I90028

AF735435

LIBRARY

TEMPLE UNIVERSITY LIBRARIES
TYLER LIBRARY
Please return this book as soon as you have finished with it.
In order to avoid a fine it must be returned by the latest date

TEMPLE UNIVERSITY LIBRARY

Artists' Pigments

A Handbook of their History and Characteristics

VOLUME 1

Robert L. Feller, Editor

(TYL)
ND
1510
A77
1986
V.1

The right of the University of Cambridge to print and sell all manner of books was granted by Henry VIII in 1534. The University has printed and published continuously since 1584.

CAMBRIDGE UNIVERSITY PRESS
Cambridge
London New York New Rochelle
Melbourne Sydney

NATIONAL GALLERY OF ART
Washington

1262100

Copyright © 1986. Board of Trustees, National Gallery of Art. All rights reserved. This book may not be reproduced, in whole or in part, in any form (beyond that copying permitted by Sections 107 and 108 of the U. S. Copyright Law and except by reviewers for the public press), without written permission from the publishers.

The hardcover edition of this book is published by the
Press Syndicate of the University of Cambridge
The Pitt Building, Trumpington Street,
Cambridge CB2 1RP
32 East 57th Street, New York, N.Y. 10022, USA
10 Stamford Road, Oakleigh, Melbourne 3166,
Australia
© Cambridge University Press 1986
Printed in Great Britain
by Balding + Mansell Limited
Designed by Susan Lehmann, Washington, D.C.

Library of Congress Cataloging-in-Publication Data
Main entry under title:
Artists' pigments.
Bibliography: p.
1. Pigments. 2. Artists' materials. I. Feller,
Robert L.
ND1510.A77 1985 751.2 85-28349
ISBN 0-521-30374-5 hard cover
ISBN 0-89468-086-2 paperback

Artists' Pigments

A Handbook of their History and Characteristics

Contents

Preface

There is growing recognition of the importance of technical examination in the study of works of art and in their care and preservation. In this light the National Gallery of Art takes great pleasure in offering this volume which describes the history and characteristics of ten pigments that have played a major role in the history of painting. Scholars have written extensively on the history of individual pigments and dyes. In the last several decades conservation scientists have published an increasing number of articles on the scientific and technical methods for identifying and characterizing artists' colorants. Rarely, however, have these two aspects of the study of pigments, their history and their scientific analysis, been brought together in one publication.

The need for this handbook was brought to my attention by Rutherford John Gettens in 1971 when he asked if the National Gallery of Art would oversee the preparation of a series of monographs on pigments, a project of international importance which he had been pursuing since 1966. With a grant from the National Endowment for the Arts matched with funds generously provided by the Ciba-Geigy Corporation and the David Lloyd Kreeger Foundation, we gladly accepted Mr. Gettens' challenge, assigning the responsibility to the distinguished chemist, Robert L. Feller, then director of the National Gallery of Art Research Project at Mellon Institute, Carnegie-Mellon University, Pittsburgh.

Close collaboration on the project with our admired colleague at the Freer Gallery of Art, John Gettens, was warmly anticipated. However, his untimely death in 1974 left the full responsibility for the completion of the project with the National Gallery of Art. Robert Feller's patient perseverance as editor and author is clearly evident in the quality of this volume. We are most grateful to him for his effort and to the fourteen contributing scholars for their dedication to the project through many years. Our special appreciation goes also to the National Endowment for the Arts and the Ciba-Geigy Corporation, and the David Lloyd Kreeger Foundation for their generous support.

J. CARTER BROWN
Director, National Gallery of Art

Foreword

One is frequently asked: why were these specific pigments selected for publication? The reason for the preparation of a particular monograph is often circumstantial. In collecting material for this book we have chosen, as did Rutherford J. Gettens before us, to rely on colleagues, wherever they may be located in the world, who have had the need or the occasion to develop special expertise and in-depth knowledge concerning a particular pigment or class of pigments. Much of the information that the authors are able to supply is based on original laboratory and bibliographic research. Descriptions of certain distinctive microscopic characteristics of Indian yellow, aureolin, and cadmium yellow, for example, came from hitherto unpublished investigations. Details on the mineral character of green earth were drawn from the specialized literature of geochemistry. Clarification of longstanding errors concerning the insect source of kermes carmine has been accomplished only this past year. Thus, it has been the search for the most advanced authorities — many of whom are still actively engaged in research — and the task of persuading these colleagues to find the time to prepare an extensive report on one area of their expertise, that has made the process of assembling these monographs such a slow one, extending over two decades of effort first by John Gettens and then by ourselves. Nonetheless, we hope that the thoroughness and unique character of the monographs will more than justify the time devoted to their preparation. We are proud to be able to continue the tradition set by Gettens in the series of monographs that he had published during the years from 1966 to 1974 in *Studies in Conservation*, the journal of the International Institute for Conservation. The director of the National Gallery of Art, J. Carter Brown, has supported the preparation of this present volume in continuance of that tradition and has formulated a vigorous program to continue this scholarly undertaking.

Work on the project was made possible by grants from the National Endowment for the Arts, together with matching grants from the Ciba-Geigy Corporation and the David Lloyd Kreeger Foundation. Of the many representatives of the National Endowment for the Arts who have been most helpful and understanding throughout the undertaking, John Spencer, director of Museum

Programs at the time the work was initiated, deserves our special appreciation. While assistant director of the National Gallery of Art, Charles Parkhurst encouraged the project through many years of correspondence with the authors and the persistent search for appropriate illustrations and bibliographic citations. The authors themselves must be especially thanked for their understanding and cooperation in the effort to assemble a set of ten monographs before publication. Many of the manuscripts were completed some years ago, yet each author graciously agreed that the handbook should not appear until it could include all ten pigments. Therefore, commercial sources cited by the authors may not be current. The support of the Andrew W. Mellon Foundation and the many valuable services of the professional and administrative staff at the Mellon Institute have also helped to make this undertaking possible.

Many of the x-ray powder-diffraction patterns were kindly provided by Elisabeth FitzHugh, using the x-ray diffraction equipment at the laboratory of the Freer Gallery of Art, Smithsonian Institution, Washington. Ms. FitzHugh also edited and reviewed the tables of diffraction data. Michael Bayard kindly reviewed all sections on optical crystallographic properties as well as those sections dealing with characteristics observable with a polarizing microscope. With his collaboration, an appendix was prepared reviewing terminology used in the characterization of pigment particles with the minerological microscope. Ruth Johnston-Feller provided many helpful criticisms and suggestions in addition to contributing an appendix on the measurement and description of color.

Particular thanks and appreciation for her exceptional dedication and special talents go to Mary Curran who served as editorial assistant and general manager throughout the project. Mrs. Curran provided extensive supplementary references and technical information which insured that the subsections of each contribution were suitably complete and uniform. It was she who insisted upon including the basic information on toxicity. Finally, we wish to thank Sandra Melzer who with much patience and outstanding skill typed the manuscripts through numerous drafts.

ROBERT L. FELLER
Research Center on the Materials of the
Artist and Conservator, Mellon Institute, Pittsburgh

Introduction

The technical examination of historic and artistic works is essential to understanding the history and character of the work and to the selection of appropriate methods for care and treatment. Among the materials to be found in easel paintings — wood, canvas, adhesives, paint vehicles, varnishes, and pigments — the latter perhaps have the greatest story to tell. Throughout history, pigments and dyes have varied in composition, quality, source, and popular usage. If one can become knowledgeable and skilled in interpreting the subtle differences introduced by such factors, much can be learned about the time, place, character, and circumstances of a given work. The accompanying monographs provide information that will be helpful in the characterization of ten pigments of major importance in the history of painting. As indicated in the bibliography at the end of this section, extensive monographs on nine other pigments, prepared under the leadership of the distinguished conservation scientist, R. J. Gettens, are also available to assist the scholar in the technical study of this essential component of the painter's oeuvre.

EXAMINATION SEQUENCE

The preliminary stage in the technical examination of a painting or painted object usually involves the use of methods that do not necessitate the taking of samples. Much can be learned by inspection with the aid of a low-powered binocular microscope. Other techniques that do not involve the taking of samples make use of reflectance spectrophotometry, analysis by x-ray fluorescence, and inspection by ultraviolet, infrared, and radiographic photography. After these initial examinations, consideration can be given to the utility of taking microscopically small samples from inconspicuous areas of the work. Portions of these samples are usually first examined for their pigment content with the aid of a polarizing microscope. Microchemical tests of both the pigment and vehicle can also be carried out under the microscope. On the basis of information thus gained, advanced analytical techniques may be applied to the remainder of the sample to determine the elements and crystalline compounds present. Typically, these procedures involve the use of an electron-beam microprobe, x-ray diffraction spectrometer, and scanning electron microscope with x-ray fluorescence capabilities. The supplementary data permit a decision to be made concerning the advisability of taking additional samples and the necessity for using even more elaborate techniques, such as neutron activation analysis, mass spectrometry, and radiocarbon dating. At each stage in the investigation, the examination and evaluation should be a joint effort of the analyst, conservator, and curator-connoisseur. One of the chief objectives of this series is to provide, for the pigments discussed, the essential information needed in each analytical step described above.

PURPOSE OF EXAMINATION

There are at least four different reasons for wanting to learn something about the pigments that an artist has used. One purpose in undertaking technical analyses is to provide an objective description of the artist's way of working. What were the basic set of pigments employed? Did the artist use relatively pure pigments of one particular color — green for example — or was the hue based on a mixture of colorants such as yellow and blue? Perhaps the final effect was achieved by building up a number of layers of paint, each differing in pigment composition. As evidenced by their pigmentation, were the paints essentially opaque or transparent?

Pigments may also be characterized for purposes of restoration — to determine what is original, to repair damaged areas, and to compensate for missing portions of a painted surface. One may wish to match the original pigments as closely as possible or to determine whether there have been changes in color over the course of time. In addition, pigment analysis may help in deciding if two passages are by the same hand and if either or both have a distinctly different character from that of the initial work.

A third reason for analyzing the pigments used is for purposes of conservation, that is, to devise techniques necessary for the care and preservation of the work. Pigments may be potentially sensitive to the deleterious effects of light, heat, or gaseous pollutants in the atmosphere. One must also determine if any chemical treatments being considered are likely to be harmful to the identified colorants.

Finally, technical examination of pigments aids in authentication and in assigning a probable date to a work of art. The analyst will wish to know as much as possible about the history of manufacture and the dates at which a particular pigment was introduced. Analysis can then reveal whether a pigment is characteristic of those of recent production or more representative of types manufactured many years ago. Investigation can also determine if a certain pigment is likely to have been available to an artist at a particular time and place.

PRECISION OF CHARACTERIZATION

Before undertaking an extensive technical examination, one should clearly define the primary purpose of the investigation. Depending upon the purpose, the required degree of analytical precision can be inferred. The usual question posed to the analyst is simply: what is the pigment used in this particular area of the painting? But the analyst must ask in turn: how precise an answer is needed? There are at least three levels of specificity at which an identification can be made. One may only need to know the general chemical type or class of the substance — simply that a pigment is a yellow or brown mineral earth color, a chromate yellow, a green earth, a lead white, or a zinc white. Such an answer is usually all that is necessary to verify the safety of a proposed method for cleaning a painting or to account for discoloration.

A higher level of precision is required to identify the variety or subspecies of the chemical type. One may be interested in determining, for example, whether a chrome yellow is the barium or the strontium salt, whether the zinc white is of the acicular or a nodular-shaped variety, or whether the green earth is a celadonite-glauconite type or a related silicate mineral such as cronstedtite. Identification of the chemical, crystallographic, and morphological variety of a pigment is often helpful in building up evidence concerning its possible age or, more simply, in determining whether the pigment has the character of an older type rather than that of a relatively modern variety.

The most precise level of characterization is required to establish correspondence or identity. There may be a need to prove that a pigment in one part of a work corresponds with — or is identical to — a pigment used in another part. Many forensic investigations require this high degree of precision to demonstrate that the pigment in a container or on a particular tool is identical to one found in another artifact.

The level of precision desired governs the elaborateness of the analytical procedures selected. The effort to establish the identical character of two samples, for example, frequently requires the determination of the number and proportion of "trace elements," often regarded as accidental contaminants or impurities. Neutron activation and x-ray fluorescence analysis are advanced methods appropriate for this purpose. Mass spectrometry, useful in determining the ratio of lead or sulfur isotopes present in certain pigments, provides another precise, objective way of characterizing the mineral source of pigments that contain these elements.

ORGANIZATION OF THE MONOGRAPHS

The material in each monograph has been organized to be of assistance to readers having a wide range of interests in the subject of artists' pigments. Section 1.0 in each chapter describes the nomenclature and general character of the pigment; section 2.0 reviews the history of its use. Under section 3.0, matters of particular concern to the artist are considered: color, permanence, compatibility, as well as certain painting and handling qualities. Section 4.0, on composition, describes the known chemical and physical varieties or sub-species of the pigment, particularly those variations in character that may arise owing to differences in source or methods of preparation. Common adulterants and impurities are also noted. Descriptions of the technical methods and procedures that can be employed to characterize and identify the pigments, information of primary concern to the analyst, are found in section 5.0. Finally, section 6.0 cites significant references to the pigment's use by a particular artist or in a specific work.

THE ESSENTIAL QUESTIONS

This handbook is part of an ongoing effort to assist the artist, curator-connoisseur, and conservator in answering three recurring questions about pigments found in artistic and historic works: what is the general chemical type of coloring agent present; is the pigment of a variety normally found in works executed at the particular time and place to which the piece is assigned; and is the pigment identical to that used in another area of the work or on another object? The first question can usually be answered unequivocally. The second and third, unfortunately, sometimes cannot. Nonetheless, the contributors to this volume have sought to point the way. One trusts that their efforts will encourage others to continue the search for more confident answers to the latter two questions, matters that so frequently intrigue the scholar and challenge the analyst.

BIBLIOGRAPHY

Monographs on artists' pigments published under the general title, "Identification of Materials of Paintings," R. J. Gettens, series editor, in *Studies in Conservation*, The journal of the International Institute for the Conservation of Historic and Artistic Works (London). Plans are underway to issue these as volume 2 of this series, under the editorship of Dr. Ashok Roy.

Azurite and Blue Verditer — R. J. Gettens and E. W. FitzHugh, 11 (1966), 54–61.

Ultramarine Blue, Natural and Artificial — J. Plesters, 11 (1966), 62–91.

Lead White — R. J. Gettens, H. Kühn, and W. T. Chase, 12 (1967), 125–139.

Lead-tin Yellow — H. Kühn, 13 (1968), 7–33.

Smalt — B. Mühlethaler and J. Thissen, 14 (1969), 47–61.

Verdigris and Copper Resinate — H. Kühn, 15 (1970), 12–36.

Vermilion and Cinnabar — R. J. Gettens, R. L. Feller, and W. T. Chase, 17 (1972), 45–69.

Malachite and Green Verditer — R. J. Gettens and E. W. FitzHugh, 19 (1974), 2–23.

Calcium Carbonate Whites — R. J. Gettens, E. W. FitzHugh, and R. L. Feller, 19 (1974), 157–184.

SELECTED REFERENCES FOR FURTHER READING

H. Wagner, *Die Körperfarben*, 2nd ed. (Stuttgart, 1939).

H. Kittel, ed. *Pigmente. Herstellung, Eigenschaften, Anwendung* (Stuttgart, 1960).

R. J. Gettens and G. L. Stout, *Painting Materials, A Short Encyclopedia* (1942; New York, 1966), 91–181.

W. H. Madison, *Unit 7, White Hiding and Extender Pigments*, Federation Series on Coatings Technology, Federation of Societies for Paint Technology (Philadelphia, 1967).

K. Wehlte, *The Materials and Techniques of Painting*, trans. U. Dix (New York, 1967), 51–168.

W. R. Fuller and C. H. Love, *Unit 8, Inorganic Color Pigments*, Federation Series on Coatings Technology, Federation of Societies for Paint Technology (Philadelphia, 1968).

J. G. Mone, *Unit 9, Organic Color Pigments*, Federation Series on Coatings Technology, Federation of Societies for Paint Technology (Philadelphia, 1968).

W. Stoy, E. T. Usowski, L. P. Larson, D. Passigli, W. H. Byler, R. Evdo, and W. von Fisher, *Unit 10, Black and Metallic Pigments*, Federation Series on Coatings Technology, Federation of Societies for Paint Technology (Philadelphia, 1969).

S. M. Alexander, "Towards a History of Art Materials — A Survey of Published Technical Literature in the Arts," in *Art and Archeology Technical Abstracts Supplement*, vol. 7, no. 3 (part I), no. 4 (part II) (1969); 8, no. 1 (part III) (1970).

Colour Index, vol. 3, 3rd ed. (Bradford, Yorkshire, England, and Research Triangle Park, North Carolina, 1971).

T. C. Patton, *Pigment Handbook*, vol. 1 (New York, 1973).

R. R. Myers and J. S. Long, eds., *Treatise on Coatings*, vol. 3, *Pigments Part 1* (New York, 1975).

R. D. Harley, *Artists' Pigments; c. 1600–1835, a Study in English Documentary Sources*, 2nd ed. (London, 1982).

Throughout this publication certain commercial materials are cited in order to specify accurately the particular experimental procedures that have been employed. Such identification does not imply recommendation or endorsement by the authors or the publisher nor does it imply that the products identified are necessarily the best available for the purpose.

Indian Yellow

N. S. BAER, A. JOEL, R. L. FELLER, and N. INDICTOR

NOMENCLATURE IN VARIOUS LANGUAGES

English: Indian yellow
German: Indischgelb
Russian: инаийский желтый
Indian: Purree, piuri
French: Jaune indien
Italian: Giallo indiano
Spanish: Amarillo indio
Dutch: Indisch geel

1.0 INTRODUCTION*

Indian yellow was probably introduced into India from Persia in the fifteenth century (Chandra, 1949b), and was known under several designations, among them *purree*, *peori* or *piuri*, and *Monghyr puri*. Derived from the urine of cows that had been fed mango leaves (*Mangifera indica Linn.*), the pigment was used mainly for watercolor and tempera-like paints. It was also occasionally used in the West as a glazing color in oil or in underpainting (Wehlte, 1975). The empirical chemical formula is given as $C_{19}H_{16}O_{11}Mg{\cdot}5H_2O$ by Gettens and Stout (1966).

According to Heaton (1947), the Indian government prohibited the pigment's manufacture on humane grounds in 1908 (see 2.3). Indian yellow has fallen into disuse and today is only of historic interest. It is listed as Natural Dye CI No. 75320 in the *Colour Index* (1971).

1.1 Brief Definition of Pigment

The principal colorant compound in the pigment is based on the yellow crystalline magnesium salt of euxanthic acid. To create Indian yellow, the cow urine was evaporated and the resultant precipitate formed by hand into balls which were exported as crude pigment, later to be washed and refined by the pigment supplier.

1.2 Current Terminology

The names of the pigment at the present time are as noted above.

2.0 HISTORY

2.1 Archaic, Obsolete, and Misrepresentative Names

Names given to Indian yellow in the past have been: piuri, peori, purree, pioury, purrea Arabica, Hardwari peori, peri rung, peoli, Monghyr puri, gogili and gaugoli. According to Maerz and Paul's *Dictionary of Color*, the pigment is also called snowshoe yellow (Kelly and Judd, 1976). The designation Indian yellow has at one time or another been applied to cobalt yellow.

NORBERT S. BAER, professor, Conservation Center, Institute of Fine Arts, New York University.

ABRAHAM JOEL, Fine Arts Conservation, 96 Spring Street, New York, N.Y.

ROBERT L. FELLER, director of the Research Center on the Materials of the Artist and Conservator, Mellon Institute, Carnegie-Mellon University, Pittsburgh.

NORMAN INDICTOR, professor of chemistry, Brooklyn College, City University of New York, adjunct professor, Conservation Center, Institute of Fine Arts, New York University.

*Much of the material in this chapter was presented in 1972 at the IIC Lisbon Congress on the Conservation of Paintings and the Graphic Arts (Baer, Indictor, and Joel, 1972).

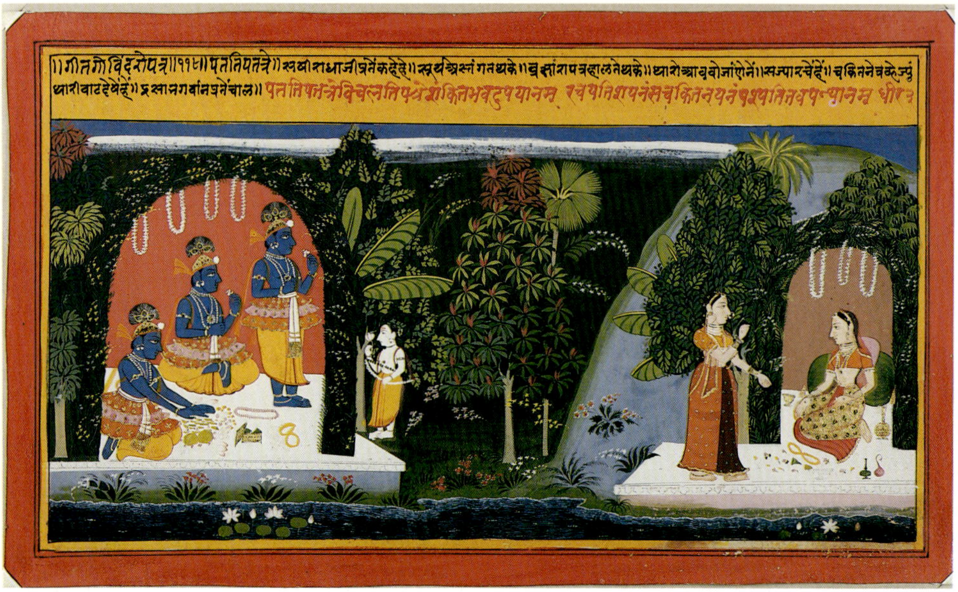

A

Fig. 1. Illustration from the Gīta Govind, Mewār Style, Rājpūt School, India, 1725–1735, Detroit Institute of Arts, 71.317. 22 x 40 mm. A. As seen under normal light. B. Viewed under ultraviolet radiation, as recorded on Kodak Ektachrome 50 with 2A Wrattan filter. C. Infrared radiation, Kodak high-speed infrared 4143 with Wrattan No. 25 filter, 3200°K bulbs. D. X-ray radiograph, 5MA/30kV for 60 seconds, Kodak Type M Industrial film. Photograph courtesy of Founders Society Detroit Institute of Arts.

2.2 **History of use**

Indian yellow was most popular in India. Moti Chandra (1949b) is of the opinion that the pigment was not employed in the miniatures of the palm-leaf period (thirteenth–fourteenth centuries A.D.); no Sanskrit text makes reference to it. He suggests that the colorant was introduced into India in the fifteenth century, probably from Persia. Chandra (1949a) states that peori could be obtained from Jaipur, where it was known as *gogili*, an Indianized form of the Persian implying "cow earth." At the beginning of the fifteenth century peori was used in miniatures on paper (Chandra, 1949b). Based on notes found on drawings, Khandalavala (1958) indicates widespread use of "peoli or gaugoli" in Pahari drawings during the period from the sixteenth to the nineteenth centuries.

Elisabeth FitzHugh's survey of the occurrences of the pigment in Indian and Persian paintings at the Freer Gallery of Art, Washington (private communication, 1976), indicates that its use appears to be limited to Indian paintings. The earliest date of the pigment's occurrence could perhaps be late sixteenth century, and its presence is limited almost exclusively to the Mughal period (late sixteenth–nineteenth centuries) (see fig. 1).

Although considerable amounts of Indian yellow are known to have reached Europe, the literature documenting its use either in the East or in the West is surprisingly sparse. Harley (1982) cites documentation (Dewhurst, 1784–1787; Fielding, 1830, 1842, 1844; Gartside, 1805; Henderson, 1806) indicating "that some English painters used Indian yellow in the late eighteenth century." She also notes Bouvier's claim (1829) to have been one of the first continental artists to use this pigment. Munkert (1905) tends to support this date, saying that the pigment had been used in Germany for about seventy-five years. Kühn (1969) cites two examples of the use of Indian yellow in nineteenth-century paintings (see 6.0).

The use of Indian yellow as a dye — "l'acide euxanthique dans teinture et la fabrication des couleurs" — was reported by Wagner (1859–1860). The method employed for dyeing

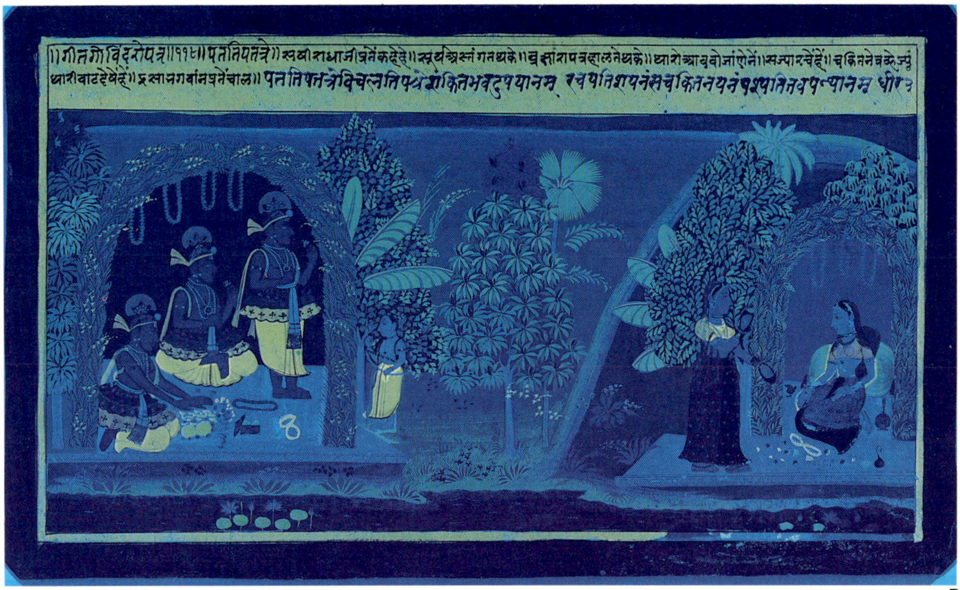

B

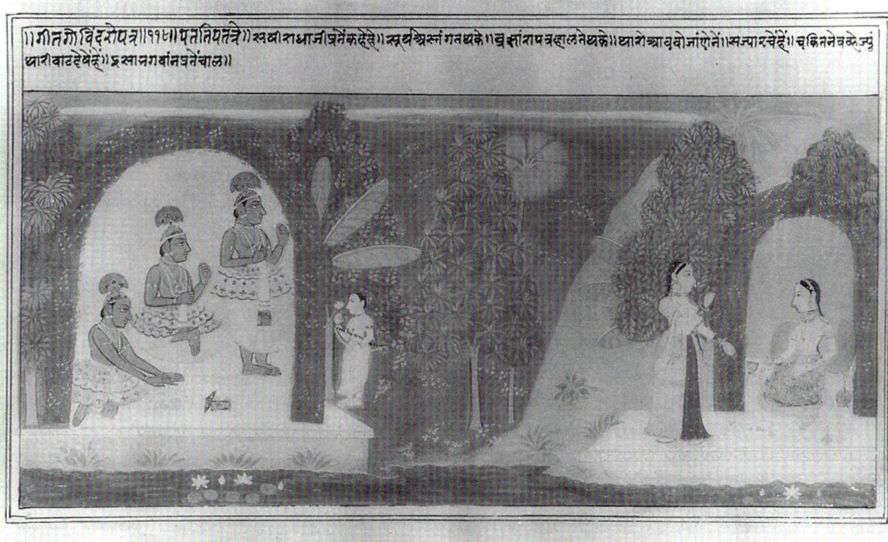

C

was simply to dissolve the pigment in hot water and dip the cloth in the solution. However, the objectionable odor imparted to the fabric soon negated this practice (Watt, 1892). Indian yellow was also said to have been used for coloring doors and railings (M'Cann, 1883).

2.3 **Terminal Date**

Heaton (1947) reported the original act prohibiting the further manufacture of the Indian yellow in 1908. An attempt to locate this document by searching the Acts of the Lieutenant Governor of Bengal in Council from 1902 to 1909 was unsuccessful. The Acts that may have provided sufficient grounds for the prohibition of the manufacture of Indian yellow are those directed toward the prevention of cruelty to animals (*Bengal Acts*, 1866/1869; *Government of India Acts*, 1890).

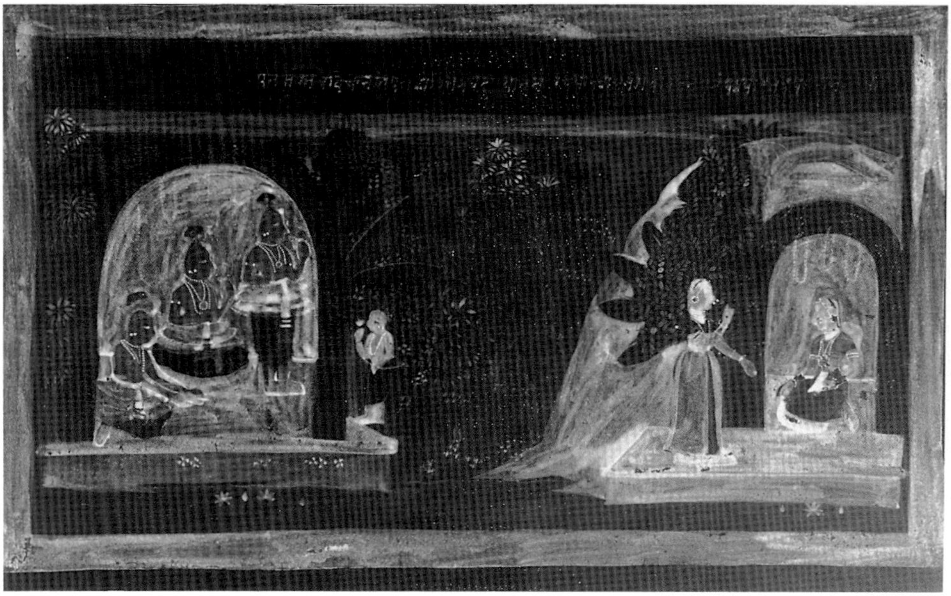

D

Act II of 1866 Bengal Acts.
An Act to provide for the better regulation of the Police within the suburbs of the Town of Calcutta.

Section XL — Whoever, within such limits as shall be from time to time defined by the Commissioner of Police, with the sanction of the said Lieutenant-Governor, in any public street, road, thoroughfare, or place of public resort, shall commit any of the following offences, shall be liable to a fine not exceeding fifty Rupees:
Clause 16 — Cruelty to Animals. Whoever shall cruelly beat, ill-treat, abuse, or torture, or shall cause or procure to be cruelly beaten, ill-treated, abused or tortured, any animal.

Act IV of 1866 Bengal Acts.
An Act to amend and consolidate the provisions of Act XIII of 1856 (for regulating the Police of the Towns of Calcutta, Madras, and Bombay) and of Act XLVIII of 1860 (to amend Act XIII of 1856).

Section LXVII — Whoever cruelly beats, ill-treats, abuses, or tortures, or causes or procures to be cruelly beaten, ill-treated, abused, or tortured, any animal, shall, for every such offence, be liable, on summary conviction before a Magistrate, to a fine not exceeding one hundred Rupees, and, in default thereof, to imprisonment, with or without hard labour, from term not exceeding three months.

Act I of 1869 Bengal Acts.
An Act for the Prevention of Cruelty to Animals.

I. The word animal shall be taken to mean any domestic or tamed quadruped . . . .
II. Every person who shall cruelly and wantonly beat, ill-treat, abuse, torture, overdrive, or overload, or cause to be beaten, ill-treated, abused, tortured, overdriven, or overladen, any animal, shall be liable to a fine which may extend to one hundred Rupees.
III. Section LXVII of Act IV of 1866, and Clause 16 of Section XL of Act II of 1866 . . . are hereby repealed.
IX. This Act shall extend to the Town of Calcutta, and to the suburbs of the Town of Calcutta . . . .
X. It shall be lawful for the Lieutenant-Governor of Bengal, by an order published in the Calcutta Gazette, to extend this Act to any city, town, station, cantonment, village, district, or portion of a district to be mentioned and defined in such order; and from time to time, by any order published as aforesaid, to revoke, vary, amend, or alter any such order.

Act III of 1869 Bengal Acts.
An Act to enable Police Officers to arrest without warrant persons guilty of Cruelty to Animals.

I. Every Police Officer may arrest without a warrant any person committing, in his view, any offence against the said Act I of 1869.

Act XI of 1890 Government of India Acts.
An Act for the Prevention of Cruelty to Animals.
Act XI of 1890 was passed by the Governor General of India in council and extended to the whole of British India and gave power to the local governments who may "by notification in the official Gazette, extend, on and from a date to be specified in the notification, the whole or any part of the rest of this Act to any such local area as it thinks fit." Pertinent portions of this Act read:

Section 2, sub-section (1) — Animal means any domestic or captured animal.
Section 3 — If any person in any street or in any place, whether open or closed, to which the public have access, or within sight of any person in any street or in any such other place, — (a) cruelly and unnecessarily beats, overdrives, overloads or otherwise ill-treats any animal . . . (c) . . . he shall be punished with fine which may extend to one hundred Rupees, or with imprisonment for a term which may extend to three months, or with both.

The Acts of Bengal, traced over the years 1866 to 1869, indicate that the increasing pressure was brought to bear against violators of the act. However, the possibility that the production of Indian yellow continued even after the passage of the Government of India Act XI of 1890 must be considered. Only gradually did production terminate, due perhaps to a combination of several factors. The escalating consequences of violating the acts or a decrease in demand for the pigment would certainly have been relevant. With a decrease in demand the milkmen would not have needed to feed the cows on mango leaves exclusively, thereby preventing the premature deaths of these animals. The local authorities may have condoned preparation of the pigment, knowing no harm would come to these animals. Thus production may have been carried on past the date of passage of the prohibition. For this reason it seems improbable that a certain date for the end of production of Indian yellow can be specified. Wehlte (1975) states, however, that the pigment has not been available commercially since 1921.

3.0 GENERAL PIGMENT PROPERTIES

3.1 Color and Spectral Reflectance

Indian yellow yields a beautiful clear yellow color and a light powdery texture of greater body and saturation in masstone than gamboge (Field, n.d., p. 157). This description is supported by Eibner (1905): "the pure pigment has an incomparably beautiful, deep and luminescent gold yellow in a shade which is achieved with no other pigment." Parry and Coste (1902), chose, on the other hand, not to consider Indian yellow as fine a color as gamboge.

Measurement of the color of Indian yellow is complicated by its fluorescence in the visible range. Moreover, as the curves in Fig. 2 indicate, the pigment's fluorescence is stimulated most strongly by radiation in the visible range, 435 nm (Feller, 1968). Fig. 3 illustrates the spectral reflectance curve as measured by a Kollmorgen KCS-10 Color-Eye®. This abridged spectrophotometer illuminates the sample with an incandescent lamp of color temperature 2857°K and measures the reflectance relative to barium sulfate at individual wavelengths, as transmitted through narrow band-pass filters. Thus, the reflectance curves should include the fluores-

Fig. 2. Behavior of Indian yellow under ultraviolet radiation, showing: A. fluorescence (relative intensity of emitted radiation) when activated by 435 nm radiation, and B. the relative intensity emitted at 535 nm when the sample is irradiated by various wavelengths from 300 to 500 nm. After Feller (1968). Sample courtesy of Winsor & Newton.

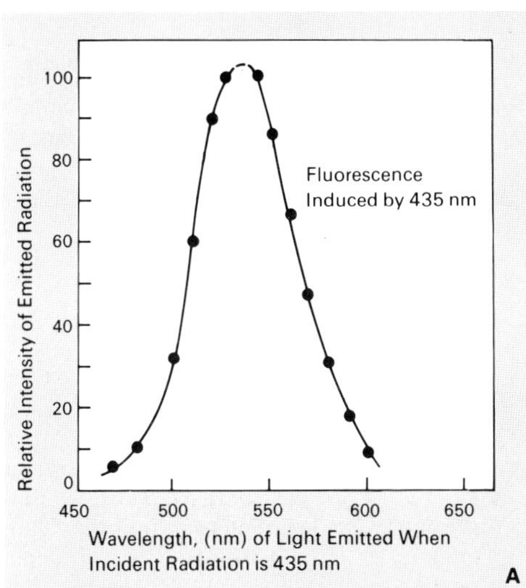

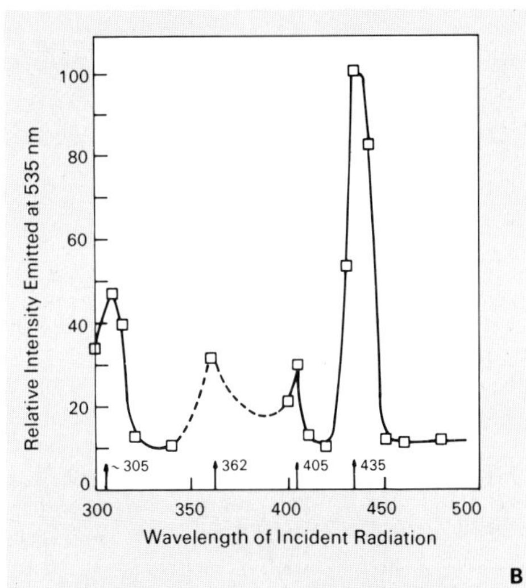

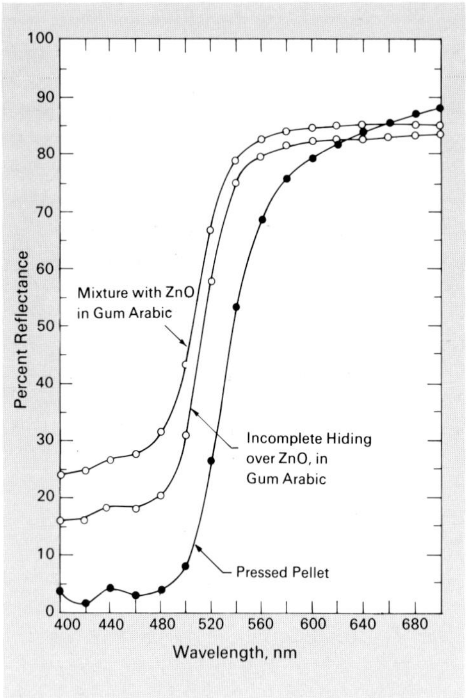

Fig. 3. Spectral reflectance of Indian yellow as measured using polychromatic illumination on a Model D-1 Kollmorgen Color-Eye® abridged spectrophotometer. (Fluorescence is thus included.) Measurements courtesy of Catherine W. Bailie and Ruth Johnston-Feller, Research Center on the Materials of the Artist and Conservator, Mellon Institute, Pittsburgh.

cence stimulated by 435 nm radiation. A reflectance curve of Indian yellow was reported by Barnes (1939) during his early investigation of artists' pigments.

The pigment specimens assembled at the Conservation Center, N.Y.U., for these investigations are listed in table 1. The Munsell notations of the masstone color of six specimens are given in table 3.

In the series of paints prepared to test lightfastness (table 2), it was noticed that the paint's initial acidity or alkalinity influenced the color, acid causing a tendency toward a paler yellow (see 3.5). The colors in this series of six paints ranged in Munsell Notation from 3-9Y/8.7-9.1/8-10 (hue/value/chroma). The NBS-ISCC color name (Kelly and Judd, 1976) for a hue in this range is light yellow.

On Indian paintings the pigment has also been identified by microscopy in mixtures with other pigments. It was most commonly found with lead white when a bright yellow was desired. Indigo was added to produce green; vermilion or hematite were added with varying amounts of lead white to give a dark brown, a pinkish yellow, or a pinkish brown flesh tone. Red lead was mixed with Indian yellow to give an orange-yellow (FitzHugh, private communication).

3.2 Hiding Power and Tinting Strength

In samples of Indian yellow taken from paintings and examined at the Freer Gallery of Art, Washington, the yellow particles were uniformly clear and transparent (FitzHugh, private communication, 1976). The refractive index of the particles is sufficiently close to that of Canada balsam (N = 1.53) that at first glance there often appears to be only a yellow stain. From the behavior of the Becke line it appears that the refractive index is indeed higher than that of Canada balsam. Gettens and Stout (1966) report the refractive index of Indian yellow as 1.67. The yellow particles are anisotropic (see 5.1).

The low refractive index means that preparations in oil and varnish are translucent; however, a bright color with good hiding can be obtained in the usual aqueous binders. The tinting strength of this pigment was considered by Wehlte (1975) to be very good but its hiding power extremely low. He believed that Indian yellow was invaluable for retouching paint losses on works with yellowed varnish.

3.3 Permanence

Indian yellow was included in the palette of lightfast artists' pigments (Normalfarben) established about the turn of the century by the German Society for Rational Painting Techniques (Munkert, 1905). Tests of its lightfastness that will be described below indicate that in a traditional water-dispersed binder the pigment is indeed more lightfast than alizarin. Authors have traditionally suggested that it may not be as stable in oils, perhaps undergoing reactions with the vehicle (Church, 1915, p. 175).

In order to confirm the pigment's light stability, paints were prepared in various vehicles by the National Gallery of Art Research Project, Mellon Institute, Pittsburgh. A sample of genuine Indian yellow, kindly provided by Winsor & Newton, was used. The samples were exposed for 205 and 384 hours in an Atlas Electric

Table I. **SAMPLES EXAMINED (Labeled Indian Yellow)**

Sample Number	*Source*	*Other Information*
I	Doerner Institute, Munich [Gelb 3.04.1]	Purchased from Günther & Wagner, Hannover, c. 1955 (Kühn, personal communication, 1972)
II	Doerner Institute, Munich [Gelb 3.04.2]	Purchased from Winsor & Newton, London, c. 1955 (Kühn, personal communication, 1972)
III	Forbes Collection [3.14.4 (372)]	Label marked Rowney
IV	Conservation Center, New York University	Purchased from Fezandie & Sperrle, New York, 1971; artists' dry color, Acid yellow 23 in Aluminum Lake
V	Forbes Collection [3.14.3 (373)]	Label marked Roberson, London; 1914 (EWF)
VI	Conservation Center, New York University	Date of acquisition unknown
VII	Conservation Center, New York University	Winsor & Newton, London; purchased 1972; watercolor; prepared from a mixture of organic pigments (Winsor & Newton)
VIII	Conservation Center, New York University	Winsor & Newton, London; purchased 1972; oil color; prepared from a mixture of organic pigments (Winsor & Newton)

Due to the limited supply of several of the samples, not all of the analytical procedures discussed in the text were applied to the complete series. Some specimens could only be studied as mounted slides.

Devices 600 WRC xenon-arc fadeometer with Pyrex-glass filter. The results, summarized in table 2, indicate that the pigment tends to be at least as lightfast as alizarin, particularly in an aqueous dispersed medium such as gum arabic, which may be close to the type of glue or gum-based vehicle in which it was traditionally employed. After an exposure of 384 hours, the British Standard (BS1006:1971) blue-wool cloths, exposed at the same time, had faded as follows: Cloth No. 7 to GGS 4.2 (Geometric Grey Scale Contrast 4.2), No. 6 to GGS 3.0, No. 5 to GGS 2.0 and No. 4 to GGS 1.2 (BS2662, 1961). These results indicate that the exposure was considerable. In view of the degree of fading of the alizarin and of the BS1006:1971 blue-cloth controls, it seems fair to state that Indian yellow in gum arabic compared in lightfastness to BS1006:1971 Class 5-6 (Feller, unpublished).

When an ultraviolet filter that sharply cuts off about 400 nm (Rohm and Haas Plexiglas® UF1, now UF4) was placed over the sample during exposure in the Atlas Electric Devices 600 WRC Fadeometer (Pyrex-glass filter), the net color change was reduced slightly in four of the

Table 2. **FADING OF INDIAN YELLOW PAINTS IN XENON ARC FADEOMETER, PYREX GLASS FILTER. COLOR DIFFERENCE (ΔE) IN CIE L*a*b* UNITS**

Test paint	*ΔE after 205 hrs.*	*ΔE after 384 hrs.*
Control (Alizarin in Rhoplex® AC33 polymer emulsion paint, 94% TiO_2)	14.6	20.8
Indian yellow in various mixtures:		
Gum arabic	2.6	10.3
Gum arabic (with added zinc white)	6.9	12.5
Acryloid® B-72 methylacrylate/ethyl methacrylate copolymer	8.4	28.2
Rhoplex® AC-33 methylacrylate/ethyl methacrylate copolymer emulsion (alkaline)	18.5	34.6
Schweitzer 5712 vinyl emulsion adhesive (acidic)	25.1	39.1

samples, perhaps fading only about 80% as much as when unprotected. Scarcely any effect was noticed in the mixtures with zinc and titanium white, possibly because these pigments themselves filter ultraviolet radiation.

Since Indian yellow fluoresces, and the fluorescence is reduced upon exposure to light, measurement of the color changes upon fading will contain uncertainties unless all aspects of the measurement procedure are carefully considered. Although the data in table 2 may contain a degree of uncertainty owing to the problem of fluorescence, it is nonetheless considered that the pigment's lightfastness relative to that of alizarin and the BS1006:1971 blue cloths is fairly indicated.

Field (n.d., p. 158) stated that Indian yellow resisted the sun's rays with singular power in watercolor painting; yet in ordinary light and air or even in a book or portfolio, the beauty of its color was not lasting. The explanation of this inconsistency eludes us. Church (1915, p. 176) also claimed that, as a watercolor, Indian yellow retained its hue unimpaired when exposed to diffused daylight. However, he further stated that "sunlight very slowly bleaches it; the hue it acquires being somewhat brownish."

In Field's notebooks from 1809, currently at the Courtauld Institute of Art, London (Harley, 1979), is a sample of Indian yellow and some brief notes in Field's hand. We are grateful to Dr. Harley for the following transcriptions of these passages and for her information that folio 338 (Field, 1809) bears a watercolor sample of Indian yellow.

Folio 338. *Indian Yellow* of Mr. Newman in lumps of a powdery, soft, light, spongy texture, which appear to have been in a liquid state. Semi-transparent, pure, weakened by acids, brightened by alkali and rather greened thereby, equally unaffected (owing the time in which Cochineal lake was destroyed) by the Sun and Jakes. Weakened but not discoloured by much longer continuance in the sun. In oil, dried nearly as soon or sooner than the oil and teints — faded in thin glazing even in the shade, totally destroyed by the sun — in body and teint it stood rather better and was not discoloured.

Folio 337v. This pigment is said to be produced from the wine of the (buffalo/camel after feeding on Mangoes) in India. Its principal use is to follow Gamboge in water colour being rather deeper or stronger but not more durable. Yet it is not to be depended on so much as Gamboge. The first executed specimens in my Chromatics having faded in two or three years shut in the Book.

At a considerably later date, Eibner (1905) had this to say:

The unfavorable judgments as to the light-fastness of Indian yellow, which were mentioned at the beginning are unfounded. I have applied pigment, which was produced purely from Indian piuri, with gum arabic on water-colour paper in different intensities of shade. Then I exposed these sheets partially without protection and partially under glass in a room with a glass roof. For two years, I have not observed the slightest decrease in intensity of the shade or any change in the color value.

Since a pigment which withstands an exposure of a year can no longer be counted among the poor pigments, it may be said that the true Indian yellow corresponds extensively to the requirements with respect to light-fastness which are to be reasonably set for an organic pigment. The exposure tests are being continued. As a result of the afore-said tests, it can already be concluded that Indian yellow is useful as a water-colour.

3.4 Compatibility

Plenderleith (1937) states that Indian yellow "may become brown, emitting a volatile agent which influences adjacent colours and may even stain adjacent pages in a book of miniatures. A green colour containing this pigment may become deep blue."

Field (n.d., p. 158) considered that in oil Indian yellow "is exceedingly fugitive both alone and in tint. Owing probably to its alkaline nature it has an injurious effect upon cochineal lakes and carmine when used with them. As lime does not injure this colour, it may be employed in fresco, according to its powers."

As a general rule, Church (1915, p. 176) believed that Indian yellow would undergo no change upon admixture with any permanent pigment, nor would it be affected by sulfur compounds. The authors are not able to affirm or deny these statements of historical interest based on direct experience. Yet they seem reasonably confirmed for either the colorant itself or mixtures with stable pigments in traditional temperalike vehicles, as found in Indian paintings in the Freer Gallery of Art, Washington (FitzHugh, private communication). Church, however, reported observing both unchanged and changed specimens when Indian yellow was prepared in linseed oil.

3.5 Chemical Properties

The working properties of the pigment may be related to the structure. The glucuronic acid portion of the molecule confers upon it considerable water solubility; the euxanthone portion permits compatibility with oily media. The color of Indian yellow is discharged in excess

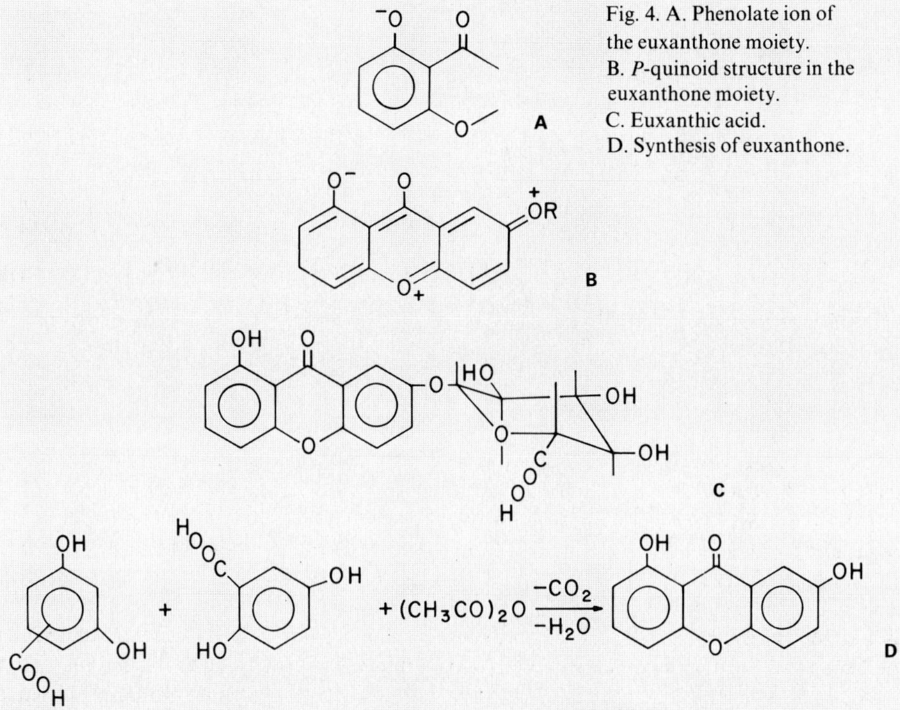

Fig. 4. A. Phenolate ion of the euxanthone moiety.
B. *P*-quinoid structure in the euxanthone moiety.
C. Euxanthic acid.
D. Synthesis of euxanthone.

aqueous acid and may be regenerated by aqueous alkalinity. This reversibility suggests that the color arises either from the highly conjugaded phenolate anion, of the euxanthone moiety (fig. 4A) or possibly from potentially *p*-quinoid structure also in the euxanthone moiety (fig. 4B).

Treatment with hydrochloric acid — removing the bases with ammonium carbonate and reprecipitating with hydrochloric acid — precipitates euxanthic acid (m.p. 154–156°C with decomposition [Church, 1915]) (fig. 4C).

3.6 Oil Absorption and Grinding Qualities

The oil absorption given by Doerner (1949) is up to 100%; that by Wehlte (1975) is approximately 40%.

3.7 Toxicity

Little has been written with authority about the toxicity of this obsolete material. However, since Indian yellow is a product of cow urine, undoubtedly it needed to be handled with reasonable standards of cleanliness.

4.0 COMPOSITION

4.1 Chemical Composition

The principal constituent of Indian yellow is a mixture of the calcium and magnesium salts of euxanthic acid (fig. 4C) (Riffault, et al., 1874; Eibner, 1909; Meyer and Jacobson, 1920; Weber, 1923; Kittel, 1960). The calcium or magnesium salt usually occurs as the basic hydrate. The euxanthone moiety is the active principal of mango leaves, and in the metabolic process, it is converted into soluble salts attached to the oxidized carbohydrate moiety. The best grades of Indian yellow contained up to 65% euxanthic acid (Wehlte, 1975); the poorest grades contained only about 34% (Zerr and Rübencamp, 1908).

An abundance of publications indicate the early chemical interest in Indian yellow, its structure and synthesis (Erdmann, 1844; Stenhouse, 1844; Schmid, 1855; Baeyer, 1870; Salzmann and Wichelhaus, 1877; Graebe and Ebrard, 1882; Graebe, 1883, 1889; Van Scherpenberg, 1918; Wiechowski, 1923; Mayer,

Table 3 **COLOR,[a] ULTRAVIOLET FLUORESCENCE, AND ELEMENTAL COMPOSITION[b] OF SPECIMENS OF INDIAN YELLOW[c]**

Sample number	*Munsell[f] designation*	*Ultraviolet fluorescence*	*% Weight loss on drying*	*% C*	*% H*	*% Ash[d]*
I	7.5YR 7/14	orange-yellow	N.D.[e]	N.D.	N.D.	N.D.
II	2.5Y 7/10	yellow	N.D.	N.D.	N.D.	N.D.
III	10YR 7/12	yellow	none	36.97	3.99	5.0
IV[g]	2.5Y 7/12	red-orange	none	9.81	4.71	41.94[g]
V	2.5Y 8/12	yellow	3.2	40.66	4.03	17.03
VI	10YR 7/12	yellow	6.7	44.73	4.17	14.39

a. Hue, value/chroma as per Munsell designation (Munsell 1929–1970).
b. Percentages based on sample dried to constant weight at 105°C.
c. See table 1 for descriptions.
d. Residue obtained at 900–1000°C.
e. N.D. denotes: no data.
f. Determined under "Tensor" incandescent illumination.
g. Sample is not Indian yellow.
Elemental analyses, Schwartzkopf Microanalytical Laboratory, Woodside, New York.

1924; Reindel and Anschutz, 1947). The structural connection between Indian yellow and the naturally occurring yellow pigments in the family of compounds known as xanthones has been made by Karrer (1947).

The chemical structure of Indian yellow has been proved by hydrolytic degradation. The mild action of mineral acids generates euxanthic acid from the essentially magnesium salt. More exhaustive hydrolysis (heat, concentrated acid) gives euxanthone and glucuronic acid. Proof of structure of the euxanthone involved its synthesis from β-resorcylic acid, hydroquinone carboxylic acid, and acetic anhydride. In this reaction dehydration and decarboxylation occur simultaneously (see fig. 4 D).

Several samples of Indian yellow specimens were submitted for carbon, hydrogen, and ash analyses. The results are reported in table 3. Included among the samples was one modern specimen known commercially as Indian yellow. The variability of the carbon, hydrogen, and ash analyses for the samples, all of which gave satisfactory infrared spectra, implies that significantly different amounts of magnesium salts are present in each. (Emission spectra of these samples all show Mg to be the principal metal present, *vide infra.*) The samples most likely also contain bound water in some form. The theoretical percentages for $C_{19}H_{16}O_{11}Mg$ are C 51.32%, H 3.63%, ash as MgO 9.07%; the theoretical percentages for $C_{38}H_{30}O_{20}$ Mg are C 54.93%, H 3.64%, ash as MgO 4.85%. Comparison of these figures with the results reported in table 3 reveals the high variability in purity associated with samples of the pigment.

A fine sample of piuri, reported by Graebe and Ebrard in 1882, analyzed as follows: euxanthic acid 51.0, magnesium 4.2, calcium 3.4, silicic acid and alumina 1.5, volatiles 39.0 — total 99.1 (Perkin and Everest, 1918). The compositions of five samples, apparently from M. M. Lefranc et Cie of Paris, as given by Parry and Coste (1902), are listed in table 4. The authors state that as the quality goes down the proportion of magnesium euxanthate steadily decreases and that of euxanthone increases.

4.2 Sources

The source of Indian yellow was once the subject of much speculation. Some thought it to be from a deposit formed in the urine of the camel (Field, 1809; Bachhoffner, 1837), elephant, or buffalo; others, an intestinal concretion (Watt, 1892). Some went so far as to identify the pigment's origin as "the urine of certain varieties of serpents" (Hilier, 1957). However, in the late nineteenth century it was Mukharji (1883–1884) who clearly established that Indian yellow was made from the urine of cows fed exclusively on mango leaves (*Mangifera indica,* Linn).

4.3 Preparation

In 1883 Mukharji visited the village of Mirzapur near Monghyr, where he found a sect of gwalas (milkmen) who manufactured the pigment. (The

Table 4. **COMPOSITION OF FIVE SAMPLES OF INDIAN YELLOW, APPARENTLY FROM M. M. LEFRANC ET CIE OF PARIS ABOUT 1902 (Parry and Coste, 1902)**

Euxanthic acid	72.3%	70.96%	4.35%	9.3%	33.34%
Euxanthone	0.0	1.12	2.80	7.40	34.00
Mg	5.35	4.88	4.85	4.60	3.70
Ca	1.75	2.43	2.61	3.33	3.70

district of Monghyr is today located in northeast Bihar [Seltzer, 1962].) He described how the urine was collected in small earthen pots, cooled, and then concentrated by heating over a fire in earthen vessels to precipitate the pigment. After straining the liquid through a cloth, the gwalas would make the sediment into a ball, dried first over a charcoal fire and then in the sun. The pigment was sold in this crude form to local merchants (chiefly marwaries) who would even advance money to the milkmen for its manufacture. The marwaries paid Rs 1 (1 sh. 8d) per pound and shipped the pigment to Calcutta and Patna (Mukharji, 1883–1884) where the price was marked up from 100 to 200% (Watt, 1892). De Puyster (1919) says that it was sold in various brands marked A to G, the latter being the poorest in quality.

One cow is said to have provided three quarts of urine, resulting in 2 oz. (56 grams) of peori per day. It is said that the annual production of this substance was as much as 10,000 to 15,000 pounds, but Mukharji considered this estimate too high. The trade was profitable, for M'Cann (1883) records that one or two dealers in Indian yellow reached an income level requiring tax payments (Watt, 1892).

The balls (fig. 5), weighing 2 to 4 oz. and having a yellow interior with a brown or greenish exterior (Watt, 1892), were shipped to Europe where manufacturers washed and purified them, separating the yellow and greenish portions. The material was then formed into yellow tablets which were redried and used as a base for water or oil color (Hilier, 1957).

Church (1915) described a procedure for purification. The crude imported material was thoroughly powdered, then washed with boiling water until the liquid filtered from it had no color. This removed the brown impurities. The washed pigment was left in contact with a saturated solution of salammoniac for a day or two, this enriched its color. The process of washing was then repeated with hot water. The purified product produced a translucent orange-yellow color. In the mid-nineteenth century, Lejort (1855) quotes the price of Indian yellow as selling for "ne vaut pas moins de 200 francs le kilogramme."

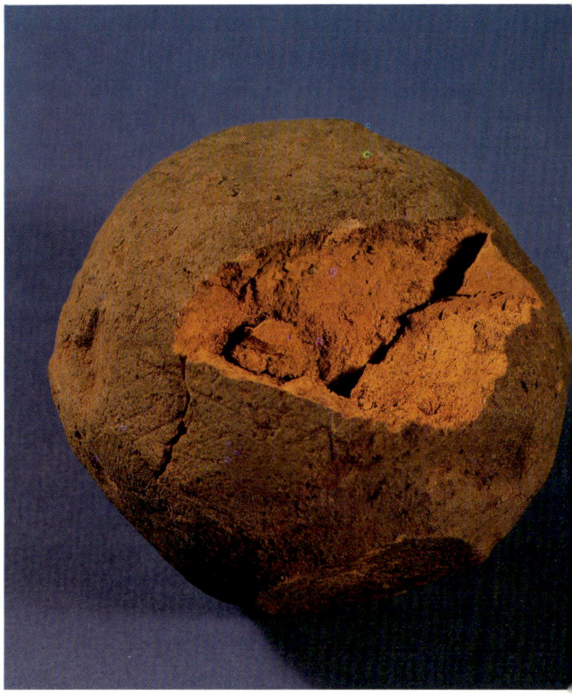

Fig. 5. Ball of Indian yellow pigment. The approximately 6-cm diameter ball has been cut open for preparation of purified pigment for technical study. The traditional golden yellow hue is thus revealed. Specimen courtesy of Winsor & Newton.

It was recorded that cows used for the manufacture of Indian yellow died after two years, but Mukharji stated the cow-keepers informed him that this was false. He himself saw cows six or seven years old which had produced peori for

Table 5. **MODERN SUBSTITUTES FOR INDIAN YELLOW**

C.I. designation	*Type*	*Chemical formula*
Pigment 40	cobalt yellow	$2K_3Co(NO_2)_6 \cdot 3H_2O$
Acid yellow 23	azo yellow	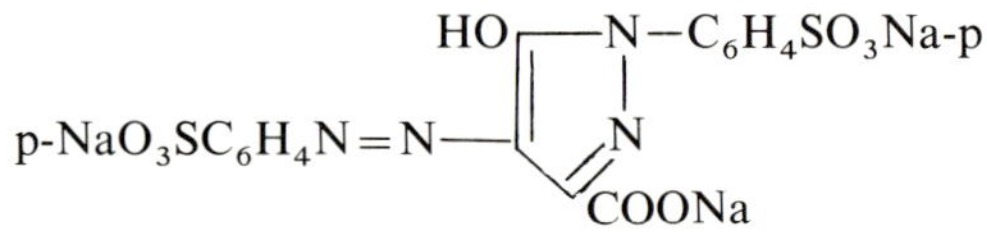
Acid yellow 63	azo yellow	$NaO_3SC_6H_4N{=}NC_6H_3\ NO_2N(NO)C_6H_3(NO_2)_2$
Acid orange 1	azo yellow	$NaO_3SC_6H_4N{=}NC_6H_4N(NO)C_6H_3(NO_2)_2$

Table 6. **CHEMICAL REACTIONS OF SUBSTITUTE INDIAN YELLOWS (*Colour Index*, 1971)**

Designation	*Solubility*	*Reactivity with Acids and Bases*
C.I. pigment 40	water — sl. soluble	Mineral acid (decomposes) Conc. NaOH (decomposes to brown oxide)
C.I. acid yellow 23	water — soluble (yellow) ethanol — sl. soluble	Conc. HNO_3 (yellow) Conc. H_2SO_4 (yellow) Conc. NaOH (redder yellow)
C.I. acid yellow 63	water — soluble (lemon-yellow) ethanol — soluble (yellow) benzene — v. sl. soluble	Conc. HNO_3 (red) changing to (brown) Conc. H_2SO_4 (red) on dilution (yellow-red) then (yellow-brown ppt.) NaOH (yellow-brown) then (brown yellow ppt.)
C.I. acid orange I	water — soluble ethanol — soluble (yellow-brown) acetone — v. sl. soluble	Conc. H_2SO_4 (red-violet) on dilution (brown-yellow) to (olive brown ppt.) NaOH (yellow-brown) with excess to (brown ppt.)

at least four years (Watt, 1892). The cows, however, appeared ill, and the villagers were reluctant to change their diets fearing a resultant drop in pigment production. The cow-keepers were labeled "cow-destroyers." On humane grounds, the production of Indian yellow was said by Heaton (1947) to have been prohibited in 1908 although it was probably in use after this (see 2.3).

4.4 Adulteration and Sophistication

Various types of organic dyestuffs have long been used to provide the artist with substitutes for Indian yellow. Azo dyes of similar hue have occasionally been employed, although they are structurally unrelated to euxanthic acid (table 5). Simple chemical tests for distinguishing some modern substitutes from genuine Indian yellow are outlined in table 6. Cobalt yellow (aureolin) was also frequently called Indian yellow in the nineteenth century (*Colour Index*, 1971) (see chapter on Cobalt Yellow, Aureolin).

Mukharji (1883–1884) mentions that piuri was also derived from a mineral which was imported from London and that it sold for only 4d per pound. The mineral could very well have been wilayiti peori, the chromate of lead which Powell (1868) distinguishes from Hardwari peori or true Indian yellow. Church (1915) noted that Indian yellow adulterated with lead chromate turns dark brown when moistened with ammonium sulfide. If samples of "Indian yellow" are incompletely bleached in acid or change from yellow to orange, lead chromate is indicated as a significant adulterant.

5.0 IDENTIFICATION — CHARACTERIZATION

5.1 Characterization by Optical Microscopy

The character of Indian yellow particles varies. One of the most striking characteristics is the presence of needle-shaped (acicular) prismatic rectangular yellow rods that tend to be bright green in certain orientations under crossed polars, the thicker shafts tending toward an olive shade (fig. 6). Extinction is parallel and under a gypsum retardation plate the prisms are seen to exhibit positive elongation. Occasionally what appear to be prisms of the same substance appear as short stubby fragments, thick enough to produce the more olive hue. When these are viewed in the vertical position under a retardation plate, one vertical section is usually yellowish, the other greenish in hue — as though the crystal prisms were twinned on the center line. The refractive index of these yellow particles is definitely above 1.52. It should be pointed out, however, that these acicular crystals are not always seen in Indian yellow from Indian paintings (FitzHugh, private communication).

An amount of colorless highly birefringent material is usually present, varying in quantity and character in different samples. Most characteristic are concretions that under the crossed polars somewhat resemble starch grains or the coccoliths in chalk. These must be clusters of radiating crystallites, for they exhibit an upright cross between crossed polars (see fig. 7). Under the gypsum retardation plate the radiating crystals are seen to exhibit negative elongation

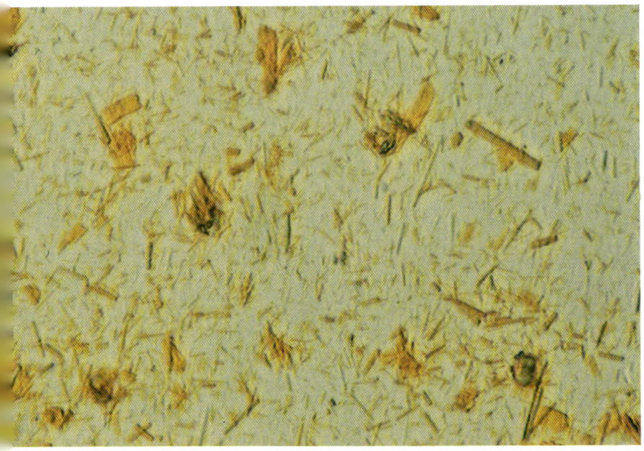

Fig. 6. Photomicrograph of Indian yellow showing needle-shaped particles, concretions, and at lower left, an olive-hued particle. Purchased from Roberson, 1914, Edward W. Forbes Collection. A. Plane polarized light in medium of refractive index 1.53 with substage shadowing.

B. Same view between crossed polars. 350x.

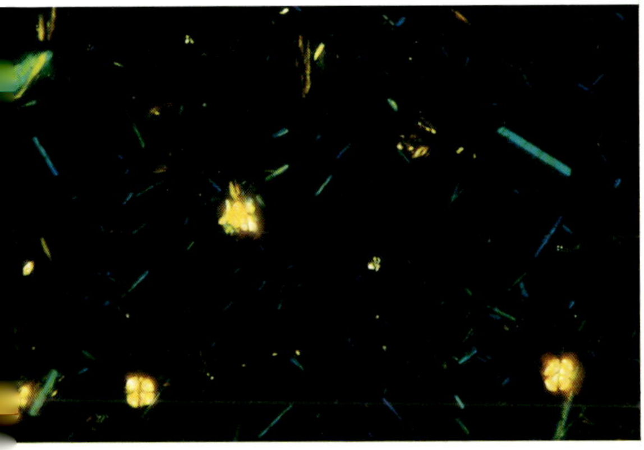

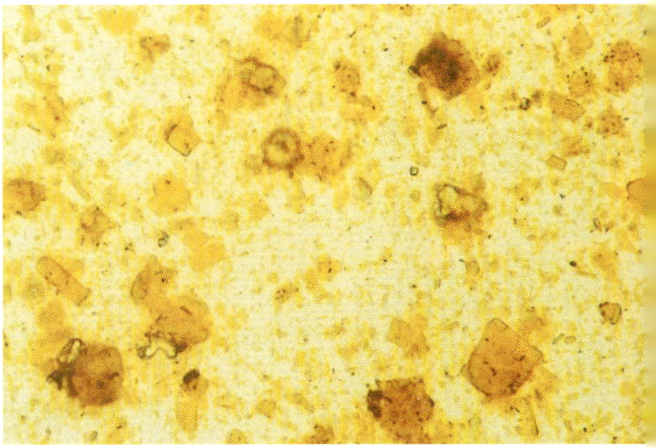

Fig. 7. Photomicrograph of Indian yellow with high concentration of quatrefoil-shaped concretions and relatively few needles. Purchased from Rowney by Edward W. Forbes, perhaps in the 1920s. A. Plane polarized light in medium of refractive index 1.53.

B. Same view between crossed polars. 350x.

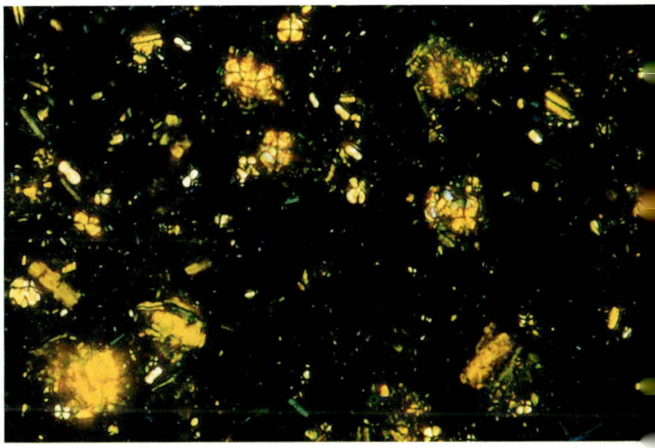

(the upper right and lower left quadrants are yellow).

Dilute acid dissolves the yellow birefringent prisms, generating what appear to be oil-in-water globules as the front of the acid solution attacks the material. Later, these globules gradually participate in the formation of fine needles that exhibit negative elongation. Perhaps these particles are euxanthic acid, for when needles of this substance are intentionally precipitated by the addition of mineral acids, they also have negative elongation.

5.2 Chemical Identification

5.21 ***Test for Magnesium with Quinalizarin.*** According to Feigl (1958, p. 225), a drop of the test solution and a drop of distilled water are placed on adjoining depressions of a spot plate and mixed with two drops of an alcoholic solution of 0.01–0.02% quinalizarin (1,2,5,8-tetrahydroxyanthraquinone). If the solution is acid, it will be colored yellow-red by the reagent. Sodium hydroxide (2N) is added drop by drop until a change to violet occurs, and then an excess is added amounting to about $\frac{1}{4}$ to $\frac{1}{2}$ of the volume then present. According to the amount of magnesium present, a precipitate of blue coloration appears; the blank remains blue-violet. The difference in shade is intensified on long standing, because the dyestuff is gradually decomposed in magnesium-free solutions owing to oxidation, whereas the colored magnesium compound is stable.

Interference by other ions is minimal; however, Feigl discusses the possibilities thoroughly.

5.22 ***Other Tests.*** If a large sample is available, Zerr and Rübencamp (1908) describe a series of tests for Indian yellow and its possible adulteration by chromate. Hydrochloric acid produces a colorless solution while depositing white flakes of euxanthic acid. Saturating this test mixture with ammonia should produce a clear yellow solution.

5.3 Instrumental Methods of Analysis

5.31 ***X-ray Diffraction.*** Two types of powder camera were used to obtain x-ray diffraction data. A Philips Powder Camera (No. 52056-B) with a 114.59 mm diameter was used to obtain the conventional Debye-Scherrer powder patterns. Because lines due to planes with very high *d* values are lost with this camera, a Guinier-De Wolff focusing camera (Guinier, 1963) with a bent crystal monochromator (Model #II-Y919; Enraf-Nonium, Garden City, New York) was also employed. As was anticipated, the Debye-Scherrer patterns for Indian yellow samples, such as that shown in fig. 8, were diffuse with frequent indications of impurities. In samples taken from Indian miniatures, the lines of ground material inadvertently mixed with the pigment tended to dominate the pattern. The spacings and relative intensities for the Indian yellow specimen that gave the best diffraction pattern (Sample VI) are given in table 7.

In a separate series of experiments, Pollack (1971) measured the intensity for diffraction peaks using a Philips diffractometer and xenon-filled proportional counter. The peaks were scanned at one-half degree two-theta per minute; the intensities listed in table 7 are peak intensities above background. Nickel filtered copper radiation was used in the diffractometer work. Table 7 contains the *d* spacings for the two additional samples. The presence of *d* spacings ranging from 5–20 Å and the absence of lines lower than 1.50Å suggests the presence of unidentified compounds, probably organic.

5.32 ***Infrared Spectroscopy.*** Infrared spectra of typical samples of Indian yellow prepared in a KBr pellet (concentration range 0.4–1.0 mg/100g KBr) are shown in fig.9. The spectra of some four or five samples displayed absorption peaks characteristic of both glucuronic acid salts and euxthanthone. Particularly complex spectra were obtained in the 6.0–8.0 microns (1650–1200 cm^{-1}) region; these are attributable to carbonyl, carboxylate anion, and C=C aromatic absorption. Broad hydroxyl absorption 3.4–2.8 microns (3000–3500 cm^{-1}) was also observed as were ether, phenol and secondary alcohol absorptions between 7.2 microns (1400 cm^{-1}) and 10.0 microns (1000 cm^{-1}). Among the several samples examined, the greatest variation was in the band at 7.35 microns (1370 cm^{-1}), which at times was missing altogether. There were also variations in intensity and resolution in bands at 6.99, 7.07, 8.62, 9.10, 9.20, and 9.40 microns (1430, 1414, 1160, 1099, 1087, and 1064 cm^{-1}). The complexity of the spectra did not permit specific assignments to individual peaks.

Treatment of Indian yellow samples with HCl and recovery through ether extraction gave material whose infrared spectra were in general similar to that of the starting material but with certain obvious differences. Although the car-

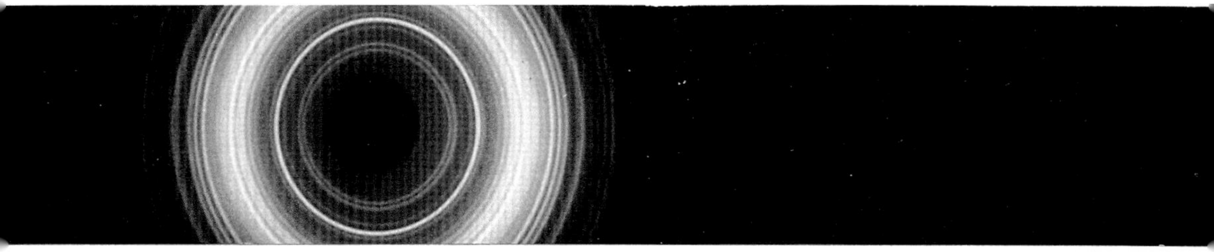

Fig. 8. X-ray powder diffraction pattern of Indian yellow. Sample from ball of pigment provided by Winsor & Newton. Photograph not to be used for measurement purposes.

Table 7. **X-RAY DIFFRACTION PATTERNS OF INDIAN YELLOW SAMPLES**

Sample VI[a, c]		*Sample VI*[b, c]		*1966 Sample*[d]		*Roberson 1910–1920*[e]		*E. W. Forbes Collection*[f] *Lump pigment. Freer Gallery of Art*	
d(Å)	*I*	*d(Å)*	*I*	*d(Å)*	*I*	*d(Å)*	*I*	*d(Å)*	*I*
—	—	—	—	20.3	12	—	—	—	—
—	—	—	—	17.7	4	—	—	—	—
—	—	—	—	—	—	15.6	10	—	—
10.0	vw[g]	9.7	vw[g]	10.0	8	10.2	12	9.72	10
9.5	vw	—	—	9.6	16	—	—	—	—
—	—	9.0	vw	9.0	43	—	—	9.03	100
7.8	vw	—	—	—	—	7.70	71	—	—
—	—	—	—	7.48	34	—	—	7.56	80
7.25	w–i	7.0	w	7.00	100	7.00	44	6.92	50
6.7	vw	—	—	6.60	37	6.60	90	6.61	80
—	—	—	—	—	—	—	—	6.07	< 10
5.95	vw	—	—	5.79	14	—	—	5.72	< 10
5.60	vw	—	—	5.50	22	5.52	16	—	—
5.4	vw	—	—	—	—	—	—	—	—
5.2	vw	5.2	vw	5.25	36	5.24	25	—	—
4.9	w–i	—	—	5.05	40	5.06	100	5.07	100
—	—	4.82	vw	4.82	34	4.79	15	4.80	< 10
—	—	—	—	4.65	10	4.63	25	4.62	< 10
—	—	—	—	4.53	12	—	—	—	—
—	—	—	—	4.41	16	4.41	49	4.42	40
—	—	—	—	4.26	11	4.25	27	4.23	< 10
4.1	w–i	4.11	i	4.06	42	4.05	23	4.00	20
3.93	w–i	3.91	w	3.90	32	3.90	24	3.88	< 10
3.78	w	3.72	vw	3.75	32	3.79	50	3.78	80b[g]
3.66	vw	—	—	—	—	—	—	—	—
3.55	vw	—	—	3.52	30	3.53	85	3.53	100
3.50	w	—	—	3.46	34	3.48	35	3.48	80
—	—	3.4	vw	3.36	5	3.41	9	—	—
3.30	vw	—	—	—	—	—	—	—	—
—	—	—	—	3.23	8	3.23	29	3.23	20
3.05	w	3.14	vw	3.035	21	—	—	—	—
3.00	vw	—	—	2.986	20	3.000	48	2.99	90b
—	—	—	—	2.870	4	2.857	9	—	—
2.83	w	—	—	2.812	20	2.812	26	2.81	40
—	—	—	—	2.764	8	2.760	11	2.75	40
—	—	—	—	2.688	16	2.684	34	2.70	60

Sample VI[a, c]		*Sample VI*[b, c]		*1966 Sample*[d]		*Roberson 1910–1920*[e]		*E. W. Forbes Collection*[f] *Lump pigment. Freer Gallery of Art*	
d(Å)	*I*	*d(Å)*	*I*	*d(Å)*	*I*	*d(Å)*	*I*	*d(Å)*	*I*
—	—	—	—	—	—	—	—	2.66	30
—	—	—	—	2.587	8	2.589	10	—	—
—	—	—	—	2.527	5	—	—	2.55	30
—	—	—	—	2.474	11	2.473	17	2.47	70
—	—	—	—	2.411	6	—	—	—	—
—	—	—	—	2.325	6	2.324	11	2.33	50
—	—	—	—	—	—	—	—	2.29	20
—	—	—	—	—	—	—	—	2.21	40b
—	—	—	—	2.120	6	—	—	2.12	30
—	—	—	—	2.077	5	2.072	12	2.08	40b
—	—	—	—	—	—	—	—	2.03	30b
—	—	1.98	w	1.956	5	1.955	5	1.96	50b
—	—	—	—	1.888	5	—	—	—	—
—	—	—	—	1.845	5	—	—	—	—

a. Guinier-DeWolff, copper tube, no filter, 10-hour exposure, 35 KV, 17 ma.
b. Debye-Scherrer, copper tube, Ni filter, 20-hour exposure, 30 KV, 15 ma.
c. See Table 1 for description of sample.
d. Phillips Diffractometer, Mellon Institute, Research Center on the Materials of the Artist and Conservator, Pittsburgh. Indian yellow 1966 sample (Pollack, 1971).
e. Phillips Diffractometer, Mellon Institute, Research Center on the Materials of the Artist and Conservator, Pittsburgh. Indian yellow Roberson 1910–1920 sample (Pollack, 1971).
f. Data kindly provided by E. W. FitzHugh, Freer Gallery of Art, Washington, Film F1841. Debye-Scherrer technique, 114.6 mm Gandolfi camera. (4 additional lines below 1.81Å)
g. Visual estimates of intensity: vw, very weak; w, weak; w–i, weak-intermediate; i, intermediate; b, broad line.

bonyl absorption region remained complex, a principal absorption band was now discernible at 5.7 microns (1750 cm^{-1}), characteristic of a free carboxylic acid. Apparently the glucuronic acid moiety is not readily hydrolyzed by warm HCl.

Drying of the KBr pellet at 105°C slightly improved the spectrum quality. Reduction of the concentration of Indian yellow in the pigment below 0.4 mg pigment/100 mg KBr led to a considerable loss of spectral detail. At 0.1 mg sample/100 mg KBr virtually no details of the spectrum were recognizable.

The high resolution (2 cm^{-1}) Fourier transform infrared spectrum (Low and Baer, 1977) reveals no major new structural details.

5.33 ***Spectrochemical Analysis.*** Arc-excited emission spectra were obtained for samples III–VI with a Jarrell Ash 21′ Grating Spectrograph. The true Indian yellow specimens (Samples III, V, and VI) gave good magnesium spectra with traces of calcium. As was expected, the aluminum lake (Sample IV) gave only strong aluminum lines. In a separate analysis, Hodge (1971) observed magnesium (major), silicon, calcium (1,000 ppm), and sodium in a specimen kindly supplied by Winsor & Newton.

5.34 ***X-radiography.*** In ordinary x-radiographs, the pigment can be expected to exhibit very low density. Fig. 1D shows a radiograph of an Indian miniature taken with Kodak Type M Industrial x-ray film, a 60-second exposure at 5 MA and 30 KV. Whereas paints that contain heavy metal elements show clearly in the radiograph, paints containing Indian yellow are highly transparent.

5.35 ***Ultraviolet Fluorescence.*** Feller (1968) (fig. 2) reported details of the spectral distribution curve of the fluorescence, indicating that the strongest emission is stimulated by 435 nm radiation. Nevertheless, fluorescence can be conveniently observed by use of a long wavelength ultraviolet lamp (365 nm). The results of the examination of a number of specimens under the usual ultraviolet lamp are presented in table 3. The modern imitation did not give the typical yellow fluorescence of true Indian yellow. If necessary, the spectral excitation and emission characteristics can be used to distinguish genuine Indian yellow from modern fluorescent dyes.

The examination of a series of Indian miniatures of known provenance, all more than 100 years old, demonstrated the liberal use of the pigment in areas of yellow, green, and orange (see 2.4 and 6.0). Modern miniatures are remarkable for the absence of the characteristic fluorescence associated with Indian yellow. Thus, examination with an ultraviolet hand lamp provides a convenient method for distinguishing potentially genuine older works from possible modern ones (FitzHugh, private communication). Fig. 1B shows the typical appearance of Indian yellow under an ultraviolet lamp, the bright yellow areas indicating the pigment.

5.4 Criteria for Positive Identification. As a mixture of materials, Indian yellow can be expected to vary slightly in character from sample to sample. The following four characteristics would make the identification highly probable: a) yellow pigment that fluoresces yellow-to-orange under long wavelength ultraviolet light; b) particles of intense yellow color when examined microscopically under plane polarized light; c) highly birefringent particles of a refractive index about 1.67, yellow in transmitted light and yellow-to-green under crossed polars, showing olive hues in the thicker particles, usually rather coarse in particle size and irregular-to-prismatic in shape exhibiting parallel extinction and positive elongation (samples from Indian paintings do not show such particles); d) all yellow particles should dissolve in dilute mineral acids. The above characteristics, accompanied by an infrared curve similar to that in fig. 9, may be considered to provide positive identification. The last is the most conclusive objective evidence available through normal analytical methods.

Although not necessarily present in every sample, the occurrence of highly birefringent colorless materials, particularly rosette-shaped concretions, would provide supplementary supporting evidence, as would a test for magnesium and an appropriate x-ray diffraction pattern. Excitation and fluorescence spectra would provide strong supporting evidence, while the de-

Fig. 9. Infrared spectra of Indian yellows. A. Roberson, 1914 sample, Edward W. Forbes Collection. B. From ball supplied by Winsor & Newton. Upper curve plotted against linear wavelength; lower curve is linear in wavenumber.

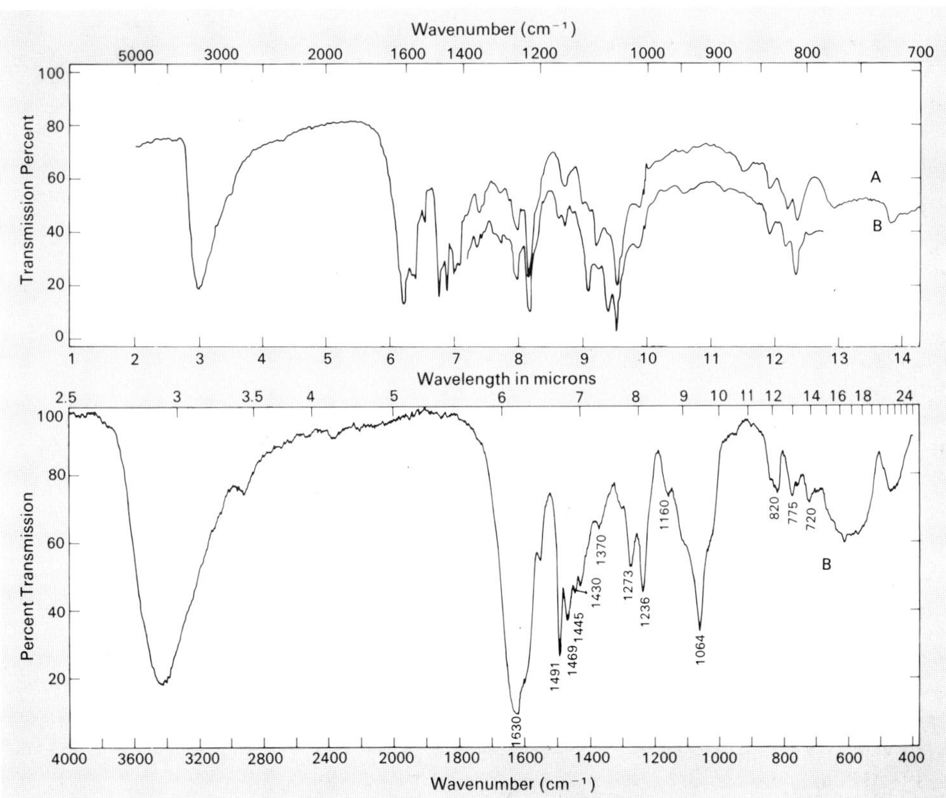

tection of euxanthic acid by chromatography or the preparation of suitable derivatives would provide excellent additional evidence.

6.0 NOTABLE OCCURRENCES

Kühn (1969) cites two examples of the use of Indian yellow in nineteenth-century paintings (Ernst Willers, *Hain bei Aniccia im Abendlicht*, 1873; August Wolf, *Madonna* after Giovanni Bellini, 1873). His reference (Kühn, 1968) to the occurrence of the pigment in *A Woman Weighing Gold*, painted c. 1662–1663 by Jan Vermeer, has subsequently been found erroneous, complicated by the unusually small sample available at the time (Feller, 1976, unpublished results).

Carlson and Krill (1978) tentatively identified Indian yellow in a Reeves and Inwood box of nineteenth-century watercolors in the Winterthur collection. Weston (1977, p. 70, Table II, R4) reported an investigation of a box of watercolors owned by Winslow Homer containing the commercial pigments of Winsor & Newton. The yellow, analyzed by emission spectroscopy is listed as "organic possibly Indian yellow or gamboge." The yellow in the watercolor box of Winslow Homer has been identified (Newman, et al., 1980) as Indian yellow using Debye-Scherrer x-ray diffraction and microscopy.

SOUTH ASIAN PAINTINGS on paper from the Freer Gallery of Art, Washington

Artist or School *Title, Date* *Accession No.*	*Method of Identification*
Mughal, Indo-Persian A *Shahnāmah* MS, late 16th–early 17th century 07.276	UV, Opt., Micr., XRD*
Mughal *Mongol Chieftain with Attendants* c. 1600 52.2	UV, Opt.
Agra *The Beloved and the Confidant* from a Rasikapriya MS 1617 31.5	UV, Opt.
Agra *The Lover Receiving a Message* from a Rasikapriya MS 1617 31.6	UV, Opt.
Mughal, School of Jahangir *Love Scene* 1605–1627 29.80	UV, Opt.
Mughal, School of Jahangir, by Hashim *Portrait of Khan Khanan* 17th century 39.50a	UV, Opt., Micr., XRD*
Mughal, School of Jahangir, by Ab'ul Hasan *Jahangir Standing on a Low Dais* 17th century 48.28	UV, Opt.
Jaipur Scenes from the *Markandeya Purana* 1799 07.602	UV, Opt., Micr., XRD*
Late Mughal *A Ruler Seated on a Terrace* 19th century 07.261	UV, Opt.
Late Mughal, same artist as 07.261 *A Prince with an Attendant* 19th century 07.621	UV, Opt., Micr., XRD*
Nepalese (?) *Portrait of a Maharajah* 19th century 07.218	UV, Opt., Micr.
Burmese or Siamese (?) *Figure of a Man with a Sword* 19th century 07.225	UV, Opt.

Analyst, Elisabeth West FitzHugh.
XRD* = x-ray diffraction; UV = ultraviolet light; Opt. = optical properties; Micr. = chemical microscopy.
*X-ray diffraction did not supply confirmation of the presence of Indian yellow. It did support microscopic evidence of the presence of pigments which were mixed with Indian yellow : lead white (07.276 and 07.602), vermilion in a brown area (39.50a), and red lead in an orange area (07.621).

ACKNOWLEDGMENTS

This project was supported in part by a grant to the Conservation Center, New York University, from the National Endowment for the Arts in Washington (Baer, Indictor and Joel, 1972). The authors are greatly indebted to Doris and Ed Wiener for a grant and the generous loan of study materials, to Elisabeth West FitzHugh for her efforts in investigating and documenting the presence of the pigment on Indian and Near Eastern paintings in the Freer Gallery of Art, Washington and to Winsor & Newton for a sample of authentic material. We also wish to thank Marjorie Shelley and James H. Frantz for their assistance with the x-ray diffraction measurements.

REFERENCES

G. H. Bachhoffner, *Chemistry as Applied to the Fine Arts* (London, 1837), 163.

N. S. Baer, N. Indictor, and A. Joel, "The Chemistry and History of the Pigment Indian Yellow," *Conservation and the Graphic Arts*, 1972 Congress of the International Institute for Conservation (Lisbon, 1972), 401–408.

A. Baeyer, "Ueber das Euxanthon und die Euxanthinsäure," *Annalen der Chemie und Pharmacie* 155, no. 3 (1870), 257–266.

N. F. Barnes, "Color Characteristics of Artists' Pigments," *Journal of the Optical Society of America* 29 (May, 1939), 208–214.

Bengal Acts (London, 1866/1869).

M. P. L. Bouvier, *Manuel des jeunes artistes et amateurs en peinture* (Paris and Strasbourg, 1832 and 1844) (1829).

British Standard 1006, Methods for the Determination of the Colour Fastness of Textiles to Light and Weathering (London, 1971).

J. H. Carlson and J. Krill, "Pigment Analysis of Early American Watercolors and Fractur," *Journal of the American Institute for Conservation* 18 (1978), 19–32.

M. Chandra, *Jain Miniature Paintings from Western India* (Ahmedabad, 1949a), 83.

M. Chandra, *The Technique of Mughal Painting* (Lucknow, 1949b), 27.

A. H. Church, *The Chemistry of Paints and Painting*, 4th ed. (London, 1915), 175–176.

Colour Index, vol. 3, 3rd ed. (Bradford, Yorkshire, England, and Research Triangle Park, North Carolina, 1971).

R. Dewhurst, "Journal, 1784–87," Manchester Reference Library, MS. 923.9 D55, 999–1,001.

M. Doerner, *The Materials of the Artist and Their Use in Painting* rev. ed. trans. E. Neuhaus (New York, 1949).

A. Eibner, "Ueber Indischgelb und Seine Ersatzmittel," *Technische Mittheilung für Malerei* 22 (1905), 174–178.

A. Eibner, *Malmaterialienkunde als Grundlage der Maltechnik* (Berlin, 1909), 195–200.

O. L. Erdmann, "Ueber das *Jaune indien* und die darin enthaltene organische Säure *(Euxantinsäure)*," *Journal für praktische Chemie* 33 (1844), 190–209.

F. Feigl, *Spot Tests in Inorganic Analysis*, 5th ed., trans. R. E. Oesper (Amsterdam, 1958), 224–225.

R. L. Feller, "Problems in Spectrophotometry," in *London Conference on Museum Climatology*, ed. G. Thomson (London, 1968),257-269.

G. Field, *Chromatography: A Treatise on Colours and Pigments* (London, n.d.), 157–158.

G. Field, "Examples and 'Anecdotes' of Pigments. Practical Journal. 1809." Courtauld Institute of Art, catalogue FIELD/6.

T. H. Fielding, *Ackermann's Manual of Colours, Used in the Different Branches of Water-Colour Painting: with an Ample Description of the Value and Properties of Each Colour* (London, 1844).

T. H. Fielding, *Index of Colours and Mixed Tints* (London, 1830).

T. H. Fielding, *On the Theory of Painting, to Which is Added an Index of Mixed Tints*, 3rd ed. (London, 1842).

M. Gartside, *An Essay on Light and Shade, on Colours, and on Composition in General* (London, 1805).

R. J. Gettens and G. L. Stout, *Painting Materials, A Short Encyclopedia* (1942; New York, 1966), 119.

Government of India Acts (London, 1890).

C. Graebe, "Ueber die Euxanthon Gruppe," *Annalen der Chemie* 254 (1889), 265–303.

C. Graebe, "Ueber Eine mit dem Euxanthon Isomere Verbindung," *Berichte der Deutschen Chemischen Gesellschaft* 16 (1883), 862–864.

C. Graebe and R. Ebrard, "Ueber Euxanthon," *Berichte der Deutschen Chemischen Gesellschaft* 15 (1882), 1675–1680.

A. Guinier, *X-ray Diffraction* (San Francisco, 1963), 149.

R. D. Harley, *Artists' Pigments; c. 1600–1835, A Study in English Documentary Sources*, 2nd ed. (London, 1982), 105.

R. D. Harley, "Field's Manuscripts: Early Nineteenth Century Colour Samples and Fading Tests," *Studies in Conservation* 24 (1979), 75–84.

N. Heaton, *Outlines of Paint Technology*, 3rd ed. (London, 1947), 139.

P. C. Henderson, *The Seasons, or Flower-garden: . . . with a Treatise, or General Instructions for Drawing and Painting Flowers* (London, 1806).

H. Hiler, *Notes on the Technique of Painting* (London, 1957), 107–108.

E. S. Hodge, *Mellon Institute Spectrographic Analysis Laboratory Report* (1971).

P. Karrer, *Organic Chemistry* (New York, 1947), chap. 42.

K. L. Kelly and D. B. Judd, *Color, Universal Language and Dictionary of Names* (Washington, 1976).

K. Khandalavala, *Pahārī Miniature Painting* (Bombay, 1958), chaps. I, IV.

H. Kittel, ed. *Pigmente. Herstellung, Eigenschaften, Anwendung* (Stuttgart, 1960), 413–414, 436.

H. Kühn, *Die Pigmente in den Gemälden der Schack-Galerie* (Munich, 1969), 60–84.

H. Kühn, "A Study of the Pigments and the Grounds Used by Jan Vermeer," *Report and Studies in the History of Art*, National Gallery of Art, Washington (1968), 155–175.

M. J. Lejort, *Chimie des couleurs* (Paris, 1855), 131.

M. J. D. Low and N. S. Baer, "Application of Infrared Fourier Transform Spectroscopy to Problems in Conservation. I. General Principles," *Studies in Conservation* 22 (1977), 116–128.

F. Mayer, "Bemerkung zu einer Arbeit des Herrn Wilhelm Wiechowski, 'Uber die Muttersubstanz des Indischgelb'," *Archiv für experimentelle Pathologie und Pharmakologie* 101 (1924), 383–385.

V. Meyer and P. Jacobson, *Lehrbuch der organischen Chemie* (Berlin, 1920), 763 764.

H. W. M'Cann, *Report on the Dyes and Tans of Bengal* (Calcutta, 1883), 92.

T. N. Mukharji, "Piuri or Indian yellow," *Journal of the Society of Arts* 32 (1883–1884), 16–17.

A. Munkert, *Die Normalfarben* (Stuttgart, 1905), 54–56.

Munsell Book of Color (Baltimore, 1929; 1970).

R. Newman, C. Weston, and E. Farrell, "Analysis of Watercolor Pigments in a Box owned by Winslow Homer," *Journal of the American Institute for Conservation* 19 (1980), 103–104.

E. J. Parry and J. H. Coste, *The Chemistry of Pigments* (London, 1902), 273–275.

A. G. Perkin and A. E. Everest, *The Natural Organic Colouring Matters* (London, 1918), 127–131.

H. J. Plenderleith, *The Conservation of Prints, Drawings and Manuscripts* (Oxford, 1937), 9.

S. S. Pollack, *Mellon Institute X-ray Diffraction Laboratory Research Report (7615–4)* (1 April 1971).

B. H. Powell, *Handbook of the Economic Products of the Punjab*, vol. 1 (Roorkee, 1868), 195.

B. de Puyster, "Use of Organic Dyestuffs for Lake Manufacture," Article IV, Kermes, Indian Yellow and Sepia, *Color Trade Journal*, 4, No. 4 (1919), 98-100.

F. Reindel and L. Anschutz, *Richter's Organic Chemistry*, vol. 4, ed. R. Anschutz (New York, 1947), 191.

M. M. Riffault, Vergnaud, and Toussaint, *A Practical Treatise on the Manufacture of Colors for Painting* (London, 1874), 417–420.

M. Saltzmann and H. Wichelhaus, "Zur Kenntniss des Euxanthons," *Berichte der Deutschen Chemischen Gesellschaft zu Berlin* 10 (1877), 1397–1403.

W. Schmid, "Ueber das Mangostin," *Annalen der Chemie und Pharmacie* 93 (1855), 83–88.

L. E. Seltzer, ed. *The Columbia Lippincott Gazetteer of the World* (New York, 1962), 1228.

J. Stenhouse, "Untersuchung einer gelben Substance, welche unter dem Namen Purree von Indien Kommt," *Annalen der Chemie und Pharmacie* (1844), 423–431.

A. L. Van Scherpenberg, *Onderzoekingen naar de constitutie van het euxanthogeen, de moederstof van hiet Indisch geel* (Amsterdam, 1918).

M. R. Wagner, "Emploi de l'acide euxanthique dans la teinture et la fabrication des coleurs," *Le technologiste* 21 (1859–1860), 122–123.

G. Watt, *A Dictionary of the Economic Products of India*, vol. 6 (London, 1892), 132–134.

F. W. Weber, Artist's Pigments, Their Chemical and Physical Properties (New York, 1923), 67–68.

K. Wehlte, *The Materials and Techniques of Painting*, trans. U. Dix (New York, 1975), 94–95.

R. C. Weston, "Pigment Analysis" in *Wash and Gouache*, ed. M. B. Cohn (Cambridge, Massachusetts, 1977), 70.

W. Wiechowski, "Uber die Muttersubstanz des Indischgelb," *Archiv für experimentelle Pathologie und Pharmakologie* 97 (1923), 462–488.

Winsor & Newton, *Notes on the Composition and Permanence of Artists' Colours, Including a Description of Oils Painting Media and Varnish* (Secaucus, New Jersey, n.d.), 8.

G. Zerr and R. Rübencamp, *A Treatise on Colour Manufacture*, trans. C. Mayer (London, 1908), 438.

Cobalt Yellow (Aureolin)

MAURA CORNMAN

NOMENCLATURE IN VARIOUS LANGUAGES

English: cobalt yellow (aureolin)
German: Kobaltgelb (Aureolin)
Russian: кобальтинитрит калия
French: jaune de cobalt (aureolin)
Italian: giallo di cobalto
Spanish: amarillo de cobalto

1.0 INTRODUCTION

Cobalt yellow, or aureolin, was introduced as an artists' pigment in 1851 and enjoyed popularity in the late nineteenth century; today it finds only limited use in the artist's palette. The name is derived from the Latin *aureus*, meaning golden. Chemically, this synthetic inorganic yellow compound is based on the precipitation of potassium cobaltinitrite, $K_3[Co(NO_2)_6]$ (Durrant and Durrant, 1970; Parkes, 1967). Almost everyone who has taken an elementary chemistry course will recall the yellow precipitate formed as a test for potassium.

Because of its moderate refractive index and its chemical properties, cobalt yellow has been more important in watercolor than in oil mediums. In modern usage, either of the terms aureolin or cobalt yellow may be used to designate the pigment. In the *Colour Index* (1971) it is listed as Pigment Yellow 40, No. 77357.

MAURA CORNMAN, assistant professor of art history and associate curator of conservation, University of Missouri, Columbia.

2.0 HISTORY

2.1 Archaic, Obsolete, and Misrepresentative Names

Almost from the outset of its use as a painting material and extending to modern times, aureolin has also been sold under the misnomer jaune indien (*Gentele's Lehrbuch*, 1909; Mierzinski, 1881) or Indian yellow (Bersch, 1901; Hofmann and Burger, 1907; Kittel, 1960). Although vaguely similar in hue, the organic colorant Indian yellow is distinctively different both optically and chemically from cobalt yellow (see chapter on Indian Yellow). Coffignier (1924) used the term aurocoline.

2.2 History of Use

Cobalt yellow was first synthesized by N. W. Fischer (1831) at Breslau during his investigation of nitrite salts. In his 1831 publication he mentioned that potassium nitrite combined with many metal nitrites, including cobalt nitrite, to form what were then considered to be "double salts," but he did not describe specific compounds until 1848. Apparently unaware of Fischer's work, Saint-Evre (1851, 1852) rediscovered the pigment in Paris. Thus, while credit for its invention has been assigned to Fischer — the pigment frequently being called Fischer's Salt (*Colour Index* 1971; Erdmann, 1866; Sadtler, 1870) — the introduction of cobalt yellow into the art world is accredited to Saint-Evre (1851) (Mayer, 1948). We find the pigment associated with his name in Tschelnitz's (1857) contemporaneous text. According to Winsor & Newton (1910) cobalt yellow was introduced in England by T. Salter in 1860.

Following the discovery of Fischer's Salt the

determination of its chemical formula was the object of considerable investigation, most notably by Stromeyer (1855), Erdmann (1866), Braun (1868), Sadtler (1870), and Rosenheim and Koppel (1898). In addition to the interest in the compound as an artists' pigment, its formation by reaction with potassium nitrite was early suggested for the detection of potassium and cobaltous ions; the reaction could also be used to separate cobalt from other metals, especially nickel (Adie and Wood, 1900; Bowser, 1909, 1910; Bray, 1909; Burgess and Kamm, 1912; Cunningham and Perkin, 1909; Drushel, 1908; Erdmann, 1866; Fischer, 1848; Fresenius, 1909; Rose, 1860; Rosenheim and Koppel, 1898).

Cobalt yellow is said to have been popularized as a pigment by the watercolorist Aaron Penley (Gettens and Stout, 1966, p. 110) and by the Leipzig artist Arthur Mühlberg (Wehlte, 1967, p. 105). It was among the pigments investigated and shown to be satisfactory by Russell and Abney in their 1888 work on the lightfastness of watercolors; at the time it apparently was a moderately popular pigment (Brommelle, 1964).

Partly owing to its high cost of production, and also to the availability of superior yellow pigments, aureolin was in vogue only a short time (Blythe, private communication, 1975; Eibner, 1909; Gettens and Stout, 1966; Wehlte, 1967; Bersch, 1905). Currently many better and less expensive yellows are available, with the result that colbalt yellow is only rarely found in modern colormen's catalogues. Although the compound has uses for the painting of glass and porcelain, where it yields a blue color on firing owing to the presence of cobalt (*Colour Index*, 1971; Kittel, 1960), it remains essentially a pigment for watercolor.

2.3 Terminal Date

The pigment is still available today, manufactured by Blythe Colours Ltd., Stoke-on-Trent, England.

3.0 GENERAL PIGMENT PROPERTIES

3.1 Color and Spectral Reflectance

Spectral reflectance curves of the pigment in oil and in watercolor mediums are shown in fig. 1. Based on these samples, the ISCC-NBS color name for aureolin in watercolor is brilliant yellow (Munsell designation 6Y/8.4/10.9); when the compound is mixed with titanium white in oil, the color name is strong yellow (Munsell designation 3Y/7.2/10). These names are also the hue designations given in Mairz and Paul (Kelly and Judd, 1976). According to Saint-Evre (1851), the pigment corresponds to the yellow in Chevreul's first circle of chromaticity.

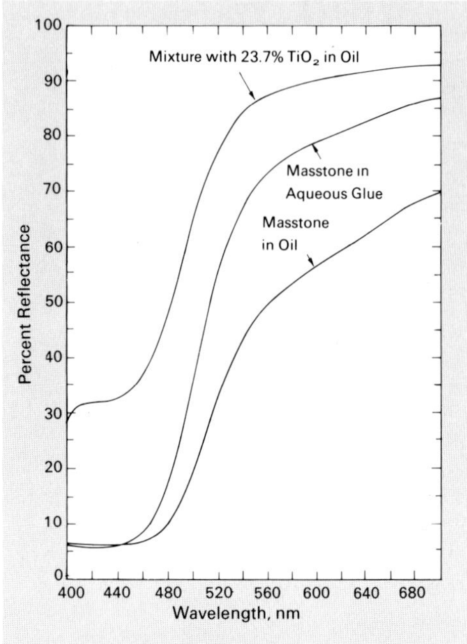

Fig. 1. Spectral reflectance curves for cobalt yellow in oil and aqueous glue.

A primrose or more greenish shade is said to have been introduced by Winsor & Newton in 1889 (Gettens and Stout, 1966, p. 110). Standage (1896) states "with cobalt blue and rose madder it forms exceedingly pure and transparent aerial greys."

3.2 Hiding Power and Tinting Strength

The pigment possesses only a moderately high refractive index, 1.72 to 1.76, a value similar to that of emerald green (copper aceto-arsenite) and blue verditer (Gettens and Stout, 1966). As a consequence of its relative transparency in oil mediums cobalt yellow is most useful when employed in watercolor, although it can also serve as a glaze in oil.

3.3 Permanence

The permanence of the pigment has been subject to question ever since aureolin's introduction as an artists' material. Experimentation over the years has indicated that cobalt yellow is given to apparently capricious reactions. Thus, the evidence in the literature concerning its permanence is contradictory — some reports suggest that it is extremely stable while others indicate that it may turn brown on exposure or contribute to the degradation of admixed organic pigments. An inconclusive picture remains, and cobalt yellow's replacement by more reliable and less expensive pigments has all but eliminated interest in further investigation of its properties and value as an artists' colorant.

Most of the older literature published concerning cobalt yellow's permanence refers to its stability in watercolor, implying that its use in oil paints was not considered a major application. Several writers state flatly that the compound is not a desirable oil pigment. Martin Fischer (1930) unequivocably declared that it should not be a part of the oil painter's palette. Laurie (1926) said that if too much oil is used the pigment will go brown. Ordinarily in the evaluations described, little or no reference is made to the process by which the particular pigment discussed was prepared; hence, there is little documentation concerning the possible presence of impurities and their effect on the compound's stability. It seems likely that impurities, present because of poor washing for example, would contribute to the inferior performance reported by some experimenters (Toch, 1911).

The pigment is said to be unaffected by atmospheric action (Bersch, 1905; Blythe, private communication, 1975) and resistant to hydrogen sulfide attack (Saint-Evre, 1852; Toch, 1911), although suspensions of it in water are slowly attacked by the gas (Church, 1915, p.169; Mellor, 1928; Stromeyer, 1855). It is sensitive to heat (*Colour Index*, 1971; Eibner, 1909; Mellor, 1928; Stromeyer, 1855), decomposing at 200°C to give cobalt oxide; hence, the pigment cannot be used in an enamel. It can, as the manufacturer suggests, be used in temperature-indicating paints. Cobalt yellow is not limefast (Eibner, 1909) and is consequently unsuitable for fresco.

From all reports, cobalt yellow is lightfast in most mediums and considered to be a permanent pigment in watercolor (Laurie, 1926). The account published by Russell and Abney in 1888 spoke highly of the compound; their specimen faded only very slowly (Brommelle, 1964). There is some indication that the pigment may decompose over time (Rose, 1860), but no modern studies have investigated the question.

While reports of the stability of cobalt yellow have varied somewhat through the years, today it is generally considered to be a relatively unstable pigment, particularly in its reactivity toward organic substances. Standage (1896) cautioned against the use of a steel spatula in blending cobalt yellow with another pigment, especially when the latter is impure. Such an action, he felt, might cause the reduction of cobalt yellow by the metal and thus accelerate its deterioration.

3.4 Compatibility

Aureolin appears to blend well with other inorganic pigments, but mixing it with organic colorants may prove disastrous. Winsor & Newton caution against its use with traditional organic pigments. The compound is known to accelerate the decomposition of such pigments as indigo, cochineal, and many lakes. In the process, cobalt yellow itself turns brown (Gettens and Stout, 1966; Toch, 1911; Winsor & Newton, 1910). The reaction with indigo was noted by Standage (1896): "Indigo is oxidized . . . by admixture with it, the yellow of the aureolin becoming brown." In watercolor, where the relative concentration of vehicle would be at a minimum, the pigment apparently is quite stable. In 1910 Blockx stated that he had been experimenting with cobalt yellow since 1879 and that several specimens in his possession had been exposed to air and summer sun for seven years and had "perfectly stood this severe test."

3.5 Chemical properties

While the results of testing are often disputed, the pigment appears indeed to be a relatively reactive substance on the basis of its reported chemical properties. It is only slightly soluble in water, 0.89 g/l at 17°C, whereas the corresponding sodium salt is readily soluble (Mellor, 1928). It is also sparingly soluble in alcohol, ether, and carbon disulfide (Saint-Evre, 1851). At 25°C the solubility in alcohol is about one-tenth that in water, 0.026 g/l in alcohol versus 0.21 g/l in water (Pierrat, 1921).

Decomposition of cobalt yellow in boiling water, with the evolution of nitric oxide fumes, leaves an alkaline, rose-red colored liquid (Mellor, 1928; Saint-Evre, 1851; Stromeyer, 1855). While the pigment resists the action of H_2S, ammonium and sodium sulfide cause immediate decomposition yielding a black residue (Mellor, 1928; Laurie, 1930). Cobalt yellow is decomposed by concentrated nitric acid and by both dilute and concentrated bases; the latter turn the remaining substance dark (Kittel, 1960; Laurie, 1930; Saint-Evre, 1851). When warmed with dilute nitric or hydrochloric acid, the pigment dissolves, producing a pink-colored solution.

3.6 Oil Absorption and Grinding Qualities

As an oil color, the pigment requires 6 parts oil to 4 parts dry powder for mixing and grinding (Uebele, 1913). The manufacturer states that severe grinding will cause the pigment to darken (Blythe, private communication, 1975).

3.7 Toxicity

The toxicity of cobalt salts ingested by mouth is low, although soluble nitrites ingested in large amounts by mouth are extremely toxic. Repeated small doses cause rapid pulse, fall in blood pressure, headache, and visual disturbances. Principal results are from experiments on animals, however, and no information with respect to humans is considered valid (Sax, 1963). If handled with care, this relatively insoluble salt should be reasonably safe; ingestion should be regarded as requiring medical advice and attention.

4.0 COMPOSITION

4.1 Chemical Composition

From the time of cobalt yellow's discovery, its composition has been in dispute. The accepted formula for the salt is $K_3[Co(NO_2)_6]$, with a degree of hydration considered to vary from 0 to 4 molecules of water, depending on the concentrations of the solutions used (Church, 1915; Rosenheim and Koppel, 1898; Sadtler, 1870). X-ray powder diffraction work by van Driel and Verweel (1936) showed the anhydrous salt to have a cubic unit cell containing four $K_3Co(NO_2)_6$. Ferrari et al. (1955) determined the unit cell edge a to be 10.505 Å. The Adie and Wood method of preparation, as noted by Church (1915), produces an especially brilliant yellow precipitate having the formula $K_2Na[Co(NO_2)_6]$ (Adie and Wood, 1900; Kittel, 1960). Blythe Colours (private communication, 1975) states that this is the formula of the substance now being marketed. From the limited information available, it appears that this compound is at least equal in pigmentary properties to the tripotassium salt and may even

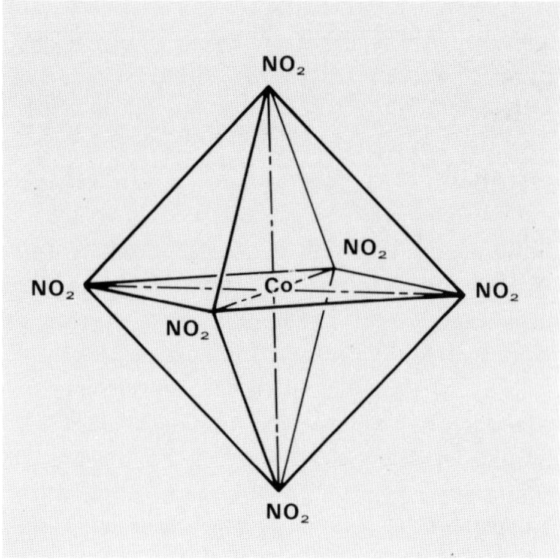

Fig. 2. Regular octahedral coordination of Co (III) by six NO_2 groups in the complex anion of cobalt yellow $[Co(NO_2)_6]^{-3}$.

be superior because it is less soluble (Adie and Wood, 1900; Church, 1915).

These and related salts may be considered derivatives of the complex acid, $H_3Co(NO_2)_6$ (Kittel, 1960). X-ray crystallographic measurements demonstrate that the molecule is stabilized by the formation of the complex anion $[Co(NO_2)_6]^{-3}$, in which six nitro groups are coordinated about the central Co (III) ion at the vertices of a regular octahedron, as shown schematically in fig. 2. Thus the earlier concept of potassium cobaltinitrite as a double salt was erroneous. The six Co-N bond lengths are identical, 2.04 Å, and the bond lengths within the NO_2 groups are N–O = 1.11 Å and O–O = 2.10 Å (Bailar et al., 1973; van Driel and Verweel, 1936; Durrant and Durrant, 1970).

Hexanitrocobaltates of this structure are strong oxidizing agents, tending to revert to the more stable cobalt (II) forms. Hence, we expect and indeed find that this complex is reactive with many substances. The distinctive yellow color of the potassium compound and other cobaltinitrites, compared with the perhaps more familiar reds, pinks, or blues of other cobalt compounds, is the result of the octahedral coordination.

4.2 Sources

There is currently only one manufacturer of the pigment, Blythe Colours Ltd., Stoke-on-Trent, England.

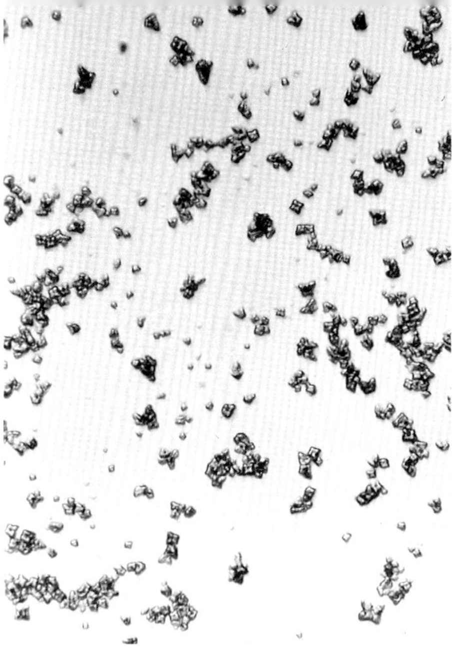

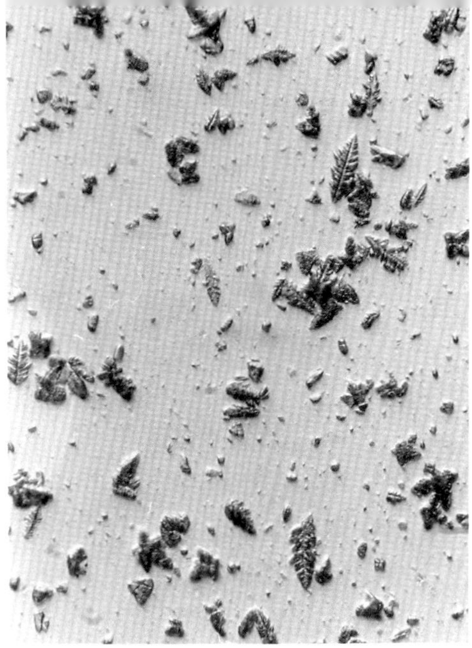

Fig. 3. Photomicrographs of (A) typical rhombohedral-type cobalt yellow, and (B) dendritic variety. Sample in mounting medium of refractive index 1.53, with substage shadowing. 450x.

4.3 Preparation

The pigment is principally produced by the acidification of solutions of cobaltous salts with acetic acid, followed by mixing with a concentrated solution of potassium nitrite. Cobalt yellow then slowly precipitates as a yellow crystalline mass (Bearn, 1923; *Gentele's Lehrbuch*, 1909; Kittel, 1960; Rose, 1860).

Developed by Hayes, another method of production directs the vapors of nitric acid, derived from its action on copper, into a solution containing cobalt nitrate and potassium, the latter preferably present as the hydroxide or acetate (*Gentele's Lehrbuch*, 1909; Mierzinski, 1881). It is necessary to add further potassium to the mixture from time to time until no additional precipitation occurs.

More recent than either of the above-mentioned methods is one probably first used by DeKoninck as a means of detecting potassium, and later published by Adie and Wood (1900) (Bowser, 1909). The yellow salt is precipitated by mixing solutions of a potassium salt and sodium cobaltinitrite, acidified with acetic acid. However, there are some problems with this method, because the relative concentrations of the solutions are critical to the salt's formation. If the potassium is in great abundance, the tripotassium salt is formed, but if the sodium cobaltinitrite is in excess, the dipotassium monosodium salt is formed. The latter is considered by the current manufacturer to be the salt produced (Blythe, private communication, 1975). However, atomic absorption analysis for sodium does not confirm this composition (see 5.33).

Following its preparation by any of the above procedures, the salt must be thoroughly washed to achieve maximum stability (Weber, 1923).

4.4 Adulteration and Sophistication

Owing to its translucent quality, cobalt yellow is often adulterated to give better hiding power and to enrich the hue. The most frequent adulterants are chrome and cadmium yellows (Church, 1915; *Gentele's Lehrbuch*, 1909; Standage, 1896). A simple means for the detection of adulterants is to place a sample in a crucible with sulfuric acid, and to heat until the acid evaporates; if the sample is free of barium and lead contaminants, no residue will be seen when water is added to the crucible (*Gentele's Lehrbuch*, 1909). Another simple test involves mixing the sample with alcohol and adding a few drops of ammonia; if the pigment contains chromate, the liquid will turn orange or yellow (Church, 1915).

5.0 IDENTIFICATION

5.1 Characterization by Optical Microscopy

Cobalt yellow crystals, although belonging to the cubic system, appear as rhombs or diamond shapes terminated with four triangular facets (see fig. 3) (Gibbs, 1874; Mellor, 1928; Saint-

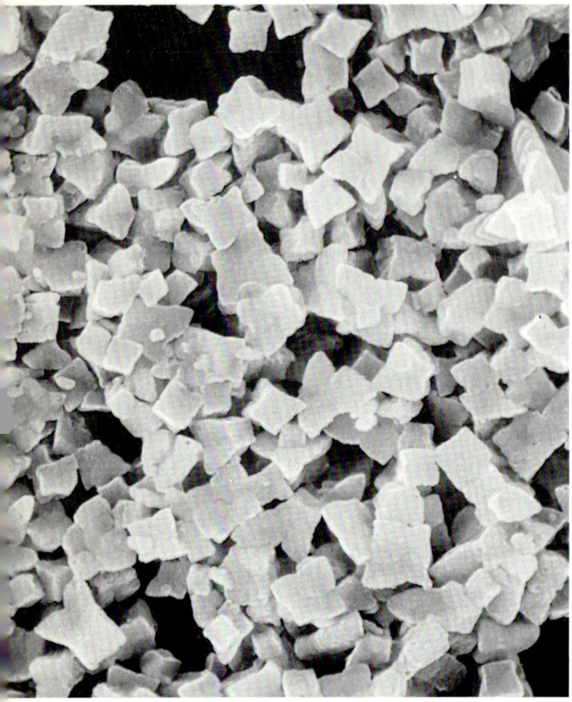

Fig. 4. Scanning electron micrograph of well-formed rhombs of cobalt yellow. 2500x. Compare with fig. 3A to see the appearance of these same particles under an optical microscope.

Fig. 5. Photomicrograph of a variety of commercial aureolin having high concentrations of dendritic particles. Samples in mounting medium of refractive index 1.52, with substage shadowing. 650x.

Evre, 1852). Fig. 4 shows the appearance of these well-formed rhombs under the scanning electron microscope. Being cubic (Bailar et al., 1973), the crystals have the distinction of being one of the few isotropic inorganic pigments. Therefore, the polarizing microscope can frequently provide an immediate indication of the presence of true cobalt yellow rather than one of its birefringent substitutes (Gettens and Stout, 1966).

Occasionally, particles of dendrite fragments can be seen under the microscope, which provides further indirect evidence that the pigment is cobalt yellow. The photomicrograph in fig. 5 indicates the appearance of one commercial product that was dendritic in character.

5.2 Chemical Identification

The presence of cobalt can be demonstrated by means of several spot tests, two of which deserve particular mention, one employing α-nitroso-β-naphthol and the other, rubeanic acid reagent (Plesters, 1956; Feigl, 1958). The procedure using the latter is as follows: the sample is heated in a porcelain crucible to burn off the organic matter, and while still warm the residue is treated with dilute HCl or HNO_3. A spot of this solution on filter paper is placed over a beaker containing ammonium hydroxide, and a drop of 1% solution of rubeanic acid in ethanol is added. The development of an orange-brown coloration indicates the presence of Co^{++}. It is advisable to run a control test with a known sample of cobalt yellow.

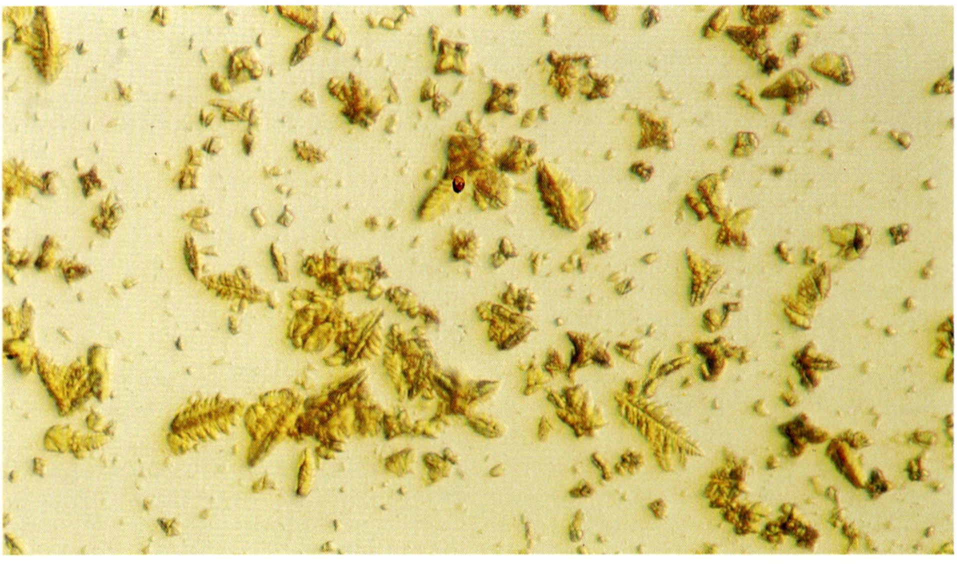

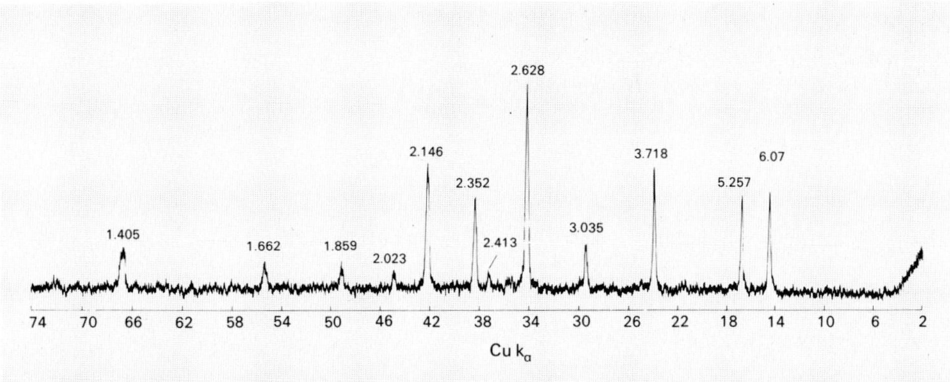

Fig. 6. X-ray diffractometer trace of cobalt yellow using Cu K_{α} radiation. The *d*-spacing in Å corresponding to the diffraction peaks is shown above the individual peak. The numbers below are the diffraction angle, 2θ.

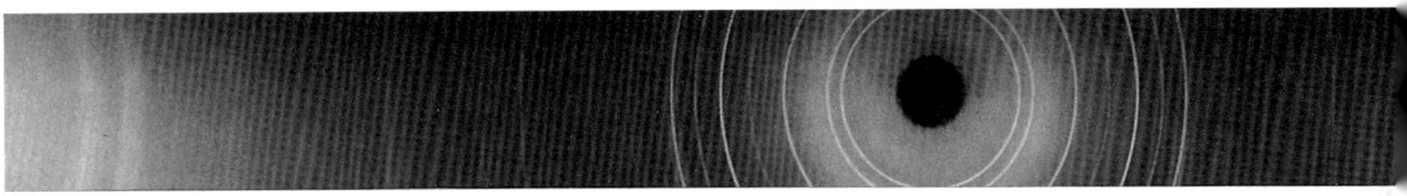

Fig. 7. X-ray powder diffraction pattern of commercial cobalt yellow pigment (Blythe Colours, Ltd.). Photograph not to be used for measurement purposes.

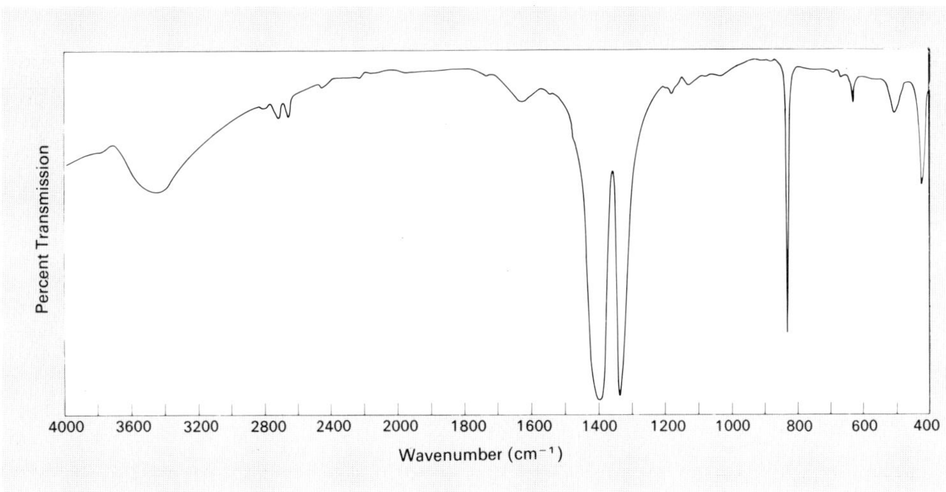

Fig. 8. Infrared absorption spectrum of typical cobalt yellow pigment.

A blue glass bead is formed when a sample is heated with borax on a platinum wire (Stromeyer, 1855; Weber, 1923).

5.3 Instrumental Methods of Analysis

5.31 ***X-ray Diffraction.*** Fig. 6 contains a diffractometer trace of Blythe cobalt yellow. The reported *d*-spacings appear in table 1. The Debye-Scherrer powder diffraction pattern of a representative sample appears in fig. 7.

5.32 ***Infrared Spectroscopy.*** The infrared spectrum of cobalt yellow, shown in fig. 8, has several sharp, distinctive absorption bands, in particular two at about 1337 and 1398 cm^{-1}, followed by a somewhat weaker one at 832 cm^{-1}. This pattern has been verified on samples

Table 1. **X-RAY DIFFRACTION DATA FOR POTASSIUM NITROCOBALTATE (III) (POTASSIUM COBALTINITRITE, CUBIC), $K_3Co(NO_2)_6$[a] (Diffractometric Data, $CuK\alpha_1$ = 1.5405 Å)**

d(Å)	*I*
6.07	30
5.257	35
3.718	60
3.035	25
2.628	100
2.412	14
2.352	45
2.1464	75
2.0234	16
1.8585	18
1.6622	20
1.4576	12
1.4046	30
1.3145	12
1.2390	10
1.1752	16
1.1204	20

[a]JCPDS Powder Diffraction File, 9-404. Courtesy of JCPDS International Centre for Diffraction Data (1982).

acquired from three artists' supply houses. Similar bands, at 1430, 1335, and 847 cm^{-1}, have been reported for the trisodium salt by Miller and Wilkins (1952).

Minor absorption bands appear at 1632, 2660, and 2720 cm^{-1} in the spectrum of the pigment. The absorption at about 3450 cm^{-1} is probably due to hygroscopic water. Miller and Wilkins (1952) report corresponding bands for the sodium salt at 1645, 2665, 2780, and 3450 cm^{-1} respectively.

5.33 ***Spectrochemical Analysis.*** The prominent lines of sodium at 5889.9 and 5895.9 Å, potassium at 7664.9 and 7698.9 Å, and cobalt at 3453.5, 3465.8, and 3529.8 Å were observed in the test samples of modern cobalt yellow burned in a direct current arc.

Atomic absorption analysis gave the following values: potassium 24.54, cobalt 13.89, sodium 0.93%, which may be compared with the respective values 25.93, 13.63, 0% for anhydrous $K_3Co(NO_2)_6$ (courtesy, E. S. Hodge, Carnegie-Mellon University, Pittsburgh). This was an unexpected result because the current suppliers believed to the best of their understanding that they were preparing the monosodium-dipotassium salt (Blythe, private communication, 1975). The detection limits are approximately 0.1, 1.0, and 0.01 ppm, respectively, for potassium, cobalt, and sodium.

Small amounts of cobalt yellow might be difficult to detect when mixed with lead white or barite because of the high x-ray absorption of lead and barium. In addition, one of the two strongest lines of cobalt yellow superimposes on the strongest line of each of these white pigments.

5.4 Criteria for Certain Identification

The identification of isotropic yellow particles with a refractive index between 1.72 and 1.76 provides a reasonably good indication of cobalt yellow. X-ray diffraction or infrared spectroscopy must also be used, particularly if the particles are too small for optical study. Either of the two latter techniques can show that the pigment is a cobaltinitrite, but some form of elemental analysis, such as emission spectroscopy, is required to establish the presence of potassium and/or sodium. If the infrared bands are weak because of contaminants or diluents, it may not be possible to distinguish cobalt yellow positively from potassium and sodium nitrate or nitrite.

Positive identification of cobalt yellow, especially distinguishing the tripotassium from the dipotassium-sodium cobaltinitrite, is best accomplished through a combination of x-ray diffraction to identify the basic compound present and elemental analysis to identify cobalt and the alkali metals.

6.0 NOTABLE OCCURRENCES

We have been able to find only one technically documented occurrence of cobalt yellow, that reported by Kühn (1969) in a work by August Wolf painted between 1874 and 1880. It is of interest to note, however, that of the forty-six watercolor artists who reported the pigments they used to Russell and Abney (1888), eighteen reported the use of aureolin. Moreau-Vauthier (1923) quotes from the artist Dagnan, who says that he occasionally employed aureolin for special subjects such as flowers and fruit.

One also finds Huish (1904), in a summary of the pigments used by thirty-two British watercolor artists of the period 1804 and 1904, reporting that half the artists used this yellow.

ACKNOWLEDGMENTS

The x-ray diffraction data were provided by Dr. Sydney Pollack formerly associated with the Mellon Institute, Pittsburgh. Atomic absorption analyses were provided by Dr. Edwin S. Hodge, also of that institution. The photomicrographs of the many morphological varieties of pigment were taken from specimens in the collection of the National Gallery of Art Pigment Bank, Mellon Institute, Pittsburgh, as were the spectral reflectance and infrared curves. A generous sample and helpful comments on the production of the pigment were kindly supplied by H. S. Clarke of Blythe Colours Ltd., Stoke-on-Trent, England.

REFERENCES

R. H. Adie and T. B. Wood, "A New Method of Estimating Potassium," *Journal of the Chemical Society, Transactions* 77 (1900), 1076–1080.

J. C. Bailar, H. J. Emeléus, R. Nyholm, and A. F. Trotman-Dickenson, *Comprehensive Inorganic Chemistry*, vol. 3 (New York, 1973).

J. G. Bearn, *The Chemistry of Paints, Pigments and Varnishes* (London, 1923).

J. Bersch, *The Manufacture of Mineral and Lake Pigments*, 3rd ed. rev., trans. A. C. Wright (London, 1901).

J. Bersch, *Malerfarben und Malmittel* (Leipzig, 1905).

J. Blockx, *A Compendium of Painting* (London, 1910), 66.

L. T. Bowser, "On the Determination of Potassium by the Cobaltinitrite Method," *Journal of Industrial and Engineering Chemistry* 1 (1909), 791–798.

L. T. Bowser, "The Speedy Detection of Potassium in Small Amounts," *Chemical News* 101 (1910), 100–101.

C. D. Braun, "Chemisch-analytische Beiträge," *Fresenius Zeitschrift für analytische Chemie* 7 (1868), 313–339.

W. C. Bray, "A System of Qualitative Analysis for the Common Elements," *Journal of the American Chemical Society* 31 (1909), 613, 614, 621, 633, 634.

N. S. Brommelle, "The Russell and Abney Report on the Action of Light on Water Colours," *Studies in Conservation* 2 (1964), 140–149.

L. L. Burgess and O. Kamm, " A Study of Cobaltinitrites and Their Application to Analytical Chemistry," *Journal of the American Chemical Society* 34 (1912), 652–659.

A. H. Church, *The Chemistry of Paints and Painting*, 4th ed. (London, 1915).

C. Coffignier, *Couleurs et peintures* (Paris, 1924), 436.

Colour Index, vol. 3, 3rd ed. (Bradford, Yorkshire, England, and Research Triangle Park, North Carolina, 1971).

M. Cunningham and F. M. Perkin, "Studies on the Cobaltinitrites," *Journal of the Chemical Society* 95 (1909), 1562–1569.

M. van Driel and H. J. Verweel, "Uber die Struktur der Tripelnitrite," *Zeitschrift für Kristallographie* (A) 95, (1936), 308–314.

W. A. Drushel, "On the Volumetric Estimation of Potassium as the Cobaltinitrite," *Chemical News* 97 (1908), 124–126.

P. J. Durrant and B. Durrant, *Introduction to Advanced Inorganic Chemistry* (New York, 1970).

A. Eibner, *Malmaterialienkunde als Grundlage der Maltechnik* (Berlin, 1909).

O. L. Erdmann, "Ueber einige salpetrigsaure Nickel- und Kobalt-Verbindungen," *Journal für praktische Chemie* 97 (1866), 385–413.

F. Feigl, *Spot Tests in Inorganic Analysis*, 5th ed., trans. R. F. Oesper (Amsterdam, 1958).

A. Ferrari, L. Cavalca, and L. Coghi, "On the Mixed Cobaltihexanitrites," *Gazzeta chimica italiana* 85 (1955), 273–287.

M. Fischer, *The Permanent Palette* (Mountain Lake Park, Maryland, 1930).

N. W. Fischer, "Ueber Stickstoffoxyd-Salze," *Poggendorff's Annalen der Physik* 21, series ii (1831), 160–163.

N. W. Fischer, "Uber die salpetrichtsauren Salze," *Poggendorff's Annalen der Physik* 14, series iii (1848), 115–125.

C. R. Fresenius, *Manual of Qualitative Chemical Analysis*, trans. H. L. Wells (New York, 1909).

Gentele's Lehrbuch der Farbenfabrikation, ed. A. Buntrock (Braunschweig, 1909), 212.

R. J. Gettens and G. L. Stout, *Painting Materials, A Short Encyclopedia* (1942; New York, 1966).

W. Gibbs, "On the Hexatomic Compounds of Cobalt," *American Journal of Science* 8, series iii (1874), 190, 286.

K. A. Hofmann and O. Burger, "Contribution to the Chemistry of Cobaltic Nitrites," *Beriche der deutschen chemischen Gesellschaft* 40 (1907), 3298–3301.

M. B. Huish, *British Water-Colour Art* (London, 1904).

JCPDS Powder Diffraction File. JCPDS International Centre for Diffraction Data (Swarthmore, Pennsylvania, 1982).

K. L. Kelly and D. B. Judd, *Color, Universal Language and Dictionary of Names* (Washington, 1976).

H. Kittel, ed., *Pigmente. Herstellung, Eigenschaften, Anwendung*. (Stuttgart, 1960).

H. Kühn, *Die Pigmente in den Gemälden der Schack-Galerie* (Munich, 1969).

A. P. Laurie, *The Painter's Method and Materials* (London, 1926).

A. P. Laurie, "The Identification of Pigments Used in Painting at Different Periods, with a Brief Account of Other Methods of Examining Pictures," *The Analyst* 55 (1930), 162–179.

R. Mayer, *The Artists' Handbook of Materials and Techniques* (New York, 1948).

J. W. Mellor, *A Comprehensive Treatise on Inorganic and Theoretical Chemistry*, vol. 8 (London, 1928), 502–503.

S. Mierzinski, *Die Erd-, Mineral- und Lack-Farben*, 4th ed. (Weimar, 1881).

F. A. Miller and C. H. Wilkins, "Infrared Spectra of Characteristic Frequencies of Inorganic Ions," *Analytical Chemistry* 24 (1952), 1253–1294.

C. Moreau-Vauthier, *Comment on peint*, ed. H. Floury (Paris, 1923), 59.

G. D. Parkes, ed., *Mellor's Modern Inorganic Chemistry* (London, 1967).

M. M. Pierrat, "Sur la solubilité de divers sels de potassium dans des mélagnes d'eau et d'alcool," *Comptes rendus* 172 (1921), 1041–1043.

J. Plesters, "Cross Sections and Chemical Analysis of Paint Samples," *Studies in Conservation* 2 (1956), 134–155.

H. Rose, "Chemisch analytische Beiträge," *Poggendorff's Annalen der Physik* 110 (1860), 411–413.

A. Rosenheim and I. Koppel, "Uber Kobaltoxydnitrite und einige Kobaltnitrocyanverbindungen," *Zeitschrift für anorganische Chemie* 17 (1898), 35–38, 41–43.

W. J. Russell and W. de W. Abney, *Report on the Action of Light on Water Colours* (London, 1888), 73–77.

S. P. Sadtler, "On the Potassiocobaltic Nitrite Known as Fischer's Salt and Some Analogous and Related Compounds," *American Journal of Science* 49, series ii (1870), 189–205.

E. Saint-Evre, "Note sur l'action de l'azotite de potasse sur le nitrate de cobalt," *Comptes rendus* 33 (1851), 166–167.

E. Saint-Evre, "Recherches sure une combinaison nouvelle du cobalt," *Comptes rendus* 35 (1852), 552–555.

N. I. Sax, *Dangerous Properties of Industrial Materials*, 2nd ed. (New York, 1963).

H. C. Standage, *The Artists' Manual of Pigments*, 3rd ed. (London, 1896), 33.

A. Stromeyer, "Ueber das salpetrigsaure Kobaltoxyd-kali," *Liebig's Annalen der Chemie und Pharmacie* 96 (1855), 218–231.

M. Toch, *Materials for Permanent Painting* (New York, 1911).

S. Tschelnitz, *Farben-Chemie* (Vienna, 1857), 56.

L. Uebele, *Paint Making and Color Grinding* (London, 1913).

F. W. Weber, *Artists' Pigments, Their Chemical and Physical Properties* (New York, 1923).

K. Wehlte, *Werkstoffe und Techniken der Malerei* (Revensburg, 1967).

Winsor & Newton, *Descriptive Handbook of Modern Water Colour Pigments* (London, 1910).

Barium Sulfate — Natural and Synthetic

ROBERT L. FELLER

NOMENCLATURE IN VARIOUS LANGUAGES

Language	*Mineral*	*Pigment — Natural, Synthetic*
English	barite	barytes, blanc fixe
French	barytite, barytine	blanc fixe, sulfate de barium
German	Baryt, Schwerspath	Barytweiss, Permanentweiss
Italian	barite	sulfato de bario
Spanish	barita	blanco fijo
Russian	барит	ланфикс баритовая сульфа́т

1.0 INTRODUCTION

For more than one hundred years, barite, the naturally occurring barium sulfate mineral ($BaSO_4$), CI Pigment White 22, and the synthetic product CI Pigment White 21 (*Colour Index* No. 77120, 1971) have provided the paint formulator with a reasonably priced white pigment of moderate hiding power. The pulverized natural mineral is traditionally called barytes in English-speaking countries; the synthetic product, blanc fixe. The natural product was first suggested as an artists' pigment about 1782, but the major introduction of both natural and synthetic materials into commerce probably took place in the period 1810 to 1820.

The principal virtues of barium sulfate white have been its chemical inertness and its nonpoisonous character. The primary uses in the paint industry have been as an extender for lead white and as a base for the preparation of lakes. Barium sulfate presently finds extensive application in industrial undercoats, particularly automotive primers. The primary use of crude barytes, however, is in oil-drilling muds.

The compound barium sulfate is also encountered in paints and colorants known as "lithopones." Lithopone is a white pigment that is a mixture of zinc sulfide and barium sulfate made by the following double-decomposition chemical reaction:

$$BaS + ZnSO_4 = ZnS + BaSO_4.$$

There is also a "cadmopone" type based on coprecipitated CdS and $BaSO_4$. Thus, the investigator of decorative paints may expect to encounter $BaSO_4$ in three principal types of white pigment: the natural mineral, the synthetic pure substance, and the synthetically prepared mixture of $BaSO_4$ and ZnS known as lithopone.

ROBERT L. FELLER, director of the Research Center on the Materials of the Artist and Conservator, Mellon Institute, Carnegie-Mellon University, Pittsburgh.

2.0 HISTORY

2.1 Archaic, Obsolete, and Misrepresentative Names

2.11 ***The Mineral.*** The names barite (mineral) and barytes (pigment) are derived from the Greek word *barys* meaning "heavy." It is said that Agricola described the mineral in the sixteenth century. Other citations of the period refer to *Lapis Solaris* (a name used by Casciorolus in 1602) and to *Lapis Bononiensis* (used by Licetus in 1640). An outline of the early history of the mineral can be found in Mellor (1923). Fine examples of barite are shown in fig. 1.

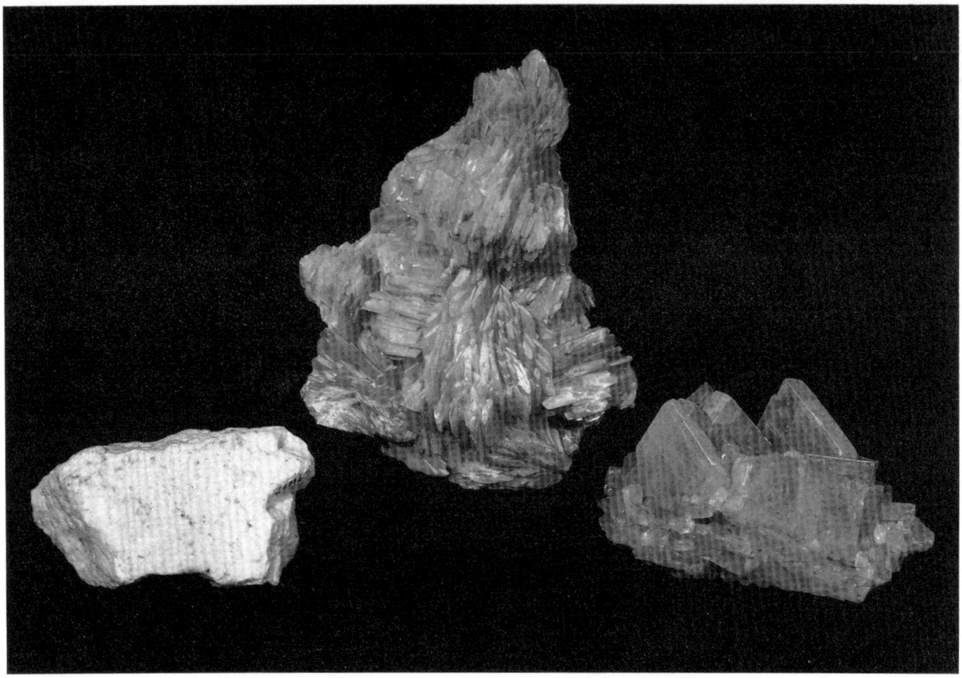

Fig. 1. Examples of Barite. The large specimen in the center is from Marienberg, Saxony. The blocky form on the right is from Felsöbanya, Hungary. Specimens courtesy the Carnegie Museum of Natural History, Carnegie Institute, Pittsburgh. The fine-grained white specimen on the left is a high-grade white mineral often used by pigment manufacturers; this sample is from a German source, probably Meggen.

In the eighteenth century Werner gave the mineral the name *Schwerspath* (heavy spar), and Bergman called it *Terra ponderosa vitriolata.* It was also known as Bolognian spar and *spath pesant. Cauk* (or *cawk*) was the name used by the Derbyshire miners for the white, opaque aggregates of the mineral familiar to the workers in those mines since at least 1668 (Percy, 1870). *Tiff* was also used, apparently by French-speaking miners. In the great barite mining area of Washington County, Missouri, a crossroads that once marked the village of Tiff still is known by that name.

The terms baryta and baryta white are encountered in the nineteenth-century literature. Coffignier (1925) mentions baratine and Tyrol white.

It was Scheele, in 1774, who first identified barium as a new element occurring in magnesia nigra (Urdang, 1942). In the early 1800s the name barytes may have referred to the oxide. This usage, found in Parkes' (1816) discussion of the substance, is apparently the meaning of the early chemical symbols shown in fig. 2 (Coward, 1927; Caven and Cranston, 1928). For example, the triangular-shaped symbols of Hassenfratz and Adet, with appropriate letters inside, are used for silica, alumina, magnesia, and lime as well as "barytes."

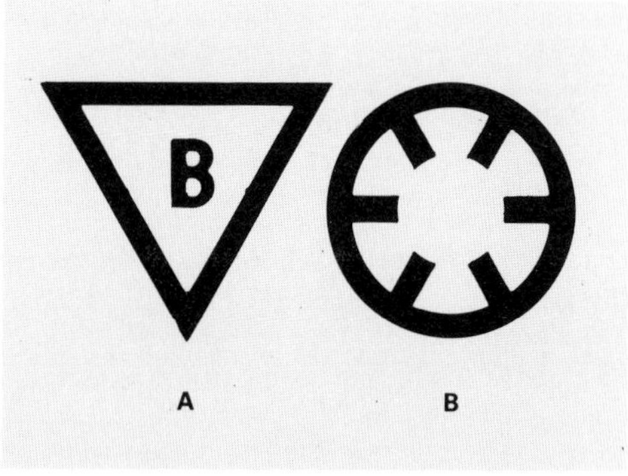

Fig. 2. Early chemical symbols for barytes taken from Caven and Cranston (1928). A. After Hassenfratz and Adet. B. After Dalton. The symbols, although labeled "barytes," probably represented barium oxide.

2.12 ***The Synthetic Product.*** Precipitated (synthetically prepared) barium sulfate is traditionally referred to as *blanc fixe*, a name said to have been proposed by Kuhlman (Coffignier, 1924, p. 87). The early popular names employed for barium sulfate in artists' colors were process white and permanent white. We also find references to an "elegant and permanent white" for watercolors sold in London around 1832 under the name mineral white. At the end of the nineteenth century, enamel white, *schneeweiss, mineralweiss,* and *neuweiss* were also familiar names (Mierzinski, 1881). Baryta is an obsolete term that may have been used both for the synthetic and the natural substance.

One must be on guard in reading the older literature because zinc white at times was also called "permanent white" (Toch, 1911, p. 150).

2.2 History of Use

2.21 ***Barytes.*** The history of the use of barite as an artist's pigment formally begins in 1782 with Guyton de Morveau's report of investigations into alternatives to lead white (Harley, 1982). It is true that the natural mineral was known in the sixteenth and seventeenth centuries, but there is no indication that barium sulfate was considered a possible pigment until de Morveau's investigations. Reference to permanent white appeared shortly thereafter in the sixth edition of Bowles' *Art of Painting in Water-Colours*, printed in 1783 (Harley, 1982). Hence, the earliest probable date for the introduction of natural barium sulfate into paints seems to be 1782–1783, close to that of the early studies on zinc white by Guyton de Morveau.

The natural mineral was not always available in a high quality deposit of white mineral, and because of this, its extensive use in paints apparently did not begin prior to the early nineteenth century. The widespread practice of adulterating lead white with barytes at that time has been noted by Leuchs; lead white thus adulterated was called Venice white at 1/1 barite/lead white, Hamburg white at 2/1, and Dutch white at 3/1 (Leuchs, 1825; Ure, 1853).

Production from mineral sources in the United States began in Virginia in 1845 and in Missouri in 1860 (Hill, 1917). Nevertheless, a booklet published by the DeLore Baryta Company (Clark, 1920) states that not until after the Civil War was the use of barite in paints fully appreciated in the United States. The composition of some of the paints of this period is cited by Heckel (1928) and by Holley and Ladd (1908). The pigment also found its way into wallpapers, linoleum, oil cloth, flannel, and shoddy cloth (Mellor, 1923, p. 802).

In 1937, over half the natural mineral consumed in the United States (350 million pounds) went into the manufacture of lithopone via the intermediate, barium sulfide. However, with the decline in the application of lithopone that has taken place in succeeding years, this use of the mineral has been reduced to minor importance (Patton, 1973).

Almost three-quarters of the modern production of barite goes into oil well drilling muds, only about 4% into paint. A major use in paints today is in the formulation of industrial undercoats, especially automotive primers. The chemical inertness of barytes recommends its use in designing chemically resistant coatings that must withstand the attack of acids, alkalis, and corrosive gases.

2.22 ***Blanc Fixe.*** It may at first seem curious that in an industrial manufacturing process, natural barium sulfate is regularly converted to barium sulfide, only to be reconverted into barium sulfate. The reason is simple; by this procedure a very pure product can be produced from relatively impure starting materials. *The Cyclopedia*, published in the period 1810–1824 (Rees) stated, for example, that native barytes was never found pure; it is probably more correct to say that pure grades are found only rarely. The *Cyclopedia* entry discusses a number of methods of preparing barium sulfate cheaply; in particular, a process published by Dr. Hope in the *Edinburgh Transactions* of 1793 is mentioned.

The use of blanc fixe probably remained almost entirely limited to watercolors until the 1820s when production may have been stimulated to provide an adulterant for lead white. *A Practical Essay on the Art of Flower Paintings* (Burgess, 1811) mentions a Constant White being prepared by a Messrs. Newman and Parkes. Nonetheless, five years later Samuel Parkes (1816) stated that the pigment had "not yet been used much in the arts except by limners as a most excellent water colour." The author added further that many years ago Mr. Hume had discovered a method of making a color from this earth "sold under the name *Hume's Permanent White*." Harley (1982, p. 175) also mentions the pigment having been brought forward by a Mr. Hume of Long Acre.

There is extensive mention of the synthetic product in the technical literature of the period 1825–1827. According to Rabaté (1964), Kuhlman's manufacture of blanc fixe began about 1830. As noted above, Coffignier (1924) states that it was Kuhlman who proposed the name blanc fixe.

A product developed in America by Henry M. Toch in 1895 was called Lake Base. Maximillian Toch (1925) reports that in 1905 not over 100 tons of Lake Base per year were sold to the paint industry, but that this figure rose to 3,000 tons in 1915. As an interesting detail, he mentions that between 1910 and 1915, the Navy experimented with a change in the formula of the famous "Battleship Grey," finally developing a composition based on 45% ZnO, 45% blanc fixe, and 10% lamp black. Doerner (1949) suggests the use of blanc fixe as an inert filtering aid for the purification of oils. This chemical compound may also be occasionally encountered in another aspect of modern conservation science; it has been suggested that barium sulfate can be precipitated within porous and friable stone to facilitate the preservation of buildings, monuments, and sculpture (Sayre, 1970).

2.3 Terminal Date

Both the natural and the synthetic mineral are still being used today as white pigment and as a base for lakes; however, the use of blanc fixe has declined sharply since 1950. The *Raw Materials Index* (1975) lists five manufacturers for barytes, three for blanc fixe.

3.0 GENERAL PIGMENT PROPERTIES

3.1 Color and Spectral Reflectance

Native barium sulfate is often of poor color owing to impurities. Hence, in early industrial practice, ultramarine and undoubtedly other blue pigments were frequently added to subdue the pinkish cast, caused principally by iron oxides (Heaton, 1947, p. 93). However, excellent white deposits exist. The author has also encountered commercial samples of blanc fixe with added ultramarine.

The synthetic product can be made with such purity that it is currently the working standard white for the reflectance spectrophotometry of colored paints, textiles, and other industrial products (Moser, 1973). A curve for the reflectance of barium sulfate relative to that of magnesium oxide is shown in fig. 3. More than thirty years ago, Vesce (1946) published the reflectance curves for both the natural and synthetic pigment in the range from 400 nm to 940 nm. Both products are also widely used for the white in photographic papers, largely because their very low solubility minimizes the possibility that impurities, which may be coprecipitated in the formation of the product, will later be released in sufficient concentration to react with the silver salts in the emulsion (Russell, 1958).

Partly because of its low hiding power the color of barium sulfate can be easily shifted by the addition of dyes and toners; it is commonly shifted toward the blue in photographic papers to offset the yellowness of the gelatin.

3.2 Hiding Power and Tinting Strength

Owing to its moderate refractive index, barium sulfate does not hide well in oil-based paints, a fact noted in the eighteenth century by Guyton de Morveau (Harley, 1982) in his original discussion of the possibility of using the mineral as a pigment. Consequently barium sulfate found its greatest early application in watercolors and other traditional water-based paints such as those involving casein and glue binders. The early study of Stutz (1926) on the transmittance of various pigments revealed that barium sulfate transmits ultraviolet radiation from 400 to below 300 nm, practically without diminution.

3.3 Permanence

The natural and synthetic barium sulfate pigments are highly stable toward color change, both with respect to exposure to light and to interaction with other pigments. Because of their poor hiding power in oil paints, however, paints formulated with these pigments may tend to increase noticeably in translucency with age. That is, they may show a greater tendency to develop pentimenti than will paints based on lead, zinc and titanium white. Oil paints containing barium sulfate may also tend to chalk as a consequence of the deeper penetration of visible and ultraviolet radiation, relative to that in paints formulated with whites of higher hiding power (Stutz, 1926) (see remark on barium sulfate's presence in a blanched paint under 6.0 Notable Occurrences).

3.4 Compatibility

Extremely inert, barium sulfate is chemically compatible with all other pigments.

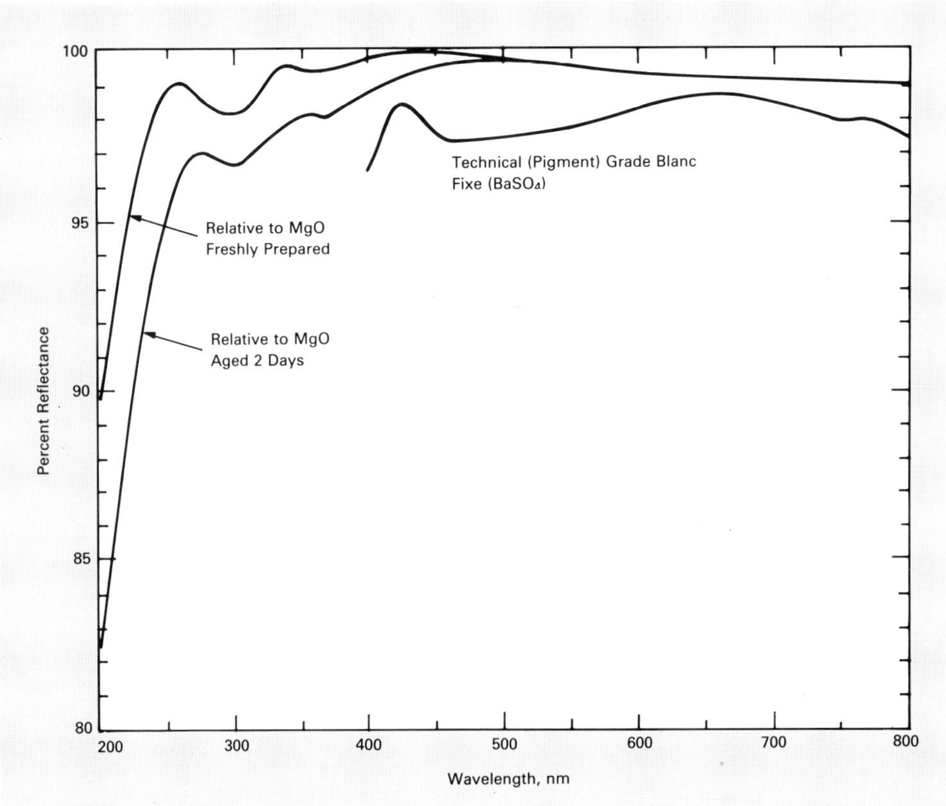

Fig. 3. Spectral reflectance of synthetic barium sulfate relative to that of magnesium oxide: technical and spectral reflectance grades (after Moser, 1973).

3.5 Chemical Properties

Barium sulfate is inert to both acidic and alkaline environments. Its solubility in strong acids is only slight (0.01 to 0.02%). Because it is not darkened by sulfide in the atmosphere, barium sulfate was popularly introduced as a "permanent white" for the artist. To obtain sufficient water soluble material for chemical tests, one usually fuses the sulfate with carbonate or chlorides (see 5.21).

3.6 Oil Absorption and Grinding Qualities

According to Patton (1973) natural barium sulfate is an attractive extender pigment owing to the following qualities: chemical inertness, reasonably high refractive index, low abrasiveness, ready wettability by vehicles, easy grinding (dispersing) properties, and low oil absorption. The *Barium Bibliography* (1961, Paint reference No. 32, p. 244) gives the oil absorption of blanc fixe as 14–21 lbs. of oil per 100 lbs. of pigment. The *Raw Materials Index* (1975) gives the oil absorption of barytes as being between 7–13. Bearn (1923) cited a value of 8.5.

3.7 Toxicity

Partially owing to its extreme insolubility in water (less than 3 ppm), the compound is not poisonous. Thus, the traditional antidote for swallowing a toxic, relatively soluble barium-containing substance is to drink a solution of sodium sulfate thereby converting the metallic ion to the insoluble sulfate. Barium sulfate (USP XVIII) is widely utilized in medical examinations as an opaque contrast medium for x-ray purposes; its low solubility results in its passing harmlessly through the body (Patton, 1973). The compound's lack of poisonous qualities was one of the chief reasons for its introduction as a white for the artist.

4.0 COMPOSITION

4.1 Chemical Composition

The natural mineral used as a white pigment is essentially pure barium sulfate (at least 94%

$BaSO_4$; see 4.33 on Pigment Grades). Silica and iron compounds, the latter particularly giving rise to color in the natural pigment, are perhaps the major impurities to be expected, but varieties of barytes are also encountered that contain significant quantities of fluoride. One notes, for example, a study made more than forty years ago demonstrating that heavy spar containing fluorspar was less effective in preventing rusting of iron during outdoor weathering than was the mineral containing silica (Wolff and Zeidler, 1937). Coffignier (1924) reported over thirty chemical analyses of barytes from a variety of sources, pointing out that some minerals could be expected to be high in silica, others high in fluoride. Typical analyses are given in Table 1.

Table 1. **SOME CHEMICAL ANALYSES OF BARYTES (COFFIGNIER, 1924)**

	Italian	*German*	*French*	*Spanish*
Barium Sulfate	90.10	95.15	92.55	98.97
Loss on ignition	0.18	0.38	0.24	0.64
Calcium Fluoride	—	1.95	3.90	—
Iron Oxide	0.25	0.40	0.20	0.20
Alumina	0.45	0.45	0.70	0.20
Silica	8.90	1.65	2.35	—

4.2 Sources

4.21 ***Barytes.*** Barite is found in many places in Germany: Saxony, the Harz, and Meggen in Westphalia (Stock, 1925). The Minden Mine in Hunterhausen is well known (Muddiman, 1942). *Gentele's Lehrbuch* (1909) mentions Lohr am Main, Thüringen bei Konigsee, and Waldshut. Prior to World War II Germany provided half the world supply of barite, primarily from Meggen, the renowned source discovered in 1845. Bearn (1923, pp. 27–31) stated unequivocally that the best barytes came from Germany. It is said that the mineral from the Hartz is contaminated chiefly with quartz (Kittel, 1960).

Important deposits occur at Felsöbànya in Hungary, Kapnik in Roumania , and at Meis and Schemnitz in Czechoslovakia. Barite is found at Kitzbühel in the Austrian Tyrol and in France, especially in the Limagne district in Puy-de-Dôme (Nédéy, 1955). Large deposits occur in Castile and Andalusia in Spain. Bearn (1923), pp. 27–31) reports that the Spanish product was often associated with a certain proportion of sulfate or carbonate of lime. There are many long-known localities in England: the Dufton lead mines in Westmorland; Alston Moor, Frizington, Cleater Moor, and elsewhere in Cumberland; the mineral veins of Cornwall; and the Settlingstones mines, Northumberland (Hurst, 1906).

Native barytes occurs in large quantities in China, and in the early twentieth century it was used there chiefly, in purified form, for mixing with other pigments (Watson, 1930).

In the United States, barite has been found in Connecticut at Cheshire, New Haven County; in New York at De Kalb in St. Lawrence County, at Pillar Point and near Chaumont in Jefferson County. The mineral is also associated with strontianite in limestone near Schoharie in Schoharie County, New York. Commercial deposits occur in veins and residual deposits in Paleozoic limestones in the southern Appalachian and central states, especially Georgia, Missouri, Tennessee, Kentucky, Virginia, North Carolina, and Alabama. The mineral is common in the lead ores of Missouri and Wisconsin. Deposits are also found in New Mexico near Barton and elsewhere in Bernalillo County, and in various districts in Dona Ana, Socorro, and Grant Counties; in Nevada; and in Yuma County, and elsewhere in Arizona (Dean and Brobst, 1955). According to Toch (1925) the sources in Virginia and Kentucky contained more than 1% lime and fluoride. It is reported that production began in Fauquier and Prince William Counties, Virginia in 1845, and in Missouri in 1860 (Williams, 1949). Figure 4 illustrates the simple method of barite surface mining practiced in Washington County, Missouri, about the time of World War I (Clark, 1920).

In the years 1964–1968, the world output of barite remained relatively stable. Approximately 60% of the mineral consumed in the United States was supplied from domestic sources in 1968; imports made up the difference, mainly from Ireland, Mexico, Canada, and Peru (Patton, 1973). Five states account for nearly all the United States' domestic barite production (Missouri, Nevada, Arkansas, Georgia, and Alaska). Washington County, Missouri, and the Ouachita Mountains of Arkansas have been cited as the principal barite-producing areas of the United States.

4.22 ***Blanc Fixe.*** Prior to 1900, all barium-sulfate-coated photographic paper used in the United States was imported, but in that year the first U.S.-based coating operation was begun (Russell, 1958). Before this, blanc-fixe-coated paper was manufactured by such concerns as Felix Schoeller in Germany, Rives and Montgolfier in France, and Wiggins-Teape in England. From 1900 until World War I the blanc fixe used in manufacturing baryta-coated paper in the United States was still primarily imported from Germany, England, or France. At the time of World War I, however, it was considered desirable to start manufacturing blanc fixe at home. This venture was pursued for about ten years but was later dropped in favor of importing a more suitable product from Germany. Then, in the early 1930s, manufacture was resumed in the United States and has continued up through the present without further interruption.

The use of blanc fixe is currently diminishing. The annual United States' consumption of precipitated barium sulfate as a pigment declined from over 30 million pounds in 1950 to an estimated less-than-10 million in 1968 (Moser, 1973).

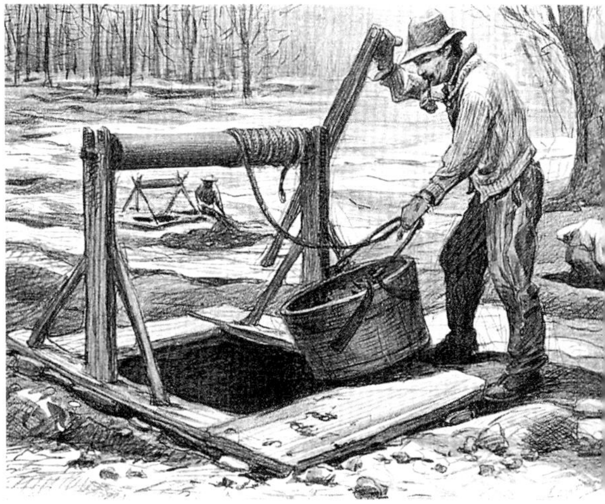

Fig. 4. Surface mining of barite in the early 1900s at Potosi, Missouri (from Clark, 1920).

Fig. 5. Cleaning ore with the "rattle box" (from Clark, 1920).

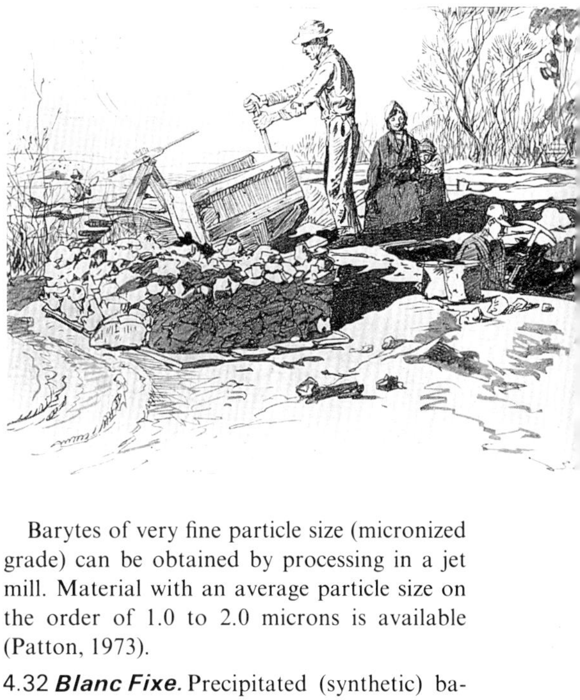

4.3 Preparation

4.31 ***Barytes.*** A booklet published by De Lore Baryta Company (Clark, 1920) is extensively illustrated with reproductions of paintings and drawings describing the mining practices in Washington County, Missouri, following the turn of the century. Figs. 4 and 5, taken from that publication, show typical surface-mining operations at the time in the region of Potosi, Missouri; fig. 5 shows the simple "rattle box" used to remove superficial debris. A description of the mining practices in Shropshire at about the same date has also been published (Malehurst Barytes Co., 1926).

The manufacture of barytes begins with a sorting of the coarsely mined barite mineral according to color. The coarse product is then crushed and wet-ground to obtain the required degree of fineness (Patton, 1973). The ground barytes is levigated, sometimes with water, sometimes by a jet of air. Traces of iron discoloration may be reduced by "bleaching," which usually consists of treatment with hydrochloric or sulfuric acid. After being bleached, the particles are again washed with water to remove residual acid; they are then dried, screened, and packaged. Trace amounts of ultramarine blue are occasionally added to improve whiteness.

Barytes of very fine particle size (micronized grade) can be obtained by processing in a jet mill. Material with an average particle size on the order of 1.0 to 2.0 microns is available (Patton, 1973).

4.32 ***Blanc Fixe.*** Precipitated (synthetic) barium sulfate is made by the reaction of a soluble barium salt with a soluble sulfate. To begin the manufacture, barytes is conventionally heated with coal, leading to the formation of barium sulfide and the gaseous products, carbon monoxide and sulfur dioxide. The barium sulfide so formed is dissolved in water, the solution filtered, and the filtrate converted to the sulfate by the addition of a solution of sodium sulfate. The precipitated barium sulfate is filtered, washed, dried, and packed for shipment (Moser, 1973).

Another manufacturing method employs barium chloride, obtained as the hydrated crystal. This is put in solution, filtered, and then reacted with sulfuric acid under suitable conditions. A third possible method involves the use of the insoluble barium carbonate, which may be dissolved in hydrochloric acid and then reacted with sulfuric acid to form the barium sulfate. Patents representing a variety of procedures for the preparation of blanc fixe are briefly described by Kittel (1960, pp. 165–171).

4.33 ***Pigment Grades.*** *Barytes*: There are essentially four grades of barytes. The three principal types are: (1) coarsely ground (< 20-mesh), employed in the glass industry; (2) finely ground (< 325-mesh), used mainly in the formulation of oil-well drilling muds; and (3) finely ground, bleached (< 325-mesh), utilized in paint and similar applications where freedom from color is an important factor (Patton, 1973). There is also a fourth grade of very fine particle size: micronized. Individual companies often prepare a wide range of pigments; for example, the Deutscher Baryte-Industry, Bad Lanterberg, at one time offered twelve classes varying in degree of fineness from 15 to 2 micrometers (Kittel, 1960). The *Raw Materials Index* (1975) lists six suppliers, each offering three or four varieties. Fig. 6 shows the particle-size distributions that are typical of coarse barytes, pigment grade, and the finest grade of blanc fixe (Moser, 1973; Sachtleben Chemie, 1975 and 1976).

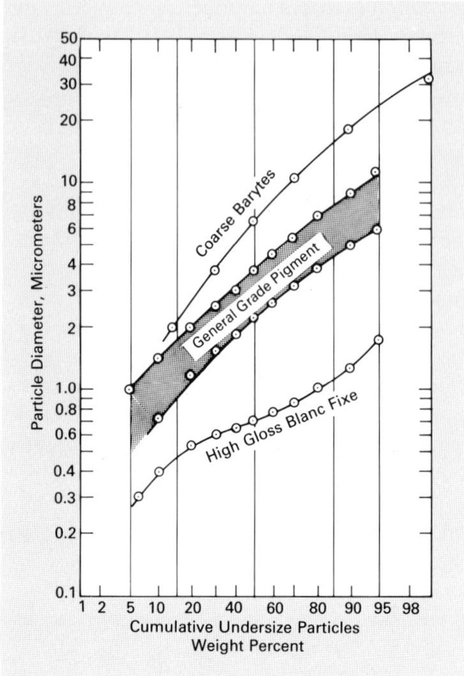

Fig. 6. Approximate cumulative particle size distribution for typical barium sulfate pigments (after Moser, 1973, and Sachtleben Chemie, 1975 and 1976).

An American standard specification covering barytes and blanc fixe pigments, ASTM D602-42 (Reapproved, 1955), calls for 94% minimum barium sulfate content in barytes, 2% maximum content of free silica (in quartz, clay, or other foreign matter), and 0.05% maximum ferric oxide. Water and other volatiles must not exceed a maximum 0.5%; matter soluble in water must be less than 0.2% (Patton, 1973).

The pH value for uncontaminated barium sulfate is close to 7.0, but the standard ASTM specification permits an acid reaction down to a pH value of 3.5 (Patton, 1973).

Blanc Fixe: Moser (1973) states that the synthetic product is manufactured in four principal grades: pigment, photographic paper, pharmaceutical, and reflectance standard. The particle-size range is 0.1–5.0 microns with an average of 0.5–2.0 microns (Bradley, 1942, p. 433).

4.4 Adulteration and Sophistication

Barytes traditionally has been a very low-cost product. Rarely are other materials added to change its properties, although one can expect to find calcium carbonate in the cheaper fillers. Heaton (1947, p. 96) also mentions a product called "Blancophone," an extender made by coprecipitating barium sulfate and calcium carbonate in the proportion of about 70/30; it was used mainly in water-based paints and to some extent in the preparation of lakes. The addition of ultramarine to improve color has been mentioned earlier.

When mixed with pigments other than white, barium sulfate "cannot be regarded simply in the light of an adulterant, since by its use in, for example, Brunswick green, the tones of the associated pigments (chrome and blue) are thereby much enhanced, and at the same time a paint is produced of much superior working properties and at considerably less cost" (Bearn, 1923, p. 29). For other specific purposes, such as providing increased resistance to corrosive attack, barium sulfate's presence in paints is recognized to be specifically beneficial. Early forms of titanium dioxide were often caused to crystallize and coalesce on a base of blanc fixe (Heaton, 1947, pp. 91–96).

The investigator, when examining tiny specimens of paint, must always be alert to the possibility that he is dealing with lithopone ($BaSO_4/ZnS$) rather than blanc fixe. Lithopone may contain 30 to 60% zinc sulfide combined with barium sulfate.

5.0 IDENTIFICATION — CHARACTERIZATION

5.1 Characterization by Optical and Scanning Electron Microscopy

5.11 ***Optical Crystallography.*** Barite is orthorhombic, dipyramidal and tabular parallel to the base. The crystal habit in the natural mineral, with respect to the particular faces that are developed, varies considerably as many standard mineralogical texts illustrate. However, the appearance is generally tabular with prismatic development less frequent, the basal plane usually being strongly developed (Milner, 1922). Cleavage is perfect parallel to the base (001) (see fig. 7), good on (010), and poor parallel to the prism (211).

The optic plane is (010), with the acute bisectrix perpendicular to (100); biaxial positive with $2V = 36–40°$; (010) yields a flash figure (Heinrich, 1965), but this is not observed in the usual tablets, as fig. 7 indicates. The dispersion is $r < v$ weak. The refractive indices are moderate, birefringence low; $\alpha = 1.634–1.637$, $\beta = 1.636–1.638$, $\gamma = 1.646–1.648$. Birefringence as seen in the usual tablets is low, $\gamma - \beta = 0.010$.

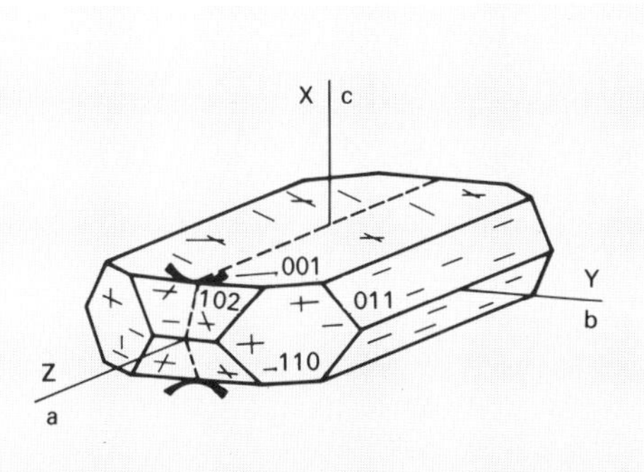

Fig. 7. Optical and crystallographic characteristics of a common crystal habit of barite.

Between crossed polars, the usual cleavage fragments in barytes are pale gray in appearance. With a first-order-red-plate, the blue-violet/orange change, particularly the orange color, is reasonably characteristic. In the usual fine particles of blanc fixe, the above optical properties can scarcely be observed; only a low degree of birefringence under crossed polars, or a faint glow owing to edge depolarization, may be apparent.

Moser (1973) states that the rhombic crystals grow along the (100) plane and that the short and long axes are the b and c crystallographic axes respectively. An elongation of the particles of blanc fixe can be seen under the scanning electron microscope (see fig. 10). In one special product manufactured in Germany, regular-shaped rhombs can be observed under the optical microscope (Sachtleben Chemie, 1975). However, normally the rhombs are irregular in shape.

5.12 ***Optical Microscopy.*** Toch (1925) and the Malehurst Barytes Company (1926) have published photomicrographs illustrating the appearance of barytes. Kittel (1960) has photomicrographs of three types of blanc fixe: large lump-shaped synthetic "Kugelspat," that could be mistaken for the natural mineral, material of fine particle-size, and a more rarely produced needle-shaped variety that he claims is the monoclinic modification. The author (Feller, 1971) has published photomicrographs at 800x to illustrate the normally expected difference between barytes and blanc fixe. Pictures of typical particles are shown in fig. 8.

The natural product is usually recognized by (a) the general rectangular or blocky character of many of the tablets as well as their relatively large size and rounded edges, (b) the low birefringence, and (c) the intermediate character of the refractive index, which results in a good contrast in Canada balsam and Permount®. Under crossed polars, many of the cleavage fragments 5–20 micrometers in diameter are seen to have rather sharp edges, the consequence of conchoidal fracture. Particles below 5 micrometers tend to have many of the above characteristics obscured, but their low birefringence is still strikingly reminiscent of that of the larger particles. The principal characteristics can usually be seen even in lakes; particularly apparent are the relief and low birefringence. The colorant in lakes is usually clustered primar-

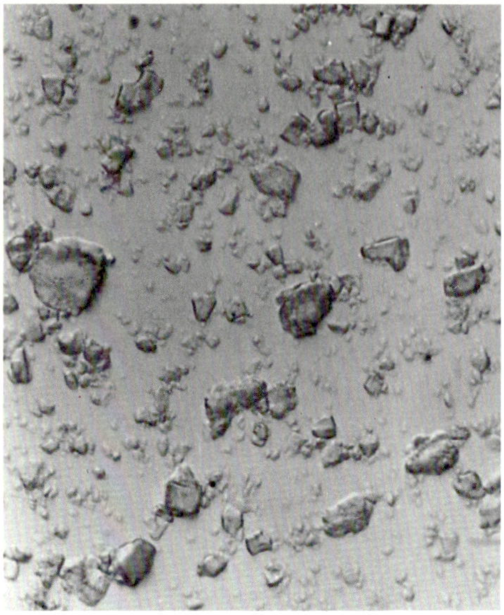

A

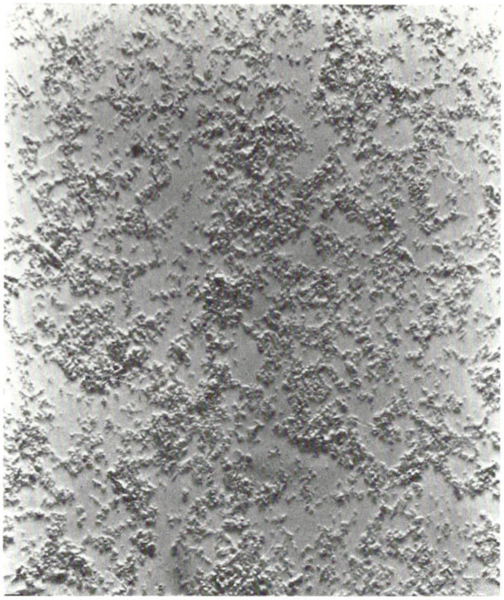

B

Fig. 8. Photomicrographs of typical (A) barytes and (B) blanc fixe pigments. The barites sample is Foam A, National Lead DeLore Division, 1966; the blanc fixe, from Chemetron, 1971, National Gallery of Art Pigment Bank, Mellon Institute, Pittsburgh. 450x with substage shadowing.

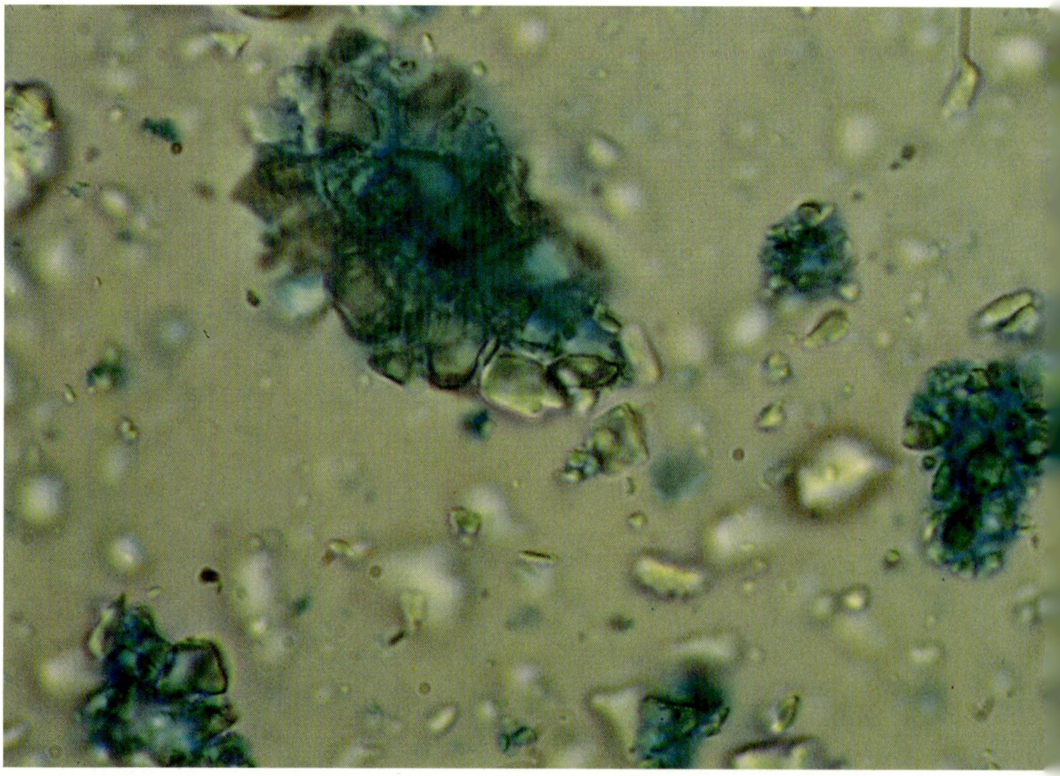

Fig. 9. Appearance of a typical lake pigment; in this case, a blue organic colorant struck on barytes. Mounted in medium of refractive index 1.53 with substage shadowing 650x.

ily on the surfaces and at the edges of the barite particles (fig. 9).

The recognition of blanc fixe of particle size below 2 micrometers poses problems. The pigment can easily be overlooked in mixtures with other pigments, although advantage can sometimes be taken of barium sulfate's insolubility in strong acids; zinc or lead white, if present in a mixed white, dissolve away. The compound's birefringence is frequently undetectable. However, the effect of its intermediate refractive index may sometimes alert the microscopist to the presence of blanc fixe because the tiny particles can act as miniature lenses, focusing the transmitted light alternately in shadows and bright spots when the microscope objective is raised or lowered. This may be observed even though the characteristic contrast may not be apparent owing to the fineness of the particles. Lithopone ($BaSO_4/ZnS$) should exhibit a higher birefringence and profile in Canada balsam owing to the zinc sulfide constituent, but it would take experience to recognize and to distinguish lithopone and blanc fixe under the microscope.

5.13 ***Electron Microscopy.*** Figs. 10A and B are scanning electron photomicrographs of blanc fixe particles at 5000 and 20,000x. An electron microscope equipped with an x-ray fluorescence analyser can be used to detect the barium in barium sulfate and possibly zinc in some of the neighboring particles, alerting the investigator to the presence of intimately coprecipitated zinc sulfide (in lithopone) or of added zinc white. Titanium, lead, and calcium should also be sought, for barium sulfate is frequently used in combination with other whites.

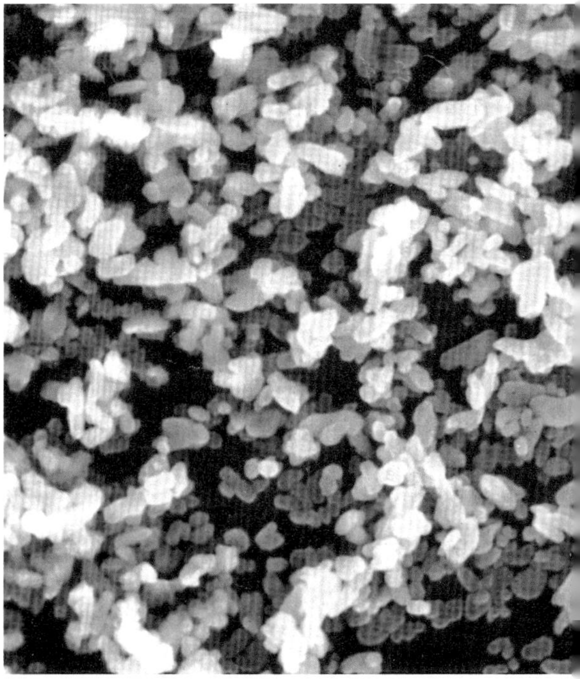

Fig. 10. Scanning electron micrographs of blanc fixe at (A) 5000x and (B) 20,000x. Chemetron Blanc Fixe No. 1, 1971, National Gallery of Art Pigment Bank sample B-377, Mellon Institute, Pittsburgh.

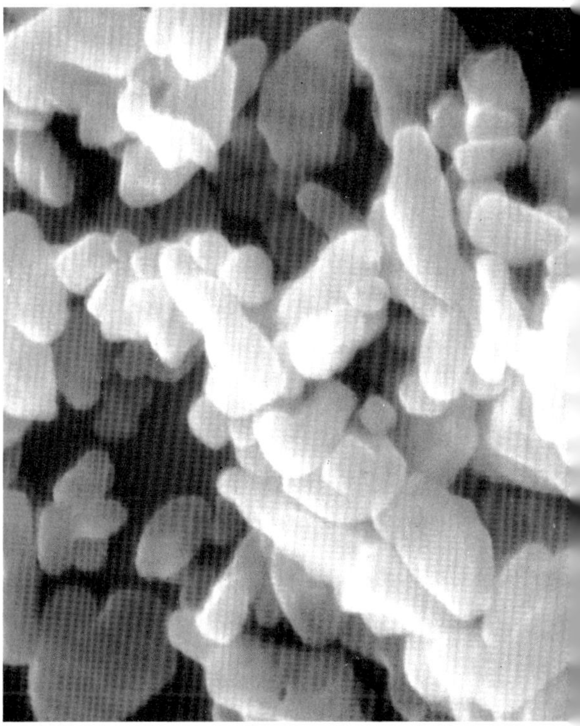

5.2 Chemical Identification

5.21 ***Conversion to Soluble Salts.*** Barium sulfate is so insoluble that in order to obtain sufficient ions in solution for microchemical tests, it must be converted to a carbonate, chloride, or sulfide. A tiny portion of the material to be tested can either be first ignited to burn off any vehicle (probably the preferable procedure) or treated with 1 or 2 drops of 10% sodium carbonate solution, evaporated to dryness, and then ignited in a microcrucible. The cooled mass is treated with a drop of dilute nitric acid to put the barium carbonate into solution. Alternatively, Feigl (1958, p. 476) suggests that the bits of sample can be mixed with 0.5 g of ammonium chloride and heated until fumes are no longer evolved, after which the sample can be dissolved in a few drops of water prior to testing. The formation of barium chloride in this manner will lead to the volatilization of zinc and cadmium chlorides if salts of these metals have been present.

5.22 ***Sodium Rhodizonate Test for Barium*** Barium and strontium ions form red to reddish-brown precipitates with sodium rhodizonate; calcium does not. The usual method of using this reagent is in the form of filter-paper test strips prepared by impregnating the papers with a dilute solution of sodium rhodizonate, drying them in a vacuum over concentrated H_2SO_4, and storing them in the dark. When carrying out tests with these treated papers, one may draw the following conclusions. If no brown fleck is formed when a drop of the unknown solution is placed on the treated test paper, the absence of both barium and strontium is indicated; the formation of a brown fleck may mean that either ion is present or that both are. If the fleck disappears on treatment with dilute hydrochloric acid (1:20), only strontium is present. (Strontium rhodizonate is soluble in dilute hydrochloric acid.) If the fleck turns red, barium is present (Feigl, 1958, p. 217). The possibility that strontium may also be present in the latter instance must not be discounted, because dissolution of a strontium portion of the brownish red precipitate in the dilute acid may scarcely be noticed.

Alternatively, a drop of the neutral or slightly acid test solution may be placed on a piece of untreated filter paper, and a drop of a 0.2% aqueous solution of sodium rhodizonate added. According to the amount of barium or strontium present, a more or less intense red-brown stain is formed (Feigl, 1958). Limit of detection: 0.2 to 0.5 micrograms of barium.

Barium may be detected by this reagent in the presence of 50x as much molar concentration of strontium. In pigments, strontium is usually present only as a trace material, but if there is the possibility that a significant amount of the element may be present, the sample should be further examined. This is most readily done by emission spectrographic or x-ray fluorescence analysis. Neutron activation analysis can also be used.

5.23 ***Tests for Sulfate.*** A highly sensitive test for sulfate is based on the ability of barium sulfate to entrap potassium permanganate upon precipitation from solution. While this procedure usually is employed as a test for sulfates, it can also be used as a test for barium sulfate itself, providing no other sulfate is present. The test is performed by converting the mineral first to barium carbonate (through fusion with sodium carbonate) and then reprecipitating the $BaSO_4$ after first adding the permanganate. Next, one acidifies the solution and adds hydroxylamine hydrochloride to decolorize the brownish precipitate of manganese dioxide that may form (Feigl, et al., 1964).

Feigl (1958, p. 315) also suggests that, to detect sulfate in the absence of insoluble sulfides or free sulfur, the sample can be reduced to the sulfide by intimately mixing it with charcoal, and four times the bulk of $NaKCO_3$, sulfur free. The mixture is fused in a magnesia spoon with a blowpipe and alcohol flame and while still warm is transferred to a glass tube and dissolved in a drop of water. The sulfate having been reduced to the sulfide by the charcoal, an iodine-azide test for sulfides can be made (see description of latter test in the chapter on Cadmium Yellow, Oranges and Reds.)

5.3 Instrumental Methods of Analysis

5.31 ***X-ray Diffraction.*** The *d*-spacings and intensities of the principal x-ray diffraction lines of barium sulfate are given in table 2. Those characteristic of beta-zinc sulfide (sphalerite structure), likely to be present in lithopone are given in table 3. Typical Debye-Scherrer powder diffraction patterns are shown in fig. 11. Pattern D is the beta (sphalerite) form which Wood (1930) said was the type of the zinc sulfide of lithopone. It can be seen that the presence of zinc sulfide is practically undetectable in the pattern of lithopone ($BaSO_4/ZnS$) because only two lines, at 1.912 Å and 1.561 Å, are distinctive of the zinc sulfide component alone. Other zinc sulfide lines coincide with those of barium sulfate. In x-ray diffractometer traces, a doublet at 3.101 Å ($BaSO_4$) and 3.123 Å (β-ZnS) should be detected if the sample is lithopone rather than blanc fixe.

Although natural barium sulfate pigment gives the same diffraction pattern as the synthetic, one can expect to find some broadening of the lines, particularly in an x-ray diffractometer trace; also possibly some impurities could be detected in the barytes.

Table 2. **X-RAY DIFFRACTION DATA FOR BARIUM SULFATE (BARITE)**[a] **(Diffractometric data, CuK α_1 = 1.5405 Å, 26°C)**

d(Å)	*I*	*d*(Å)	*I*
4.44	17	1.947	<1
4.34	36	1.930	7
3.90	57	1.857	16
3.77	12	1.787	3
3.576	31	1.760	9
3.442	100	1.754	9
3.317	67	1.726	5
3.101	97	1.723	6
2.834	53	1.681	7
2.734	16	1.673	14
2.726	47	1.669	10
2.481	14	1.636	8
2.444	2	1.593	8
2.322	15	1.590	7
2.303	6	1.534	18
2.281	7	1.526	11
2.209	27	1.495	3
2.120	80	1.474	10
2.104	76	1.457	3
2.056	23	1.426	8

[a] JCPDS Powder Diffraction File 5-0448, Courtesy of JCPDS International Centre for Diffraction Data (JCPDS 1982).

Table 3. **X-RAY DIFFRACTION DATA FOR BETA-ZINC SULFIDE**[a] **(Present in Lithopone, $BaSO_4/ZnS$, CuKα_1 = 1.5405 Å, 26°C)**

d(Å)	*I*
3.123	100
2.705	10
1.912	51
1.633	30
1.561	2
1.351	6
1.240	9
1.209	2
1.1034	9
1.0403	5

[a] Sphalerite (cubic) structure. Stable below 1020°C. JCPDS Powder Diffraction File 5-0566, Courtesy of JCPDS International Centre for Diffraction Data (JCPDS 1982).

5.32 ***Infrared Spectroscopy.*** Barium sulfate has a distinctive and sharp set of absorption bands in the infrared that permit it to be readily differentiated from other white pigments. Most distinctive are the strong bands at 1075, 1120, and 1185 cm^{-1} and the small sharp band at 985 cm^{-1} (*Infrared Spectroscopy*, 1980; Tarte and Nizet, 1964; Newman, 1979). The natural and synthetic barium sulfates cannot be distinguished conclusively by infrared although, as can be seen in fig. 12, the absorption bands of the natural product tend to be broader. The two cannot be distinguished from lithopone ($BaSO_4/ZnS$) unless the spectra is examined in the region of wave numbers lower than 400 cm^{-1} (wavelengths longer than 25 nm) (Nyquist and Kagel, 1971). Fig. 13 indicates the distinctive bands in this region that one may expect to find in lithopone. If zinc oxide rather than zinc

Fig. 11. X-ray powder diffraction patterns using 114.6 mm diameter Gandolfi camera. A. Blanc fixe Chemetron, B. Barytes DeLore Foam A. C. Lithopone Sherwin Williams 60N. D. Beta zinc sulfide, mineral sphalerite, Spain, USNM 134545 (Department of Mineral Sciences, Smithsonian Institution). Patterns courtesy of the Freer Gallery of Art, Washington. (Photographs not to be used for measurement purposes.)

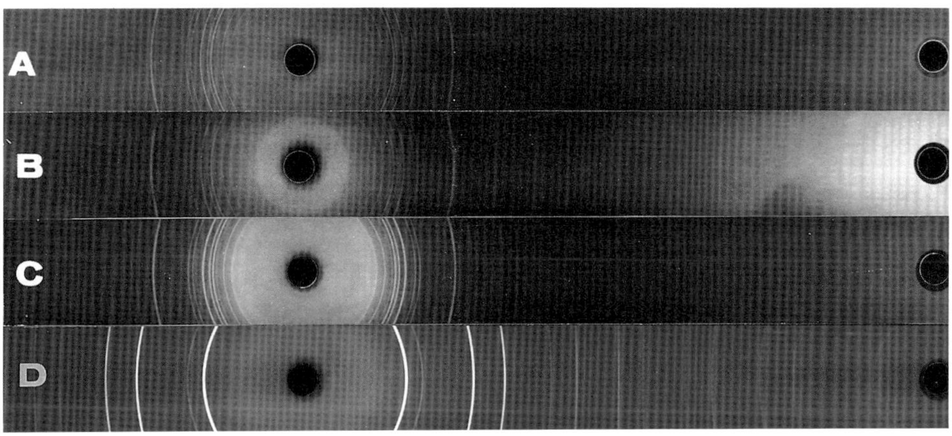

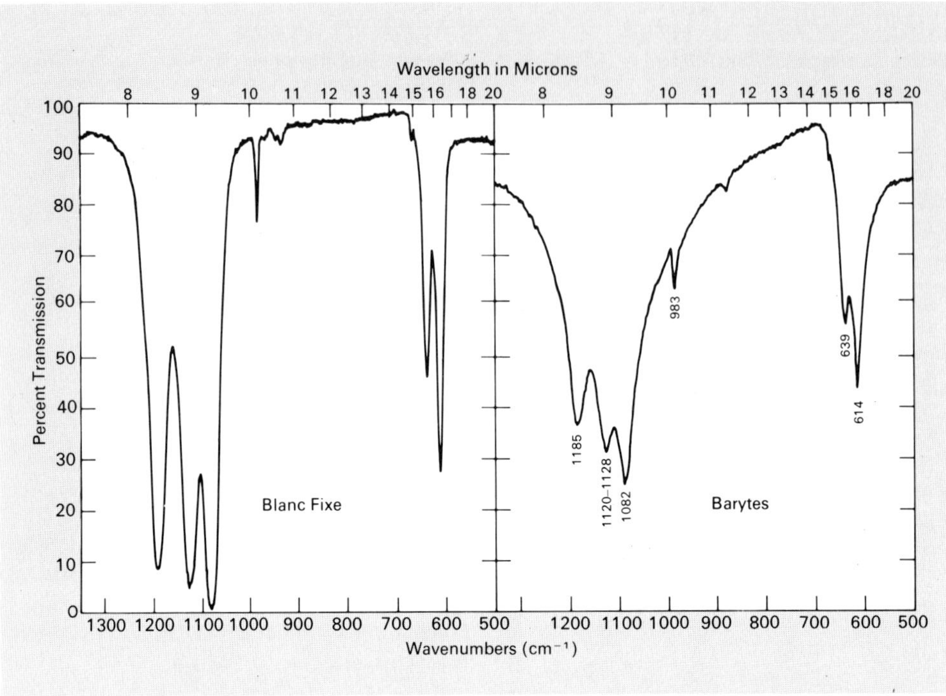

Fig. 12. Principal absorption bands of barium sulfate pigments in the infrared spectrum. Note broadening of the bands in the natural product, barytes.

sulfide were present, the distinctive absorption band would fall between 500 and 400 cm^{-1} (see chapter on Zinc White). Because the infrared spectrum of barium sulfate might be confused with that of basic lead sulfate, a control should be run if the latter is suspected on the basis of an emission spectrographic or microscopic examination.

5.33 ***Spectrochemical Analysis.*** The principal lines in the arc emission spectrum of barium are at 5535.55, 4554.04, 5777.67, and 4934.09 Å in descending order of intensity. The sensitivity is on the order of 1 to 5 ppm. The barium line at 4554.04 is freer from interference than are 5535.55 and 4934.09, both of which are prone to interference from iron (Ahrens and Taylor, 1961). There are also intense lines for barium in the ultraviolet at 3345.02, 3302.59 and 3282.33 Å. The principal sulfur line, at 4815.51 Å, is only 80% as strong as the two strongest barium lines, and there are two other fairly strong lines for sulfur at 4694.13 and 4162.70 Å. The green color of the flame, indicating the presence of barium, is often readily noticeable.

Analysis of the minute samples of barium in the air by atomic absorption spectroscopy has been described by Magyar (1971).

The main x-ray fluorescence lines for barium are $K\alpha$ at 32.1 kV and $K\beta$ at 36.6 kV (Hanson, 1973).

5.34 ***X-Radiography.*** The presence of barium, a heavy metal, results in the white pigment dense to x-rays, with a mass absorption coefficient of 210 cm^2g^{-1} for Cu$K\alpha$ radiation. Because of its great chemical purity and inertness and high density to x-rays, the pigment is used extensively for medical radiography of the alimentary canal.

5.4 Criteria for Positive Identification

Although, like all crystalline compounds, barium sulfate gives a distinctive x-ray diffraction pattern, care must be taken to differentiate the Debye-Scherrer pattern of barium sulfate from that of lithopone ($BaSO_4/ZnS$) (See 5.31).

Owing to the insolubility of the sulfate, there are no simple microchemical tests that can be made and x-ray diffraction will perhaps usually be relied upon for positive identification of barium sulfate. An infrared curve frequently can provide supporting evidence.

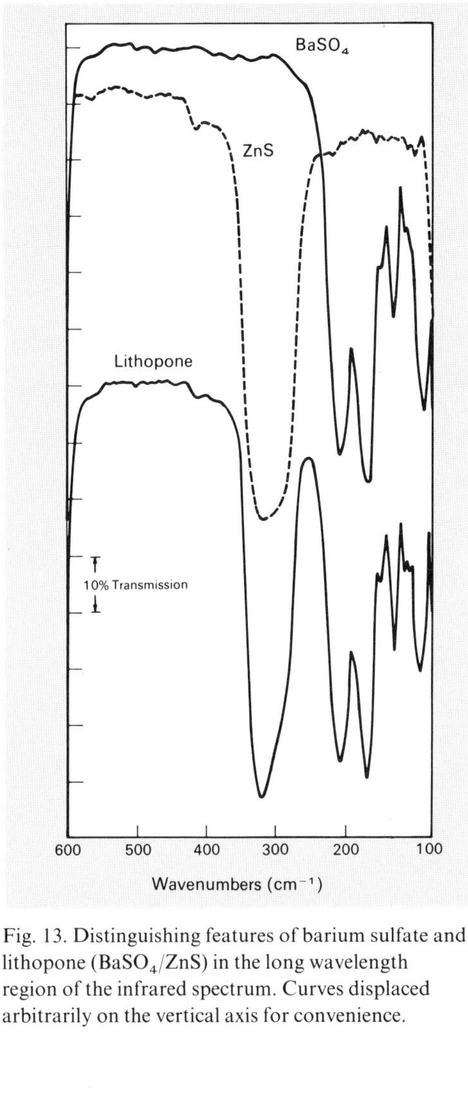

Fig. 13. Distinguishing features of barium sulfate and lithopone ($BaSO_4/ZnS$) in the long wavelength region of the infrared spectrum. Curves displaced arbitrarily on the vertical axis for convenience.

Under the microscope, the natural pigment is usually recognized by its particle shape, profile, refractive index, and characteristic low birefringence. Even so, some coarsely ground synthetic products could be confused with the natural material (see Kittel, 1960). The colors under a sensitive-tint retardation plate, particularly the orange hue, are reasonably distinctive. Finely divided barite and blanc fixe are much more difficult to recognize. Emission spectographic or x-ray fluorescence analysis is absolutely necessary to detect not only barium but the presence or absence of zinc, lead and titanium.

The analyst must continually be alert to the difficulty in differentiating between lithopone ($BaSO_4/ZnS$) and blanc fixe. Positive identification can be made only by x-ray diffraction, although the absence of zinc or the presence of only very minor amounts of zinc would minimize the possibility of the sample being lithopone. The infrared spectrum in the region 400–100 cm^{-1} would also be helpful to detect significant quantities of zinc oxide or sulfide. The detection of significant quantities of zinc is not a sufficient basis for establishing the presence of lithopone because zinc white (zinc oxide), rather than zinc sulfide, could be present.

Barium sulfate is widely used in mixture with other white pigments. Titanium barium whites and titanated lithopones are also manufactured that may contain 15 to 30% titanium dioxide (Moser, 1973); hence, considerable caution must be exercised before the analyst can claim positive characterization of a barium-sulfate-containing white.

6.0 NOTABLE OCCURRENCES

Barium sulfate can be expected to occur rather widely, particularly in paints made after the 1820s and throughout the 1900s. Nonetheless, because there are no simple microchemical tests for this very insoluble pigment and because museum scientists in the past have not regularly had x-ray diffraction at their disposal, positive identifications have been rare.

Kühn (1969), in his examination of 140 nineteenth-century paintings in the Schack-Galerie, found barium sulfate in almost 60% of the grounds and in better than 20% of the whites, greens, blues, yellows, and reds.

Barytes and cerulean blue were found in a blanched area of Léger's *The City, Fragment, Third State, 1919* (Philadelphia Museum of Art, Arensberg Collections 50-134-124) (Siegl, 1967) and in a green in Picasso's *Guitar* of 1919 (Museum of Modern Art, New York) (Feller, 1967).

Barytes, possibly used in lake pigments, has been observed in World War I airplane camouflage colors (Toelle, et al., 1972). Blanc fixe was an important ingredient in "Battleship Grey" (Toch, 1925).

In forgeries, Gettens (1954) found lakes on barytes in vermilion-hued colors on Chinese ceramics. Riederer (Fleming, et al., 1971) found the pigment in some purportedly Etruscan paintings on terracotta.

ACKNOWLEDGMENTS

The author gratefully acknowledges the measurements and valuable comments of Dr. Sidney Pollack regarding the crystallographic and x-ray diffraction data and of Dr. Gerald Carlson regarding the infrared spectra. Scanning electron micrographs were prepared by Martin Haller. The kindness of Peter S. Schenck of the former De Lore Division, National Lead Company, in providing us in 1966 with a copy of the 1920 booklet by Allen W. Clark is much appreciated, as is the kindness of R. W. Unwin of the National Lead Company for permission to reprint two of the illustrations. Mr. Frank D. Davis of the Frank D. Davis Company, Los Angeles, California, kindly supplied samples of barytes including types used in oil-drilling mud.

The author very much appreciates the helpful comments in 1967 of Dr. F. Greenaway of the Science Museum, South Kensington, England, concerning John Dalton's symbols for barytes which appeared on a wall chart used by Dalton in his lectures in Scotland in the years around 1807; the original was destroyed by enemy action in 1940. This chart appeared on the cover of *Chemistry in Britain* 2, no. 12 (December 1966).

Elisabeth West FitzHugh provided the author with an x-ray diffraction pattern of β-sphalerite, the variety of zinc sulfide said to occur in lithopone.

REFERENCES

L. H. Ahrens and S. R. Taylor, *Spectrochemical Analysis*, 2nd ed. (Reading, Massachusetts, 1961), 205–209, 358.

Barium Bibliography, paint reference no. 32 (New York, 1961), 244.

J. G. Bearn, *The Chemistry of Paints, Pigments and Varnishes* (London, 1923).

J. J. Bradley, Jr. "White and Colorless Pigments of Low Refractive Index (LRI)," in *Protective and Decorative Coatings* 2nd ed., vol. 2, J. J. Mattiello ed. (New York, 1942), 428–434.

J. C. Burgess, *A Practical Essay on the Art of Flower Painting* (London, 1811), 42.

R. M. Caven and J. A. Cranston, *Symbols and Formulae in Chemistry* (London, 1928), 2.

A. W. Clark, *The Story of Barytes* (St Louis, 1920).

C. Coffignier, *Couleurs et peintures* (Paris, 1924), 87–96.

C. Coffignier, *Colours and Varnishes*, trans. A. Harvey (London, 1925) 27.

Colour Index, vol. 3, 3rd ed. (Bradford, Yorkshire, England, Research Triangle Park, North Carolina, 1971).

H. F. Coward, "John Dalton: The Early Years of the Atomic Theory as Illustrated by Dalton's Own Notebooks and Lecture Diagrams," *Journal of Chemical Education* 4 (1927), 23–27.

B. G. Dean and D. A. Brobst, *Barite Deposits in the United States* (Washington, 1955).

M. Doerner, *The Materials of the Artist and Their Use in Painting*, trans. E. Neuhaus (New York, 1949), 102.

F. Feigl, *Spot Tests in Inorganic Analysis*, 5th ed., trans. R. F. Oesper (Amsterdam, 1958).

F. Feigl, I. T. de Souza Campos, and S. L. Dalto, "New Applications of the Wohlers Effect in Spot Test Analysis," *Analytical Chemistry* 36 (1964), 1657–1659.

R. L. Feller, "Analysis of Pigments" in *American Painting to 1776: A Reappraisal*, ed. I. M. G. Quimby (Charlottesville, Va., 1971), 327–344.

R. L. Feller, "Barytes Found in Blanched Paint," *Bulletin, International Institute for Conservation—American Group* 8, no. 1 (1967), 10.

S. J. Fleming, H. Jucker, and J. Riederer, "Etruscan Wall-paintings on Terracotta: A Study in Authenticity," *Archaeometry* 13 part 2 (1971), 143–167.

Gentele's Lehrbuch der Farbenfabrikation ed. A. Buntrock (Braunschweig, 1909): vol. 1, 125–127 (barytes); vol. 3, 127–137 (blanc fixe).

R. J. Gettens, "True and False Vermilion on Early Chinese Ceramics," *Far Eastern Ceramic Bulletin* 6, no. 1 (March, 1954), 16–27.

V. F. Hanson, "Quantitative Elemental Analysis of Art Objects by Energy-Dispersive X-ray Fluorescence Spectroscopy," *Applied Spectroscopy* 27, (1973), 309–334.

R. D. Harley, *Artists' Pigments; c. 1600–1835, A Study in English Documentary Sources*, 2nd ed. (London, 1982).

N. Heaton, *Outlines of Paint Technology*, 3rd ed. (London, 1947).

G. B. Heckel, *The Paint Industry* (St. Louis, 1928).

E. W. Heinrich, *Microscopic Identification of Minerals* (New York 1965), 131.

J. M. Hill, *Barytes Deposits in the United States* (Washington, 1917), 169–179.

C. D. Holley and E. F. Ladd, *Analysis of Mixed Paints, Color Pigments and Varnishes* (New York 1908).

G. H. Hurst, *Painters' Colours, Oils, and Varnishes: A Practical Manual*, 4th ed. (London, 1906).

Infrared Spectroscopy—Its Use in the Coatings Industry (Philadelphia, 1980), Spectra 789, 792.

JCPDS Powder Diffraction File, JCPDS International Centre for Diffraction Data, (Swarthmore, Pennsylvania, 1982).

H. Kittel, ed. *Pigmente. Herstellung, Eigenschaften, Anwendung* (Stuttgart, 1960).

H. Kühn, *Die Pigmente in den Gemälden der Schack-Galerie* (Munich, 1969).

J. C. Leuchs, *Anleitung zur Bereitung aller Farben und Flüssigkeiten* (Nuremberg, 1825).

M. Magyar, "Analysis of Barium Sulfate by Atomic Absorption," *Environmental Science and Technology* 5, no. 4 (1971), 358–359.

Malehurst Barytes Company, "The Mining and Refining of Barytes," *The Industrial Chemist* (May 1926), 213–220.

J. W. Mellor, *A Comprehensive Treatise on Inorganic and Theoretical Chemistry*, vol. 3 (London, 1923), chap. 24.

S. Mierzinski, *Die Erd-, Mineral- und Lack-Farben*, 4th ed. (Weimar, 1881), 369–375.

H. B. Milner, *An Introduction to Sedimentary Petrography* (London, 1922).

F. H. Moser, "Barium Sulfate, Synthetic," in *Pigment Handbook*, vol. 1, ed. T. C. Patton (New York, 1973), 281–288.

E. W. Muddiman, "Witherite," *Journal of the Oil and Colour Chemists Association* 25, no. 265 (1942), 127–142.

Nédéy, G., "Sulfate de baryum naturel," *Peintures, pigments, vernis* 35 (1955), 529–532.

R. Newman, "Some Applications of Infrared Spectroscopy in the Examination of Painting Materials," *Journal of the American Institute for Conservation* 19, 1 (1979), 42–62.

R. A. Nyquist and R. O. Kagel, *Infrared Spectra of Inorganic Compounds* (New York, 1971), Spectra 403, 443.

S. Parkes, *The Chemical Catechism* (New York, 1816), 129–131.

T. C. Patton, 'Barium Sulfate Natural," in *Pigment Handbook* vol. 1, ed. T. C. Patton (New York, 1973), 275–280.

J. Percy, *The Metallurgy of Lead* (London, 1870), 537.

H. Rabaté, "Sur l'ordre chronologique d'apparition des principaux constituents solids des peintures et des preparations assimilées," *Peintures, pigments, vernis* 40, no. 10 (1964), 601–604.

Raw Materials Index, Pigment section (Washington, 1975).

A. Rees, *The Cyclopedia*, vol. 4 (Philadelphia, 1810–1824).

R. E. Russell, "Barium sulfate," in Tappi Monograph Series no. 20, *Paper Coating Pigments* (New York, 1958), 19–26.

Sachtleben Chemie, "Barium Sulphate for Extenders and Fillers in Coatings and Plastics," *Polymers, Paint and Colour Journal* (18 June 1975), 533–538.

Sachtleben Chemie, "The Effects Exerted by the Particle Size of Barium Sulphate Extenders in Enamels and Emulsion Paints," *Polymers, Paint and Colour Journal* (3 November 1976), 852–862.

E. V. Sayre, "Direct Deposition of Barium from Homogeneous Solution within Porous Stone," *New York Conference on Conservation of Stone and Wooden Objects*, IIC, part 1 (New York, 1970), 115–118.

F. Siegel, "Blanching of Paint in a Léger," *Bulletin, International Institute for Conservation — American Group* 8, no. 1 (1967), 8–9.

E. Stock, "Heavy Spar," *Farben-Zeitung* 31 (1925), 341–342.

G. F. A. Stutz, "The Testing of Paint Pigments for Transparency to Ultraviolet Radiation in ZnO," *Journal Franklin Institute* 202, (1926), 89–99.

P. Tarte and G. Nizet, "Etude infrarouge de quelques composés du type K_2SO_4 et $BaSO_4$," *Spectrochimica acta* 20 (1964), 503–513.

M. Toch, *Materials for Permanent Painting* (New York, 1911).

M. Toch, *The Chemistry and Technology of Paints*, 3rd ed. (New York, 1925).

A. D. Toelle, H. D. Hastings, and B. F. Hardesty, "Project Butterfly Dart III," *Journal of the Society of World War I Aero Historians* 13, no. 4 (1972), 325–341.

G. Urdang, *The Apothecary Chemist Carl Wilhelm Scheele* (Madison, 1942).

A. Ure, *A Dictionary of Arts, Manufacturers and Mines*, vol. 1 (London, 1853), 122.

V. C. Vesce, "Spectral Characteristics of Pigments in the Visual and Infrared Bands," *Protective and Decorative Coatings*, vol. 5 ed. J. J. Mattiello (New York, 1946), Chap. 4.

E. Watson, *China, The Maritime Customs, II. The Principal Articles of Chinese Commerce (Import and*

Export), Special Series, no. 38, 2nd ed. (Shanghai, 1930).

F. J. Williams, "Barium Minerals," in *Industrial Minerals and Rocks*, 2nd ed., eds. S. H. Dolbear et al., (New York, 1949), 77–94.

H. Wolff and G. Zeidler, "Investigations on the Use of Heavy Spar in Paint and Varnish Industries," *Paint and Varnish Production Manager* 16, no. 2 (1937), 7, 8, 10, 12.

W. A. Wood, "An X-ray Examination of Lithopones," *Journal of the Society of Chemical Industry* (11 July 1930), 300–301T.

BIBLIOGRAPHY

J. Bersch, *The Manufacture of Mineral and Lake Pigments*, 3rd ed. rev., trans. A. C. Wright (London, 1901).

H. S. Booth, E. F. Pollard, and M. J. Rentschler, "Reactions in Inert Fused Substances," *Industrial and Engineering Chemistry* 40 (1948), 1981–1982.

D. Brewster, *The Edinburgh Encyclopedia*, vol. 5 (Philadelphia, 1832), 684.

M. Cremer, "Blanc Fixe Micro, a New, Multi-purpose Barium Sulfate Extender," *Polymers Paint Colour Journal* 172 (1982), 560–563, 565, 566, 568–570.

G. Deniges, "Microchemical Test for Barium with Sodium Tungstate (Abstract)," *Analyst* 55 (1930), 156; from *Annales de chimie analytique* 11 (1929), 365.

G. Genin, "Sulfate de baryum precipite," *Peintures, pigments, vernis* 35 (1959), 578–581.

H. Gruszczyk, "Geological Conditions of the Occurrence of Barite," *Instytut geologiczny Warsqy prace*, no. 59 (1970), 9–30 Review in *Chemical Abstracts* 77 (1972), 64504g.

R. B. Ladoo and W. M. Myers, *Nonmetallic Minerals*, 2nd ed. (New York, 1951).

K. V. S. N. Raju and M. Yaseen, "The Effect of Part Replacement of Titanium Dioxide by Barytes on the Properties of Alkyd Paints," *Journal of the Oil Colour Chemists' Association* 65 (1982), 401–408.

L. Rossi and R. G. Coronato, "Differential Analytical Study of Calcium, Barium, Strontium, and Lead Sulfates," *Revista de la Asociacion bioquimica Argentina* 5, 14, 17–24; "Characterization of Calcium, Barium and Strontium," (1940), *Revista de la Asociacion bioquimica Argentina* 5, 14 (1940), 27–32 (see *Chemical Abstracts* 35, 403[8]).

R. M. Santmeyers, "Barite and Barium Products, Part 1 — General Information," *U.S. Bureau of Mines Information Circular 6221* (Washington, D.C., 1930).

R. M. Santmeyers, "Barite and Barium Products, Part 2, Barium Products," *U.S. Bureau of Mines Information Circular 6223* (Washington, D.C., 1930).

W. Withering, "Experiments and Observations on the Terra Ponderosa," *Philosophical Transactions of the Royal Society of London* (1784), 293–311.

Cadmium Yellows, Oranges and Reds

INGE FIEDLER and MICHAEL A. BAYARD

NOMENCLATURE IN VARIOUS LANGUAGES

English: cadmium yellow
German: Cadmiumgelb
Russian: желтый кадмий
French: jaune de cadmium
Italian: giallo di cadmio
Spanish: amarillo de cadmio

1.0 INTRODUCTION

The yellow compound cadmium sulfide was first recommended as an artists' pigment in 1818 by its discoverer, Stromeyer (1819), but its use on a reasonably wide scale had to await the commercialization that began in the mid 1840s. Cadmium yellow is fundamentally cadmium sulfide, CdS, although various chemical and physical modifications have been developed in order to produce shades ranging from very light lemon yellow to medium orange. There are two principal types of the yellow: the pure compound and the lithopone variety, consisting of coprecipitated cadmium sulfide and barium sulfate.

Greenockite and hawleyite are natural cadmium sulfide minerals which frequently occur as a yellow coating on sphalerite (zinc sulfide) (fig. 1). Although the statement has been made that artists have found use for greenockite for nearly two thousand years, the authors know of no evidence to support the claim (Curtis & Wright, 1954; Patton, 1973).

The hue of the sulfide is altered toward orange and red by the inclusion of an increasing amount of selenium. A further development in the manufacture of these hues has been the introduction of the cadmium/mercury sulfides. Because of their close chemical relationship, the oranges and reds are included herewith, although the principal emphasis will be on the yellow.

1.1 Brief Description of Pigment

The cadmium sulfide class of pigments includes the cadmium yellows and light oranges (CdS), the light cadmium yellows which contain added zinc in solid solution with CdS [(Cd, Zn)S], and the sulfoselenide deep oranges and reds [Cd(S, Se)]. These can be extended with the inert pigment $BaSO_4$ to give the lithopone varieties. A variant substitutes mercury for some of the cadmium to yield cadmium/mercury oranges and reds [(Cd, Hg)S]. The pure cadmium sulfide yellow is designated CI Pigment Yellow 37, No. 77199, while the yellow cadmium lithopone is CI Pigment 35, No. 77117. The sulfoselenide orange is CI Pigment Orange 20, and the end member (CdSe) is Pigment Red 108, both CI No. 77196, although the latter is also given CI No. 77202. The mercury/cadmium oranges are CI Pigment Orange 23, while the red varieties are listed as CI Pigment Red 113, both CI No. 77201. The CI designations for the oranges and reds include the lithopones (*Colour Index*, 1971).

INGE FIEDLER, Conservation scientist, Department of Paintings and Objects Conservation, Art Institute of Chicago.

MICHAEL A. BAYARD, Research microscopist and consultant, Particle Data Laboratories, Elmhurst, Illinois.

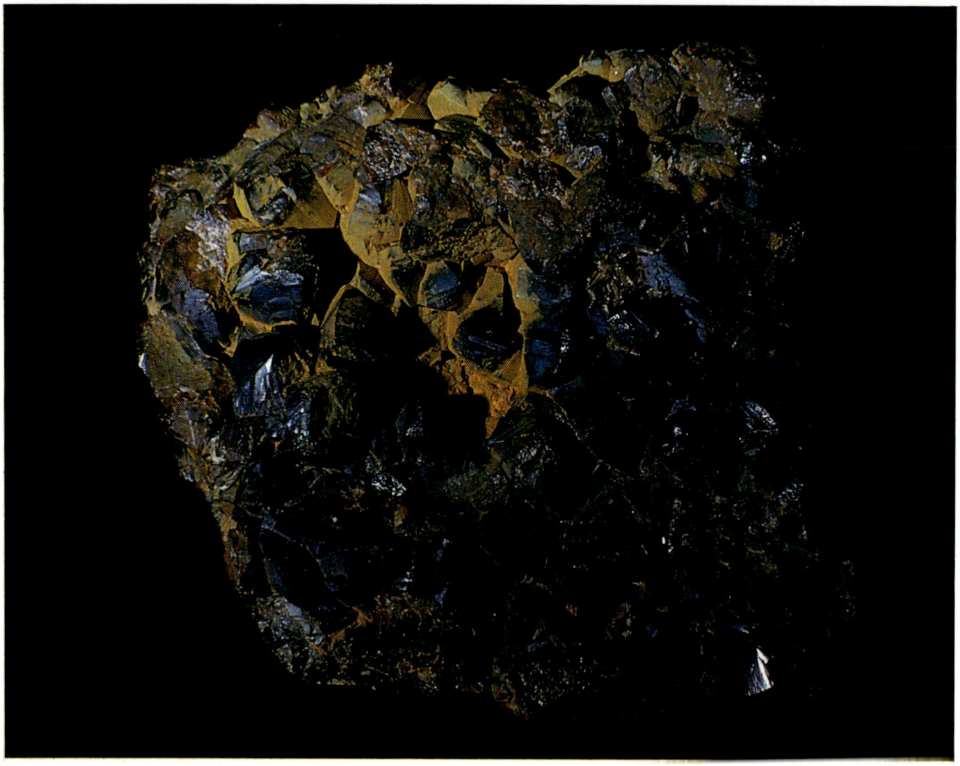

Fig. 1. Typical light coating of greenockite on surface of other mineral (here sphalerite). Specimen, approximately 8 x 6 x 4 cm, courtesy of Carnegie Museum of Natural History, Carnegie Institute, Pittsburgh.

1.2 Current Terminology

Cadmium sulfide is a relatively recent addition to the artists' palette; as a result the pigments are usually referred to by their generic names rather than by fanciful terms: cadmium yellow primrose, lemon, light, medium, and dark; and cadmium oranges and reds in light, medium, and dark shades. Names introduced in the nineteenth century, Aurora yellow and jaune brilliant, are still used by Winsor & Newton, Inc. With barium sulfate as an extender, the pigment is termed cadmium lithopone (or cadmopone) or described as a cadmium-barium shade, while trade names, such as Cadmolith (Glidden), are also in use.

2.0 HISTORY

The observation by Stromeyer in 1817 that a sample of zinc carbonate from the Salzgitter zinc factory in Germany gave an iron-free oxide with a yellow color led to the discovery of a new element. Because of its close association with zinc, the element, a metal, was named cadmium, derived from cadmia fornacum, the designation applied to the furnace used in smelting zinc ore (calamine or cadmia). Concurrently, Hermann, Roloff, Meissner, and Karsten also noted the existence of this yellow substance in association with various processes for producing zinc products (Budgen, 1924; Gregory & Burr, 1926; Weeks & Leicester, 1968). Cadmium is not found in a free state and only rarely occurs as a relatively pure cadmium bearing mineral such as greenockite or hawleyite. The commercial source of the metal is the refining of zinc ores that contain cadmium as an impurity. Precipitation of the yellow-colored sulfide from solutions of cadmium salts by the use of hydrogen sulfide was one of the chemical phenomena first observed involving the newly discovered element (Gay-Lussac, 1818; Curtis & Wright, 1954).

2.1 Archaic, Obsolete and Misrepresentative Names

From the time of its introduction as an artists' pigment, cadmium sulfide has been, with few exceptions, known as cadmium yellow. At first this designation referred essentially to a single shade, comparable to a medium yellow of today. Later the term cadmium orange was used to refer to a darker hue. Because of the desire for lighter shades, jaune brilliant or brilliant yellow, a mixture of lead white and cadmium yellow, was offered soon after the commercialization of cadmium sulfide. At that time the addition of a white pigment was the only way to achieve a lighter color. Doerner in 1949 described brilliant yellow as "a very light mixture of cadmium yellow and either Cremnitz or zinc white; it is durable but wholly superfluous."

Other names for cadmium yellow were: cadmia, radiant yellow, daffodil, Aurora yellow (introduced in 1889), Orient yellow (the latter two being trade names of Winsor & Newton), and neutral orange (Standage, 1886; Parry & Coste, 1902; Mayer, 1975). Church (1915, p. 166) described daffodil as being "a variety of cadmium sulfide prepared at a red heat and containing a small quantity of magnesia." He described Aurora yellow as "a bright, beautiful pigment consisting essentially of cadmium sulfide, more opaque than most other varieties of cadmium . . . and (it) possesses a pure yellow hue. Its stability is greater than many other varieties of this pigment." Neutral orange, he stated, was "a mixture of cadmium yellow and Venetian red."

A number of mixtures using cadmium yellow can presently be found; examples are: jaune brilliant (cadmium yellow and lead white, sometimes with vermilion added) and cadmium green (cadmium yellow and viridian). An imitation Naples yellow hue is available as either a mixture of cadmium sulfide, ferric oxide, and zinc oxide for watercolors or cadmium yellow, lead white, light red, and yellow ochre for oil paints (according to Winsor & Newton, n.d.). Because of toxicity, true lead antimonate (Naples yellow) has been largely replaced by the cadmium sulfide based substitute (see chapter on Lead Antimonate Yellow).

2.2 History of Use

Cadmium yellow, under the name of sulphuret of cadmium, was suggested for use as an artists' pigment by Stromeyer shortly after his discovery of the metal. Referring to experiments done in 1818, he had this to say: "This sulphuret, from its beauty and the fixity of its color, as well as from the property which it possesses of uniting well with other colors, and especially with blue, promises to be useful in painting. Some trials made with this view gave the most favorable results" (Stromeyer, 1819). Though this early reference indicated the possibilities, little use could be made of the suggestion until suitably large sources of the metal became commercially available. The metal was being produced about 1829 in Upper Silesia (the main producer of cadmium until the mid 1840s), but the small quantity of cadmium metal and its sulfide then available was, according to R. Wagner (1872) and Budgen (1924), mainly employed in type-metal, fusable alloys, and as a blue flame colorant in fireworks. The artist had to wait for the expanded commercialization in the 1840s.

Parnell (1846, p. 429) provides an interesting comment on the early interest in this colorant: "Sulphuret of cadmium may be fixed on silk by first impregnating the goods with chloride of cadmium and afterwards passing them through a weak solution of hydro-sulphate of potash or soda. It is easy to perform this operation." Further on Parnell notes that "the facility with which silk may be dyed by the process above mentioned, leads us to believe that if cadmium was to become more common, its sulphuret would be employed in paintings as well as in dyeing."

There is very little information available about the early history of the pigment. Furthermore, most of this information concerns its manufacture rather than application. Melandri is said to have introduced the pigment into the technique of oil painting in 1829 (Munkert, 1905; Eibner, 1909; de Wild, 1929). Reference has also been made to limited use in France and Germany beginning that same year (Kelley, 1970). According to Harley (1982) cadmium yellow was not found in any English color lists before 1835. Bachhoffner (1837) in a paragraph on cadmium states: "Its oxide is of an orange color, and would no doubt form a good pigment did there exist a sufficient supply of the metal." (Oxide in this case was a misidentification of the sulfide). Apparently cadmium was imported into the United States in 1842, but no information is given as to amount or use (Kelley, 1970). Harley (private communication) reports that George Field, in his *Practical Journal 1809*, had a watercolor sample of cadmium yellow on folio 385, with the comment "cadmium yellow works well — acts slightly on knife — not more changed than the paper by suspending in damp and foul air more than five months." Although the folio is not dated, on its reverse Field made a notation on native ochres, dated November 28,

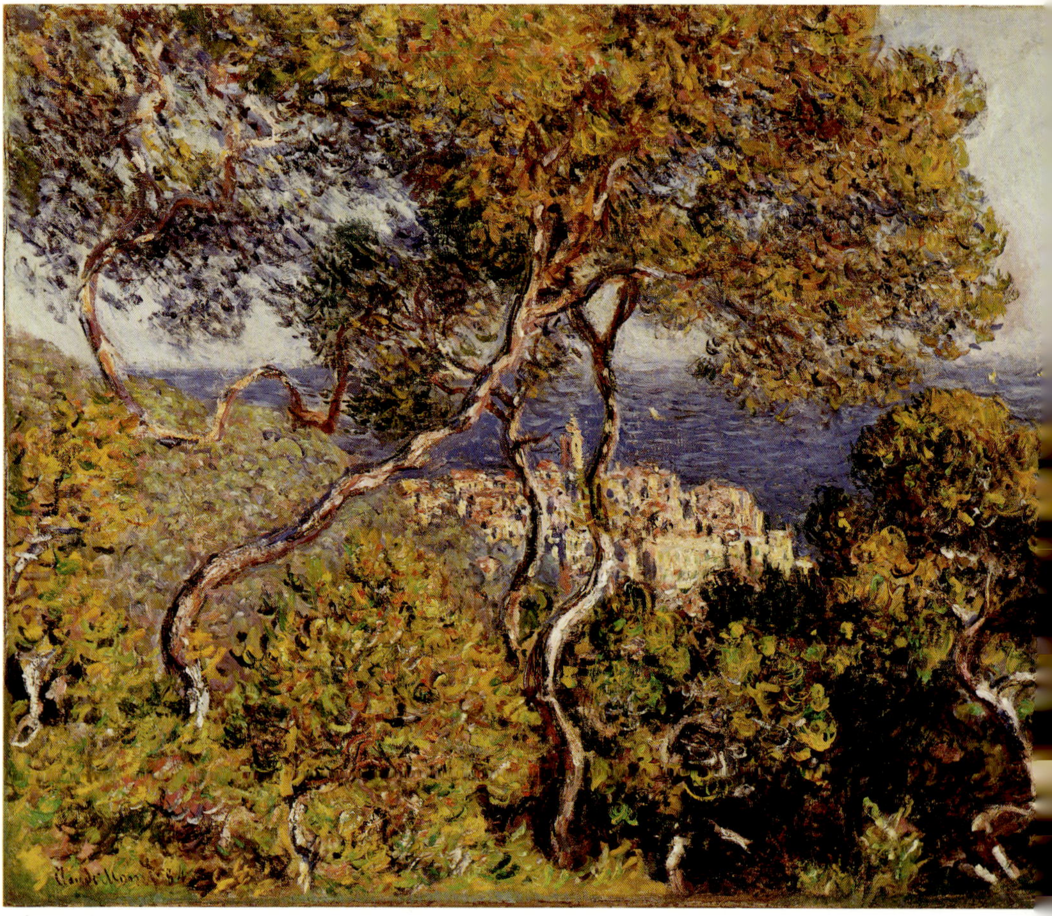

Fig. 2. Claude Monet, *Bordighera*, painted 1884; oil painting showing the use of cadmium yellow (Potter Palmer Collection, Art Institute of Chicago, 1922.426).

1844. This suggests that the cadmium sample predated 1844. There are several authors (Laurie, 1914; F. W. Weber, 1923; Kelley, 1970) who allude to the pigment's commercial introduction in England by 1846. We know that it was shown by Winsor & Newton at the 1851 Exhibition held at the Crystal Palace in London (*Great Exhibition* . . ., 1851).

The scarcity of the pigment apparently continued for many years. An American artist (*Handbook of Young Artists* . . . 1865) mentions cadmium sulfide in a footnote as a matter of current interest. In 1870, Field had this to say: "The metal cadmium affords, by precipitation with sulphuretted hydrogen, a bright warm yellow pigment, which passes readily into tints with lead white, appears to endure light, and remains unchanged in impure air; but the metal from which it is prepared being hitherto scarce, it has been as yet little employed as a pigment, and its habits are, therefore, not fully ascertained" (Field, 1870).

Exact statistics on the early production of the metal were not kept, though Chizhikov (1966) refers to less than 250 pounds as the maximum world production for any year prior to 1871. Wagner (1872) gives the more optimistic figure of 700 pounds of metal for Europe in the same years. Even if the higher figure were correct, the supply would still have been limited. Metallurgy, medical uses and pyrotechnics competed with the pigment trade for the available metal.

In a Devoe and Co. catalog of 1878, published in New York City, cadmium yellows and cadmium orange cost fifteen dollars per pound as compared to chrome yellow at seventy cents per pound, a twenty-fold difference in price. Guignet (1888) considered cadmium "a perfect color if not so expensive." Apparently the pigment appealed to Monet, for we find it used in a number of his works, including the painting titled *Bordighera* of 1884 (fig. 2). Cadmium pigments from various manufacturers (Weber, 1890) are summarized in table 1.

A modest 300 pounds of the metal (and no sulfide) were produced in the United States in 1906; by 1911, 2,392 pounds of cadmium sulfide were manufactured. Nonetheless, owing to its continued high cost, cadmium yellow remained limited in its use by artists until around 1917, when the amount employed in industry began to increase rapidly. Manufacturing techniques also improved, thus removing most of the problems associated with the early pigment. Instead of a material "variable in trustworthiness" as stated by Hunt in 1880, cadmium sulfide became by the 1920s a reproducible, permanent, bright pigment available in a wide variety of shades and compatible with most other materials. As in other cases, the artist benefited from the fallout of industrial technology. By 1922, production had reached 134,774 pounds, the price had come down to one dollar and thirteen cents a pound, and the pigment was currently used for such mundane applications as paint for streetcars, a colorant for soap, and an ingredient in rubber (Budgen, 1924; Toch, 1925).

The lowering of the cost of the pure material and the introduction of cadmium lithopones perhaps were the most important factors in bringing cadmium pigments into widespread industrial use. In the United States, Marston, in 1921, patented the first lithopone as a cadmium yellow with about 5% admixture of barium sulfate. By 1927, cadmium lithopones with over 60% sulfate were produced at about half the cost of the pure pigment (Ward, 1927). A wide variety of shades were available, and their tinting strength, heat and chemical resistance were only slightly less than those of the pure substance. With the price one-half to one-third that of pure cadmium sulfide, the lithopones provided better hiding power on a cost basis. Both types of pigment were widely employed in the ceramics, rubber, and glass industries and as colorants in the new plastics; however, only the pure pigments were recommended for use by artists. According to Chizhikov (1966), total use of the metal in Great Britain increased from 375 tons in 1943 to 1,040 tons in 1958; pigment production for the same period increased forty-fold, from 4.9 tons to 207 tons.

Present day use of cadmium pigments is concentrated mainly in the plastics industry,

Table 1. **CADMIUM PIGMENTS LISTED IN F. WEBER AND CO. TRADE CATALOGUE OF 1890**

F. Weber	*DR. FR. Schoenfeld and Co.*	*Newman*	*Winsor & Newton Catalog No.*
Oil Colors (Class H)			
No. 380. Cadmium yellow, light	Cadmium No. 1 citron Cadmium No. 2 light		780 Cadmium yellow pale
No. 381. Cadmium yellow, medium	Cadmium No. 3 medium Cadmium No. 4 deep		781 Cadmium yellow deep
No. 382. Cadmium yellow deep	Cadmium No. 5 orange		782 Cadmium orange
No. 383. Cadmium orange			Also: 789 Orient yellow 793 Aurora yellow
Watercolors (moist and cakes)			
Cadmium yellow Cadmium orange	Cadmium yellow, light Cadmium yellow, deep Cadmium orange	Cadmium yellow (In cakes)	Cadmium yellow Cadmium yellow, pale Cadmium orange Aurora yellow
Pastels			
Cadmium yellow No. 1 Cadmium yellow Nos. 2, 3 Cadmium yellow Nos. 4, 5, 6			

with about 75% of the approximately 9,000 ton annual (world) production devoted to this application. The ceramics and surface coatings industry account for another roughly 10% each. Under ceramics is included cadmium's use as a colorant for metal enamels and in glasses; for instance, the red glass used for railway, marine, and other signalling purposes is colored with $CdSe_{0.33}S_{0.66}$ suspended in the glass. The remaining 5% of production is spread among such applications as cellulose laminates for work surfaces, printing and silk-screen inks, cosmetics, leather, and rubber (Cadmium Association, 1978). At present, the cadmium yellows, oranges, and reds are still one of the most important classes of pigments.

2.3 Terminal Date

The pigment continues in use. At the present time, there are five manufacturers in the U.S.A. listed in the *Raw Materials Index* (1975).

3.0 GENERAL PIGMENT PROPERTIES

3.1 Color and Spectral Reflectance

Though it is a single chemical entity, cadmium sulfide can be made in various shades ranging from very light yellow to deep orange. At one time it was thought that there might be a range of polysulfides from CdS to CdS_5, that impurities were the main cause of color variance, or that different polymers of CdS formed the different shades (Buchner, 1887; Eibner, 1909). These ideas seemed partially verified by some early density measurements such as those conducted by Church (1915), but the difficulty of achieving accuracy on such finely divided material was not appreciated at that time. It is now known that the main factor influencing the color of cadmium sulfide, as with many pigments, is the particle size, with the lighter shades having the smaller size distribution. In addition, color is affected by crystal structure, crystal perfection, particle shape and size, state of agglomeration, and the presence of other materials (Bersch, 1901; Milligan, 1934). In older samples, Weiser (1928) states that imperfect crystals 0.1 to 1 micrometer in diameter are yellow, while those 4 to 7 micrometers (usually consisting of agglomerates) are bright orange. Modern material usually consists of fine to very fine particles, with most in the submicrometer range. This is the result of precise control during precipitation and calcination.

Pure cadmium sulfide provides a yellow pigment of high chroma, with the introduction of zinc sulfide in solid solution forming a series of pigments with decreasing redness. A maximum of about 25 mole percent zinc sulfide is used (Huckle, et al., 1966). Sulfoselenide oranges and reds form a second series of pigments based on cadmium sulfide. They have the same hexagonal structure but with replacement of some of the sulfur by selenium. As selenium increases, deeper reds are produced, up to a maximum useful cadmium selenide content of 50 mole percent. Fig. 3 shows the family of reflectance curves which can be obtained from this series of pigments, published by Huckle, et al. (1966). In addition, Curtis and Wright (1954) show the location of the cadmium sulfide and sulfoselenides on the CIE chromaticity diagram. Johnston and Feller (1965) have published the reflectance, scattering, and absorption coefficients of cadmium red light, and Barnes (1939) reported spectral reflectance curves for cadmium yellow, orange, and red.

A third series of pigments is based on solid solutions of mercuric and cadmium sulfides: cadmium/mercury sulfide. These were developed despite earlier published work that reported CdS-HgS mixed crystals to be black and

Fig. 3. Typical spectrophotometric reflectance curves of cadmium sulfide and sulfoselenide pigments (after Huckle, et al., 1966).

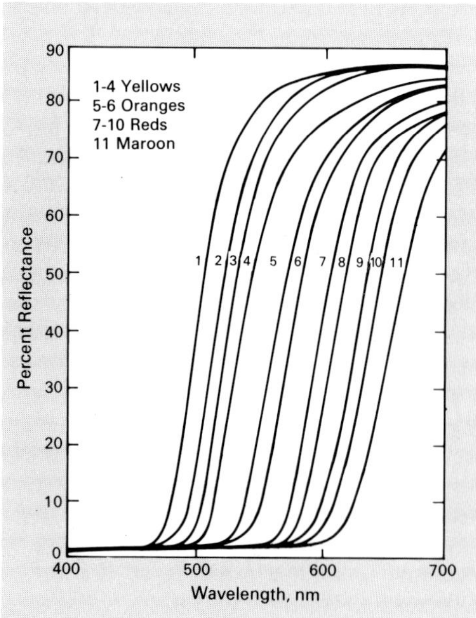

lacking in useful color value. In present commercial use the maximum amount of mercuric sulfide is about 20 mole percent for the deep maroon. The pigments became available in 1955, and the original patentee, Hercules (now Ciba-Geigy Corporation), gave them the trade name Mercadium® (Huckle, et al., 1966). The General Color Company, a division of H. Kohnstamm and Company, called them Cadmere® pigments; this line has been discontinued. Moore (1973) has published spectral reflectance curves for cadmium/mercury sulfides and a table giving the relation of hue to composition.

Lithopone extensions of all of the above mentioned pigments may be regarded as a fourth variety.

Typical Munsell values for light yellow are: 8.OY/9.2/9.0 (hue-8.OY, value-9.2, and chroma-9.0); dark yellow 9.OYR/7.8/9.0. The dominant wavelength varies from 575 nm in the light yellows to 587 nm in the medium oranges. The present day lemon yellows and darkest oranges obtained without the addition of selenium extend in dominant wavelength from about 550 nm to 595 nm. In the light yellow, the purity is 81.9% and in the medium orange, 86.9% (Barnes, 1939). With an ICI illuminant type C, the tristimulus values of masstones range from X = 71.58, Y = 76.66, and Z = 10.83 (for light yellow) to X = 51.94, Y = 42.18, and Z = 4.90 (for medium orange).

3.2 Hiding Power and Tinting Strength

The combination of high refractive index and small particle size imparts good hiding power and light scattering ability to the pure yellows and oranges. The lighter shades are somewhat more opaque because of the smaller particle size. As would be expected, the lithopones possess considerably poorer hiding power and lower tinting strength, though the presence of barium sulfate does not materially affect the spectral reflectance curve, owing to the relatively low refractive index of the latter. In terms of cost for a given degree of hiding power, the lithopones are competitive with the more chemically pure types.

3.3 Permanence

William Holman Hunt (1906, p. 456) considered that

> Cadmium indeed at the best is very capricious, and if trustworthy, as many good authorities declare it to be, it is only so when exceptional care is spent on its preparation. Some specimens painted on a trial canvas in 1860, had in 1880 sunk to the colour of dirty beeswax, and some prepared by Mr. Dawson (and therefore above suspicion as to its genuineness), soon after became greatly vitiated, while chromes put on the trial canvas in 1860 are still incomparably superior in brilliancy to cadmium, lemon and strontian yellow of the same date.

Certainly the stability of cadmium pigments in the early days was dependent on the care and knowledge utilized in their preparation. The medium yellow (pure cadmium sulfide) of the past was usually quite stable; however, in the early preparations of light shades the presence of free sulfur was said to be common. Exposure to moist air was thought to cause bleaching owing to the formation of sulfurous acid which could act on the cadmium sulfide to form the colorless sulfate (Friend, 1910). Pigments containing cadmium oxalate or carbonate also faded when exposed to light and air. Certain lighter pigments of the past, though supposedly free of sulfur, would sometimes bleach with the formation of hydrates and sulfates owing to the presence of small sulfur crystals with highly reactive surfaces. Calcining tended to eliminate this action. Darkening was another common transformation with time, usually caused both by uncombined sulfur and the presence of iron, copper, lead, or arsenic. Even 0.1% of one of these impurities would cause noticeable change. Thus, much of the poor reputation of past cadmium yellows was due to impurities, additives, and adulterants rather than to any inherent instability in the cadmium sulfide (Church, 1915). Martin Fischer (1930) considered the cadmiums to be suitable for *The Permanent Palette*. They were also included in the German list of permanent pigments, "Die Normalfarben," established in the late nineteenth century (Munkert, 1905).

By the 1940s, the pigments' stability had been greatly improved. Toch (1946, p. 100) reports his experience upon examining a bright yellow indicator, placed on top of a gas tank, which pointed the way to an air field. "It was painted with lemon cadmium," he said, "not mixed with oil, but with a spar varnish so that each succeeding rain storm would wash away any dust or dirt that might be collected. At the end of six months a close examination showed the lemon cadmium in as brilliant a condition as the day it was applied." Toch considered that this probably would not have been true of any other yellow pigment. He had indeed been informed, he said, that the Department of Commerce had adopted the use of cadmium yellow as an airplane signal.

Outdoor exposure can lead to bleaching by oxidation of the sulfide to sulfate, usually caused by the combined action of air, moisture, and ultraviolet radiation. Any two of the above factors do not necessarily produce severe bleaching, and an impervious vehicle will minimize or eliminate this effect. In addition, there is a tendency for paints formulated with the pigment to chalk; this behavior can be minimized by the use of a suitable binder (Curtis & Wright, 1954). Sulfoselenide reds, especially the deeper varieties, are more resistant to weathering than are the yellows. The mercury-cadmium reds are not as resistant, though they are used extensively for plastics and auto upholstery (Huckle, et al., 1966). All cadmium pigments possess good resistance to alkalies and as such can be employed in water-dispersed polymer emulsions. Because of calcining they display excellent heat resistance at the temperatures of enamel baking and plastic processing (Patton, 1973).

3.4 Compatibility

Lead- and copper-based pigments are the main source of incompatibility with cadmium yellows. Authors such as Church (1915) and Standage (1886) refer to formation of black sulfides of lead or copper, leading to gray, green, or brown discoloration. Darkening in the presence of lead pigments was the slower process. Such interactions were undoubtedly more prevalent in older pigments. The most rapid darkening apparently occurred with emerald green (Schweinfurt green), a form of copper acetoarsenite. Hunt (1880) refers to this as "an example of a well-known misalliance, cadmium and emerald green. When put on, the tint was brighter than this at the side. It is now as it was within two weeks of its first application — nearly black." Cadmium yellow with a small amount of sulfur, prepared by the authors, reacted immediately to form a dark brown substance upon heating with emerald green (fig. 4); modern commercial samples did not react immediately, though this does not preclude long-term interaction. The nominal reactions with emerald green and lead chromate are as follows:

$$4CdS + Cu(C_2H_3O_2)_2 \cdot 3\,Cu(AsO_2)_2 = 4\,CuS + Cd(C_2H_3O_2)_2 + 3\,Cd(AsO_2)_2$$

$$CdS + PbCrO_4 = CdCrO_4 + PbS$$

The type of medium influences the rate of reaction, the most rapid being in watercolors and the slowest in oil. The vehicle itself could cause change if the oil were "rancid" with

Fig. 4. Microscopic evidence of reaction between cadmium yellow and emerald green. The cadmium sulfide was freshly precipitated and uncalcined. There is darkening of the cadmium yellow particles due to formation of copper sulfide, with the reaction taking place only on the cadmium grains because they are less soluble than the copper acetoarsenite. Aroclor 5442 mount; refractive index 1.66. 500x.

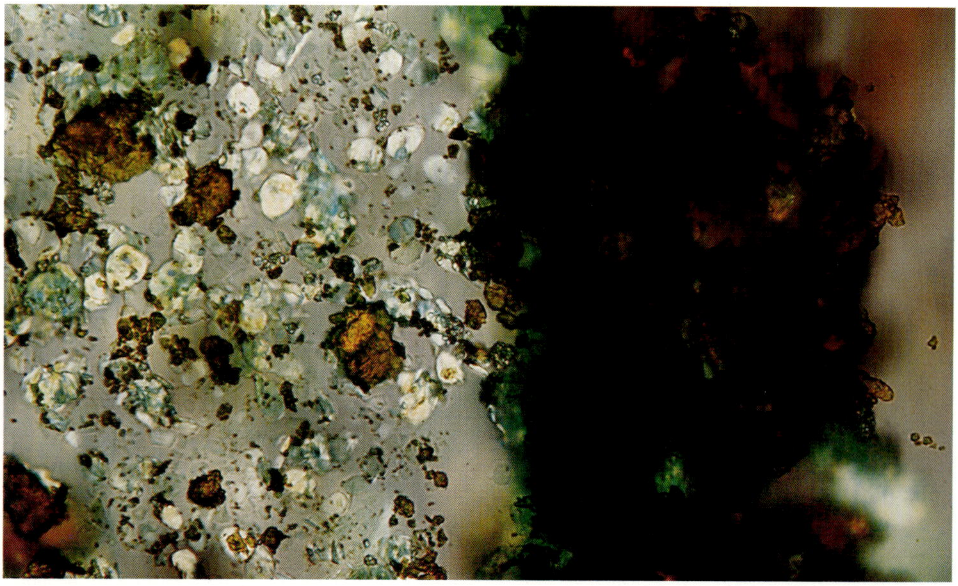

organic acids present. Interactions through use of iron palette knives are also said to have caused darkening owing to the formation of iron sulfide.

The amorphous form of cadmium yellow used by Monet sometimes darkened considerably in areas where it was mixed with chrome yellow. Macroscopically, areas with the mixture appeared as a yellow-brown (ochrelike) discoloration, darker than the surrounding yellows. Optical microscopy and electron probe analysis indicated that the darkening occurred only when the chrome yellow and cadmium yellow were in close contact; the unaltered yellow areas of these paintings did not reveal as close an association of the two components. The vehicle used may have some influence on the changes, although further studies on the mechanism of incompatibility are needed.

In the past, cadmium yellow was not recommended for use in fresco painting. This was particularly true of the lighter shades, since these contained cadmium oxalate or carbonate or perhaps a surfeit of free sulfur. In the alkaline fresco medium, reactions would occur that led to rapid fading of the light yellows. The normal medium yellow would not be subject to this, but the poor reputation of the cadmiums may have restricted their use (Laurie, 1911; F. W. Weber, 1923).

Not all past material was incompatible, as is illustrated by jaune brilliant, a reasonably stable pigment based on a mixture of cadmium sulfide with lead white. Even here, caution was advised, because in some cases this light pigment was not prepared from pure cadmium sulfide (Hurst, 1917). Although much of the incompatibility in early samples was attributable to impurities, very little problem remained with commercial pigments by the mid 1920s.

3.5 Chemical Properties

Cadmium sulfide is a partially covalent compound with a stoichiometric composition of 77.8% Cd and 22.2% S. Formation energy amounts to 34 K cal/mole. At normal temperatures the compound is essentially inert to air, humidity, hydrogen sulfide and sulfur dioxide. It readily dissolves in concentrated hydrochloric acid and in dilute or concentrated nitric acids with the evolution of hydrogen sulfide, forming cadmium chloride or cadmium nitrate. Solubility in water is very low, a saturated solution at room temperature containing 0.2 ppb (parts per billion). Alkalies, acetic acid, and temperatures up to 300°C have little or no effect on cadmium sulfide. Above 300°C gradual oxidation occurs in air with conversion to oxide and sulfate. Concentrated solutions of SO_2 (sulfurous acid), ammonium chloride, and acid chlorides and iodides will dissolve the compound (Chizhikov, 1966; Budgen, 1924; *Handbook of Chemistry and Physics*, 1970–1971).

The zinc-cadmium system must also be considered; up to 25% of many of these pigments consist of zinc sulfide in solid solution with the cadmium. Chemically and physically the two elements behave similarly. The presence of zinc in the cadmium lattice increases its solubility and decreases its acid resistance, though only marginally; solubility in water will still be in the parts per billion range (Huckle, et al., 1966).

Another consideration is the presence of up to 62% barium sulfate in the cadmium lithopones. The water solubility of barium sulfate is on the order of 0.0002%. While the compound is slightly soluble in most strong acids, it is otherwise inert, with no appreciable changes on exposure to air, humidity, hydrogen sulfide, or sulfur dioxide (see chapter on Barium Sulfate). Temperatures up to 300°C and alkalies likewise have little effect.

3.6 Oil Absorption and Grinding Qualities

Oil absorption ranges from 18% for the pale yellows to 22% for the pale orange (Patton, 1973). The *Raw Materials Index* (1975) lists oil absorption for Pigment Yellow 37 at 19.5 to 26%; Pigment Orange 20 at 28 to 32%. For lithopones (Pigment Yellow 35) the values given are 14 to 20%. The low oil absorption for barium sulfate, 8 to 16%, is reflected in the decreased oil requirement for lithopones. Partially because of this relatively low absorption value, lithopones readily disperse into most binders and plastic systems. The cadmiums do not retard drying, exhibit no abnormal settling tendencies, and may be regarded as slow but reliable driers (Toch, 1911; Curtis and Wright, 1954).

3.7 Toxicity

Most reported cases of cadmium poisoning involve ingestion of soluble cadmium salts, inhalation of cadmium vapor, or ingestion of acid foodstuffs which have had contact with cadmium-plated articles. Curtis and Wright in 1954 had not been able to trace any case of poisoning which had arisen through the ingestion of paint or pigmented plastic. Fortunately, the pigments are sufficiently insoluble to be safe for most uses. Rat feeding tests showed tolerances of 5 grams per kg in single doses and 1 gram per kg per day in 44-day feeding tests (Huckle, et al., 1966). The LD_{50} in rats was greater than 10 g/kg which compares with a value of 13.7 g/kg for ethyl alcohol and 5.14 g/kg for boric acid (*Merck Index*, 1976). As long as they are properly encapsulated in a plastic matrix, cadmium colorants are permitted in toys and food containers by the French government and several other European countries (Dunn, 1974).

Plastic water pipes formulated with cadmium-containing pigment have passed strict heavy-metal ion migration tests carried out by the National Water Council in the United Kingdom (Dawes, 1978).

At the present time the toxicity of cadmium sulfide is being investigated by the Dry Color Manufacturers' Association (1978), and the Art Hazards Project, Center for Occupational Hazards, Incorporated, New York City (private communication). The toxicity rating of most of the cadmium pigments for contact with the skin is considered not significant; the rating of cadmium/mercury sulfides is moderate (McCann, 1979). A possible risk does exist if cadmium coated (enameled) metal or ceramic is heated in excess of 700°C. At brazing or welding temperatures, cadmium vapor could be evolved in relatively high concentrations. Recommendations are being considered for warning labels on the paint to protect users of cadmium pigments from a suspected though not yet proven hazard.

4.0 COMPOSITION

Cadmium pigments can be divided into three major categories based on composition. These are the pure sulfide and zinc-modified CdS yellows and light oranges, the sulfoselenide deep oranges and reds, and the cadmium-mercury reds. Lithopones of these categories can be considered another variation.

4.1 Chemical Composition

4.11 *Cadmium Sulfide.* Cadmium sulfide occurs in two crystal forms, hexagonal, α (density 4.8261), and cubic, β (density 4.835). The densities are those based on calculations from x-ray diffraction data. The normal form which results from precipitation by hydrogen sulfide is the cubic one as demonstrated by Böhm and Niclassen in 1924. At an early date, an amorphous form, considerably softer than the crystalline ones was considered to exist. Investigations of Allen and Crenshaw (1912) reported the presence of much amorphous material. However, the morphology visible under the microscope in small particles would not have permitted these workers to characterize the internal crystal order with confidence.

Modern x-ray diffraction indicates that most pigments are at least partially crystalline. All forms can coexist at normal temperatures, though heating slowly converts the cubic into the hexagonal. Densities given for cadmium pigments in the older literature range from about 3.9 to 4.8; this range is due in part to the extreme variability of substances labeled cadmium yellow which sometimes contained materials such as zinc oxide, cadmium carbonate, and cadmium oxide in addition to the sulfide. Also, undoubtedly, experimental errors in measuring density on such finely divided materials led to a spread in values.

In many present-day light yellows the substitution of zinc for some of the cadmium results in a range of densities from about 4.40 to 4.82 (Hercules, 1976). Chemically, there can be a complete series of substitutions ending with pure zinc sulfide (density 3.98), though a maximum of 25% zinc is employed in commercial pigments (Huckle, et al., 1966).

4.12 *Cadmium Selenide and Sulfoselenides.* The sulfoselenides result from replacement of some of the sulfur by selenium in the hexagonal cadmium sulfide lattice. The terminal density of this system (CdSe) is 5.81; the densities of the intermediate materials fall between this and the 4.82 value of the pure sulfide. The crystal-lattice dimensions also change as selenium is substituted. Since sulfur and selenium are close congeners in the periodic table, the chemical properties of the two combined with cadmium are almost identical with those of the pure sulfide. Properties such as refractive index, crystal structure, hiding power, particle size, and morphology are all similar.

4.13 *Cadmium Lithopones.* Cadmium lithopones are usually made by coprecipitation from solutions containing cadmium sulfate and barium sulfide. Equimolar quantities would yield a pigment containing 38.2% cadmium sulfide and 61.8% barium sulfate by weight. As with other cadmium pigments, a calcining operation is then performed. Lithopones are physical mixtures rather than crystallographic ones and have densities similar to those of the pure CdS pigments. Dark orange and red lithopones can be prepared in a similar manner, with selenium included in the precipitation solutions.

4.14 ***Cadmium/Mercury Sulfides.*** In the manufacture of cadmium/mercury pigments, the initial step is a coprecipitation of cadmium and mercury sulfides in solution. If a lithopone is desired, barium is generally added at this time. Following the precipitation, a calcining operation in an inert atmosphere is employed to convert the cubic form to the hexagonal. As with the cadmium-zinc mixtures, the cadmium/mercury substances are solid solutions with variable Cd-Hg ratios and variable lattice constants. The composition range from deep orange to maroon involves 11 weight percent HgS in the orange to 26 weight percent in the maroon (Huckle, et al., 1966; Moore, 1973).

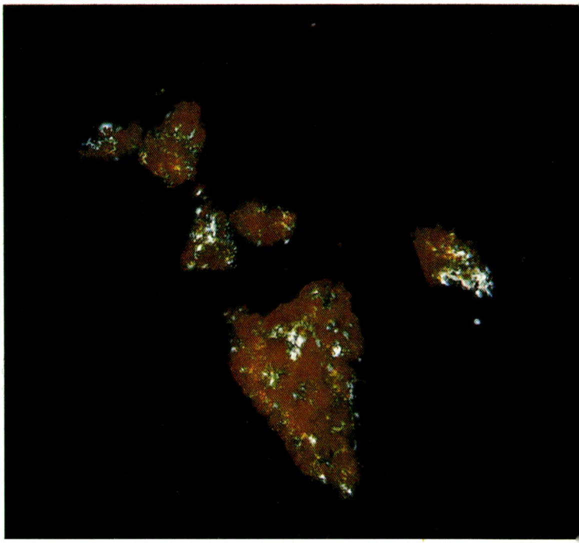

Fig. 5. Photomicrograph of greenockite found on sphalerite, between crossed polars. Grant County, New Mexico. Aroclor 5442 mount; refractive index 1.66. 500x.

4.15 ***Natural Cadmium Sulfide.*** One of the natural forms of cadmium sulfide was first noted by Connell (1840) in the Edinburgh *Philosophical Journal.* It was named greenockite, after Lord Greenock who found the mineral in the course of excavating the Bishopton tunnel near Port Glasgow. A Mr. Brown of Lanfyne apparently found a large crystal of natural cadmium sulfide in 1810 but misidentified it as sphalerite (Jameson, 1840; Weeks & Leicester, 1968).

Greenockite commonly occurs as a fine powdery coating on various zinc minerals, particularly sphalerite and wurzite (figs. 1 and 5). More rarely, quartz, smithsonite and magnetite are substrates for the mineral. It has also been discovered as larger crystals associated with zeolites. Greenockite can be found in Scotland, Bohemia, France, Greece, Bolivia, and parts of the United States.

Fig. 6. Hemimorphic pyramidal habit of greenockite. This is the ideal form, mainly seen in the large crystals of the mineral (after Dana, *A Textbook of Mineralogy*, 4th ed. rev. [New York, 1932]).

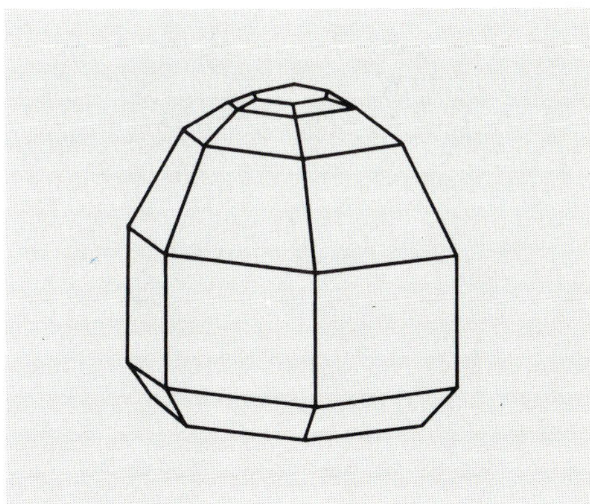

Properties of the mineral greenockite (Dana, 1944) are as follows: hexagonal crystal system, space group C6mc with $a = 4.42$ and $c = 6.724$ (fig. 13). It forms hemimorphic pyramids with a more complex positive end (fig. 6). Greenockite is brittle, with conchoidal fracture, hardness 3 to 3.5, density approximately 4.9, orange yellow to brick red streak, adamantine to resinous luster, and color yellow to orange. Optically it is uniaxial, positive for red to blue green, negative for the rest of the spectrum. It exhibits weak pleochroism, with indices of $\epsilon = 2.529$ and $\omega = 2.506$. Winchell (1951) lists a distinctive cleavage on 1010 and a dispersion of 0.023.

Several authorities claim that greenockite has been widely used as a pigment during the past 2,000 years (Curtis & Wright, 1954; Huckle, et al., 1966; Patton, 1973; Cadmium Association, 1978). However, the rarity of the mineral and the difficulty of separating it from its normally dark substrate seem to preclude this. So far as the authors know, no one has reported the finding of cadmium sulfide in paintings before the mid 1800s, after the date of its first artificial synthesis.

The cubic form of cadmium sulfide was discovered as a mineral by Traill and Boyle in 1955. It was named hawleyite after Professor Hawley of Queens University, Kingston, Canada. Like greenockite, it tends to form yellow to orange incrustations on zinc minerals and as such cannot be readily distinguished visually from the former. The properties of hawleyite are: cubic space group $F\bar{4}3m$, lattice constant $a = 5.818 \pm 0.005$ Å, sphalerite structure, calculated density 4.87, and refractive index above 2. It has been found in the Keno-Galena Hill area, Yukon Territory. Traill and Boyle (1955) believe that many specimens identified as greenockite by "casual hand-specimen examination may prove to be hawleyite when identified by x-ray diffraction."

In general, the two minerals are similar chemically and physically to their artificial counterparts. Except perhaps if the crystals are unusually large, there is little chance that morphology and optical properties could be used to differentiate natural from man-made samples. The method of deposition of the minerals apparently is similar to the precipitation utilized in manufacturing. As the cadmium minerals are normally associated with zinc, this element is a common impurity in samples scraped off a sphalerite or wurzite substrate, and there is a possibility that some zinc may be included in the crystal of the cadmium mineral. X-ray fluorescence analysis on two mineral samples removed by the authors from ZnS by scraping gave the following compositions in weight percent:

	Hawleyite Eureka, Nevada	*Greenockite Grant County, New Mexico*
Cd	66.9%	27.3%
S	24.5	18.6
Al	3.0	3.7
Si	2.2	28.8
Zn	1.4	5.3
Ca	—	4.1
Fe	—	10.1
Total	98.0%	97.9%

The sample of greenockite was impure, partially due to the difficulty of removing the mineral from its substrate.

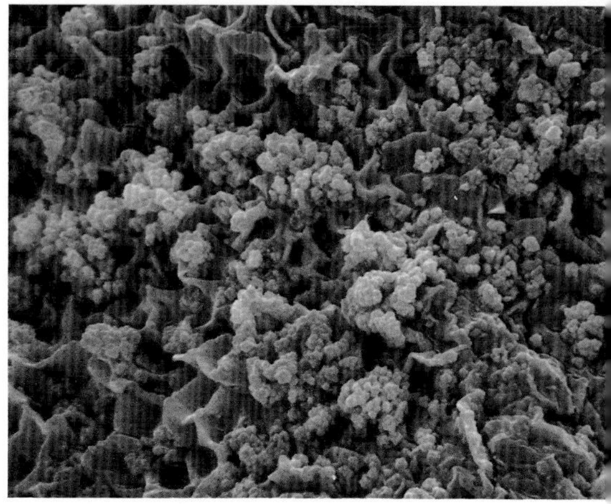

Fig. 7. A. Scanning electron micrograph of greenockite. The smaller irregular particles are the CdS mineral, while the folded background structure is that of the sphalerite substrate. Complete separation of the two was impossible in this instance. Sample from Grant County, New Mexico. 3500x.

B. Scanning electron micrograph of hawleyite. In this case almost complete separation of the mineral from its substrate was possible. There are some individual particles and agglomerates which are similar to the man-made pigments; others appear as elongated single crystals or agglomerates grown in whiskerlike or dendritic fashion. Sample from Eureka, Nevada. 3500x.

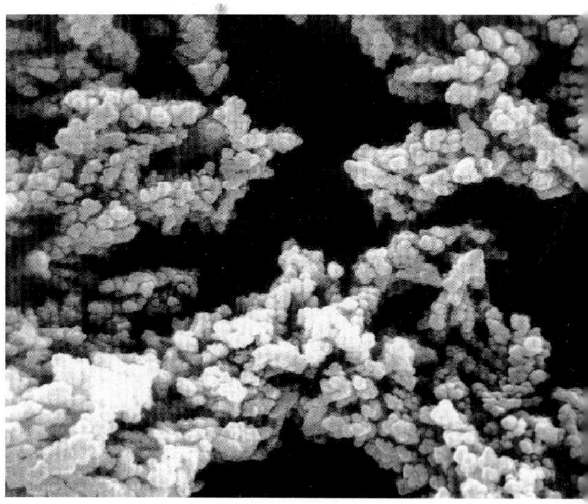

Impurities would be expected to vary greatly from sample to sample (fig. 7). Nonetheless, upon comparing the x-ray fluorescence data and optical properties of hawleyite with those of artificial samples, it would be difficult to differentiate the natural from the synthetic.

4.2 Sources

The supply of cadmium, a by-product of zinc refining, is entirely dependent upon the rate of operation of the world's zinc smëlters. The U.S.A., Canada, and Australia are the chief producers of cadmium metal.

At the present time, in the U.S.A., there are five manufacturers of cadmium yellow listed in the *Raw Materials Index* (1975): Ferro, Kohnstamm, Glidden, Harshaw, and Hercules. Kohnstamm supplies the plastics industry and Ferro the ceramics manufacturers; the other three sell pigments to the various artists' suppliers.

4.3 Preparation

Outlines of the two major methods of making cadmium sulfide — the wet and the dry process — were published by Stromeyer as early as 1819. Although there have been many refinements and improvements, the basic techniques have not changed to the present time. Modern quality control and pure starting materials now permit each method to yield reproducible colors of assured permanence and compatibility; in the past this could not always be guaranteed.

Dry Process

The dry process originally involved heating cadmium metal, oxide, or carbonate with sulfur out of contact with air at temperatures of 300–500°C. These reactions are represented in the following equations:

$$Cd + S = CdS$$

$$2\,CdO + 3\,S = 2\,CdS + SO_2$$

$$2\,CdCO_3 + 3\,S = 2\,CdS + SO_2 + 2\,CO_2$$

Either a stoichiometric mixture or one with an excess of sulfur was employed. Temperature, time of reaction, purity of materials, and oxygen abundance all had some effect on the final properties of the material. In many cases the operation was performed in stages with more materials added from time to time, and if the color did not appear "right," more heating could be arranged or more of one or another of the ingredients could be added according to the experience of the operator. Obviously, it was not a precise process in the modern industrial sense (Remington and Francis, 1954).

After completion of the reaction, a wet grinding operation was performed to give final size and shape to the particles. Both of these parameters have been demonstrated as important in determining the color attributes of the pigment, and hence, grinding had to be well controlled. Experiments done by the authors show that grinding of fusion-prepared sulfides will result in any shade from light yellow to deep orange depending on the thoroughness of the size reduction. While any shade could in theory be produced by varying the grinding, in many cases a dull pigment was the result of the experiments (Zerr & Rübencamp, 1908).

Compared to the older methods of manufacturing by the dry process, the major differences in present-day practice are the use of calcining (heating to 700–800°C in an inert or sulfur atmosphere), purer raw materials, better control over temperature and oxygen content, and refinements in the grinding operation. Calcining is probably the most important difference as it establishes the crystalline conditions needed for a bright pigment. Nonetheless, little use is presently made of the dry process.

Wet Process

The wet process precipitates cadmium sulfide by the reaction of a soluble cadmium salt with a soluble sulfide. The cadmium salts generally used are the chloride, nitrate, sulfate, or iodide, while the source of sulfur may be hydrogen sulfide, barium sulfide, sodium sulfide, or various thiosulfates. The time of reaction, temperature, concentration of reactants, impurities, pH and the rate of addition of one reactant to the other all affect the color and permanence of the resulting pigment. Representative equations are as follows:

$$CdCl_2 + H_2S = CdS + 2\,HCl$$

$$CdSO_4 + Na_2S = CdS + Na_2SO_4$$

$$CdI_2 + BaS = CdS + BaI_2$$

In the earliest practice, hydrogen sulfide was the source of the sulfide ion. For the dark shades the gas was bubbled through a porous plate in the bottom of an earthenware reaction vessel (Woulfe vessel or equivalent). The mixture was constantly stirred at elevated temperatures to ensure formation of the relatively large crystals and agglomerates needed for dark shades. The salt concentration was high, and the precipitate always in contact with the hydrogen sulfide (Remington and Francis, 1954). When lighter shades were desired the cadmium concentration was lowered, and the hydrogen sulfide bubbled through just below the surface. It was important to keep the precipitate out of contact with the gas as much as possible (Budgen, 1924). In another method used to create light hues cadmium metal was dissolved in acid at the same rate as it reacted with a stream of hydrogen sulfide bubbled through the resulting solution. Because of the acidity, the low concentration of cadmium salt, and the necessarily minimal contact of the precipitate with the sulfide gas, a light shade (material with small particle size) resulted. The precipitated medium yellow of the past was made by the complete or almost complete reaction of a cadmium salt (generally sulfate or chloride) with sulfide ion in neutral or slightly acid-concentrated solutions.

By the 1870s, hydrogen sulfide gas was being supplanted by sodium sulfide because of the ease and precision of handling and of maintaining the desired concentrations of both the cadmium and the sulfide ions (Wagner, 1872). Cold dilute solutions gave a yellow product, while hot concentrated neutral or acid solutions yielded a darker hue which was the normal medium yellow of the time. In most cases the pigment was used without calcining, the particle size, shape, surface, and hence color and brilliance being determined entirely by the precipitation reaction. Table 2 presents an outline of the colors and crystal forms precipitated from the various salts at different temperatures and degrees of acidity (Milligan, 1934).

No subsequent heating was employed. When properly washed, made from pure starting materials, and reacted so as not to produce free sulfur, the pigments produced by this procedure were stable. Unfortunately, lighter yellows were popular, and many attempts were made to produce them without complete success. For example, pale shades could be prepared by partial precipitation from cold dilute solutions of cadmium salts or by rapid precipitation from acid solutions, although at the expense of lower stability as compared to that of the "normal" medium-yellow to light-orange pigment. Very light hues could be readily attained using cadmium nitrate and sodium sulfide or ammonium thiosulfate as reactants. These methods produced a light shade; however, having up to 20% free sulfur content, the resultant pigments were rarely stable.

Laurie (1914) mentions a light yellow based on a calcined mixture of cadmium sulfide with zinc sulfide and either zinc or magnesium oxide. Mixtures with lead white were also attempted, but the cadmium portion had to be very pure to avoid rapid darkening through formation of lead sulfide. Even the best of these would probably not stand the test of time.

Another common way in which light cadmium yellow was made in the early days of manufacture involved either cadmium carbonate or cadmium oxalate. In extreme cases over 30% oxalate was present and only 1.5% cadmium sulfide. When mixed with this much white (the oxalate or carbonate), cadmium yellow was indeed light, but it was none too stable, exhibiting a tendency to fade spontaneously. Analyses by Eibner (1909) of representative pigments were as follows:

Cd	46.5%	54.2%	34.9%	70.2%
S	1.5	6.7	9.4	19.7
C_2O_4	31.9	24.1	25.6	6.3
H_2O	14.9	10.7	9.5	3.7
Zn	—	—	16.3	0.4
Total	94.8%	95.7%	94.7%	100.3%

It is not known if the zinc in these samples was combined as sulfide or oxide, though the latter is probable. The abundance of water recorded in the analyses may have been due to the water of crystallization in the calcium or cadmium oxalate.

When commercial precipitation was finished, thorough washing and filtering was done to eliminate the soluble reaction products. Washing was accomplished with distilled or soft water, followed by filter pressing. In some cases digestion with carbon disulfide was performed to remove the last trace of free sulfur (Church, 1915).

In investigations similar to those of Milligan (table 2), involving solutions of 0.1 normality cadmium salt, the authors found that rapid precipitation from dilute acid solutions gave a light yellow; an almost red shade was obtained from more concentrated solutions (fig. 8). The rate of addition and amount of agitation proved important; slow mixing produced a darker color. It was possible to have both an almost red shade and a medium orange in layers in the same solution, acid or neutral, depending on the rate of addition of sodium sulfide. In many cases the color of the precipitate in the reaction liquid would be grossly different from experiment to experiment. Upon drying, however, a group of pigments with approximately the same light orange shade resulted. Moderately hard water yielded a darker pigment.

The present-day wet processing is an extension and refinement of older practices. The temperature, reactant purity, and other parameters that affect precipitation are known and

Table 2. **COLOR AND CRYSTALLINE FORM OF PRECIPITATED CADMIUM SULFIDE (MILLIGAN, 1934)**

Cadmium salt used at 0.1 N concentration	*Without the Addition of Acid*				*In the Presence of Added Acid*			
	At 30°C		*At 100°C*		*At 30°C*		*At 100°C*	
	Color	*Crystalline form*	*Color*	*Crystalline form*	*Color*	*Crystalline form*	*Color*	*Crystalline form*
Sulfate.......	Yellow	Hexagonal*	Yellow	Cubic	Yellow	Cubic†	Red	Cubic
Nitrate	Yellow	Hexagonal	Yellow	Hexagonal*	Yellow	Cubic†	Red	Cubic
Chloride....	Yellow	Hexagonal*	Yellow	Hexagonal	Yellow	Hexagonal	Red	Hexagonal
Bromide	Yellow	Hexagonal*	Yellow	Hexagonal	Orange	Hexagonal	Red	Hexagonal
Iodide........	Orange	Hexagonal	Orange	Hexagonal	Orange	Hexagonal	Orange	Hexagonal

*Trace of cubic beta (β) cadmium sulfide.
†Trace of hexagonal alpha (α) cadmium sulfide.

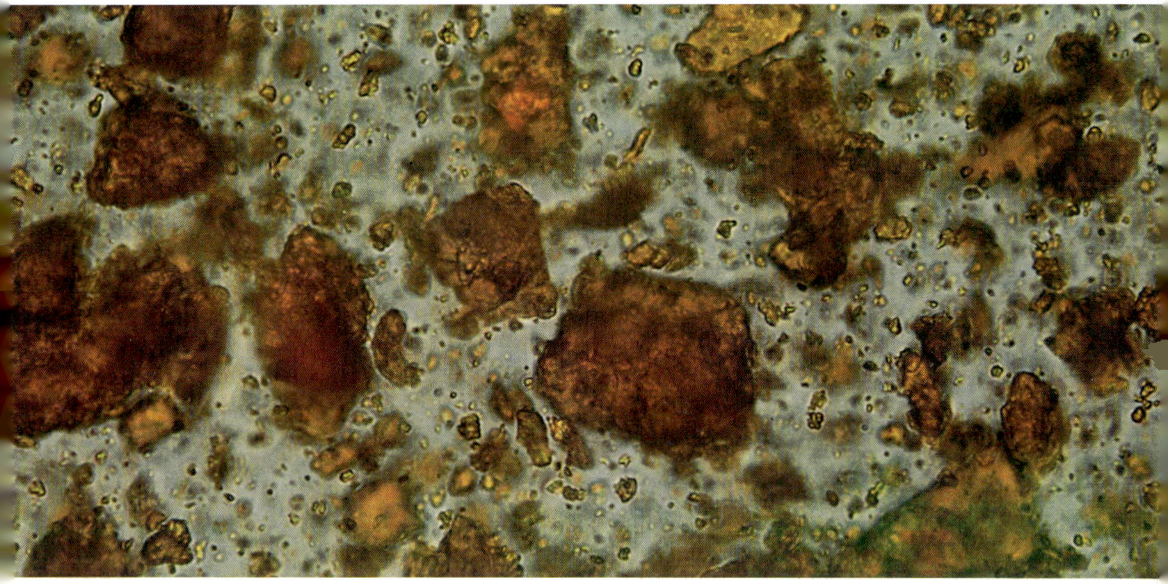

Fig. 8. Reaction product of $CdCl_2$ and Na_2S. The cadmium sulfide formed here would be similar to that initially formed in the wet process before grinding. Reaction was carried to completion using slightly acid 0.1N solutions of both reactants. Aroclor 5442 mount; refractive index 1.66. 800x.

are well controlled. As in most commercial procedures there is an abundance of proprietary processes and trade secrets. For example, in the making of the lighter shades, precipitation modifiers such as cobalt, nickel, phosphate, potassium alum, aluminum hydrate, organic esters, and sodium carbonate have been added to the reactants.

A major modification over the old practice of making light pigments is the direct addition of zinc to the cadmium sulfide lattice by simultaneous precipitation of zinc and cadmium. If a soluble zinc salt, usually with the same anion as the cadmium salt, is added before precipitation, a mixed sulfide, $Cd_{1-x}Zn_xS$, is formed with the proportion of zinc controlled by the concentration of zinc ion in solution. Up to 25% zinc is incorporated to yield the lightest shades; beyond this amount, the resulting pigment takes on a distinctly greenish hue (Patton, 1973). The zinc-cadmium solid solution is as durable as the pure cadmium sulfide. There are electrolytic processes that make use of a cadmium anode, an aluminum or sulfur-copper sulfide cathode, and an electrolyte of either sodium thiosulfate, iron sulfide in acid, or sodium chloride solution with hydrogen sulfide bubbled through. The iron component of the FeS electrolyte is not incorporated into the resulting CdS. Alternating current can also be used with cadmium electrodes in a thiosulfate solution. These processes produce a good, finely divided pigment, but at a greater cost than the more conventional processes, and therefore are not much used (Zerr & Rubencamp, 1908; Remington & Francis, 1954).

In both old and new preparations the presence of impurities must be considered. Elements such as copper, iron, lead, and nickel, which form colored or unstable sulfides, must be avoided. Thallium, even in trace amounts, promotes adverse reactions to light. The purity level of all reactants employed today is in the range of 99.6–99.8%. Attainment of this level, easily achieved by 1920, presented problems in the past (Budgen, 1924). The final factor responsible for the reproducibility and permanence of today's product is the almost universal use of calcining to enhance the brilliance of the pigment by establishing the desired crystal structure and surface (fig. 9) (Eibner, 1909; Budgen, 1924; Remington & Francis, 1954; Wagner, 1960; Taylor & Marks, 1966).

4.31 ***Cadmium Lithopones.*** Cadmium lithopone yellows developed out of economic necessity because the cost of the pure pigments was prohibitive for most industrial applications. Lithopones were introduced by Marston in 1921. By 1927, with a cost less than half that of the pure pigment, extended cadmiums were being used extensively in industry. The best product was obtained by coprecipitation; either barium sulfide was added directly to cadmium sulfate solutions, or a soluble barium salt was reacted with sodium sulfate in the solution from which cadmium sulfide was being precipitated. The reactions involved are:

$$BaS + CdSO_4 = CdS + BaSO_4$$
$$BaS + CdCl_2 = CdS + BaCl_2$$
$$BaCl_2 + Na_2SO_4 = BaSO_4 + 2\,NaCl.$$

The first method results in a product with 61.8% barium sulfate by weight. The second procedure can be made to give any amount of extender desired. Early lithopones contained only 5% barium sulfate, although modern types tend toward the 61.8% value. Even the so-called pure cadmium sulfides today sometimes contain a few percent of this extender.

Zinc sulfide can also be added as a lightener, resulting in the greenish lemon and primrose types of cadmium yellow. Typical samples of this variety analyzed as follows (Curtis & Wright, 1954):

Cadmium Sulfide	31.6%	37.1%
Zinc Sulfide	5.1%	0.8%
Barium Sulfate	63.3%	62.0%

Though the sulfide and sulfate are precipitated at the same time in the same solution, there is no crystallographic mixing of cadmium into the barium sulfate lattice or vice versa (Ward, 1927; Taylor & Marks, 1966; Patton, 1973). X-ray diffraction shows that the cadmium lithopones are strictly a physical mixture, although scanning electron microscopy reveals the mixture to be an intimate one.

While extension of pure pigments can be achieved by mechanically grinding barium sulfate with cadmium sulfide, the tinctorial strength and durability of the product is inferior to that of coprecipitated varieties, and there is some possibility of separation of the components when dispersed in a vehicle, even though the densities of the two pigments are similar (Curtis & Wright, 1954).

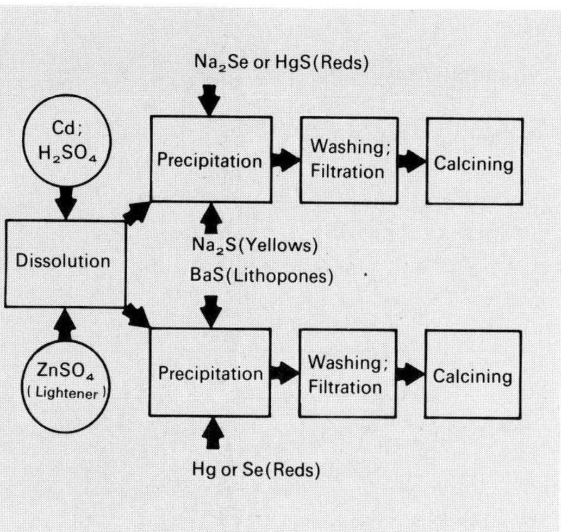

Fig. 9. General outline of present day manufacturing practice based on the wet method.

4.32 ***Cadmium Reds.*** The sulfoselenide reds were first commercialized in 1910, although this type of pigment was mentioned in a patent granted in 1892 (Gettens & Stout, 1966). Preparation originally consisted of heating cadmium sulfide, sulfur, and selenium to about 600°C in an inert atmosphere. In 1919 an improved process was developed by Bayer which gave a more consistent product and also did away with losses of expensive selenium through volatilization. This method followed the procedure used in the precipitation of cadmium sulfide, but utilized a mixture of alkaline sulfides and selenides. The precipitate from this process is yellow; the red shade develops upon calcining. Calcining must be well controlled, with a temperature of about 300°C employed; if overheated, the pigment becomes brown. The relationship between sulfur/selenium ratios and hue is given in table 3.

Table 3. **COLOR AND SELENIUM CONTENT OF CADMIUM SELENIDE REDS (after Curtis and Wright, 1954)**

Cd/Se	100/12 Orange
Cd/Se	100/15 Orange-red
Cd/Se	100/20 Light red
Cd/Se	100/25 Bright red
Cd/Se	100/30 Dark red
Cd/Se	100/33 Maroon red
Cd/Se	100/40 Dark maroon

Lithopones are made in a two-step process, as a coarse pigment is obtained if BaS/Se is substituted for the sodium salts. The cadmium is first precipitated as the carbonate, and then the lithopone formed using barium sulfides and selenides with an excess of alkali present (Heaton, 1956).

Another way of achieving orange to maroon shades involves the substitution of mercury for cadmium. This type of pigment was developed because of a shortage of selenium around 1948; in response the Imperial Paper and Color Corporation initiated research to find a substitute material (U.S. Patent 2,878,134). The group of pigments is practically identical in behavior and appearance to the sulfoselenides but is less costly (Moore, 1973).

4.4 Adulteration and Sophistication

The high cost of cadmium resulted in such extensive adulteration in the past that it was necessary to forewarn the prospective buyer to choose only pure materials and to acquire them from a reliable colorman. Much of the poor reputation of early pigments was due to the adulterants, among which were zinc oxide, zinc carbonate, zinc sulfide, zinc chromate, arsenic yellow, lead chromate, gypsum, cadmium oxide, cadmium phosphate, Indian yellow, tin sulfide, mercury and bismuth compounds, lead iodide, and barium sulfate. The first three compounds sometimes constituted as much as 50% of the material (Budgen, 1924). With the exception of the coprecipitation with barium sulfate in lithopones, the addition of these substances seldom resulted in a pigment as stable as cadmium sulfide itself. When added as lighteners, cadmium carbonate, cadmium oxalate, and free sulfur can definitely be considered as adulterants, particularly in regard to stability. These components gave rise to either bleaching or darkening depending on conditions of use (Buchner, 1887; Church, 1915).

The presence of adulterants was of such concern that it was recommended that the artist perform his own chemical tests to determine the purity of a cadmium pigment. As recently as 1927, Fox and Bowles suggested several wet chemical tests to establish the purity of CdS. The adulterants tested for were lead, arsenic, antimony, free sulfur, barium, silica, and cadmium oxalate Eibner (1909) earlier recommended the following standards: (1) dissolution in hydrochloric acid should result in a clear solution if there are no additives present, (2) the addition of ammonia or calcium chloride to the acid solution should not produce turbidity, (3) a chemical test for arsenic must be negative, (4) no extractables should be found upon boiling in water, nor should there be any acidity or alkalinity in the water that has been boiled with the pigment, (5) no water should be evolved when the sample is heated to 100°C.

Another type of cadmium yellow, cadmium chrome yellow ($CdCrO_4$), was mentioned by Bersch (1901). It was produced by a precipitation reaction between a cadmium salt and potassium chromate and was recommended as a deep yellow equal to lead chromate yet stable under atmospheric attack, though no references to its use are given. A variation, basic cadmium chromate [$Cd_2(OH)_2CrO_4$], is mentioned in the *Colour Index* (1971) as Pigment Yellow 77188.

5.0 IDENTIFICATION-CHARACTERIZATION

5.1 Characterization by Optical and Scanning Electron Microscopy

Two crystal forms (hexagonal and cubic), an amorphous type and a spherulitic modification, are encountered in cadmium pigments. The

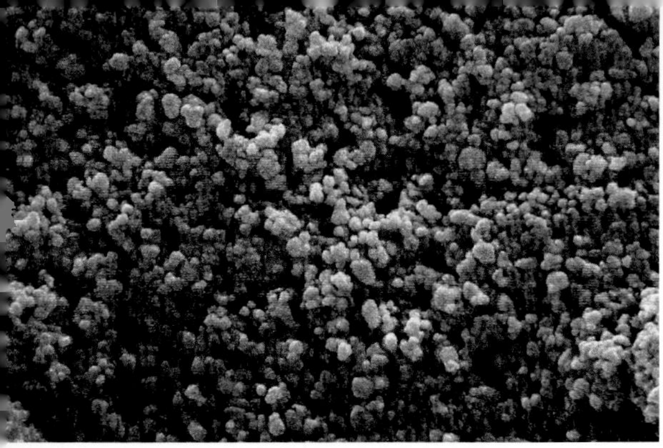

Fig. 10. Scanning electron micrographs of various cadmium sulfide pigments.
A. Cadmium Orange Extra Deep, Fezandie and Sperrle, New York. The structure is typical of the small rounded particles and agglomerates in a product of recent manufacture. The size range of individual particles is narrow (0.2–0.8 μm), and some apparent fusion of the particles can be seen. 5000x.

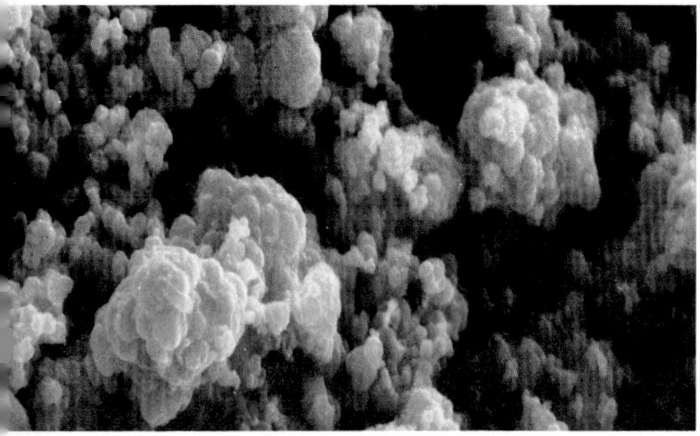

B. Cadmium Yellow Dark, Beaverhall (Edward W. Forbes Collection, Art Institute of Chicago). Some large, fused agglomerates are present, in addition to the smaller varieties. The larger ones could give rise to a pseudospherulitic appearance by optical microscopy. 5000x.

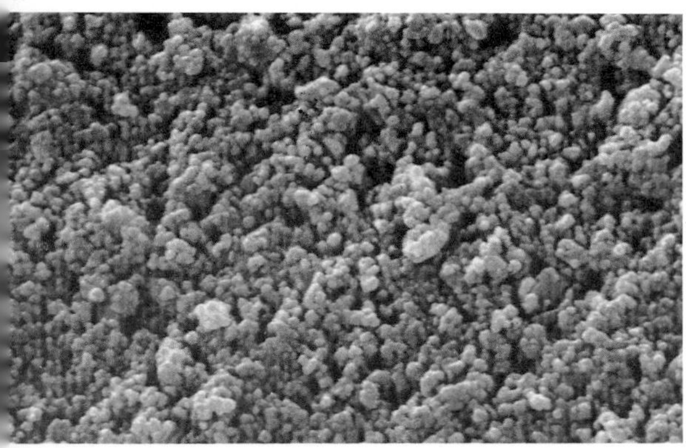

C. Primrose Cadmium Yellow X-2272 (Standard), Imperial Hercules, Inc. There is no discernible difference between this cadmium lithopone and pure CdS which has uniform particle size. 5000x.

hexagonal is the type most commonly found, but all forms have refractive indices considerably greater than 2. In general, the morphology observable by optical microscopy is nondescript due to the small particle size encountered, with the pigment particles ranging from crystallites of 0.1–0.2 micrometers in their longest dimension to single particles 3–4 micrometers in diameter. Often there is a tendency to form agglomerates which are difficult to break up. Under high magnification some samples appear to have rounded grains while others are angular, although their true shape can only be seen by electron microscopy (fig. 10 A, B, and C). When viewed in ordinary light or plane polarized light the color transmitted by the particles appears yellow to orange, depending on shade and grain size, and this coloration is visible even in the smallest grains. The small size of the individual crystallites coupled with their intense absorption color can be used as an identification key, though this can be masked by the secondary spectrum of some microscope optics. Because of their high refractive index, the Becke-line effect causes many of the symmetrical or smaller particles to have a bright center spot when the microscope is slightly defocused by increasing the distance between objective and object. When examined in crossed polars both the isotropic and anisotropic modifications usually are visible in the yellow to orange absorption color, though the lighter yellow shades may be distinctly greenish (fig. 11 A and B). The high refractive index, small size, and generally rounded shape give rise to anomalous polarization effects at the edges of the particles (edge depolarization), and in many instances only the edges of the grains will be visible.

The lithopones generally have the same appearance microscopically as the pure cadmium shades from which they were derived. The sulfoselenides and cadmium-mercury pigments appear essentially the same as the yellow sulfides, though the absorption color is necessarily shifted toward the red. Regrettably, cadmium pigments are variable with no single set of parameters applicable to all (fig. 12 A and B).

5.11 ***Hexagonal (Greenockite).*** Hexagonal cadmium sulfide (fig. 13) has two principal indices of refraction, ω 2.506 and ϵ 2.529. The birefringence is 0.023. If zinc has been added to the crystal lattice the refractive indices will be lowered depending on the amount of zinc, with

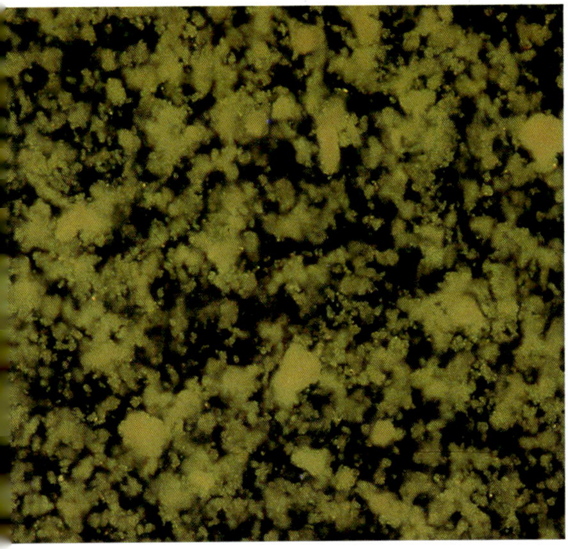

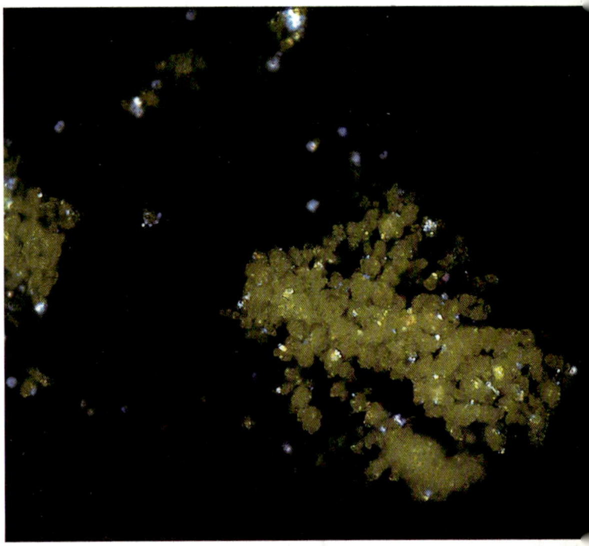

Fig. 11. Photomicrographs of typical specimens from paintings.
A. Crushing from watercolor by Alexander Calder, Untitled, Art Institute of Chicago, 1965.258. Between crossed polars these cadmium yellow particles have a distinct green coloration typical of the lighter shades. Aroclor 5442 mount; refractive index 1.66. 650x.

B. Sample from *Venice, Palazzo Dario*, by Claude Monet, Art Institute of Chicago, 1933.446, crossed polars. This is an example of a modern type of cadmium yellow used by Monet in 1908. Aroclor 5442 mount; refractive index 1.66. 650x.

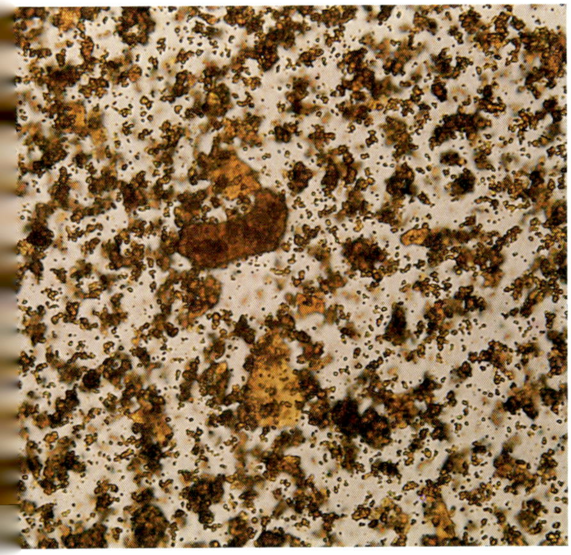

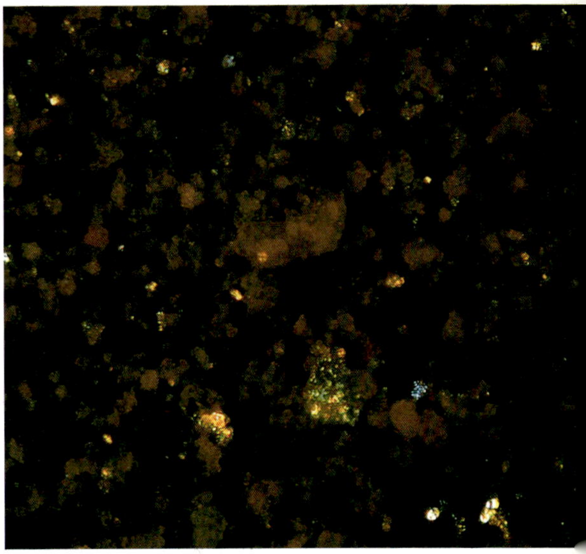

Fig. 12. A. Cadmium Yellow Dark, Beaverhall (Edward W. Forbes Collection, Art Institute of Chicago). A wide range of sizes is present, with the thinner particles appearing yellow and the thicker ones deep orange to brown. Most grains are equant and rounded, with considerable edge darkening due to high index. Aroclor mount; refractive index 1.66. 450x.

B. Same area between crossed polars. A few spherulites are present. 450x.

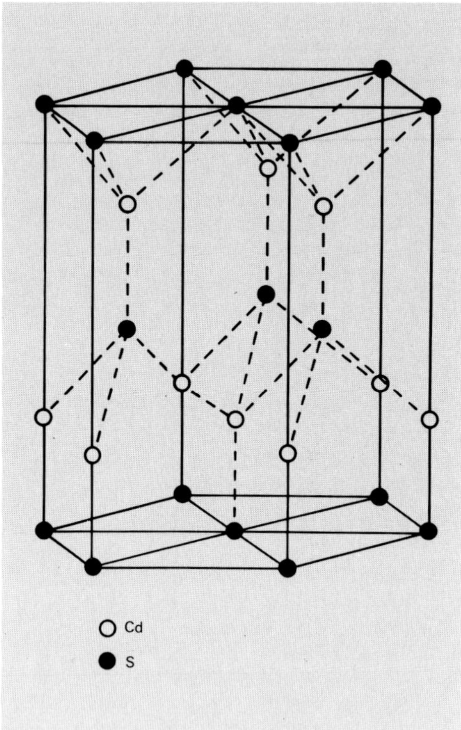

Fig. 13. Hexagonal crystal lattice of cadmium sulfide. Zinc can directly substitute for cadmium and selenium for sulfur to give mixed sulfides and/or sulfoselenides (after *Technical Notes on Cadmium, Cadmium Pigments*, Cadmium Association, 1978).

the end point of the series having indices of 2.356 and 2.378. The substance is optically positive and, depending on thickness, exhibits a yellow to orange color which masks polarization colors. It is slightly pleochroic (Dana, 1944; Winchell, 1951). Because of the small size encountered in commercial pigments, the optical properties are rarely measurable. Individual grains sometimes show extinction, though their number and the perfection of extinction is highly variable from sample to sample. The facts that even a one micrometer particle may be an agglomerate of several randomly oriented crystals, that only a short optical path is available for the relatively weak birefringence to be revealed, and that the high refractive index and small size enhance edge depolarization, all contribute to imperfect or practically nonexistent extinction. The same may be said of attempts to measure birefringence, principal indices, or optic sign. Even with large single crystals, immersion methods for index determination would entail sulfur-selenium melts, a procedure that is neither easy nor accurate (Merwin & Larsen, 1912). Between crossed polars, hexagonal particles of cadmium pigments appear mainly in the absorption color, usually with minimal change on stage rotation. The edges of the grains are considerably brighter than the centers owing to internal reflection and depolarization. There are no morphological or optical properties that would indicate the hexagonal internal order present in most samples, though x-ray diffraction readily confirms this. Slowly grown large crystals are prismatic with parallel extinction, while sublimed CdS exhibits filaments and needles, stubby prisms and twins after 1011 and 2023 (Mellor, 1923).

5.12 ***Cubic (Hawleyite).*** Though only one example of the cubic variety, hawleyite, was available for investigation, it was very similar to both the man-made pigment and the hexagonal mineral. The refractive index was above 2.0, and the small particles were bright and unchanging under crossed polars when the stage was rotated. The sample apparently tended to grow in linear agglomerates similar to whisker or dendritic growth, though this characteristic would not be adequate as a key for identification of hawleyite (fig. 14 A, B). The cubic form is not common in pigments, as verified by x-ray diffraction; only one example was found and that a mixture with the hexagonal variety.

A specimen of hawleyite examined by Traill consisted of "a mixture of massive siderite and coarsely crystalline sphalerite, with a bright yellow coating. Microscopic examination revealed that the coating was composed of a bright yellow mineral . . . admixed with lesser amounts of fine grained sphalerite and siderite. The grain-size of the hawleyite was so small that individual grains could not be discerned" (Traill & Boyle, 1955).

5.13 ***Spherulitic Forms.*** In addition to irregularly shaped particles, there are two modifications of cadmium sulfide which have spherical symmetry. These have been encountered by the authors mainly in older samples. One form, though not truly spherulitic, shows a dark cross between crossed polars not unlike that seen in starch grains (fig. 15 A, B). The particles take on the absorption color of the cadmium yellow, usually having the bright edges often seen associated with the irregular particles under crossed polars. This pseudospherulitic modifi-

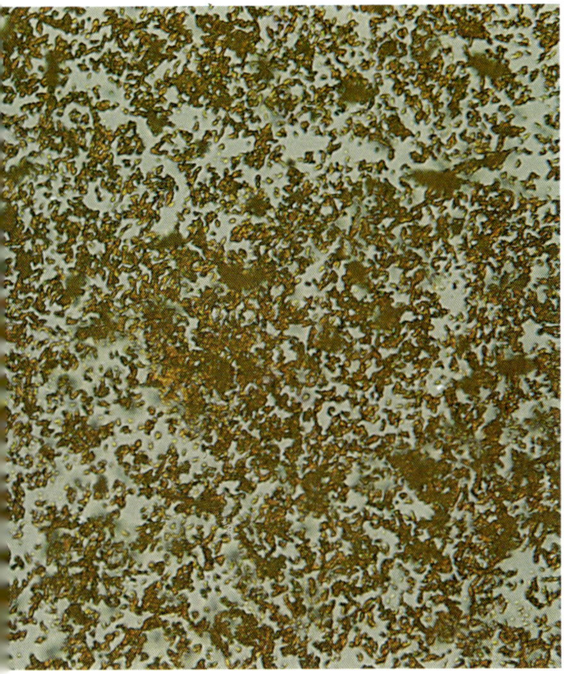

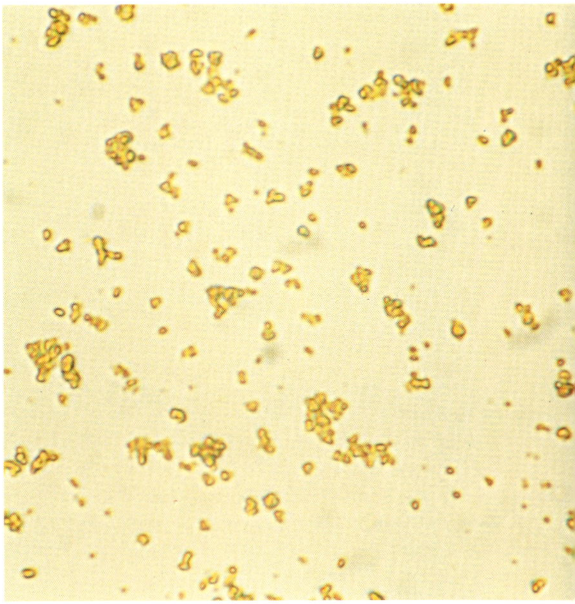

Fig. 15. A. Cadmium sulfide showing equant particles with rounded edges, a size range of <0.3 to 1.5 μm with agglomerates. Current production, Winsor & Newton, Inc. Aroclor 5442 mount; refractive index 1.66. 2000x.
B. Same sample between crossed polars. Many of the particles show crosses because of the edge depolarization associated with the pseudospherulitic modification. 2000x.

Fig. 14. A. Hawleyite from Eureka, Nevada. Some of the elongated growth modifications are present. Aroclor 5442; refractive index 1.66. 600x.
B. Same area between crossed polars. Although indicating cubic crystal structure by x-ray diffraction, this substance appears to be anisotropic. 600x.

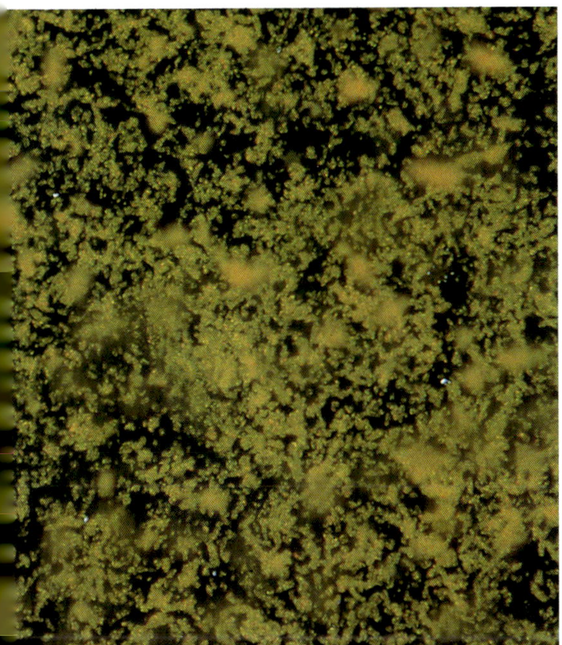

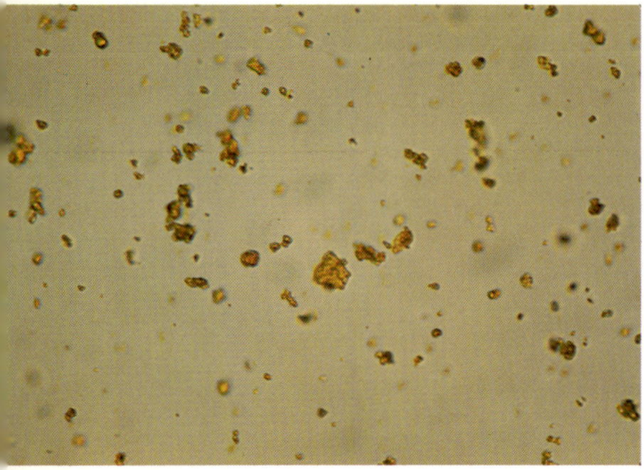

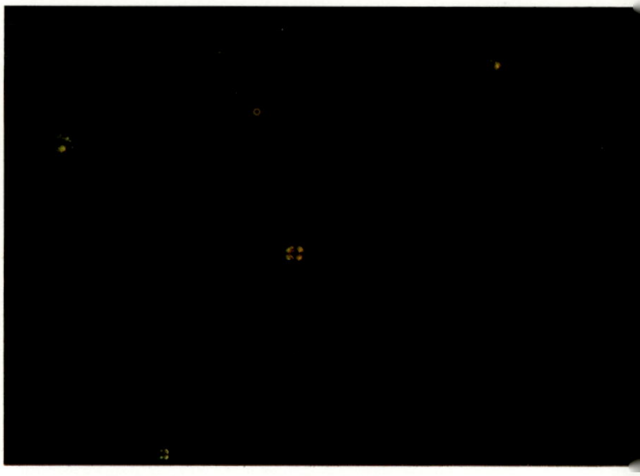

Fig. 16. A. Cadmium yellow, Roberson, 1926 (Edward W. Forbes Collection, Art Institute of Chicago), seen in plane polarized light. This sample shows many unusual modifications including large angular flakes. Aroclor 5442 mount; refractive index 1.66. 600x.

B. Same area between crossed polars. Several spherulites are present, with the two upper ones being fragments, while the center and lower ones are intact, although of different sizes. 600x.

cation may have cubic, hexagonal, or amorphous internal order, although the structure usually associated with spherulites, namely crystallites radiating from a central point, is not present. The spherulitic appearance is caused by symmetrical depolarization due to the combination of round shape and the high index of the particles compared to the mounting medium. The same phenomenon can be seen in the back lens of a high aperture objective in crossed polars or from small, high refractive index glass beads in a low index medium. The phenomenon has also been noted in cadmium reds.

Another form of cadmium yellow is truly spherulitic. In this case the hexagonal variety is found as crystallites radiating from a central point. The particles are larger than the "normal" cadmium sulfide, with a range from 4–10 micrometers. In some cases multiple intergrown spheres or fragments of spherulites are present. Under high magnification a faint radiatin line structure delineating the individual crystallites can sometimes be seen, along with a small point from which the material nucleated (fig.16A,B). In the size range found, the birefringence color is usually first order white, though this is masked by the yellow absorption color. A first-order red compensator will cause adjacent quadrants of the particle to appear in different colors (red and blue) when dealing with the true spherulites; the other spherical modification will not exhibit this characteristic.

5.14 ***Amorphous Type.*** An amorphous form of CdS has been found in many of Claude Monet's paintings that date from before 1900 (fig. 17). A variety of phenomena associated with this substance can be demonstrated with the optical microscope, with particle size and state of agglomeration markedly influencing polarization effects. Small individual particles or loose agglomerates appear the same as the cubic or hexagonal types. Larger aggregates (to 50 micrometers) have an unusual appearance. In many cases, edges of the large flat flakes are composed of small particles, loosely bound together, that appear bright between crossed polars regardless of orientation. The centers are more or less fused with no apparent optical activity. Occasional intermediate-size flakes (10–20 micrometers) appear totally isotropic. They are probably composed of the fused material found in the centers of the large agglomerates. All stages of agglomeration can be found in a single sample — from loosely associated groups of "normal" CdS through large flakes with fused centers and, finally, single pieces of the fused substance without the small particles around the edge. All particle types were proved to be CdS by electron microprobe analysis. X-ray diffraction indicated that a sample of this variety of cadmium yellow was about 10% hexagonal, with the balance amorphous. It is not known if the fusion is a result of pressure during grinding, a slow reaction with time, or a

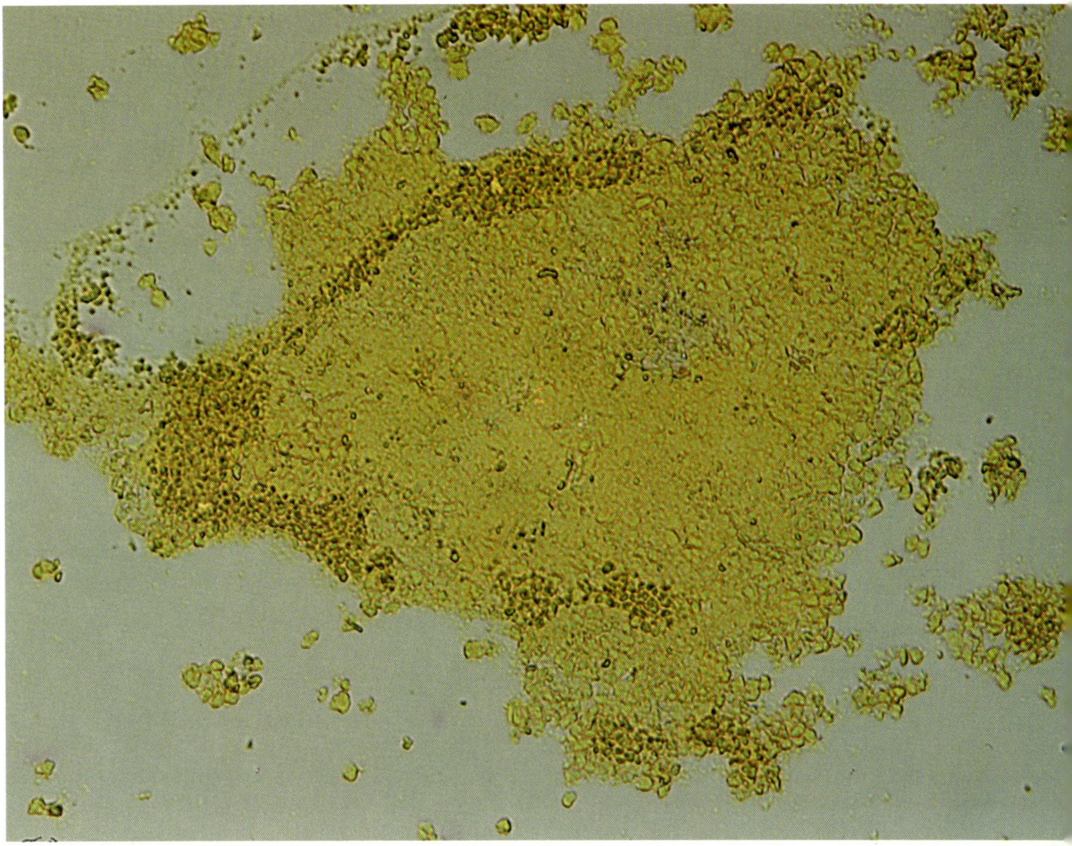

Fig. 17. Crushing of sample from Claude Monet's *The Custom House at Varengeville*, Art Institute of Chicago, 1933.1149. This shows the partially fused amorphous form of CdS used by Monet in 1897. Aroclor 5442 mount; refractive index 1.66. 600x.

result of some other step in the processing of this pigment.

Processes are known for creating amorphous varieties, either by gradual addition of cadmium oxide to molten potassium thiocyanate or by heating cadmium thiocyanate with acetic acid in a sealed tube at 250°C (Budgen, 1924).

5.15 ***Barium Sulfate.*** Barium sulfate is orthorhombic with indices of 1.637, 1.638, and 1.649 and a birefringence of 0.012 (see chapter on Barium Sulfate). Because of the small size of the particles and intimate admixture with the cadmium component, the optical properties of barium sulfate are usually completely masked in lithopones. They thus appear to have much the same high index as does the pure sulfide, and only the absorption color of the cadmium is seen in both crossed and uncrossed polars. Sometimes it is possible to see isolated 2–4 micrometer-sized particles of barium sulfate, although this is the exception unless the sample comprises physically mixed components rather than being coprecipitated. In mixtures, cadmium sulfide particles tend to agglomerate around the slightly larger barium sulfate grains. If one wishes to see the barium sulfate, acid can be used to dissolve the cadmium sulfide, leaving most of the extender behind (fig. 18 A, B, C).

5.2 Chemical Identification

Yellow, orange, or red pigments suspected of being cadmium based could consist of pure cadmium sulfide, cadmium sulfide modified with zinc, cadmium sulfoselenide, lithopones of these types, or a cadmium-mercury sulfide. The simplest approach to narrowing down the possibilities is first of all to test for sulfide and cadmium. The detection of lithopones by chemical means poses many more problems, although a simple initial test would involve dissolution in concentrated hydrochloric acid followed by

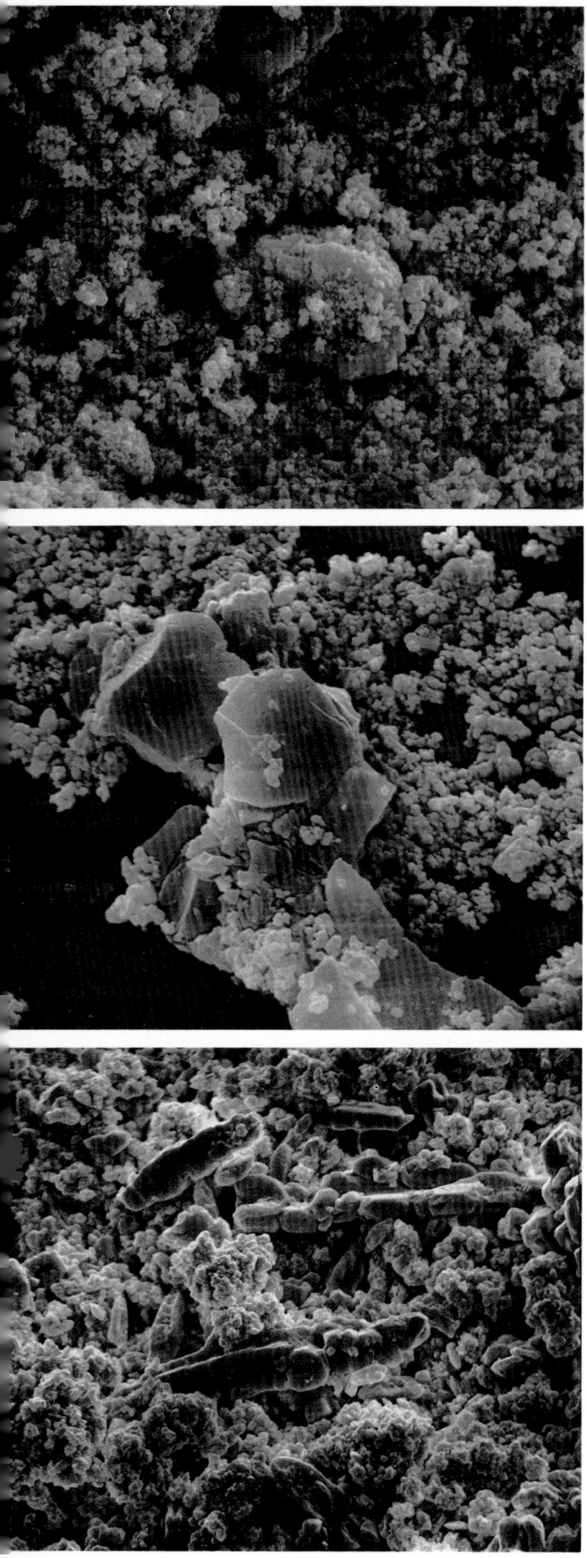

Fig. 18. Cadmium lithopones. Scanning electron micrographs.
A. An example of bimodal distribution of particles, the larger particles being $BaSO_4$. C.P. Cadmium Lemon No. 1470, Harshaw Chemical Co. 3750x.
B. Cadmium lithopone pigment. Some large particles of $BaSO_4$ are evident, though these are the exception rather than the rule. Cadmolith Primrose Pb-407, Glidden Pigments. 3500x.
C. Cadmium lithopone sample B showing barium sulfate component remaining after treatment with concentrated hydrochloric acid. Since $BaSO_4$ is not totally acid insoluble, some recrystallization has formed the larger particles. The smaller ones represent the habit found in the original pigment. 2000x.

evaporation and redissolution in dilute hydrochloric acid. The residue insoluble in the dilute acid can then be examined for $BaSO_4$. If barium is to be detected microchemically, one may fuse the sample and test for barium ions with sodium rhodizonate (see chapter on Barium Sulfate).

5.21 ***Sodium Azide-Iodide Test for Sulfur.*** Perhaps the easiest test to make is for sulfide. Gettens, et al., (1972) described a sensitive test involving the catalytic decomposition of sodium azide (NaN_3). The suspected pigment is placed in contact with a solution of sodium azide dissolved in water, often under a cover slip. The concentration of azide is not critical; from 5 to 15% will do. Iodine is then added either as a small crystal or as a solution in alcohol or aqueous potassium iodide. Bubbles of nitrogen form on the surface of the pigment particle in the presence of sulfide; in some cases a single 1–2 micrometer particle gives rise to several bubbles 10–20 micrometers in diameter. The reaction is a surface-induced catalytic one that requires no dissolution of the sample; thus, if the test is performed carefully, the pigment can be recovered for other procedures. Sensitivity is better than 10^{-12} gram because the bubbles are so much larger than the particles responsible for their generation.

5.22 ***Sulfate Test for Sulfide.*** Gettens, et al., (1972) describe the oxidation of sulfide to sulfate by bromine and its reaction with a calcium salt to form insoluble calcium sulfate (gypsum) as needles and clusters of needles (fig. 19). To perform the test the sample is placed in a 1% solution of calcium chloride or acetate and inverted over a bromine water bottle. In time the sulfide is oxidized and reacts with the calcium salt to form gypsum. Barium sulfate does not interfere owing to its low solubility; it remains as a residue.

5.23 ***Reaction with Potassium Mercuric Thiocyanate.*** The sample may be dissolved in nitric acid and then buffered with sodium or ammonium acetate; some free acid is not objectionable. A concentrated drop of potassium mercuric thiocyanate is mixed with a dilute droplet of the substance in question. If cadmium is present, colorless orthorhombic prisms separate slowly, one end pyramidal and the other beveled or truncated. When a small amount of zinc is admixed, the prisms are more irregular and slender; with still higher concentrations of zinc, "arrowheads" of a mixed cadmium-zinc compound are seen. Zinc salts alone give rapid precipitation of feathery dendritic crystals, although even a trace of cadmium will inhibit this form.

Lead, silver, copper, cobalt, and manganese can give interfering reactions with this reagent although each precipitate has a distinct morphology or color. Copper produces greenish yellow dendrites or boat-shaped crystals; cobalt yields deep blue crystals. Nickel forms yellow spherulites; manganese crystallizes with an appearance similar to that of badly formed cadmium mercuric thiocyanate, but without the beveled or pyramidal ends. Lead precipitates either granular or prismatic crystals which may interfere with cadmium. However, in the case of lead (and silver) the crystals are smaller than those formed with cadmium (Chamot & Mason, 1960).

5.24 ***Reaction with Di-p-nitrophenylcarbazide.*** A test for cadmium used by Plesters (1956) is as follows: A drop of the unknown in solution is mixed on a spot plate with a drop of sodium hydroxide solution (10%) and potassium cyanide solution (10%), a drop of reagent (0.1% alcoholic solution of di-p-nitrophenylcarbazide), and two drops of formaldehyde (40%). In the presence of cadmium a blue-green precipitate or color is formed. The reagent alone gives a violet color with formaldehyde, and hence, the color should be compared with a blank using the reagents only (Feigl, 1958).

5.25 ***Cadmium Chloride Formation.*** When reacted with hydrochloric acid, cadmium sulfide forms cadmium chloride with the evolution of hydrogen sulfide. Even on rather small samples, the odor of hydrogen sulfide frequently can be noticed. Crystallization following the hydrochloric acid treatment yields thin hexagonal plates and hexagonal rods or needles radiating out from a point. The rods are the preponderant form, with the plates appearing sparingly or not at all (figs. 20, 21, and 22). The principal refractive indices of cadmium chloride are 1.636 and 1.705. When zinc is present the indices are lowered due to the formation of a mixed chloride. Barium sulfate is dissolved only to a limited extent by the acid and is frequently apparent as 1–5 micrometer transparent particles after the sulfide goes into solution.

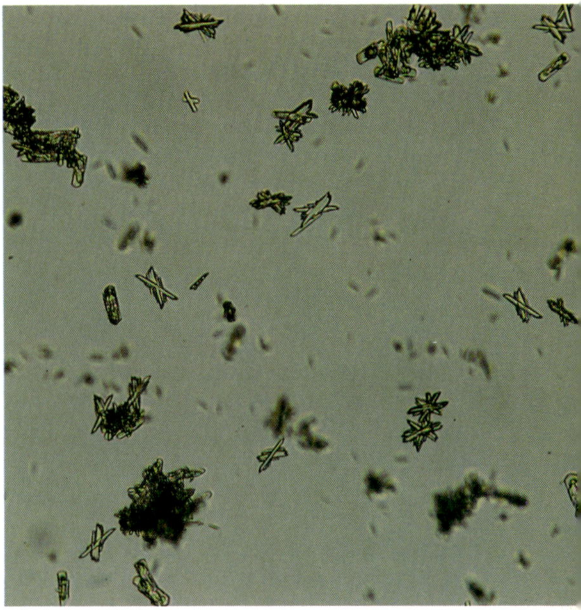

Fig. 19. Calcium sulfate crystals (gypsum) resulting from the sulfate test for sulfide. Aroclor 5442 mount; refractive index 1.66. 300x.

5.26 ***Reaction with Oxalic Acid.*** If oxalic acid is added to a moderately concentrated neutral or slightly acid drop of the material to be examined, colorless monoclinic prisms of $CdC_2O_4 \cdot 3\,H_2O$ will form if cadmium is present. They are thin with square or truncated ends, the acute angle being 60°. The extinction angle is 24°. Single crystals, X's, and radiates are the common habits. If more than 10% zinc is present in relation to the cadmium, spherulites and pseudo-octahedra with the morphology of zinc oxalate will form (Chamot & Mason, 1960).

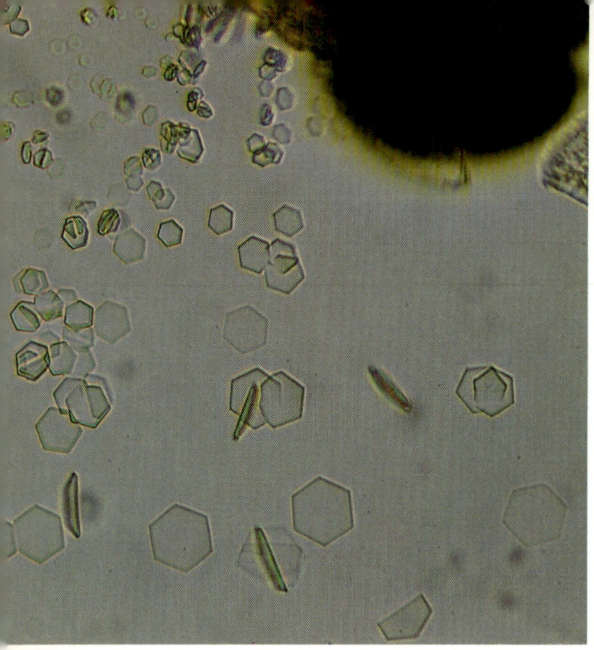

Fig. 20. Reaction product of cadmium sulfide and hydrochloric acid. The hexagonal plate morphology is distinctive, although this form is somewhat rare. Aroclor 5442 mount; refractive index 1.66. 750x.

Fig. 21. Reaction product of cadmium sulfide and hydrochloric acid. Aroclor 5442 mount; refractive index 1.66. 500x.
A. Clusters of hexagonal crystals are the common mode away from the drop edge.

B. Clusters between crossed polars.

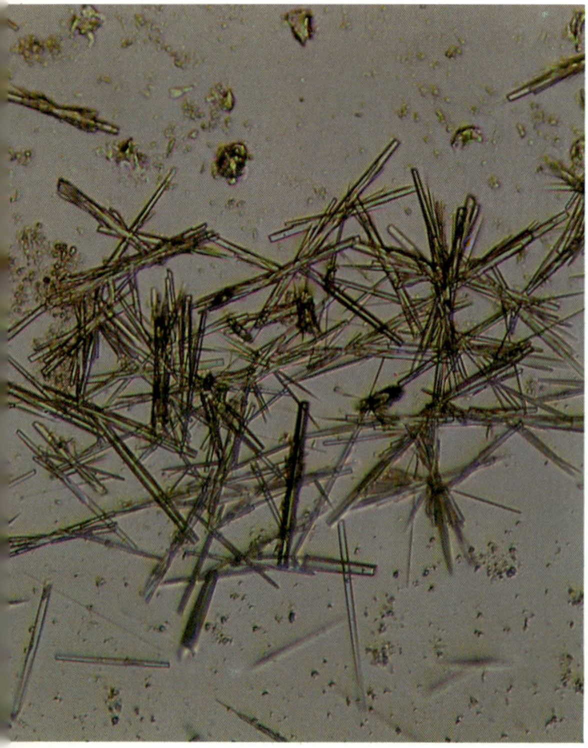

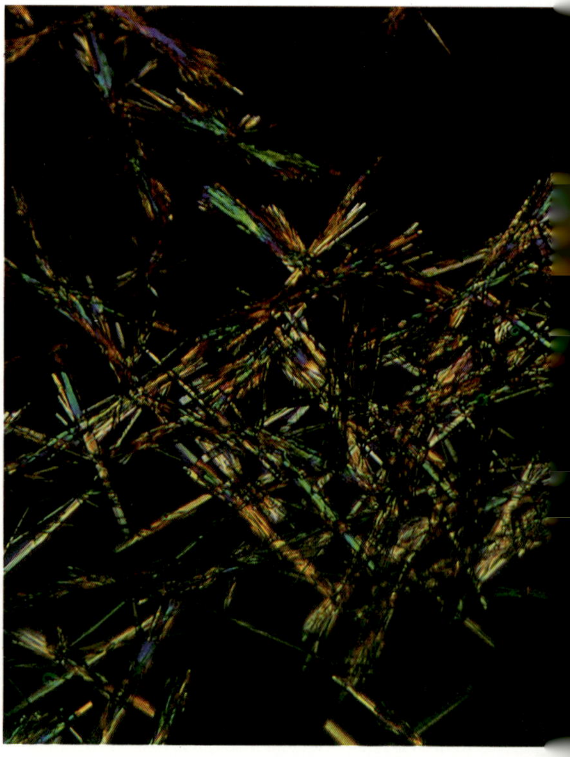

5.3 Instrumental Methods of Analysis

5.31 ***X-ray Diffraction.*** The x-ray diffraction data for the two crystal forms of CdS are given in table 4. Skinner and Bethke (1961) have reviewed the published data on the unit-cell dimensions for greenockite.

In the pigments that the authors have investigated, the hexagonal form was the preponderant one, though in varying degrees of crystal perfection from an almost glassy state with weak diffuse lines to well-formed crystals exhibiting sharp diffraction lines. In only one instance was the cubic type found and that as a mixture with about an equal amount of the hexagonal material. Though the particles are small, there is no evidence for line broadening due to size alone. In the case of the lithopones, the diffraction patterns indicate a physical mixture of hexagonal CdS and orthorhombic $BaSO_4$ with no evidence for crystallographic mixing (fig. 23). The x-ray diffraction data for $BaSO_4$ can be found in the chapter on Barium Sulfate.

A complete series ranging from pure CdS to pure ZnS can be formed as zinc is added to the

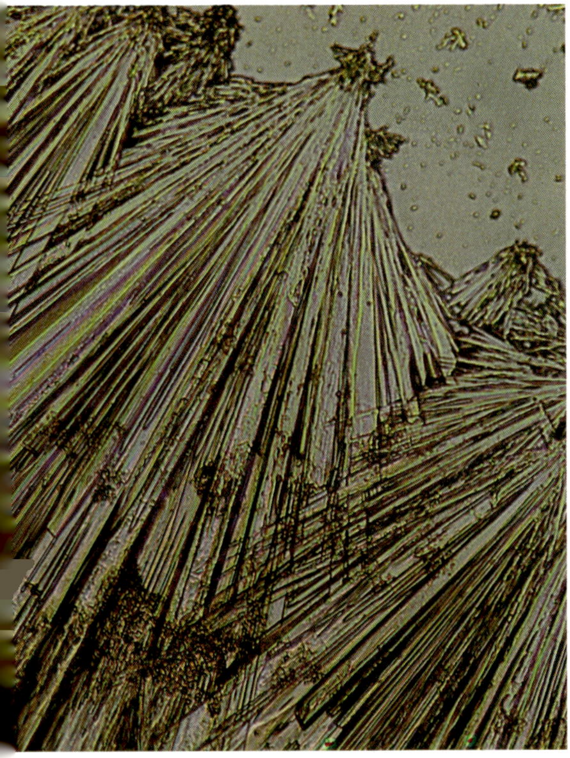

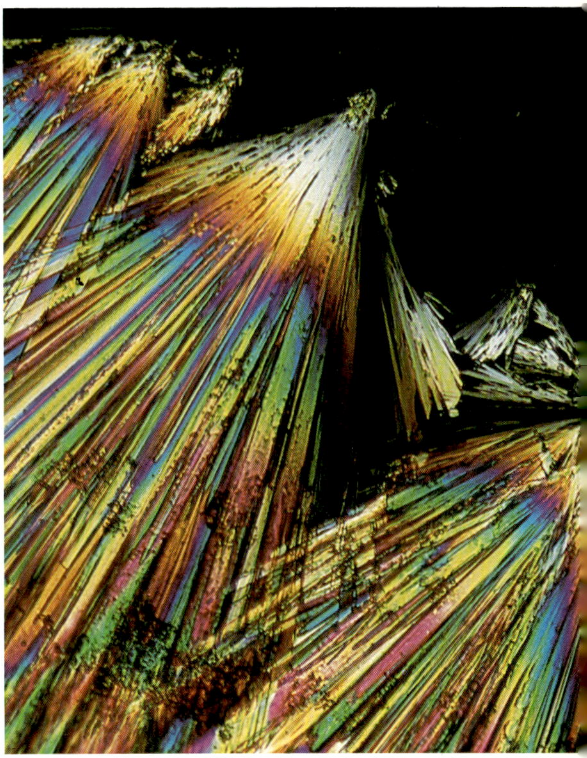

Fig. 22. Reaction product of cadmium sulfide and hydrochloric acid. Aroclor 5442 mount; refractive index 1.66. 500x.
A. Radiating prisms of cadmium chloride, the common habit near the edge of the reaction drop.

B. Same area between crossed polars.

cadmium sulfide lattice, the unit cell becoming smaller as the content of zinc increases. The change in *d*-spacing of the three major reflections of hexagonal zinc-cadmium sulfide can be used to indicate the approximate zinc content, although accuracy is low at zinc concentrations below 5% (see table 5 and fig. 24). Lattice spacings can also be used to determine whether a sample that shows zinc and cadmium by x-ray fluorescence contains the zinc as oxide, sulfide, or $Cd_{1-x}Zn_xS$ (Ulrich and Zachariasen, 1925; Smith, 1955; Hurlbut, 1957; Huckle, et al., 1966).

Pigment samples from Monet's *The Custom House at Varengeville* and Fernand Léger's *Divers on a Yellow Background* in the collection of the Art Institute of Chicago were analyzed both by electron microprobe (XRF) and x-ray diffraction (XRD), with a 57 mm camera used for the diffraction data. As can be seen in the patterns in fig. 25, the Monet pigment is an example of poorly crystallized material while that from the Léger is well formed. Principal lines found and measured, versus percentage of zinc, were:

Sample	*Principal Lines* d (Å)			*Percentage of Zinc Determined by* XRD	XRF
Monet (AIC 1933.1149)	3.57	3.35	3.17	0	0
Léger (AIC 1953.176)	3.53	3.31	3.11	12	10

Curtis and Wright (1954) compared x-ray diffraction patterns of cadmium sulfide and sulfoselenide pigments and were able to show the expansion of the lattice in the red pigment. The constitution of cadmium-selenium red pigments has been a matter of controversy in the

Table 4. **X-RAY DIFFRACTION DATA FOR CADMIUM SULFIDES**

Diffractometric Data CuKα_1 = 1.5405 Å, 25 C		*Diffractometric Data* CuKα = 1.5418 Å	
Alpha Cadmium Sulfide, Hexagonal, Greenockite[a]		*Beta Cadmium Sulfide, Cubic, Hawleyite*[b]	
d(Å)	I	*d*(Å)	I
3.58	75	3.36	100
3.36	60	2.90	40
3.16	100	2.058	80
2.450	25	1.753	60
2.068	55	1.680	10
1.898	40	1.453	20
1.791	18	1.337	30
1.761	45	1.298	10
1.731	18	1.186	30
1.679	4	1.120	30
1.581	8	1.028	5
1.520	2	0.985	20
1.398	16	0.918	5
1.3536	6	0.887	5
1.3271	12		
1.3032	8		
1.2572	12		
1.2247	2		
1.1940	8		
1.1585	12		
1.1249	8		
1.0743	6		
1.0540	2		
1.0340	4		
0.9934	4		
0.9881	6		
0.9842	6		
0.9827	6		
0.9729	2		
0.9533	10		
0.9265	2		
0.9081	8		
0.8956	2		
0.8878	2		
0.8804	4		
0.8653	<1		
0.8624	<1		
0.8315	4		
0.8166	4		
0.8158	4		

a. JCPDS Powder Diffraction File, 6-0314 with major correction.
b. JCPDS Powder Diffraction File, 10-454 with minor correction.
Courtesy of JCPDS International Centre for Diffraction Data (1982).

past, but the above study clearly indicates the existence of cadmium sulfide and selenide in solid solution rather than a separate class of compounds. Huckle, et al., (1966) demonstrated the change in lattice dimensions of the cadmium-mercury sulfide system.

The *Encyclopedia of X-rays and Gamma-Rays* (1963, p. 426) outlines a mechanism whereby ghost diffraction patterns can be formed by the interaction of CdS and copper radiation. Copper has a characteristic peak at 1.541 Å and a fairly intense continuum with a peak at about 0.5 Å. When this white radiation impinges on a cadmium sulfide sample, most of the radiation below 0.4641 Å (the K edge of cadmium) is absorbed. The "white" radiation above this range is reflected by the crystal lattice, and as would be expected with such radiation, produces a series of overlapping diffraction patterns which would normally lead to a low scattering angle diffuse halo. However, the silver absorption edge occurs at 0.4858 Å, and at this absorption edge a silver emulsion will experience a sudden increase in photographic sensitivity. The combination of these two phenomena give rise to a spectrum with an apparent incident radiation centered at 0.475 Å. The band width of the radiation is merely the difference between the two K absorption edges and is 0.0217 Å. Thus, an exact replica of the 1.541 Å radiation pattern is formed, though with spacings of 0.31 times those of the "normal" pattern.

5.32 ***Infrared Spectroscopy.*** The absorption band characteristic of cadmium sulfide occurs at 250 cm^{-1}; that of zinc sulfide occurs at 290 cm^{-1}. Because most sulfides have absorption in this range, it may be difficult to differentiate the various types with assurance. Considerable caution must be exercised in the interpretation of spectra of unknown samples. Many spectrometers in the past had poor sensitivity in the 30–40 micron region where sulfide absorption is located. Fig. 26 shows slight changes in the principal absorption region in a series of pigments of increasing selenium content. Barium sulfate shows up well even in amounts as small as 1%; details regarding its infrared spectrum can be found in the chapter on Barium Sulfate. Afremow and Vandeberg (1966) have published reference spectra of cadmium yellow lithopones, cadmium reds, and cadmium/mercury sulfides.

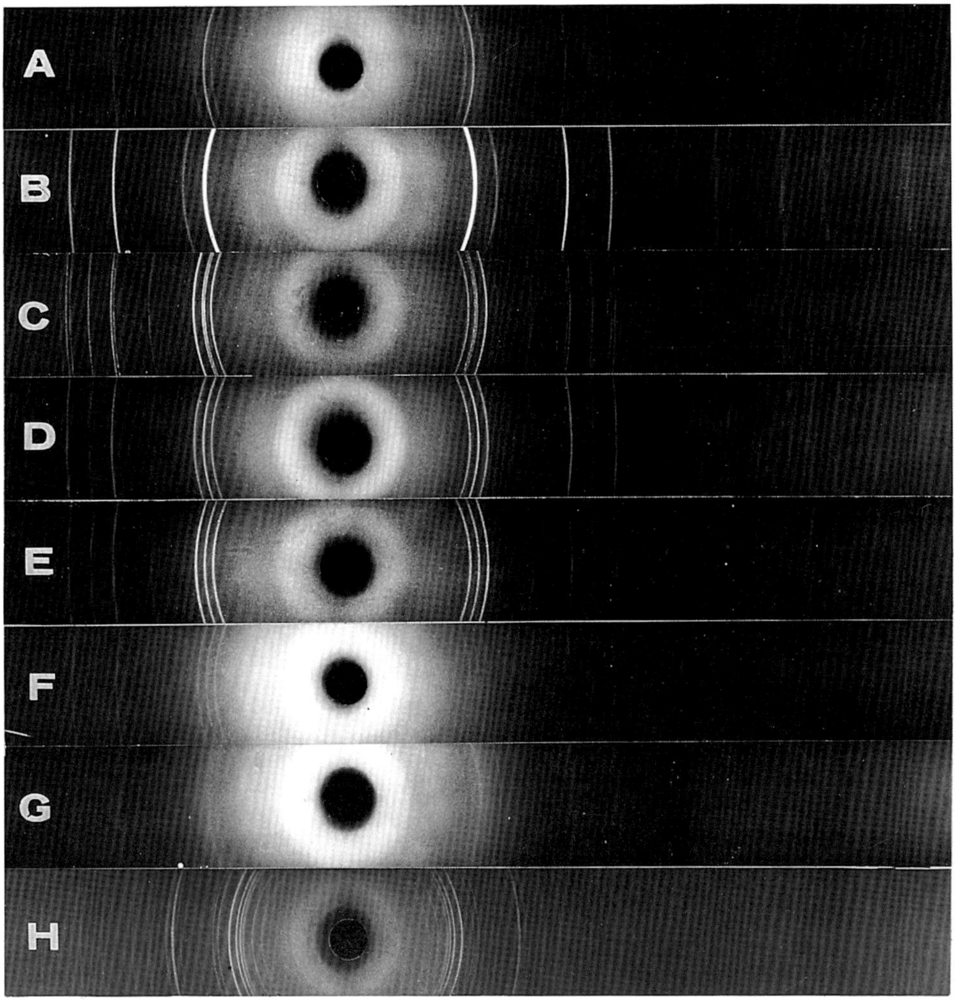

Fig. 23. X-ray powder diffraction patterns of cadmium yellows and minerals taken on a 114.6 mm Debye-Scherrer camera. A. Greenockite (hexagonal), Grant County, New Mexico. B. Hawleyite (cubic), Eureka, Nevada. C. Cadmium Yellow Deep, Fezandie and Sperrle, Inc., New York, 8 August 1977. This is a mixture of cubic and hexagonal CdS. D. Primrose Cadmium Yellow X-2822 from Imperial Pigments/Hercules, Inc. E. Cadmium Yellow C-411, Ferro Corp. (Zn approximately 20%). F. Pure Orange, Code No. 3350, Glidden. G. Cadmium Yellow Dark, Beaverhall (Edward W. Forbes, Art Institute of Chicago), an almost amorphous pattern. H. Cadmolith Primrose, Pb-407, Glidden Pigments, a physical mixture of CdS and $BaSO_4$. (D through H have the hexagonal crystal structure in varying degrees of crystal perfection.) Photographs are not to be used for measurement purposes.

Fig. 24. Values of zinc concentration plotted against change in crystal lattice dimensions in Ångstroms (Å), showing the possibility of measuring zinc abundance based on x-ray diffraction alone.

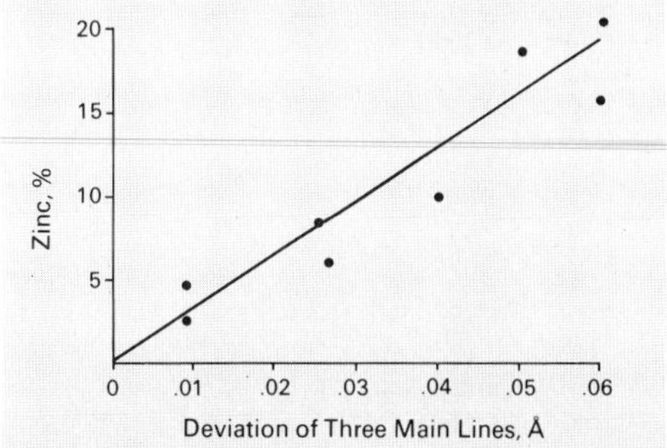

Table 5. **CORRELATION BETWEEN ZINC IN CdS LATTICE AND d VALUES FOR THREE MAJOR CdS LINES**

Sample[1]	*Major X-ray Lines*, d *(Å)*	*Percentage ZnS by X-ray Diffraction*	*Percentage ZnS by X-ray Fluorescence or from Manufacturer's Data (Corrected for Lithopone)*
1. Cadmium Yellow Code C-411, Ferro Corp.	3.51 3.30 3.10	20	21
5. Cadmolith, Primrose Pb 407 (c. 20 yrs. old), Glidden Pigments	3.56 3.35 3.15	3	5.3
6. Cadmolith, Primrose Lot-14433, Glidden Pigments	3.55 ? 3.10	? Pattern Poor	5–10 (by manufacturer)
7. Pure Cadmium Lemon, No. 3150, Glidden Pigments	3.53 3.32 3.12	12	9.8
8. Pure Primrose, No. 3050, Glidden Pigments	3.48 3.30 3.10	20+	16 (by manufacturer)
9. Pure Golden No. 3250, Glidden Pigments	3.57 3.35 3.15	3	4.5 (by manufacturer)
10. Pure Orange No. 3350, Glidden Pigments	3.56 3.34 3.14	6	2.1
12. Cadmium Yellow Light, Fezandie and Sperrle, Inc.	3.53 3.30 3.11	16	18
13. Cadmium Yellow, Medium, Fezandie and Sperrle, Inc.	3.56 3.33 3.14	5	0
14. Cadmium Yellow Deep, Fezandie and Sperrle, Inc.	3.51 3.29 3.10	20+ Mixture with Cubic	8.7
16. Cadmium Yellow Med., Permanent Pigments, Inc.	3.56 3.34 3.10	10	no data
28. Cadmium Yellow Lemon, No. 23, Harshaw (EWF Collection)[2]	3.54 3.33 3.13	9	8.1
29. Cadmium Yellow, Weber (EWF Collection)	3.54 3.31 3.12	7	6

Sample[1]	Major X-ray Lines, d (Å)	Percentage ZnS by X-ray Diffraction	Percentage ZnS by X-ray Fluorescence or from Manufacturer's Data (Corrected for Lithopone)
34. Cadmium Yellow (EWF Collection)	3.56 3.36 3.16	2	0
35. Cadmium Yellow Pale, Beaverhall (EWF Collection)	3.57 3.36 3.17	0	0
37. Cadmium Yellow, Winsor & Newton (EWF Collection)	3.57 3.36 3.16	1	0

1. Samples used throughout this study were arbitrarily numbered to allow correlation of different analytical methods listed in the tables.
2. Edward W. Forbes Collection, Art Institute of Chicago.

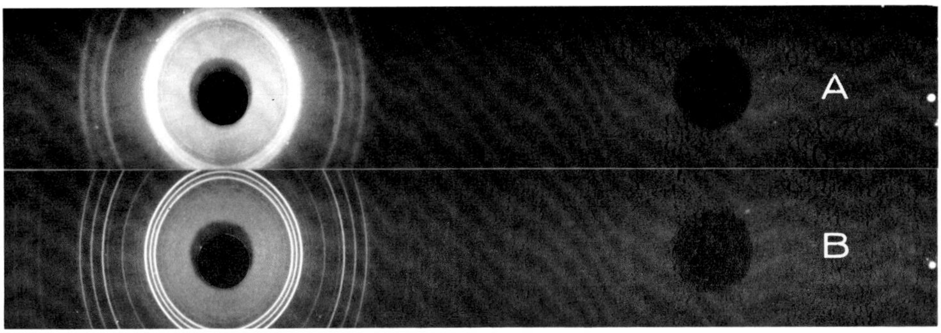

Fig. 25. X-ray powder diffraction patterns taken on a 57.3 mm Debye-Scherrer camera of samples from paintings. A. Claude Monet, *The Custom House at Varengeville*, Art Institute of Chicago, 1933.1149. Here the pattern is diffuse due to a large abundance of the amorphous form. Zinc content indicated zero both by x-ray diffraction and x-ray fluorescence. B. Fernand Léger, *Divers on a Yellow Background*, Art Institute of Chicago, 1953.176. This is an example of a well-crystallized CdS with 12% zinc; barium is present to the extent of 4–6%.

Raman spectroscopy utilizes the shift to lower energies when a monochromatic beam of light interacts with the chemical bonds and crystal structure of a substance. A weak point in the technique is the possibility of fluorescence, which is sometimes caused by impurities, completely masking all Raman information. Several samples were evaluated by both macro- and micro-Raman methods. A sample of commercial CdS modified with zinc gave a broad peak near the literature value (Ross, 1972) of 300 cm^{-1}, while sulfoselenides were shifted toward the value for pure CdSe, 209 cm^{-1} (fig. 27). However, a sample of pure CdS fluoresced enough to negate Raman, a reaction possibly caused by trace impurities, while an uncalcined sample of precipitated cadmium sulfide gave no spectrum at all. Barium sulfate was easily detected by macro-Raman in manufactured pigments at levels as low as 1%. The application of Raman spectroscopy to these materials is relatively new and more work is needed to enable it to become a reliable method of characterization.

5.33 ***Spectrochemical Analysis.*** The major visible lines are at the following wavelengths: for cadmium 6438.47 Å and for zinc 6362.35 Å; the major barium lines are all in the visible region 5777.66, 5535.55, and 4554.04 Å. The most sensitive lines in the ultraviolet region for cadmium are 3610.51, 3466.20, 3403.65, and 2288.02 Å; for zinc 3345.02, 3302.59, and

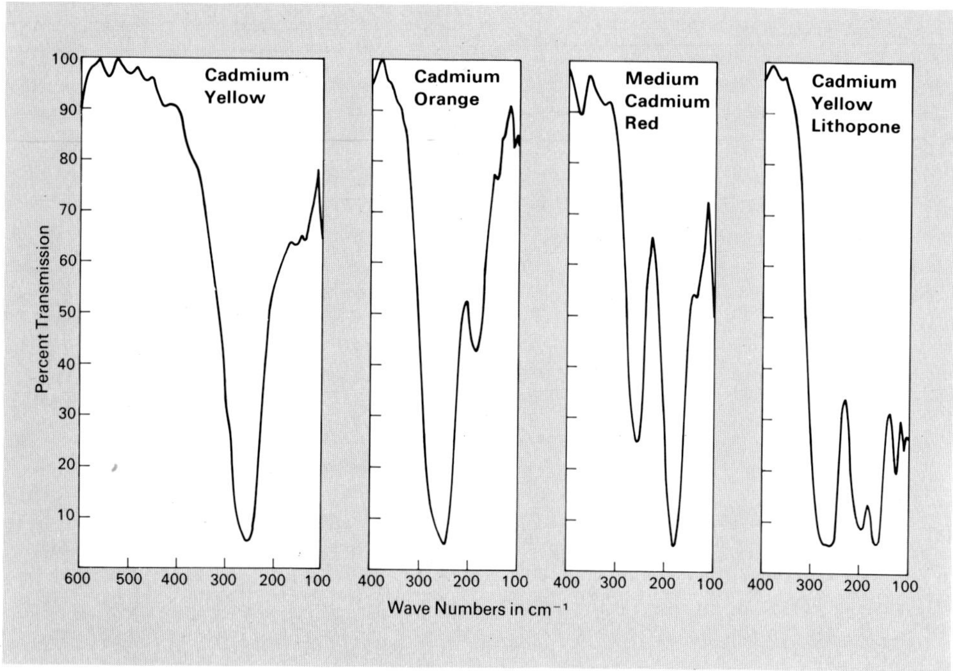

Fig. 26. Infrared spectra of cadmium sulfide and cadmium selenide pigments in the region of principal absorption. Increasing selenium content is evidenced by the band at 180 cm^{-1}, which can be masked by the presence of barium sulfate as seen in the spectrum for the cadmium lithopone. Data courtesy of the Research Center on the Materials of the Artist and Conservator, Mellon Institute, Pittsburgh.

2138.56 Å; barium lines are weak (Saidel, et al., 1955).

Crushings and sections can often provide sufficient material for emission spectrographic analyses. However, because samples from paintings will usually be limited, it is best to do all nondestructive tests before emission spectroscopy is used, because this method of analysis destroys the sample. Emission spectroscopy can be considered primarily as a semiquantitive tool, with nominal breakdown into matrix, major (1–10%), minor (0.1–1%) and trace (less than 0.1%) components. Sensitivity varies considerably for different elements. Use of the laser microprobe (a method employing a laser to vaporize 25–50 micrometer samples and to analyze the resulting light by means of a spectrometer) presents an opportunity to extend the method's applicability to smaller samples.

Emission spectroscopy is mainly a tool for detecting metallic (cationic) elements; the sensitivity for anions is low. With 10 mg samples, trace elements can easily be seen. As the sample size diminishes, however, these will not be so readily apparent. Elements detected in six commercial products are given in table 6.

5.34 ***X-Radiography.*** The mass absorption coefficients for cadmium, barium, zinc, and sulfur are plotted in fig. 28. Barium is generally considered the most absorbent material, sulfur the least. The K edges occur at 5.018 Å for sulfur, 1.283 Å for zinc, 0.4641 Å for cadmium, and 0.3314 Å for barium.

The absorption values are based on the empirical relationship $\mu = C\lambda^n$, where C is a constant unique for each absorption range of each element, λ is the wavelength of the radiation in Å, and n is an exponent unique for each range (i.e., from K to L_1; L_1 to L_2; etc.) and common to all elements. For the absorption edge ranges of the studied elements, n ranges from 2.83 to 2.65. Accuracy is expected to be about 5%.

Because of the K-edge locations, the absorption coefficients at one angstrom are: S (sulfur) 26.3, Zn (zinc) 141.0, Cd (cadmium) 74.4, and Ba (barium) 116.4. Absorption coefficients for the primary line of copper (1.542 Å) are: S-89.58, Zn-60.19, Cd-235.68, and Ba-368.76.

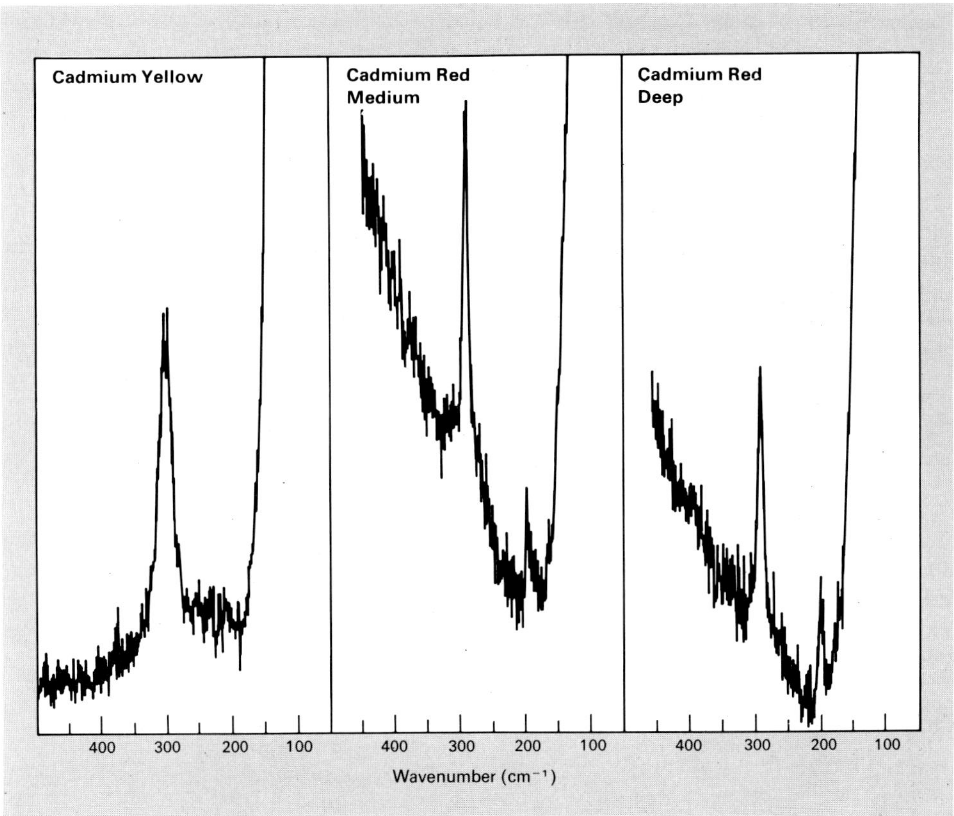

Fig. 27. Raman spectra of (A) zinc modified cadmium sulfide, (B) medium cadmium sulfoselenide red, and (C) deep cadmium sulfoselenide red.

Table 6. **ELEMENTS DETECTED BY EMISSION SPECTROMETRIC ANALYSIS**

Sample	*Matrix*	*Major*	*Minor*	*Trace*
1. Cadmium Yellow Code C-411, Ferro Corp.	Cd	Zn	Al, Mg, Ca, Si, B	Cu, Mn, Fe, Ti, Pb
5. Cadmolith-Primrose Pb-407, Glidden Pigments	Cd	Zn, Ba	Ca, Al, Si, Mg, B	Fe, Mn, Cu, Ti
7. Pure Cadmium Lemon, Code no. 3150, Glidden Pigments	Cd	Zn, Ba	Sr	Cu, Ca, Al, Pb
10. Pure Orange, Code no. 3350, Glidden Pigments	Cd	Zn, Ba	Mg, Mn, Ca, P, Si	Cu, Fe, Pb
14. Cadmium Yellow Deep, Fezandie and Sperrle, Inc.	Cd	Zn, Ba	Sr, Pb	Ca, Al, Cu, Fe
21. Kadmiumgelb hell, H. Schmincke and Co.	Cd	Zn, Ba	Sr, Al	Ca, Cu, Pb

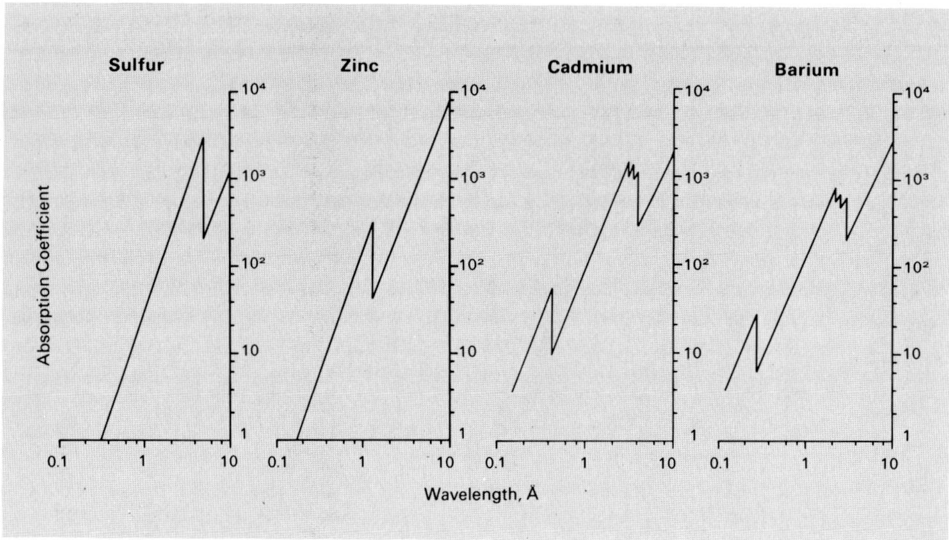

Fig. 28. Absorption curves for sulfur, zinc, cadmium, and barium between 0.1 and 10 Å. Sulfur and zinc show only the K-edge discontinuity, while cadmium and barium display the K transition and the three discontinuities at the L level.

With a molybdenum target (0.711 Å) the coefficients are: S-10.01, Zn-53.71, Cd-29.99, and Ba-46.92, while tungsten (0.24 Å) gives: S-0.32, Zn-1.724, Cd-5.852, and Ba-8.77 (*Encyclopedia of X-rays and Gamma-Rays*, 1963).

5.35 ***Neutron Activation Analysis.*** Five samples were analyzed by neutron activation analysis, with sixteen elements evaluated in two samples and eighteen in the other three. The results are reported in table 7, and the method used is summarized as follows: The matrix effect due to the presence of large amounts of cadmium in the samples was eliminated by preparing standards containing the various additional metal ions in approximately the same quantity of cadmium sulfide and then irradiating them under the same conditions as the unknowns. These were prepared from the pure metals in solution and deposited into the cadmium sulfide matrix by a micropipetter. Two different irradiations were performed: one five-minute irradiation in the pneumatic tube system of the Ford Nuclear Reactor, followed by immediate counting of the sample (or standard) to measure the elements S, Cd, Ba, Cu, Mn, Al, Na, and Ca, and a five-hour irradiation, followed by a 1 to 6 week delay before counting, for the measurements of the elements Se, Hg, Cr, Ag, Cs, Ni, Fe, Zn, Co, and Sb. Gamma rays characteristic of the radionuclides produced were measured using two matched large volume Ge(Li) detectors interfaced to a ND-4420 computer analyzer system. Sample size for this technique may be as small as one or two milligrams or less (John D. Jones, private communication).

Although the major components of cadmium pigments can be accurately analyzed by the above technique, neutron activation has its main application in the characterization or "fingerprinting" of samples based on an analysis of trace elements. The method is capable of good accuracy and, in some instances, very good solid solution sensitivity, though absolute weight sensitivity is not as high as that in x-ray fluorescence. Because of the large number of trace elements which are quantifiable, activation analysis should be a good indicator of sample identity, or lack thereof; though incidental contamination from the other pigments, vehicle, etc., however, may complicate this type of evaluation.

5.37 ***X-ray Fluorescence.*** The major cadmium lines are: in the K series 0.5643 Å (23.11 kV) and 0.4750 Å (26.09 kV), and in the L series 3.9562 Å (3.12 kV), 3.7380 Å (3.32 kV), and 3.5140 Å (3.53 kV). Because electron microprobe analysis is generally performed at voltages of 25 kV and below in order to excite x-rays from small volumes, the high energy K transitions are not of major interest analytically. The only important interference is a silver line with a relative intensity of 50 at 3.9344 Å (3.15 kV);

Table 7. **ANALYSIS OF CADMIUM BASED PAINT PIGMENTS FOR MAJOR AND TRACE ELEMENTS BY INSTRUMENTAL NEUTRON ACTIVATION ANALYSIS**

	Parts Per Million (micrograms/gram)			
Elements	*Sample No. 10 Pure Orange*[1]	*Remarks*	*Sample No. 70 C. P. Golden*[2]	*Remarks*
Sulfur	186,000 ± 900		238,000 ± 1200	
Cadmium	566,800 ± 200		799,960 ± 276	
Barium	151,830 ± 3000		not detected	less than 1800
Antimony	1.25 ± 0.1		3.16 ± 0.17	
Copper	not detected	less than 80	not detected	less than 120
Manganese	35.9 ± 1.1		13.6 ± 1.2	
Calcium	1692 ± 212		1392 ± 252	
Selenium	783 ± 86		5390 ± 593	
Mercury	8.04 ± 1.1		not detected	less than 4
Chromium	not detected	less than 15	17.2 ± 3	
Silver	not detected	less than 3	not detected	less than 3
Cesium	not detected	less than 0.02	not detected	less than 0.02
Nickel	not detected	less than 84	not detected	less than 55
Iron	not detected	less than 100	not detected	less than 80
Zinc	10,462 ± 192		2184 ± 42	
Cobalt	not detected	less than 0.16	not detected	less than 0.2

1. Pure Orange No. 3350 Glidden Pigments.
2. C. P. Golden No. 1489 Harshaw Chemical Co.

Elements	*Sample No. 26 Primrose Cadmolith*[3]	*Remarks*
Sulfur	222,000 ± 12,700	
Cadmium	235,000 ± 2,200	
Barium	426,000 ± 11,300	
Antimony	16.8 ± 0.35	
Copper	not detected	less than 22
Manganese	not detected	less than 2.5
Calcium	251 ± 45	
Selenium	37.3 ± 3.2	
Mercury	not detected	less than 5
Chromium	not detected	less than 18
Silver	not detected	less than 4
Cesium	not detected	less than 1.2
Nickel	not detected	less than 93
Iron	not detected	less than 581
Zinc	24,600 ± 220	
Cobalt	0.53 ± 0.08	
Sodium	333 ± 16	
Aluminum	176 ± 2.4	

3. Primrose Cadmolith, Sample No. 766, Chemical and Pigment Co., 1/24/34 (Edward W. Forbes Collection, Art Institute of Chicago).

Elements	*Sample No. 31 Cadmium Yellow*[4]	*Remarks*
Sulfur	233,000 ± 14,000	
Cadmium	564,000 ± 2,900	
Barium	not detected	less than 594
Antimony	2.20 ± 0.15	
Copper	46.7 ± 7.8	
Manganese	not detected	less than 4.5
Calcium	2,050 ± 136	
Selenium	29.7 ± 2.8	
Mercury	not detected	less than 4
Chromium	not detected	less than 15
Silver	not detected	less than 5
Cesium	not detected	less than 1.6
Nickel	not detected	less than 123
Iron	not detected	less than 758
Zinc	42,200 ± 373	
Cobalt	0.76 ± 0.07	
Sodium	5,598 ± 105	
Aluminum	60 ± 1.5	

4. Cadmium Yellow, Roberson 1926 (Edward W. Forbes Collection, Art Institute of Chicago).

Elements	*Sample No. 35 Cadmium Yellow, Pale*[5]	*Remarks*
Sulfur	175,000 ± 22,000	
Cadmium	508,000 ± 4,300	
Barium	19,500 ± 600	
Antimony	4,320 ± 14	
Copper	not detected	less than 640
Manganese	not detected	less than 15
Calcium	3,249 ± 282	
Selenium	41.7 ± 5.9	
Mercury	not detected	less than 9.2
Chromium	88.9 ± 7.2	
Silver	not detected	less than 6
Cesium	not detected	less than 2.4
Nickel	not detected	less than 130
Iron	not detected	less than 1850
Zinc	1,320 ± 22	
Cobalt	not detected	less than 4
Sodium	not detected	less than 1000
Aluminum	38,972 ± 344	

5. Cadmium Yellow Pale, Beaverhall (Edward W. Forbes Collection, Art Institute of Chicago).

although a crystal spectrometer will separate this from the cadmium line, there will be some overlap in an energy dispersive system. Sulfur has a simple spectrum with only the Kα1,2 transition at 5.3728 Å (2.31 kV) of analytical use. The major source of interference is lead with lines at 5.2986 Å (2.34 kV) and 5.2846 Å (2.35 kV). The zinc K line is at 1.436 Å (8.63 kV), while barium has its main analytical transition at 2.775 Å (4.47 kV) (White, 1965).

Samples smaller than the electron diffusion zone give approximately the correct Cd:S ratios because the energies, and hence diffusion volumes, for the two are similar. Zinc will not give a correct ratio in very small samples as its energy is considerably higher than that of the other two. Also, since barium sulfate present as an extender may have larger particles than the cadmium sulfide component, local variations in composition can lead to errors in determining barium sulfate concentration (Bayard, 1973).

Twenty-seven bulk cadmium pigments were analyzed by electron excited x-ray fluorescence. The weight percentages, corrected for atomic number, absorption, and fluorescence, appear in table 8. (Data on lithopones will be in error because of their large abundance of undetected oxygen.)

Sixteen of the samples contained zinc in amounts ranging from 1.4–22.5%, corresponding to its use as a lightener. Such a broad range would, in principle, allow for characterization of the shade of cadmium yellow by using the microprobe data. However, interference from zinc oxide would introduce errors. Unfortunately, it is difficult to measure the quantity of oxygen in such systems.

Reviewing the information in table 8, we can describe an average present-day cadmium yellow as ranging from pure CdS to CdS containing anywhere from a few percent up to 20% zinc. Though not containing the abundance of barium sulfate found in lithopones, so-called pure samples may still contain 1–5% barium. If a reasonably detailed characterization of a cadmium yellow sample is contemplated, zinc, barium, titanium, and aluminum concentrations should be determined as well as those of cadmium and sulfur.

5.4 Criteria for Positive Identification

Unequivocal identification of cadmium pigments can be difficult owing to the variety of materials which are encountered. In general, one would wish to ascertain the presence (or ab sence) of the following: CdS, $Cd_xZn_{1-x}S$, CdS_xSe_{1-x}, $Cd_xHg_{1-x}S$, $BaSO_4$, and possibly ZnO. The presence of adulterants cannot be ruled out in older samples (see 4.4).

Examination with an optical microscope may suggest the presence of a cadmium-based material; this analysis, combined with microchemical tests, may place a sample in the "cadmium sulfide or possibly sulfoselenide class." The sodium azide-iodide test (5.21) provides a sensitive test for sulfide. However, complete characterization requires x-ray diffraction, x-ray fluorescence, and/or emission spectroscopy, depending on the amount of sample available. Even with instrumental analysis the subtleties of the cadmium pigment system may preclude unequivocal characterization without the availability of sufficient sample and without considerable experience on the part of the analyst.

In most cases x-ray diffraction will yield positive identification of the principal compounds present, based on the crystallographic parameters of the components CdS, $Cd_xZn_{1-x}S$, and CdS_xSe_{1-x}, either with or without $BaSO_4$. While this method will permit one to be certain of the principal compounds, it requires a sample size three to four orders of magnitude larger

Table 8. **ANALYSIS OF CADMIUM PIGMENTS BY X-RAY FLUORESCENCE**

Sample	*Cd*	*S*	*Zn*	*Other*
1-Cadmium Yellow Pigment, Code 411, Ferro Corp.	59.2	24.6	14.1	
3-Cadmium Yellow, c. 1950, Winsor & Newton	57.5	35.9	3.5	Ba 1.1
4-Cadmium Sulfide, current product, Winsor & Newton	70.9	25.8		Ti 1.2
5-Cadmolith Primrose Pb 407, 20 yr. old, Glidden	29.9	25.3	1.6	Ba 41.1
7-Pure Cadmium Lemon 3150, Glidden Lot 13754	62.6	27.1	6.5	Ba 1.8
10-Pure Orange 3350, Glidden Lot 11000	64.9	26.3	1.4	Ba 5.3
12-Cadmium Yellow Light, Fezandie & Sperrle, Inc.	55.6	24.3	12.2	Ti 5.7
13-Cadmium Yellow Med., Fezandie & Sperrle, Inc.	37.5	34.3		Ti 26.1
14-Cadmium Yellow Deep, Fezandie & Sperrle, Inc.	65.3	25.4	5.8	Ba 1.4
21-Cadmium Yellow Light, H. Schminke & Co.	51.3	19.1	22.5	Ba 5.1
23-Cadmium Yellow Dark, H. Schminke & Co.	51.5	29.9	2.5	Ba 14.2
28-Cadmium Yellow Lemon #23 Harshaw (EWF)[1]	66.3	26.2	5.4	
29-Cadmium Yellow, Weber (EWF)	28.5	25.4	1.9	Ba 42.1
31-Cadmium Yellow, Roberson 1926 (EWF)	67.3	23.1	7.5	
32-Cadmium, Roberson (EWF)	81.2	16.7		
33-Cadmium Yellow Pale, Le Franc (EWF)[2]	8.2	1.6		Cr 29.0, Sr 59.1
34-Cadmium Yellow (EWF)	72.4	25.5		
35-Cadmium Yellow Pale, Beaverhall (EWF)	55.2	26.5		Al 16.2
36-Cadmium Yellow Dark, Beaverhall (EWF)	71.7	25.7	3.0	Al 1.0
37-Cadmium Yellow, Winsor & Newton (EWF)	72.5	25.5		
38-Reaction of $CdCl_2$ and Na_2S	73.3	24.7		
50-Cadmium Orange, Roberson (EWF)	73.2	23.5	0.5	Ba 0.8
67-CP Cadmium Primrose 1466, Harshaw	61.6	19.7	12.6	Ba 4.0
68-Cadmium Lemon 1470, Harshaw	59.6	21.3	5.8	Ba 11.2
69-CP Cadmium Golden 1481, Harshaw	73.2	19.2	5.5	
70-CP Cadmium Golden 1489, Harshaw	78.0	19.9		
93-Pure Cadmium Red Deep, Lot #14093, Glidden	66.2	10.5		Se 17.3, P 3.2, Ba 0.8
Theoretical Composition, CdS	77.81	22.19		

1. EWF = Edward W. Forbes Collection, Art Institute of Chicago.
2. Although labeled cadmium yellow, this pigment is largely strontium chromate.

than that needed for the electron microprobe. In the case of early pigments of the amorphous type, x-ray diffraction will be useless.

Electron microprobe analysis will give positive identification of the elements present in the major components of cadmium pigments. It will give at least semiquantitative information as to the abundance of the elements. However, it will not tell how they are combined; for example, while microprobe analysis will indicate the amount of zinc, it will not show whether the zinc is combined as a solid solution with CdS, or as the compounds ZnS or ZnO. The major advantage of the electron microprobe is the small sample size required (as small as 15 cubic micrometers), making this method particularly applicable to the examination of specimens taken from paintings or objects. Also, it is nondestructive.

Scanning electron microscopy combined with energy dispersive x-ray fluorescence will provide the same elemental information as the electron probe if there is no interference from the binder. In addition, the micro-morphological data from the SEM may give some key as to the sample origin. Sample preparation will probably require low temperature ashing to remove the medium, though interference from inorganic components may still partially obscure both elemental and morphological information. More work is needed to extend the use of this method.

Optical microscopy in conjunction with microchemistry is a reasonable approach in convincing the investigator that he is dealing with a pigment that contains cadmium and a sulfide or selenium, though it will generally not give specific information as to zinc content, the amount of barium present, or the method of chemical combination of these metallic elements. Most of the chemical methods are destructive and require a relatively large amount of sample.

Far infrared spectroscopy yields information on the metal-sulfide bonds, but this will usually not be sufficient for unequivocal identification. The barium sulfate component can be easily seen by infrared, however, even at levels of 1%. In ideal cases the sample can be removed from the potassium bromide pellet for further tests.

6.0 NOTABLE OCCURRENCES

The use of cadmium was mentioned by William Holman Hunt, who noted the fact in his diary of 24 March 1855 during the painting of the *Scapegoat.* He was not pleased with the effect of the pigment, noting: "work in default of clouds at the water of the marsh erasing and repainting it preferring to rely on a mixture with Emerald green rather than another I had used containing cadmium" (Walker Art Gallery, 1969, p. 43).

Standage (1886, pp. 98–99), in documenting the palettes of several prominent nineteenth-century English artists, listed those using cadmium as: Frederic Leighton, Lawrence Alma-Tadema, William Quiller Orchardson, John Pettie, and Colin Hunter. Church (1915, p. 290) in his chapter on selected palettes lists Samuel Palmer as painting with cadmium 1, 2, and 3. From *Portfolio* 1875–1876 he also quotes references to P. H. Calderon and P. G. Hamerton as cadmium yellow users.

Birren (1965) lists the following artists as having some form of cadmium yellow on their palette: Eugene Delacroix, Jean Baptiste Camille Corot, Paul Signac, Pierre Bonnard, André Derain, Henri Matisse, Fernand Léger, and Raoul Dufy. Kühn (1969) has reported finding cadmium yellow in paintings by the nineteenth-century Swiss artist, Arnold Böcklin in the Schack-Galerie. Analyses conducted by Richter and Härlin (1974) on the artist's own collection of dry pigments agree with the findings of Kühn.

The majority of Monet's paintings at the Art Institute of Chicago have cadmium yellow as one of the yellows used. Moisse, the dealer who supplied most of his pigments, stated that Monet employed three shades of cadmium: light, dark, and lemon (Tabarant, 1923). Because only small samples are taken from paintings and other pigments are frequently mixed with the cadmium, it was impossible to identify the shades positively. Based on morphology there were, however, at least two different forms found (see 3.2). Delbourgo and Rioux (1974), using electron microprobe analysis, identified cadmium sulfide in *La Cathédrale de Rouen* by Monet.

An example of cadmium orange may be seen in Charles Demuth's *Gladioli: Flower Study No. 4* (fig. 29), in the Art Institute of Chicago.

PAINTINGS from the Art Institute of Chicago

Cadmium Yellow

Artist or School *Title, Date,* *Collection or Location*	*Analyst and* *Method of* *Identification*
Vincent van Gogh *Still Life: Fruit* c. 1887 1949.215	Micr.[1] (Betty Engel[2]), E. Probe[3]
Vincent van Gogh *Self-Portrait* 1886–1887 1954.326	Micr. (Betty Engel[2]), E. Probe
Fernand Léger *Follow the Arrow* 1919 1953.341	Micr. (Marigene Butler[4]), E. Probe
Fernand Léger *Divers on a Yellow Background* 1941 1953.176	Micr., E. Probe, XRD[3]
Henri Matisse *By the Sea* 1904 77.1974[5]	Micr., E. Probe
Henri Matisse *Bathers by a River* 1916–1919 1953.158	Micr. (Marigene Butler[4]), E. Probe
Henri Matisse *Le Château Chenonceaux* c. 1918 76.1974[5]	Micr., E. Probe

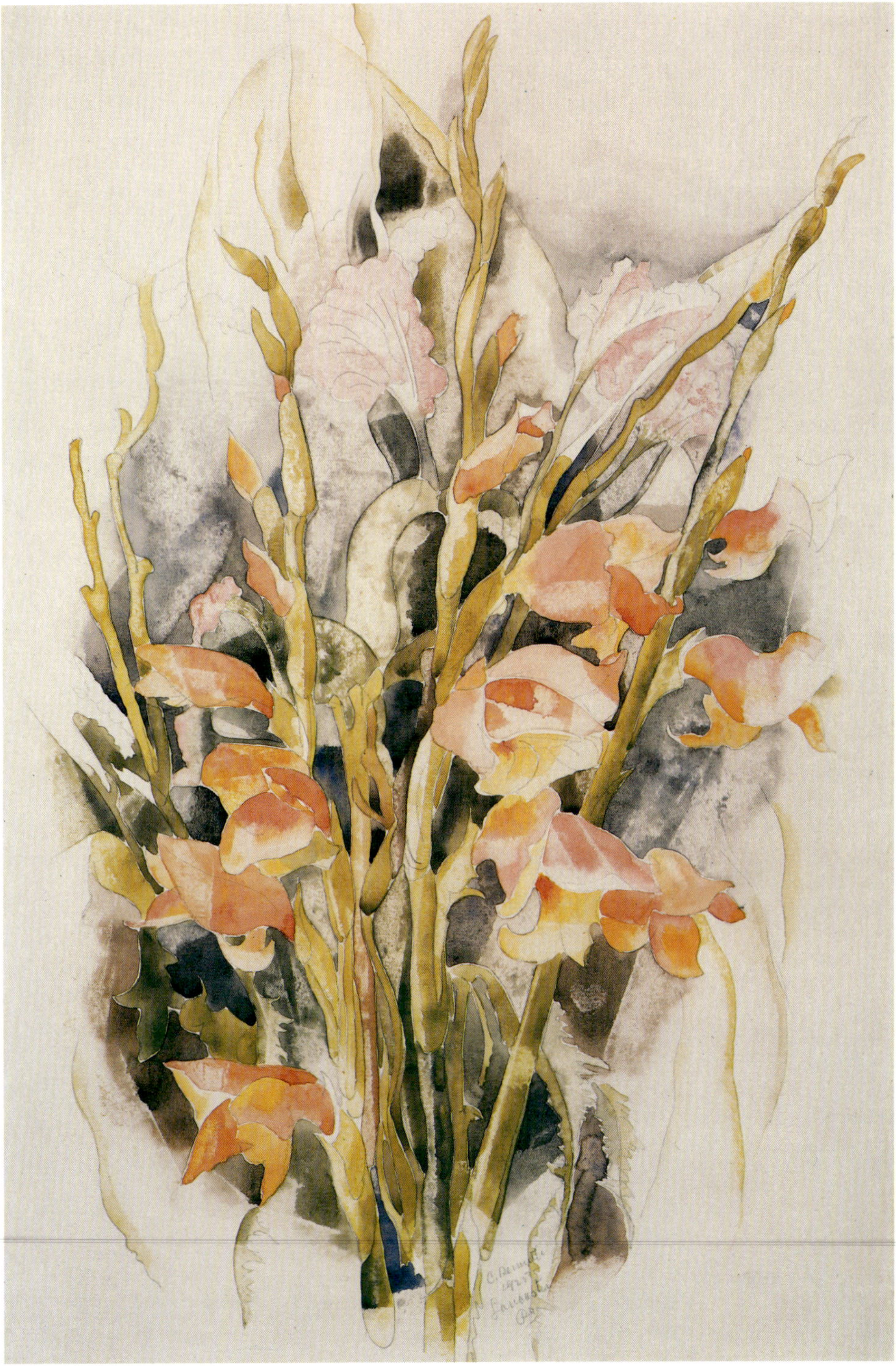

Fig. 29. Charles Demuth, *Gladioli: Flower Study No. 4*, 1925. An example of the use of cadmium orange in a watercolor (Olivia Shaler Swan Memorial Fund, Art Institute of Chicago, 1933.472).

Artist or School Title, Date, Collection or Location	Analyst and Method of Identification
Claude Monet *Monet's House at Argenteuil* 1873 1933.1153	Micr., E. Probe
Claude Monet *Still Life: Apples and Grapes* 1880 1933.1152	Micr., E. Probe
Claude Monet *The Cliff Walk, Pourville* 1882 1933.443	Micr., E. Probe
Claude Monet *Etretat* 1883 1964.204	Micr.
Claude Monet *Bordighera* 1884 1922.426	Micr.
Claude Monet *Rocks at Belle Isle* 1886 1964.210	Micr., E. Probe
Claude Monet *Torrent, Creuse* 1888–1889 1922.432	Micr., E. Probe
Claude Monet *Grainstacks, Snow, Sunset* 1891 1922.431	Micr.
Claude Monet *Grainstacks, End of Day, Autumn* 1891 1933.444	Micr.
Claude Monet *The Grainstacks in the Snow, Overcast* 1891 1933.1155	Micr., E. Probe
Claude Monet *The Custom House at Varengeville* 1897 1933.1149	Micr., E. Probe, XRD[3]
Claude Monet *Morning on the Seine* 1897 1933.1156	Micr.
Claude Monet *Vétheuil* 1901 1933.447	Micr.
Claude Monet *Vétheuil at Sunset* 1901 1933.1161	Micr.
Claude Monet *Charing Cross Bridge, London* 1901 1933.1150	Micr.
Claude Monet *Waterloo Bridge* 1903 1933.1163	Micr., E. Probe
Claude Monet *Houses of Parliament, Westminster* 1903 1933.1164	Micr., E. Probe
Claude Monet *Venice, Palazzo Dario* 1908 1933.446	Micr.
Claude Monet *Venice, San Giorgio Maggiore* 1908 1933.1160	Micr., E. Probe
Cadmium Orange	
Henri Matisse *Le Château Chenonceaux* c. 1918 76.1974[5]	Micr., E. Probe

1. Microscopy by Inge Fiedler unless otherwise indicated.
2. Betty L. Engel, 1975.
3. Electron microprobe analysis and x-ray diffraction by McCrone Associates, Chicago, Illinois.
4. Marigene H. Butler, private communication.
5. On loan to the Art Institute of Chicago from the Robert R. McCormick Charitable Trust.

WATERCOLORS AND GOUACHES from the Collection of the Art Institute of Chicago, Department of Prints and Drawings

Cadmium Yellow

Artist or School Title, Date, Collection or Location	Analyst and Method of Identification
Léon Bakst *Dieu Bleu: Pretres Agri* 1911 1920.2522	Micr.[8], E. Probe[7]
Alexander Calder *Untitled* no date 1965.258	Micr.[8], E. Probe[7]
Stuart Davis *Electric Lights and Buildings* c. 1931–1932 1973.711	Micr.[8], E. Probe[7]
Fernand Léger *Study for The City* 1919 1952.1007	Micr.[8], E. Probe[7]
John Singer Sargent *Olive Trees, Corfu* no date 1933.505	Micr.[8], E. Probe[7]
John Singer Sargent *An Artist at His Easel* 1914 1962.971	Micr.[8], E. Probe[7]

Artist or School Title, Date, Collection or Location	*Analyst and Method of Identification*
Cadmium Orange	
Léon Bakst *Boutique Fantasque: Woman in a Red Bonnet* 1917 1920.2531	Micr.[8], E. Probe[7]
Charles Demuth *Gladioli: Flower Study No. 4* 1925 1933.472	Micr.[8], E. Probe[7]

WATERCOLORS from the Fogg Art Museum, Harvard University

Cadmium Yellow

Artist or School Title, Date, Collection or Location	*Analyst and Method of Identification*
Charles Demuth *Fruit and Sunflower* c. 1924 1925.5.3 (Fogg Art Museum)	Weston[6]/XRF
Winslow Homer *Hunter in the Adirondacks* 1892 1939.230 (Fogg Art Museum)	Weston[6]/XRF

6. Weston, 1977
7. Electron microprobe analysis performed by McCrone Associates, Chicago, Illinois.
8. Micr. by Inge Fiedler unless otherwise indicated.

ACKNOWLEDGMENTS

The authors wish to acknowledge the following persons for their assistance in this project: Robert Feller and his staff for infrared analysis, color and spectral reflectance data and for invaluable advice throughout the work and Marion Weimann, Jr., for locating obscure references and for aid in gaining access to the Scanning Electron Microscope User's Lab at the University of Chicago (funded by a grant from the National Institute of Health, Biotechnology Resources Division). We extend our gratitude to Rosamond D. Harley for much of the historical information, including that about George Field's journal, and to John D. Jones, manager of the Phoenix Memorial Laboratory, University of Michigan, for his contribution on neutron activation analysis. Terry Keegan, graduate student at the University of Chicago, assisted with the x-ray diffraction analysis.

We also wish to thank the following individuals and manufacturers for providing samples and technical information: Zora Pinney; Brian J. Heath, vice president of Winsor & Newton, Inc.; Paul Tudder of Glidden Pigments; Michael J. Dunn of H. Kohnstamm and Co., Inc.; Peter J. Hesse of H. Schmincke and Co.; Marylou Hickey of The Hercules Chemical Co., The Shepherd Chemical Co., and The Harshaw Chemical Co. Additional samples from the E. W. Forbes collection were kindly provided by the Intermuseum Laboratory, Oberlin, Ohio.

Much of the work was done with the help of staff members of the Art Institute of Chicago: Ilse Hecht, research associate, translated innumerable papers on cadmium pigments; Harold Joachim kindly provided access to the prints and drawings collection; and David Chandler assisted with the examination and sampling of watercolors and gouaches. The photography department has helped with much of the graphics; in particular we wish to acknowledge the assistance of John Mahtesian, Anne Benolken, and Eugene Yates. To Alfred Jakstas, conservator, we extend our appreciation for his encouragement and support throughout this endeavor.

REFERENCES

L. C. Afremow and J. T. Vandeberg, "High Resolution Spectra of Inorganic Pigments and Extenders in the Mid-Infrared Region from 1500 cm^{-1} to 200 cm^{-1}, *Journal of Paint Technology* 38 (1966), 169–202.

E. T. Allen and J. L. Crenshaw, "The Sulphides of Zinc, Cadmium and Mercury; Their Crystalline Forms and Genetic Conditions," *American Journal of Science* 34 (1912), 341–396.

G. H. Bachhoffner, *Chemistry as Applied to the Fine Arts* (London, 1837), 117.

N. F. Barnes, "Color Characteristics of Artists' Pigments," *Journal of the Optical Society of America* 29 (May 1939), 208–214. See also N. F. Barnes, "A Spectrophotometric Study of Artists' Pigments," *Technical Studies in the Field of the Fine Arts* 7 (1939), 120–130.

M. A. Bayard, "Application of Electron Microprobe to the Analysis of Free Particulates," *Microprobe Analysis*, ed. C. A. Anderson (New York, 1973), 323–348.

J. Bersch, *The Manufacture of Mineral and Lake Pigments*, 3rd ed. rev., trans. A. C. Wright (London, 1901), 153–154, 439–440, 462.

F. Birren, *History of Color in Painting* (New York, 1965), 42–74.

J. Böhm and H. Niclassen, "Uber amorphe Niederschläge und Sole," *Zeitschrift für anorganische Chemie* 132 (1924), 7.

G. Buchner, "Ueber Schwefelcadmium und ii ber die vershiedenen Cadmiumfarben des Handels," *Chemiker-Zeitung*, ed. G. Krause 11, no. 72 (7 September 1887), 1087–1089, and 11, no. 73 (11 September 1887), 1107–1109.

N. F. Budgen, *Cadmium: Its Metallurgy, Properties and Uses* (London, 1924), 100–105.

Cadmium Association, *Technical Notes on Cadmium, Cadmium Pigments* (London and New York, 1978).

E. M. Chamot and C. W. Mason, *Handbook of Chemical Microscopy*, vol. 2, 2nd ed., 7th printing (New York, 1960), 137–148.

D. M. Chizhikov, *Cadmium*, trans. D. E. Hayler (Oxford, 1966).

A. H. Church, *The Chemistry of Paints and Painting*, 4th ed. (London, 1915), 162–167.

Colour Index, vol. 3, 3rd ed. (Bradford, Yorkshire, England, and Research Triangle Park, North Carolina, 1971).

A. Connell, "Chemical Examination of Greenockite, or Sulphuret of Cadmium," *The Edinburgh New Philosophical Journal* (Edinburgh, 1840), 392–395.

P. J. Curtis and R. B. Wright, "Cadmium Pigments," *Journal Oil and Colour Chemists' Association* 37 (1954), 26–43.

J. D. and E. S. Dana, *The System of Mineralogy*, vol. 1, 7th ed., rev. by C. Palache, H. Berman, and C. Frondel (New York, 1944), 228–230.

B. Dawes, "Cadmium Pigments — Are They Worth It," *European Plastics* 5, no. 12 (1978), 32.

S. Delbourgo and J.-P. Rioux, "Contribution à l'étude de la matière picturale des impressionnistes," *Annales du Laboratoire de recherche des Musées de France* (1974), 34–42.

F. W. Devoe, *Trade Catalogue* (New York, 1878).

M. Doerner, *The Materials of the Artist and Their Use in Painting*, rev. ed. trans. E. Neuhaus (New York, 1949).

Dry Color Manufacturers' Association, *Criteria for Warning Labeling on Artists' Paints Containing Cadmium* (Arlington, Virginia, 1978).

M. J. Dunn, "Cadmiums — Valuable Colours for Plastics," presented at the Canadian Society of Plastics Industry (Quebec City, 1974).

A. Eibner, *Malmaterialienkunde als Grundlage der Maltechnik* (Berlin, 1909), 128–143.

Encyclopedia of X-rays and Gamma-rays, ed. G. Clark (New York, 1963).

B. L. Engel, "A Study of the Materials and Techniques Used by Vincent van Gogh," (Ms., Intermuseum Laboratory, Oberlin, Ohio, 25 May 1975).

F. Feigl, *Spot Tests in Inorganic Analysis*, 5th ed., trans. R. E. Oesper (Amsterdam, 1958).

G. Field, "Examples and Anecdotes of Pigments. Practical Journal 1809" (Courtauld Institute of Art, University of London, 1809), catalogue FIELD/6.

G. Field, *The Rudiments of Colours and of Colouring*, rev. by R. Mallet (London, 1870), 82.

M. Fischer, *The Permanent Palette* (Mountain Lake Park, Maryland, 1930).

J. J. Fox and T. H. Bowles, *The Analysis of Pigments, Paints and Varnishes* (London, 1927), 35, 65–66, 91–95.

J. N. Friend, *An Introduction to the Chemistry of Paints* (London, 1910), 38–39.

Gay-Lussac, "Sur un nouveau Métal (le Cadmium)," *Annales de Chimie et de Physique* 8, ser. 2 (1818), 100–101.

R. J. Gettens, R. L. Feller, and W. T. Chase, "Vermilion and Cinnabar," *Studies in Conservation* 17 (1972), 58–59.

R. J. Gettens and G. L. Stout, *Painting Materials, A Short Encyclopedia* (1942; New York, 1966).

Great Exhibition of the Works of Industry of All Nations, 3 vols. (London, 1851).

J. C. Gregory and M. S. Burr, "Cadmium and its Compounds," in *A Text-Book of Inorganic Chemistry*, vol. 3, part 2, ed. J. N. Friend (London, 1926), 159.

C.-E. Guignet, *Fabrications des couleurs* (Paris, 1888), 111–112.

Handbook of Chemistry and Physics, 51st ed., ed. R. C. Weast (Cleveland, 1970–1971).

Handbook of Young Artists and Amateurs in Oil Painting (New York, 1865), 51.

R. D. Harley, *Artists' Pigments: c. 1600–1835, a Study in English Documentary Sources*, 2nd ed. (London, 1982).

N. Heaton, *Outlines of Paint Technology*, 3rd ed. (London, 1956), 136–138.

Hercules, Incorporated, *Characteristics of Hercules Imperial Pigment Colors* (Wilmington, Delaware, 1976), 7.

W. G. Huckle, G. F. Swigert, and S. E. Wiberley, "Cadmium Pigments — Structure and Composition," *Industrial and Engineering Chemistry*

Product Research and Development 5, no. 4 (December, 1966), 362–366.

W. H. Hunt, *Pre-Raphaelitism and the Pre-Raphaelite Brotherhood*, vol. 2 (New York, 1906), 455–456.

W. H. Hunt, "The Present System of Obtaining Materials in Use by Artist Painters, as Compared with that of the Old Masters," *Journal of the Society of Arts*, vol. 28 (London, 1880), 485–499.

C. S. Hurlbut, Jr., "The Wurzite-Greenockite Series," *American Mineralogist* 42 (1957), 184–190.

G. H. Hurst, *Dictionary of Chemicals and Raw Products Used in the Manufacture of Paints, Colours, Varnishes and Allied Preparations*, 2nd ed., rev. (London, 1917), 72–73.

R. Jameson, ed., "Notice of Greenockite, A New Mineral Species of the Order Blende," *Edinburgh New Philosophical Journal*, vol. 28 (Edinburgh, 1840), 390–392.

JCPDS Powder Diffraction File. JCPDS International Centre for Diffraction Data (Swarthmore, Pennsylvania, 1982).

R. M. Johnston and R. L. Feller, "Optics of Paint Films: Glazes and Chalking," *Application of Science in the Examination of Works of Art*, Proceedings of the Seminar, 7–16 September, 1965, Museum of Fine Arts (Boston, 1967), 86–95.

W. K. Kelley, "Cadmium Colors in Coatings Industry," *American Paint Journal Convention Daily* 55, no. 15 (1970), 22–23.

H. Kühn, *Die Pigmente in den Gemälden der Schack-Galerie* (Munich, 1969).

A. P. Laurie, *The Materials of the Painter's Craft* (Philadelphia, 1911), 136–137, 311.

A. P. Laurie, *The Pigments and Mediums of the Old Masters* (London, 1914), 16, 51.

M. McCann, *Artist Beware* (New York, 1979), 136–137.

R. Mayer, *A Dictionary of Art Terms and Techniques* (New York, 1975), 24–25, 272.

J. W. Mellor, *A Comprehensive Treatise on Inorganic and Theoretical Chemistry*, vol. 4 (London, 1923), 595.

Merck Index, 9th ed., ed. M. Windholz (Rahway, New Jersey, 1976).

H. E. Merwin and E. S. Larsen, "Mixtures of Amorphous Sulphur and Selenium as Immersion Media for the Determination of High Refractive Indices with the Microscope," *The American Journal of Science*, ed. Edward S. Dana, 4th series, vol. 34 (1912), 42–47.

W. O. Milligan, "The Color and Crystal Structure of Precipitated Cadmium Sulfide," *Journal of Physical Chemistry* 38 (1934), 797–800.

E. L. Moore, "Cadmium/Mercury Sulfides," *Pigment Handbook*, vol. 1, ed. T. C. Patton (New York, 1973), 395–399.

A. Munkert, *Die Normalfarben* (Stuttgart, 1905), 47–53.

E. A. Parnell, *A Practical Treatise on Dyeing and Calico-Printing* (New York, 1846), 429.

E. J. Parry and J. H. Coste, *The Chemistry of Pigments* (London, 1902), 180.

T. C. Patton, "Cadmium Sulfide Pigments," *Pigment Handbook*, vol. 1 (New York, 1973), 389–393.

J. Plesters, "Cross-sections and Chemical Analysis of Paint Samples," *Studies in Conservation* 2 (1956), 148–149.

Raw Materials Index, Pigment section (Washington, 1975).

J. S. Remington and W. Francis, *Pigments: Their Manufacture, Properties, and Use* (London, 1954), 117–120, 199.

E. L. Richter and H. Härlin, "The Pigments of the Swiss Nineteenth-century Painter Arnold Böcklin," *Studies in Conservation* 19 (1974), 83–87.

S. D. Ross, *Inorganic Infrared and Raman Spectra* (London, 1972), 99–101.

A. N. Saidel, V. K. Prokofiev, and Raiski, *Tables of Spectrum Lines* (Berlin, 1955), 539–545.

B. J. Skinner and P. M. Bethke, "The Relationship between Unit-cell Edges and Composition of Synthetic Wurtzites," *American Mineralogist* 46 (1961), 1382–1398.

F. G. Smith, "Lattice Constants of Cadmium Sulfide," *American Mineralogist* 40 (1955), 696–697.

H. C. Standage, *The Artists' Manual of Pigments* (Philadelphia, 1886), 35–36, 45–46, 73–75.

F. Stromeyer, "New Details Respecting Cadmium," *Annals of Philosophy*, trans. from *Annalen der Physik* 14 (Leipzig, 1819), 269–274.

Tabarant, "Couleurs," *Le bulletin de la vie artistique*, ed. Bernheim-Jeune (Paris, 1923), 287–290.

C. J. A. Taylor and S. Marks, eds., *Pigments, Dyestuffs and Lakes*, Oil and Colour Chemists' Association, Paint Technology Manuals, pt. 6 (New York, 1966), 192–193, 324–325.

M. Toch, *The Chemistry and Technology of Paints*, 3rd ed. (New York, 1925), 73.

M. Toch, *Materials for Permanent Painting* (New York, 1911), 102.

M. Toch, *Paint, Paintings and Restoration*, 2nd ed. (New York, 1946), 100.

R. J. Traill and R. W. Boyle, "Hawleyite Isometric Cadmium Sulfide, A New Mineral," *The American Mineralogist* 40 (1955), 555–559.

F. Ulrich and W. Zachariasen, "Uber die Kristallstruktur des α und β CdS, sowie des Wurtzits," *Zeitschrift für Kristallographie* 62 (1925), 260–273.

R. Wagner, *A Handbook of Chemical Technology*, 9th ed. trans. and ed. W. Crookes (London, 1872), 82.

H. Wagner, in *Pigmente*, 3rd ed., ed. H. Kittel (Stuttgart, 1960), 196–201.

Walker Art Gallery, *William Holman Hunt* (Liverpool, March-April 1969; Victoria and Albert Museum, London, May-June 1969).

H. W. D. Ward, "New Cadmium Pigments," *The Chemical Trade Journal and Chemical Engineer* 80 (1927), 59–60.

F. Weber and Co., *Catalogue of Artists' Materials and Draughtsmen's Supplies* (Philadelphia, 1890).

F. W. Weber, *Artists' Pigments, Their Chemical and Physical Properties* (New York, 1923), 28–31.

M. W. Weeks and H. M. Leicester, *Discovery of the Elements*, 7th ed. (Easton, Pennsylvania, 1968), 502–509.

H. B. Weiser, *The Colloidal Salts* (New York, 1928).

C. R. Weston, "Pigment Analysis. Appendix B," in M. B. Cohn, *Wash and Gouache* (Fogg Art Museum, Cambridge, Massachusetts, 1977), 67–70.

E. W. White, G. V. Gibbs, G. G. Johnson, Jr., and G. R. Zechman, Jr., *X-ray Wavelengths and Crystal Interchange Settings for Wavelength Geared Curved Crystal Spectrometers*, 2nd ed. (University Park, Pennsylvania, 1965).

A. M. de Wild, *The Scientific Examination of Pictures*, trans. L. C. Jackson (London, 1929), 61–62.

A. N. Winchell, *Elements of Optical Mineralogy*, 2nd ed., pt. 3 (New York, 1951).

Winsor & Newton, *Notes on the Composition and Permanence of Artists' Colors Including a Description of Oils Painting Media and Varnishes* (Secaucus, New Jersey, n.d.).

G. Zerr and R. Rübencamp, *A Treatise on Colour Manufacture*, trans. C. Mayer (London, 1908), 125, 152–153.

BIBLIOGRAPHY

H. S. Clarke, "Cadmium Pigments: A Non-toxic Group of Pigments," *Proceedings Society of Plastics Engineers Annual Technical Conference New Orleans* (1979), 300–306.

D. J. Heiser, "Cadmium Pigments," *Plastics Compounding* 2, no. 5 (1979), 85.

D. J. Heiser, "Current Regulatory Status of Cadmium Pigments," *Color Research and Application* 6, no. 4 (1981), 237.

F. Nakahara, "Effects of Ammonium Salts on the Change of Coprecipitates of Cadmium Sulfide and Mercury (II) Sulfide during Heat Treatments — Studies on Mercury Cadmium Sulfide Pigments VII, *Journal of the Japan Society of Colour Material (Shikizai Kyokaishi)* 54, no. 3 (1981), 123–130.

F. Nakahara, "Effects of Ammonium Salts on the Change of Coprecipitates of Cadmium Sulfide and Mercury (II) Sulfide during Heat Treatments — Studies on Mercury Cadmium Sulfide Pitments VIII, *Journal of the Japan Society of Colour Material (Shikizai Kyokaishi)* 54, no. 4 (1981), 188–192.

Red Lead and Minium

ELISABETH WEST FITZHUGH

NOMENCLATURE IN VARIOUS LANGUAGES

Artificial Pigment	*Mineral*
English: red lead	minimum
German: Mennige or Bleimennige	Mennige
Russian: свинцовци	сурик
Chinese: ch'ien tan or tan-fen	
French: minium or minium de plomb	minium
Italian: minio or minio di piombo	minio
Spanish: minio	minio or azarcon, from Arabic *zarqūn* (Levey, 1962)
Japanese: entan	

1.0 INTRODUCTION

Red lead, CI Pigment Red 105, No. 77518 (*Colour Index*, 1971), was one of the earliest pigments artificially prepared and is still in use today. The naturally occurring mineral minium (from the Latin name *minium*) was used as a pigment only at an early date, if at all.

1.1 Brief Definition of Pigment

Red lead is lead tetroxide (Pb_3O_4), the same chemically and crystallographically as the mineral minium.

ELISABETH WEST FITZHUGH, Conservation Scientist, Freer Gallery of Art, Smithsonian Institution, Washington, D.C.

1.2 Current Terminology

Red lead is the name in common use today and according to the unabridged Oxford English Dictionary, was known at least as early as about 1450. The term minium was used for the artificial material from classical times and is still occasionally employed. See 2.1 for other terms that are currently in use.

2.0 HISTORY

2.1 Archaic, Obsolete, and Misrepresentative Names

Pliny the Elder, the Roman writer of the first century A.D., applied the name *minium* to cinnabar, the naturally occurring form of the pigment vermilion (mercuric sulfide) and he called red lead *minium secondarium* (Bailey, 1929). Cinnabar was often adulterated with red lead and the name minium was used for the mixture. Gradually, however, it came to be applied to red lead alone as recorded by Agricola in *De re metallica* (Hoover & Hoover, 1950). Both Pliny and Dioscorides, the Greek writer of the same date, used the term false sandarach for red lead because of its resemblance in color to sandarach. The latter was their term for realgar, the naturally occurring red arsenic sulfide. Vitruvius, the Roman author of the first century B.C., referred to red lead as sandarach (Hoover & Hoover, 1950). Other terms applying to the pigment or to mixtures containing it were also used by classical writers: *sandyx* was a mixture of red lead and red ochre; *syricum* or *siricum* could be synonyms for red lead or litharge (Smith & Hawthorne, 1974), for sandyx, or for a mixture of sandyx and synopis (a red iron oxide).

Minien is a term for red lead found in a fifteenth-century Strasbourg manuscript (Borradaile, 1966). *Menning* was used by Agricola in the sixteenth century. Paris red and Saturn red (*Pariserrot, saturnine, rouge de Saturn, rosse di Saturno*) were sometimes used for red lead but also appear to be synonyms for orange lead (see below) (Bearn, 1923, p. 114; Dunn, 1973). De Massoul lists Saturnine red as an expensive variety of minium purified by repeated washing with water (Harley, 1982).

The pigment is also called red lead oxide and its equivalent in various languages: German, *Saturnrot Bleioxid*; French, *plomb oxidé rouge*; Italian, *piombo ossidato rosso*. The designation *bleizinnober* has also been used. In both French and Russian *minium de fer* and *iron minium* (Malevanny, 1968) refer not to red lead but to ochre. The mineral crocoite ($PbCrO_4$) is sometimes referred to as red ore of lead.

Orange lead or orange mineral (in French *mine orange*) can be a synonym for red lead, but more commonly this term refers to the finest, or purest, red pigment prepared from lead white and containing less free litharge than ordinary varieties (Weber, 1923). D. V. Thompson (1936) defines minium as orange lead and says that the medieval pigment is paler and more orange than the modern material. Oguchi (1969) states that when Japanese red lead is prepared from lead white it is called orange mineral or orange lead. The name orange mineral was first used in the United States in the nineteenth century and the pigment was described as being made from lead white (Candee, 1967). Orange lead is said to have a finer texture and lighter color than minium and has been used with organic reds in printing inks to give opacity (Champetier & Rabaté, 1956). It has been used with dyes and other materials, including lead acetate and calcium sulfate, to make *rouge universel* or *rouge vermilion*, and it has served as a base for aniline colors (Bearn, 1923, p. 117).

2.2 History of Use

A Chinese text of the early fifth century B.C. refers "unambiguously to the manufacture of lead compounds from the metal" (Schafer, 1955). The term *ch'ien tan*, literally "lead cinnabar," was used in the Han Dynasty (second century B.C.–second century A.D.), indicating the pigment was prepared artificially from lead (Schafer, 1955). Thus it is certain that the manufactured product was known at this date. There appears to be no evidence that the natural mineral was ever used in China.

Vitruvius says that red lead was first produced from lead white accidentally in a fire and notes: "A much better result is obtained in this way than from the natural substance which is produced from the mines" (Forbes, 1955). This seems to indicate that the artificial material replaced the naturally occurring mineral in the classical world at an early date.

The use of chemically prepared pigments such as Pb_3O_4 does not appear to have preceded the introduction of metals (Multhauf, 1967). The earliest use of red lead thus probably dates to the time of the development of lead metallurgy in both China and the Near East (Schafer, 1955). Mass production of metals in the Near East through ore smelting began in the fifth and fourth millenia B.C.; lead was produced extensively in Anatolia by 2300–1800 B.C. (Wertime, 1973). It has been stated that red lead was known in ancient Mesopotamia (Forbes, 1955), although direct evidence is lacking. According to R. C. Thompson (1925), the word *s-ā-s-u* on Assyrian cuneiform tablets means red lead, and the substance is described as being made by the roasting of lead white.

The earliest objects on which red lead has been found in Egypt were uncovered at the Greco-Roman site of Hawara (Forbes, 1955; Petrie, 1889). It has been identified on Roman Fayum portraits of the second to fourth centuries A.D. (Buck & Feller, 1972; Roy, 1979). It was not reported in a set of dry pigments from Pompeii although the author listed cerussa usta (burnt lead white) (Augusti, 1967). The Greek writer Theophrastus (fourth–third centuries B.C.) does not mention its preparation (Caley & Richards, 1956), but later classical writers have described the pigment. Vitruvius, whose description of the manufacture of red lead from lead white is noted above, was apparently the first to record this process (Hoover & Hoover, 1950). Both Dioscorides and Pliny describe the preparation of red lead from lead white, and Pliny also discusses the heating of unspecified lead ores to produce the pigment.

In ancient Chinese texts we find the same confusion in nomenclature between minium and cinnabar as is found in the works of classical authors. However, minium was recognized as an artificial pigment. Oguchi (1969) suggests that artificial minium was introduced into China from India or the Near East in the late Han period (third century A.D.), but it appears that some Chinese were well aware of the chemical changes necessary to produce the pigment. The Chinese alchemical writer Ko Hung was scornful of those who could not understand these changes, and in 320 A.D. he wrote, "The ignorant do not believe that minium and lead white are products of the transformation of lead, and may not know that a mule is the offspring of a donkey and a horse.... There are things that are as clear as the sky, yet men prefer to sit underneath an upturned barrel." (Needham, 1974). In the Sung dynasty (tenth–thirteenth centuries) a special government factory supplied minium to official painters and decorators (Schafer, 1955). Schafer suggests that minium was imported into India from China at this period, because the Indians called it "China flour" (*cinapishta*).

Red lead has been identified on wall paintings and paintings on silk from Tun Huang in central Asia, dating from the tenth to the eleventh centuries. It has been found on Chinese wood and stone sculpture of an earlier date, but no occurrences are presently known on later Chinese paintings, where vermilion is the common red pigment. It has been identified on wall paintings in central China of the fifth–ninth centuries A.D. (Gettens, 1938b), and of the sixteenth century (Gettens, 1938c). Further west it has been found on sixth-century Buddhist wall paintings in Afghanistan (Gettens, 1938a). The darkening of red lead on these wall paintings is discussed in 3.3.

In Japan the roasting of "black lead," in open pans, to obtain "red lead" is described in an eighth-century document, indicating that red lead was prepared from lead or its ores by heating (Takamatsu, 1879–1880). Red lead pigment, packaged in paper and preserved in the Shōsō-in, the eighth-century warehouse in Nara, has a high litharge content, varying from 70.90 to 92.20%; the conversion process from litharge was clearly inefficient at that date (Yamasaki, 1959a). (References to litharge in this chapter are to painters' litharge, yellow PbO, unless otherwise indicated. See 4.3.) The earliest known use of red lead in Japan is on the wall paintings of the Horyuji temple, dating from the seventh–eighth centuries. From that date to the present it has been used in Japan on a variety of objects, including sculpture, paintings on silk and paper, and prints.

Red lead has been identified on Indian fifteenth- to nineteenth-century miniature paintings and on eighteenth-century Indian wall paintings and sculpture. It has also been reported on Nepalese-Tibetan *thang-ka* paintings of the late eighteenth or early nineteenth century. The pigment has been found on Persian fourteenth- to seventeenth-century miniatures (see fig. 1) and on Turkish miniatures of the sixteenth and seventeenth centuries (see 6.0 for details). The preparation of red lead from lead white has been described by the chief librarian of Shah Abbas, the sixteenth-century Persian emperor (Welch, 1981). Orange lead, prepared from lead white, is said to have been used by sixteenth-century Mughal painters, and good quality minium is still manufactured in Jaipur (Chandra, 1949). Wulff (1966) illustrates a modern Persian lead oxidizing furnace used to prepare lead oxides for ceramic glazes. These furnaces could equally well be used to make lead oxide pigments. There are three sets of modern Persian pigments in the Freer Gallery Laboratory collection. They are used in present-day miniature painting and are supposedly "traditional" pigments. All three sets include red lead, labelled *surandj*. The pigment in two of the sets is described as Russian made. However, no reference has been found to the use of the pigment in Russia; on Scythian sites in South Russia various iron oxide pigments have been extensively described, but no red lead was found (Malevanny, 1968).

The Persian alchemist and physician of ninth- to tenth-century Baghdad, al-Razi, recognized the two oxides of lead, minium and litharge, as artificially prepared substances (Multhauf, 1967). An eleventh-century Arabic manuscript on bookbinding lists red lead as an ingredient of various inks and dyes (Levey, 1962). The pigment has been found on a fourteenth-century Arabic manuscript in the collection of the Freer Gallery. In the Mappae Clavicula, a tenth- and twelfth-century medieval manuscript which records the technology of earlier periods, preparation of red lead is achieved by treating metallic lead with vinegar and then heating the lead white thus formed (Smith & Hawthorne, 1974). Theophilus, the eleventh-century medieval scholar, describes roasting lead white while stirring to produce minium on a scale appropriate to a small factory (Hawthorne & Smith, 1963). A fifteenth-century German treatise on the dyeing of textiles describes the manufacture

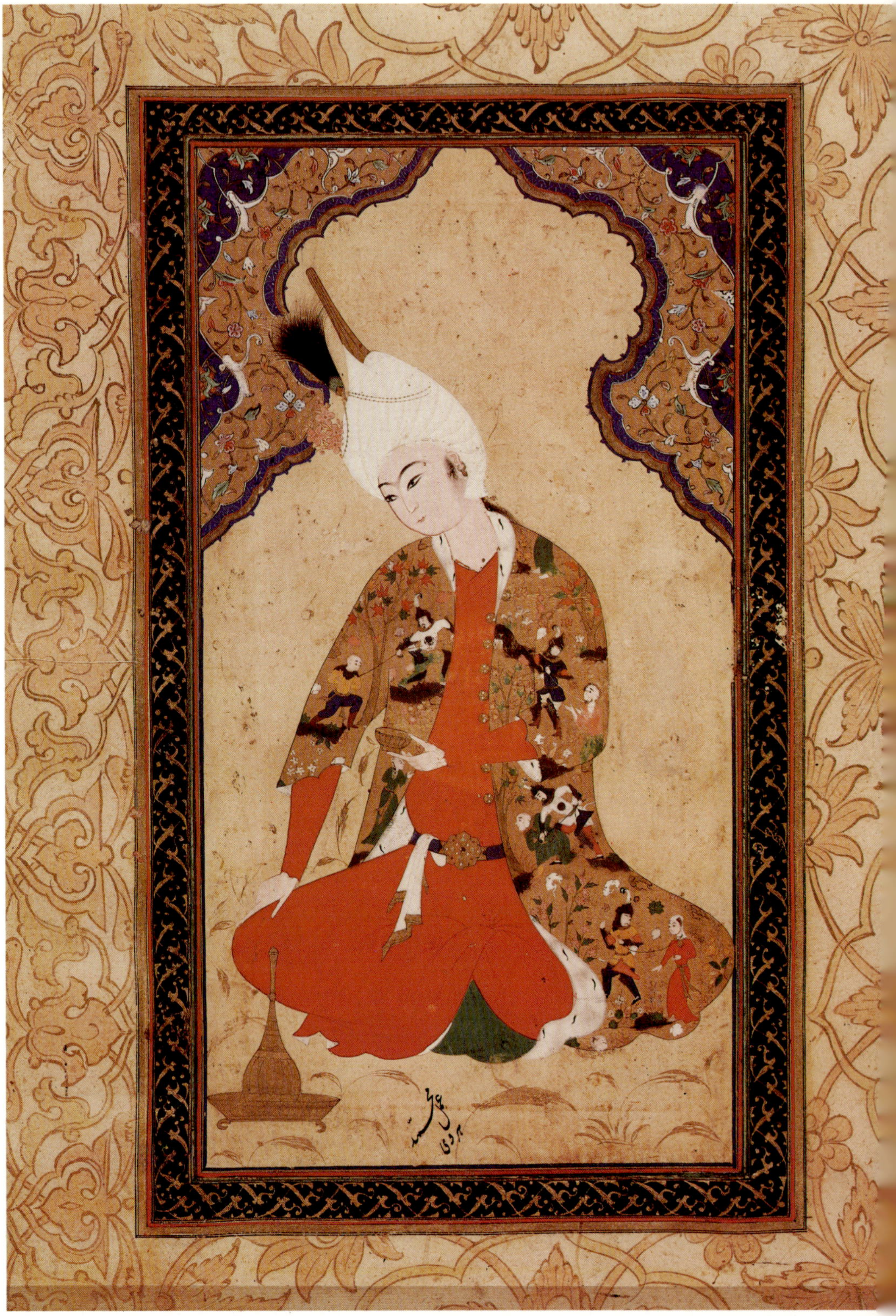

of minium by simple melting and oxidation of lead (Radosavljevic, 1972). Agricola in the sixteenth century knew that red lead was an artificial product (Hoover & Hoover, 1950). Slavic, Armenian, and Azerbaijan recipes of the fifteenth to eighteenth centuries involve the making of the pigment from lead metal, which is first converted to lead white by treatment with vinegar and sal ammoniac (Radosavljevic, 1972).

Red lead was used widely in medieval manuscripts, both alone and with vermilion (D. V. Thompson, 1936); the word miniature comes from the Latin verb *miniare*, to write with minium (Loumyer, 1908). Laurie (1910) says it was used on illuminated manuscripts and in miniature painting in Europe from the eighth century on, and that it was used in Byzantine manuscripts beginning in the seventh century. The pigment has been found on polychrome sculpture and panel paintings from the twelfth, thirteenth, and fifteenth centuries, and occasionally on mural paintings. It has been reported on English sixteenth-century miniature paintings, where it was mixed with lead white, vermilion, and ochre to form the flesh tones (Murrell, 1976).

Red lead is mentioned frequently in the literature on easel painting, but was not highly recommended (Harley, 1982). It is not a common pigment on European easel paintings, but it has been found on works dating from 1300 to 1900 in the Doerner Institute (Kühn, 1973). It was identified in the bole of a fifteenth-century Italian painting (see fig. 2) and in a scattering of other paintings up through the nineteenth century. Its occurrence on a painting by Renoir from 1876 (Butler, 1973) may be only as an adulterant since it is mixed with vermilion. Red lead was not popular with artists in the nineteenth century, but it has been found in three collections of nineteenth-century pigments: a set of painting materials owned by J. M. W. Turner (N. W. Hanson, 1954), a German set of pigments (Richter & Härlin, 1974a), and a collection of pigments used by the Swiss painter Arnold Böcklin (Richter & Härlin, 1974b). It has also been identified on one of Böcklin's paintings (Kühn, 1969). Red lead was important as a decorators' paint in England and was made on an industrial scale in the seventeenth century directly from lead. In the eighteenth century it was manufactured both from lead and from litharge (Harley, 1982).

Fig. 1. Persian miniature, *Portrait of a Prince*, Muhammad Harawī, Safavid period, mid-sixteenth century. Pigment of orange-red robe is red lead. Freer Gallery of Art, Washington, 37.8.

Red lead, imported from England, was available to painters in America from 1738 on. The Wetherill Company in Philadelphia first manufactured the red pigment in 1809, and the orange in 1811 (Candee, 1967). It is of interest that the price of the superior product, orange lead, was about twice that of the red (Candee, 1967). As in England, the pigment was probably used more as a decorators' paint than an artists' pigment, although there is little evidence on this point. It has been found on six out of one hundred American paintings of the eighteenth-century Hudson Valley school (E. Quandt, Private communication). Red lead is mentioned as an artists' material in late eighteenth- to early nineteenth-century Quebec (Levenson, 1983). Like lead white, it has been recommended for protecting the back of a canvas (Mayer, 1970).

Red lead has been used over vermilion to give a decorative effect. Cennini (D. V. Thompson, 1960) describes its use as hatching, or as decoration over vermilion to give the effect of silk cloth. Interestingly enough, a similar technique is quite common on Japanese paintings of the Ukiyo-e school (seventeenth–nineteenth centuries).

2.3 Terminal Date

Although still produced for a number of purposes, red lead is not widely used today as an artistic or decorative pigment. Lead tetroxide finds important applications in pottery glazes and in glass manufacture (*Colour Index*, 1971). Perhaps its best-known use is as an ingredient in anticorrosive paint on iron and steel. The first patent involving the addition of red lead to a protective coating for metal was registered in 1852 (Smith, 1978). For many years the pigment's presence was common in the priming coat used for this purpose. It would be found either alone or, more recently, mixed with other pigments such as siliceous iron oxide or magnesium silicate. Today increased use is made of other corrosion-inhibiting pigments such as basic lead silico-chromate, zinc yellow (zinc chromate) and basic zinc chromate (Fuller, 1968; Eikhoff, 1973); see chapter on Chrome Yellow and other Chromate Pigments. In 1973 the total consumption of zinc chromate was

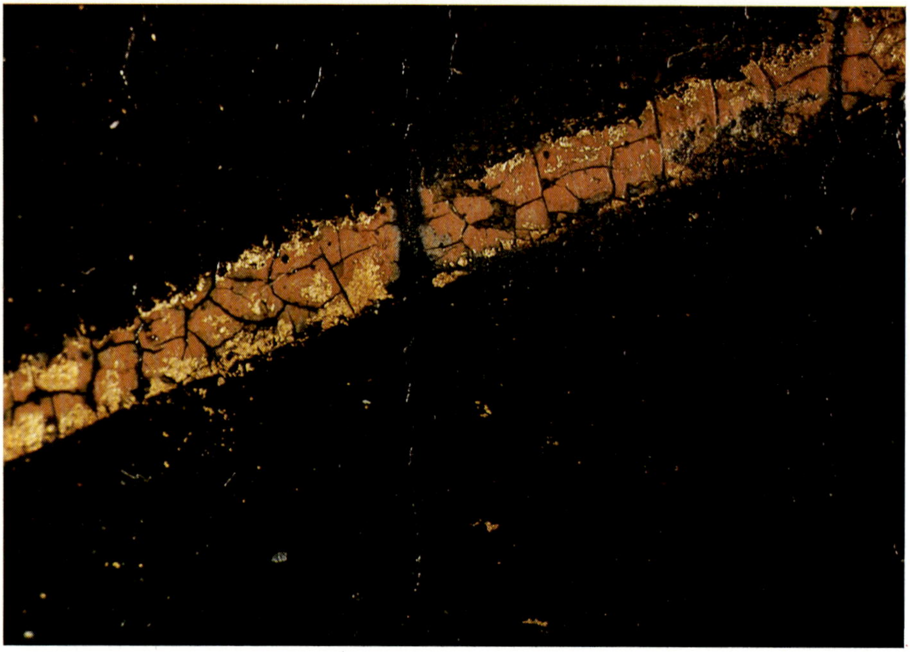

Fig. 2. Cross section, approximately 15x. Bole containing red lead and calcium sulfate, from *Madonna and Child Enthroned with Four Saints*, Andrea di Giusto, first half of fifteenth century. Los Angeles County Museum of Art, A.5805,47.1. Courtesy of Norman Muller.

8,000 tons and of red lead, 7,000 tons (*Kline* . . . , 1975).

3.0 GENERAL PIGMENT PROPERTIES

A dense, fine-textured pigment, red lead contributes excellent body and good hiding power to drying oil paints, although its tinting strength is low. The pigment has a tendency to darken in watercolor and wall paintings, but in oil mediums it is quite stable. Its anticorrosive properties on iron are of major importance in the commercial market (Dunn, 1946). Red lead exerts a powerful action on drying oils, especially if there is a certain litharge content (Taylor & Marks, 1966). Oil-based paints possess good adhesion, low moisture absorption, and low permeability to water vapor and oxygen (Dunn, 1973).

Many of the good physical and mechanical properties of red lead in linseed oil can be attributed to the formation of lead soaps. These soaps are formed by chemical combination of the pigment with free fatty acids, present in the linseed oil. They produce crystalline aggregates that mechanically reinforce the paint film so it is tough and elastic and adheres well to metal surfaces. As the linseed oil ages, more fatty acids are freed to continue this reaction. Photomicrographs of these lead soaps have been published (Dunn, 1945; Perera & Heertjes, 1971).

Like most lead-containing compounds, the pigment is poisonous. This fact was early known; Dioscorides noted that red lead was more poisonous than vermilion and described measures taken to prevent poisoning. Its toxicity and the possibility of its darkening in some circumstances (see 3.3) are regularly cited as disadvantages of the pigment's use.

Red lead, acting as a repellent to silverfish, has recently been identified on orange end papers of Chinese books dating to as early as the fourteenth century (Reese, 1978).

3.1 Color and Spectral Reflectance

Red lead is usually more orange than red. Slight differences in color among the various types can be related to the amount of unchanged litharge present and to particle size. The pigment is No. 77578, pigment red 105 in the *Colour Index* (1971).

Curves of red lead's spectral reflectance have been published (Barnes, 1939; Oguchi, 1969;

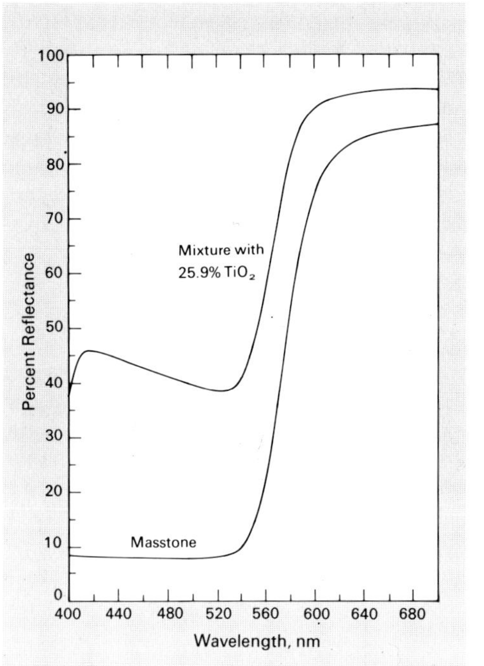

Fig. 3. Spectral reflectance of red lead in oil.

Dunn, 1973, 1975), and the dominant wavelength, purity, and luminosity have been reported (Yamasaki, 1959a). Spectrophotometric curves were also published by Kenjo and Toishi (1964) in their study on the darkening of red lead. Fig. 3 gives typical curves. The masstone of Munsell Notation, 0.8 YR/6.40/13.36, may be designated as a vivid reddish orange (Kelly & Judd, 1976), while the mixture with titanium white (at 25.9% TiO_2 by weight) may be properly designated as a light yellowish pink. Dunn (1973) states that the fumed type is not as deep a red as the furnace type, but instead is more orange. This variation can be seen from the curves that accompany Dunn's remarks.

Red lead absorbs strongly in the ultraviolet from 200 to 500 nm and reflects in the infrared region. Dunn (1973, 1975) has published spectral reflectance curves of red lead and photographs of a dozen lead pigments in ultraviolet, visible, and infrared.

3.2 Hiding Power and Tinting Strength

Because red lead has a high refractive index, N(Li) = 2.42, it has good hiding power and provides relatively opaque paints in oil. The tinting strength is good (Wehlte, 1975).

The particle size varies greatly. The fume type, probably the finest modern red lead, has a relatively high percentage of particles under 1.0 μm in diameter and no particles larger than 8–10 μm (Dunn, 1973). A sample of Japanese pigment dating to the eighth century A.D., probably prepared by heating litharge, had particles ranging from 5–10 μm in diameter (Yamasaki, 1959a). The four varieties of red lead shown in the scanning electron micrographs (fig. 4) vary in particle size from 0.2 to 5 μm. Dunn (1975) gives a table of the hiding power of various lead pigments showing furnace red lead to be 55, fumed red lead 70, and lead white 20–25 ft^2/lb.

3.3 Permanence

The unsuitability of red lead as a pigment on frescoes and in watercolor has been noted. In the early fifteenth century, Cennino Cennini (D. V. Thompson, 1960) mentioned that red lead on wall paintings turned black with exposure to air. Indeed, numerous examples of darkened pigment can be cited.

Dark red lead has been reported on several medieval manuscripts. Laurie (1910) noted its occurrence on the eighth-century Irish Lindisfarne Gospels. In a seventeenth-century English color catalogue in the London library the color sample of red lead had reportedly gone black (Harley, 1982). Whether this is due to the formation of lead sulfide or to another cause is not known. The pigment is common on Japanese paintings of the Ukiyo-e school (seventeenth–nineteenth centuries). On such works in the Freer Gallery of Art, Washington, the orange color is usually retained, probably because the paintings were originally mounted as scrolls and were therefore rolled up and somewhat protected. However a few occurrences of brown red lead have been noted on these paintings (see fig. 5). Darkened red lead is also present on an Indian fifteenth-century miniature (FGA 59.1). No black lead dioxide could be identified in samples from these paintings either in microscope slides or by x-ray diffraction.

Some examination of darkened red lead on wall paintings has yielded more concrete results. On wall paintings from three widely scattered sites — in central China, Kizil in Chinese Turkestan, and in Afghanistan — Gettens (1938a, b, c) noted that red lead had changed to a color he described as "chocolate brown," and suggested that this was due to the formation of brown lead dioxide. This suggestion has been

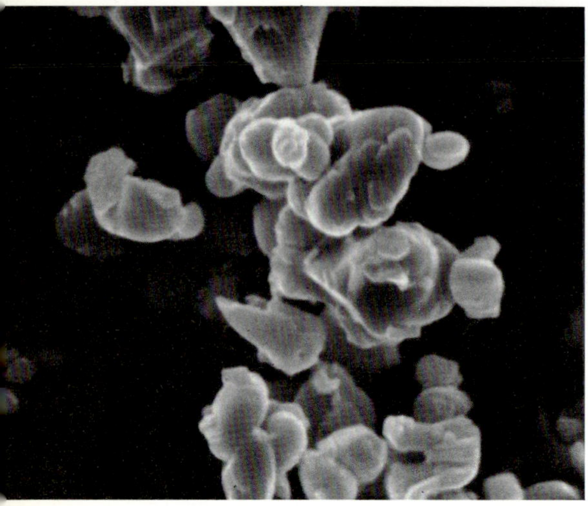

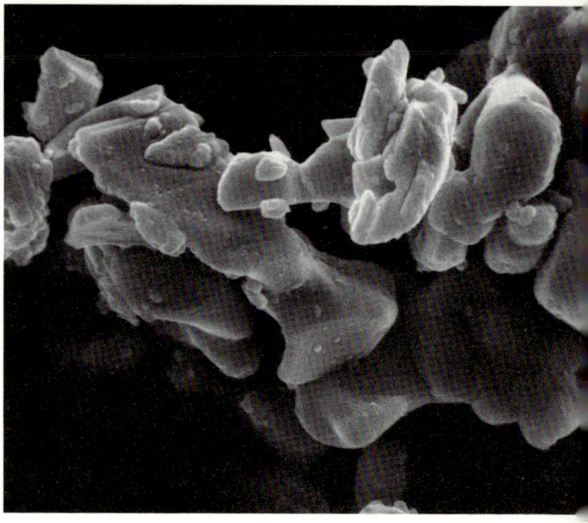

Fig. 4. Scanning electron micrographs of red lead, courtesy of Walter R. Brown, National Museum of Natural History, and John Winter, Freer Gallery of Art, Smithsonian Institution, Washington. Captions by John Winter.

A. "Furnace" red lead, National Lead Co., 1936. 7500x. Rather irregular particles, no well-defined habit, edges slightly rounded. Mostly in size range 0.5–3 μm. Except for those of one or two particles at the top, the surfaces do not appear to be fractured. Aggregation only moderate.

B. "Fumed" red lead, National Lead Co., 1936. 7500x. Characteristics of individual particles similar to those in A, except that particles are smaller, 0.3–1 μm. However, this characteristic is partly offset by greater aggregation. Aggregates also appear "tighter," with individual particles being physically joined in several places, rather than simply adhering together.

C. Yashiro No. 8, *tan.* 7500x. Particles are larger than those in A and B, mostly 2–5 μm with only a few smaller. They are irregular and angular particles with rounded edges and little evidence of crushing or fracturing. Many particles are joined to form larger ones (perhaps crystallographic twinning). Some particles perhaps show more regular habit than those in A or B.

D. Turner palette No. 21. 7500x. Wider range of particle sizes than in A, or B, or C, 0.2–3 μm. Except for greater number of larger particles, looks very much like B. Particles irregular and angular, with rounded edges, quite a lot of aggregation. No particular evidence of crushing or fracturing.

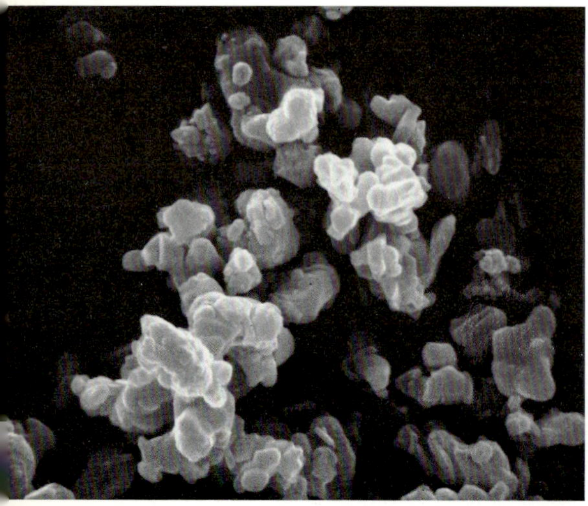

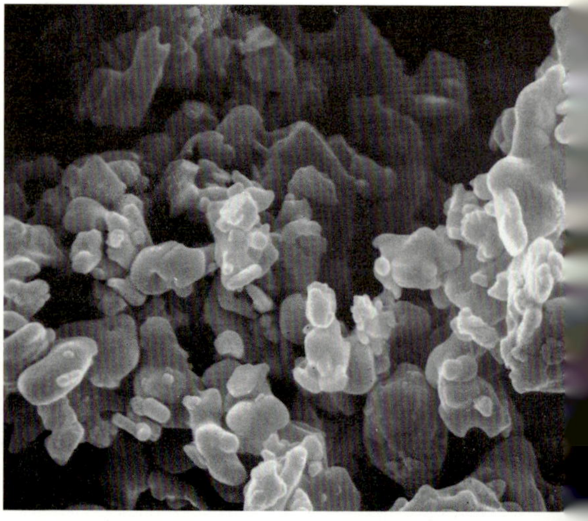

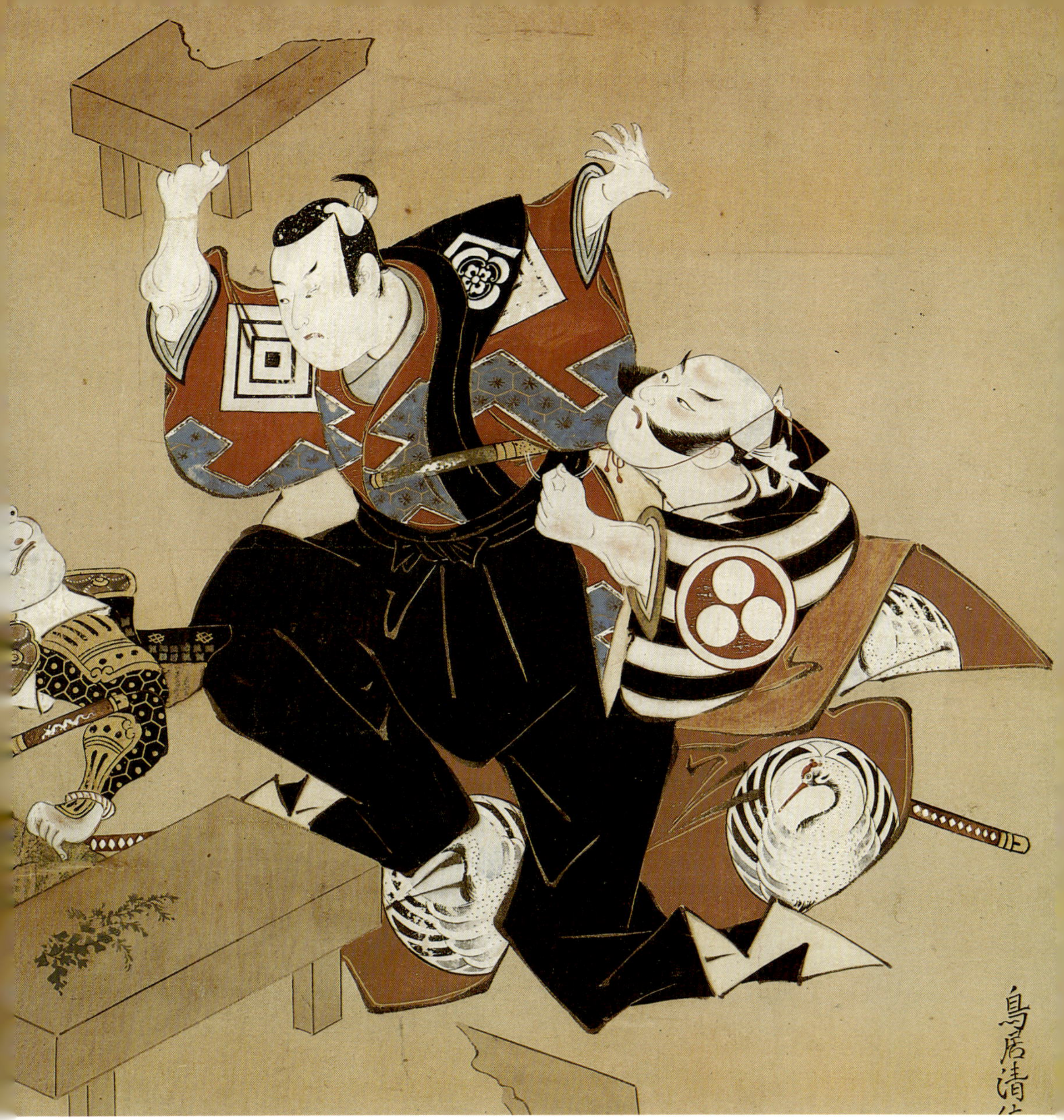

Fig. 5. *Three Actors Playing a Scene*, Torii Kiyonobu I (1664–1729), Japanese, detail. The red-brown robe of actor, lower right, is darkened red lead. Freer Gallery of Art, Washington, 98.14.

confirmed by recent examination of another wall painting fragment from Kizil. As identified by x-ray powder diffraction the darkened red lead on this fragment is clearly a mixture of red lead and black lead dioxide (see table 4 and fig. 17).

Red lead, partially or completely transformed to black lead dioxide, has also been observed on several Swiss wall paintings of the thirteenth to the seventeenth centuries. PbO_2 was identified by x-ray diffraction (H. Kühn, private communication, 1976). Some black-colored particles, with red lead, have been observed microscopically in samples of darkened pigment from early medieval wall paintings from Eastern Turkestan, although PbO_2 could not be identified by x-ray diffraction (Riederer, 1977). In Japan, wall paintings of the seventh and tenth centuries show red lead that has altered to a brown color (Yamasaki, 1959b).

In films prepared from a glue medium darkening can be produced artificially by exposure to light and high humidity (Gettens & Stout, 1966). Test panels with a kaolin priming (the ground used in Chinese wall paintings) on a hardboard support were exposed to 90% relative humidity at 40°C; within five days the red color had become "as brown as the specimens from the Tun Huang caves" (Gettens, unpublished). Previously it had been thought that the chocolate brown flesh color of the Buddhas on central Asian wall paintings was original (Warner, 1938). Weber (1923) states that dry powdered red lead turns black on exposure to light. He suggests, however, that this is not a reaction produced by oxidation since it will occur in an evacuated vial exposed to light. Kenjo and Toishi (1964) tested Pb_3O_4 in atmospheres of oxygen, carbon dioxide, and nitrogen, in darkness and ultraviolet light, with and without glue medium, measuring the resulting changes by spectrophotometer. Slightly more darkening was evident in oxygen than in the other gases; slight lightening was noted in carbon dioxide and nitrogen. The presence of glue seemed to protect the red lead in oxygen and carbon dioxide, but not in nitrogen. Litharge was likewise found to be protected by the glue medium. *Nikawa*, the animal glue used, is the binder found in Japanese paintings.

The darkening of lead tetroxide may not be a simple matter of alteration to PbO_2. As noted in 4.3, the chemistry of the various lead oxides is complex. The black PbO_2 itself is not particularly stable; it decomposes slowly due to the action of light and is decomposed by low heat — α-PbO_2 starts decomposing at 270°C, β-PbO_2 at 400°C (Pascal, 1963). If PbO_2, a strong oxidizing agent, forms, it may react with the medium or other organic materials present. The mechanism of the formation of this black discoloration is still incompletely understood.

Various authors have pointed out that red lead in linseed oil medium can alter to a lighter color. Exposure to the elements — sunlight, rain, and atmospheric CO_2 — can cause the formation of the basic lead carbonate, lead white, to give a chalky surface (Eikhoff, 1973). Augusti (1965) has suggested that the red pigment, although listed, was not among the specimens found at Pompeii because it had altered to lead carbonate. Whitened red lead has been reported on paintings in Japanese sixteenth-century temples (Emoto, 1974).

Another defect of the pigment is its susceptibility to attack by hydrogen sulfide or by sulfide-containing pigments, leading to the formation of black lead sulfide (PbS).

3.4 Compatibility

The incompatibility of red lead and orpiment (As_2S_3) was noted by medieval writers (Smith & Hawthorne, 1974). Nonetheless, in oil and in synthetic resin vehicles the pigment is considered to be relatively safe from darkening. Wehlte (1975) does not recommend its use with sulfur-containing pigments such as cadmium sulfide, vermilion, or ultramarine.

3.5 Chemical Properties

When heated strongly, red lead decomposes to form litharge (PbO) (Parkes, 1961). When heated gently in air, it changes to a dark reddish brown and then violet, but its red color is regained on cooling (Bearn, 1923, p. 115). The mineral minium has a fusibility of 1 (i.e., it will fuse in a candle flame at about 525°C), and it gives off oxygen in a closed tube (Palache et al., 1944).

Red lead is soluble in dilute acids. After treatment with dilute nitric acid and rapid evaporation, characteristic colorless dendritic crystals of lead nitrate form (see figs. 6A and 6B). Sometimes brown lead dioxide forms with the lead nitrate; de Wild (1929) points out that this is most readily observed by reflected light. Application of concentrated nitric acid will also cause formation of brown PbO_2 (Plesters, 1956). After slow evaporation of the nitric acid solution, lead nitrate may be observed in the form of highly refracting octahedra (figs. 6A and 6C). It has been noted that the presence of a reducing agent like sugar or alcohol, with or following the addition of nitric acid, will cause the brown lead dioxide to dissolve (Champetier & Rabaté, 1956), which can constitute a confirmatory test. The addition of oxalic acid to the nitric acid solution will also dissolve the PbO_2. While red lead is soluble in acetic acid, de Wild (1929) says that PbO_2 will form when the pigment is treated with acetic acid and moderate heating.

Dilute (3N) HCl causes simultaneous formation of white $PbCl_2$ and liberation of Cl_2. It has been suggested (Parkes, 1961) that lead dioxide may form intermediately, a reaction that will also proceed with concentrated HCl. When red lead is dissolved in dilute HNO_3, and dilute HCl is then added, colorless or white crystals of lead chloride can be observed in the form of irregular prisms, feathery dendritic masses, or most characteristically, tong- or cross-shaped crystals, like X's with slightly curving arms (Chamot & Mason, 1940; see fig. 7). Dilute

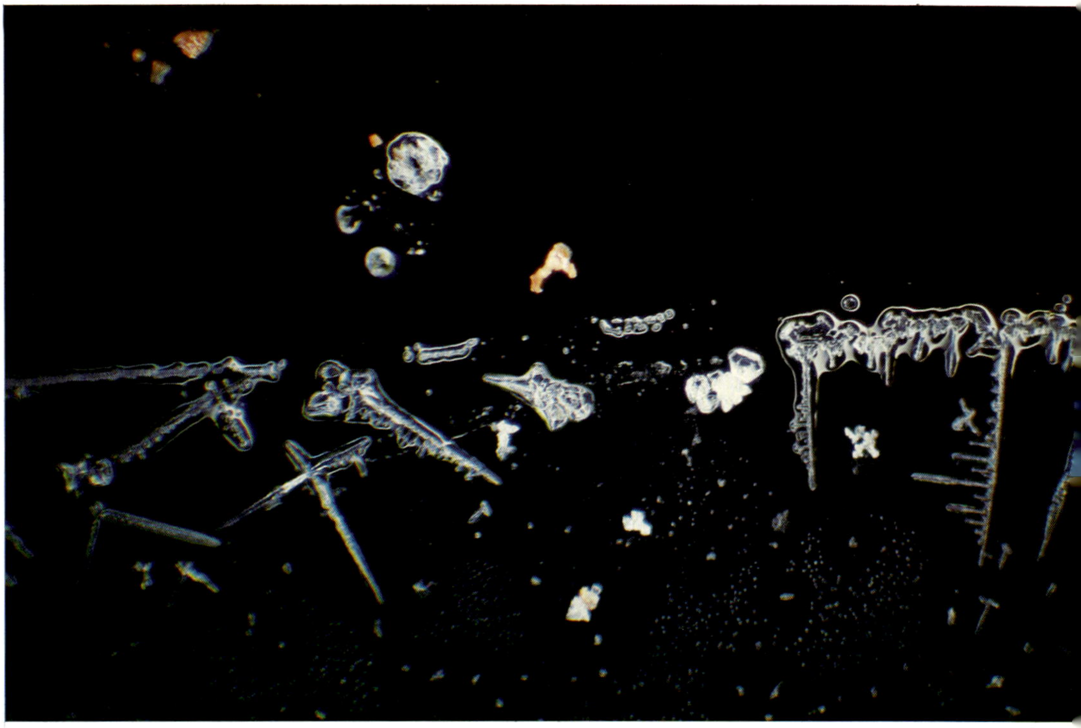

Fig. 6. Microchemical tests for lead.
A. Test for red lead with dilute nitric acid. Both dendritic and octahedral lead nitrate can be seen. Test was made on several particles of red lead from a Buddhist stone sculpture, sixth century, Freer Gallery of Art, Washington, 13.27. One large particle and one small one remain; the largest dimension of the large particle is about 0.07 mm. Dry preparation by reflected light. 125x.

B. Lead nitrate, dendritic, by transmitted light. Dry mount. 60x.

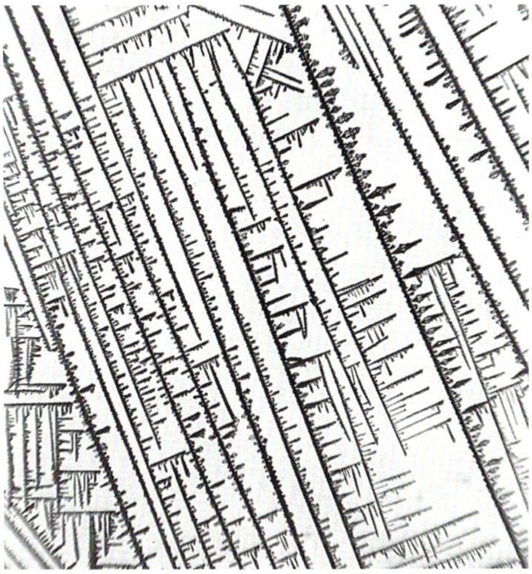

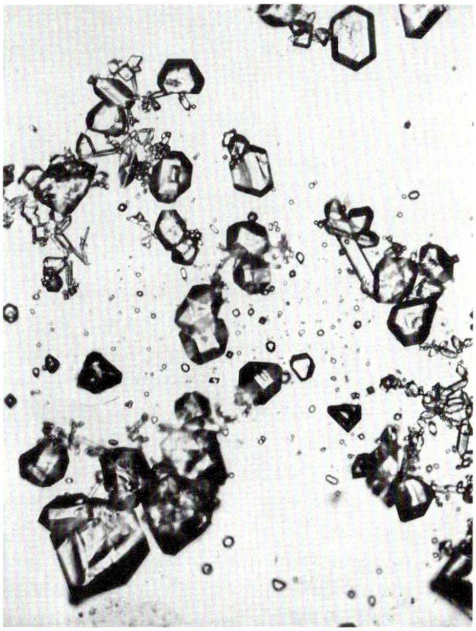

C. Lead nitrate, octahedral, by transmitted light. Wet mount without cover glass. 155x. Photo W. T. Chase. Figures 6B and C reprinted by permission from *Studies in Conservation* (Gettens, Kühn & Chase, 1967).

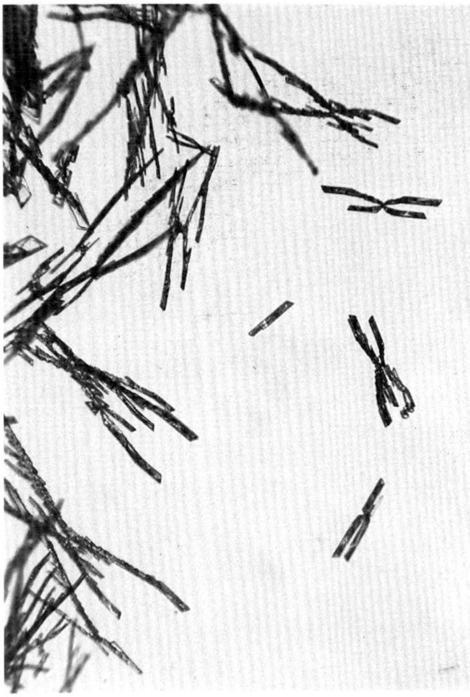

Fig. 7. Lead chloride, by transmitted light. Wet mount without cover glass. 60x. Reprinted by permission, from *Studies in Conservation* (Gettens, Kühn & Chase, 1967).

H_2SO_4 causes the precipitation of heavy granular $PbSO_4$.

Red lead is not attacked by sodium hydroxide (Champetier & Rabaté, 1956).

3.6 Oil Absorption and Grinding Qualities

The oil absorption of red lead pigments tends to be low. The *Raw Materials Index* (1975) indicates that typical oil absorption is 6 to 7.5%. Although red lead mixes well with oil, paints made with the 85% grade must be used within a few hours of mixing; otherwise, the paint will thicken owing to soap formation to such an extent as to become unusable (Mattiello, 1942).

The measured density of artificial lead tetroxide is 8.9–9.2, compared with a calculated density of 8.925 (Roberts et al., 1974).

3.7 Toxicity

Because of its lead content, the pigment has a high toxicity (Wehlte, 1975).

4.0 COMPOSITION

4.1 Chemical Composition

Red lead is lead tetroxide, Pb_3O_4. It belongs to the tetragonal crystal system. Its structure consists of chains of $Pb^{IV}O_6$ octahedra sharing opposite edges. These chains are linked by the Pb^{II} atoms, each with a pyramidal arrangement of three O atoms (Wells, 1975; see fig. 8).

Fig. 8. Crystal structure of Pb_3O_4. a. Presentation in three-dimensional space. b. Plan. From A. F. Wells, 1975.

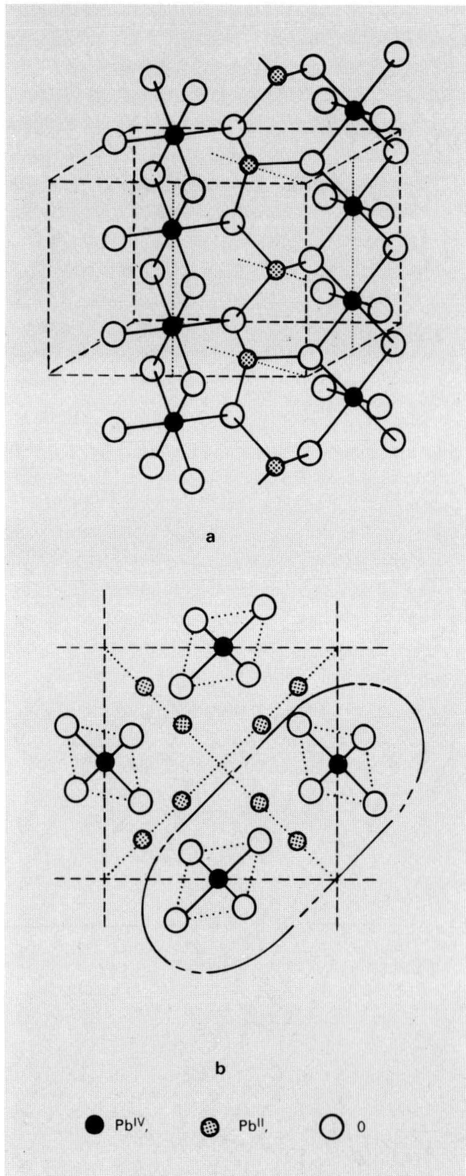

Dunn (1973) states that the pigment is presently sold in grades of 98, 97, 95 and 85% Pb_3O_4. The balance is principally litharge, the presence of which is considered desirable for many uses. There are two major types of red lead available in these ranges of composition: furnace and fume (see 4.3 Preparation). The latter tends to provide finer particle sizes, as indicated in the particle-size distribution curve published by Dunn (1973). A grade known as "orange mineral," made by the old method of roasting lead white, is still available although no longer much in demand. In the early part of the twentieth century, most of the red lead was the 85% grade, which because of its reactivity with the vehicle had to be mixed on the site. With the high cost of labor the demand for premixed paint grew and eventually a 98% grade was marketed (Dunn, 1973).

4.2 Sources

The mineral minium is found in small amounts in many localities, usually as an alteration product of galena or cerussite (Palache et al., 1944). Both Forbes (1955) and Petrie (1889) suggest that a naturally occurring red oxide of lead was the source of the pigment of Greco-Roman date found at Hawara in Egypt. However, there seems to be no mention of the mineral red lead in ancient Egyptian texts (Harris, 1961). Except for these suggestions, there is little indication that the natural mineral ever was used as a pigment. In the section discussing history of use (2.2), reference is made to the manufacture of red lead in China as early as the fifth century B.C.

There remain today many manufacturing sources of this valuable pigment. Dunn (1973) cites three current sources of supply in the United States: Eagle-Picher, Hammond, and National Lead.

4.3 Preparation

The preparation of the red pigment from lead white was known in Greek and Roman times. However, in fifteenth-century Germany (see 2.2) lead metal was the starting material in a two-stage process in which litharge was the intermediate product. From the Far East descriptions of the manufacture of red lead from what was evidently the metal are known much earlier. Although such early descriptions of making red lead from lead metal imply its preparation from litharge, this process as such was not mentioned until the eighteenth century in England (Harley, 1982).

It must be borne in mind that the yellow painters' material known as litharge, frequently used as a drier in oils, is not the same crystallographically as the mineral litharge. Mineral litharge is red PbO, of the tetragonal crystal system. Painters' litharge is identical to the mineral massicot, yellow PbO, which is orthorhombic. Massicot as a pigment name can refer to any lead-based yellow, most commonly lead-tin yellow (Harley, 1982).

Litharge (PbO) can be heated in air to produce red lead, the speed of the reaction depending on the fineness of the PbO (Pascal, 1963). The simple lead-roasting muffle furnace shown in fig. 9 is typical of that used in the West in the late nineteenth and early twentieth centuries. In

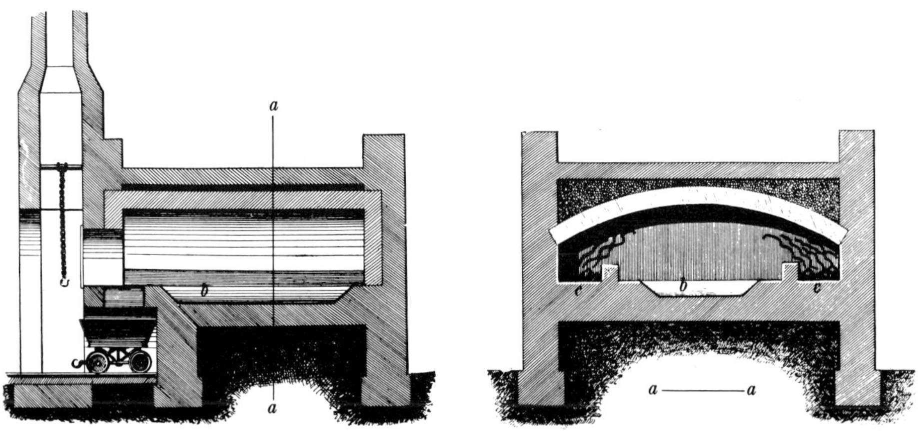

Fig. 9. Basic structure of a reverberatory furnace for converting lead to dross and finally to red lead. Here illustrated is the drossing furnace. Lead metal to be converted is placed at b; fireplaces at c. After Zerr and Rübencamp (1908).

the past, temperatures ranging from 300 to 500°C have been given for this process (above 550°C only PbO is stable). At present a range of 450–470°C is used, and the product is called furnace red lead. Pigment prepared in this way contains a certain amount of litharge. To obtain pure Pb_3O_4, dilute acetic acid can be used to remove the PbO that may be present (Wells, 1975). In practice this procedure is not followed in the paint industry; the presence of some PbO, which is a powerful drier, is believed advantageous. As noted above, the Pb_3O_4 content of commercial red lead varies from 85 to 98%; the composition of the balance, when listed, is given as PbO (*Raw Materials Index*, 1975). It is generally agreed that the transformation of litharge to red lead is not a simple process and that there are many complex intermediate oxides (Wells, 1975; Gillibrand & Halliwell, 1972).

It is reported (Pascal, 1963) that finely divided lead can oxidize to minium at room temperature, the process being accelerated by light and humidity. Lead monoxide can also transform to minium at room temperature; in one instance this took place within eight years.

The second common manufacturing method today produces a material known as fume red lead. In this process atomized molten lead is instantly heated in an oxidizing atmosphere to approximately 1800°C. The condensed material from this heating is collected in a series of chambers, fume red lead being the final and finest product collected (Dunn, 1973).

Other preparation methods exist, which were not commonly applied in commerce (Bearn, 1923, pp. 114–115). Lead sulfate can be calcined with sodium carbonate and sodium nitrate to produce red lead. In the so-called nitrate process, metallic lead is fused with sodium nitrate to produce lead oxide and sodium nitrite; the latter is washed away and the remaining lead oxide is treated as above. Two noncommercial methods of forming minute crystals of red lead have been reported: lead carbonate can be fused in a mixture of sodium and potassium nitrates, or lead oxide or carbonate can be heated in hot concentrated potassium hydroxide solution (Palache, et al., 1944). A pigment of very fine particle size can be obtained by vaporizing lead in an arc between two electrodes and oxidizing it in a current of oxygen (Champetier & Rabaté, 1956).

Red lead can be prepared, as in antiquity, by heating lead white ($2PbCO_3 \cdot Pb(OH)_2$) in air between 425 and 430°C, with the release of water and carbon dioxide. Although higher and lower temperatures can be used, transformation appears to be most rapid and efficient in this range (Brown & Nees, 1912). As already noted, the pigment produced by this process is called orange lead.

4.4 Adulteration and Sophistication

As mentioned, the principal impurity found in conjunction with red lead is litharge. This is not an intentional adulterant, but is a result of the manufacturing process.

The pigment has long been used as an adulterant in other reds, notably vermilion. This practice dates at least from the time of Pliny. The transfer of the name *minium* from cinnabar to red lead, largely because of this practice, has already been noted by Cennini (D. V. Thompson, 1960). In seventeenth-century England, and later in America, vermilion was adulterated with red lead, and methods used to detect this have been described (Harley, 1982).

In the past, the pigment has also been doctored with organic dyes to give it greater brilliance (Doerner, 1949). One common product was known as "American vermilion," a mixture of red lead and red lakes (Gettens, et al., 1972). A red pigment on Chinese pottery figurines of questionable origin (Gettens, 1954) was found to consist of vermilion, red lead, barite, and modern dyes.

5.0 IDENTIFICATION — CHARACTERIZATION

5.1 Characterization by Optical and Scanning Electron Microscopy

Red lead is weakly but plainly birefringent, and its refractive index is high, N(Li) = 2.42. Anomalous blue-green interference colors seen with crossed polars are highly characteristic and often manifest themselves as "a peculiar iridescence" (Gettens, 1954). Examples can be seen in figs. 10, 11, 12, and 13. The pigment characteristics of a high refractive-index, orange color by transmitted light, blue-green color between crossed polars, and solubility in dilute nitric acid make the identification of red lead almost certain by simple light microscopy.

The mineral minium, belonging to the tetragonal crystal system, occurs as an alteration product of galena (PbS) and other lead minerals and is commonly associated with galena, cerussite ($PbCO_3$), massicot (PbO), and limonite

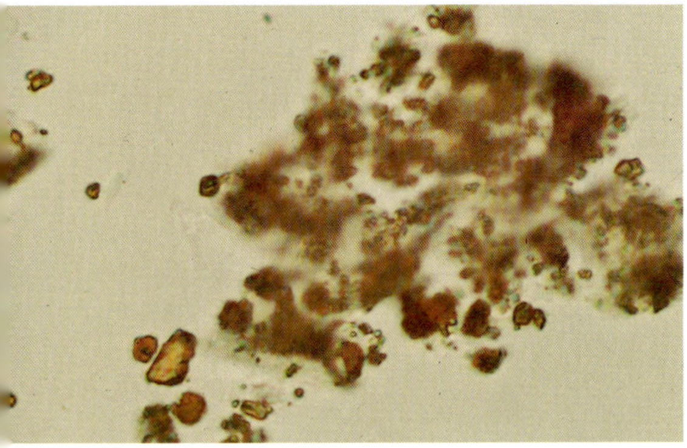

Fig. 10. Red lead on orange sticks of fan, Japanese painting *A Boy with a Fan*, Ukiyo-e, late seventeenth century, Freer Gallery of Art, Washington, 05.96. 1200x. Oil immersion objective, mounted in Canada balsam, refractive index 1.53. A. In plane polarized light.

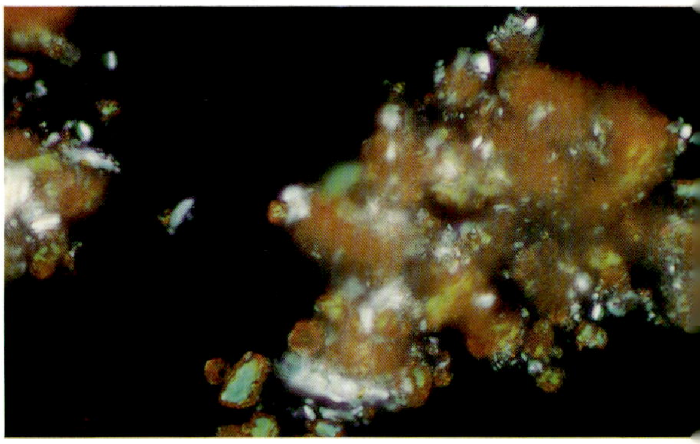

B. Same field between crossed polars.

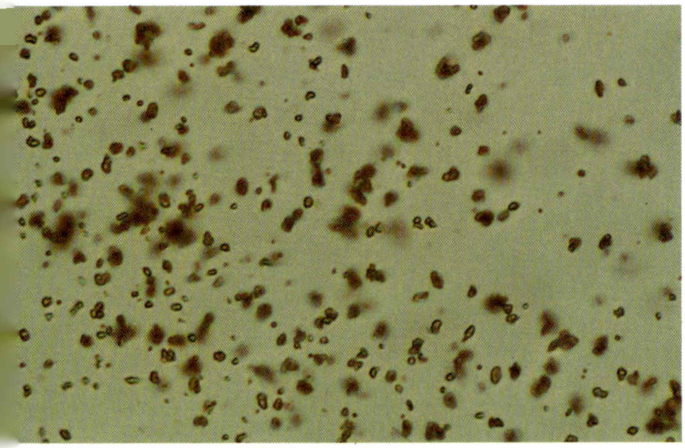

Fig. 11. "Fumed" red lead. National Lead Co., 1936. 1200x. Oil immersion objective, mounted in Canada balsam, refractive index 1.53. A. In plane polarized light.

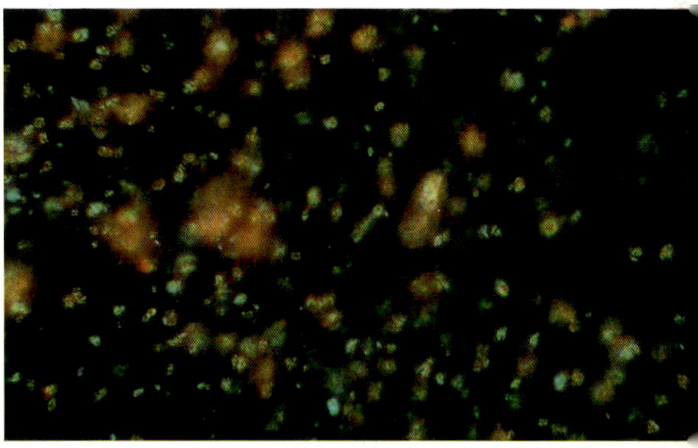

B. Same field between crossed polars.

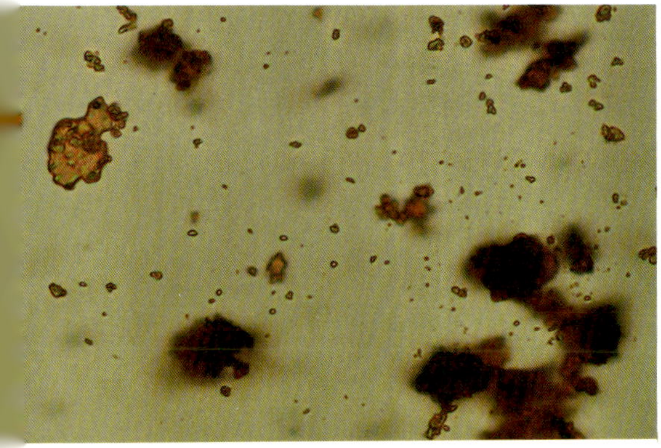

Fig. 12. Red lead, Turner palette No. 21 (see J. W. Hanson, 1954). 1200x. Oil immersion objective, mounted in Aroclor 5442, refractive index 1.66. A. In plane polarized light.

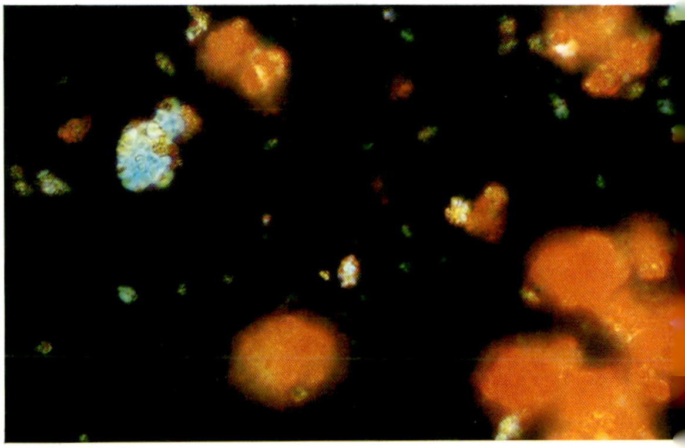

B. Same field between crossed polars.

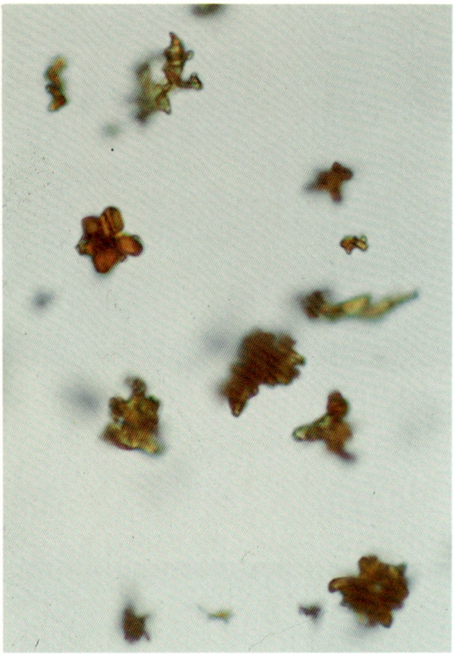

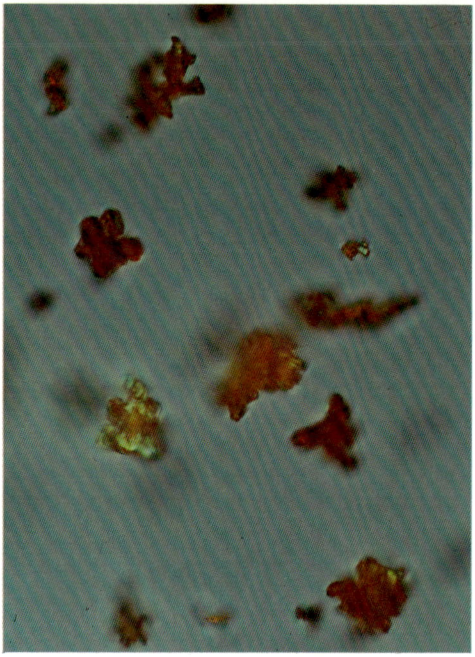

Fig. 13. Red lead, modern Japanese pigment, *tan*, Yashiro No. 8. 1200x. Oil immersion objective, mounted in Canada balsam, refractive index 1.53. A. In plane polarized light.

B. Same field between partially crossed polars.

C. Same field, rotated 50°, between partially crossed polars.

D. Same field as C, between crossed polars.

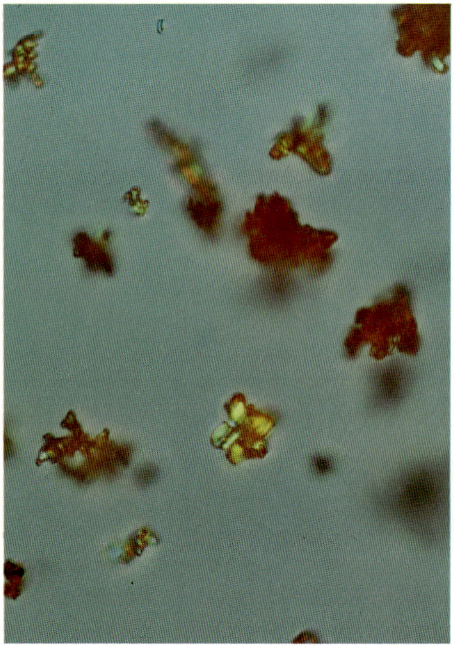

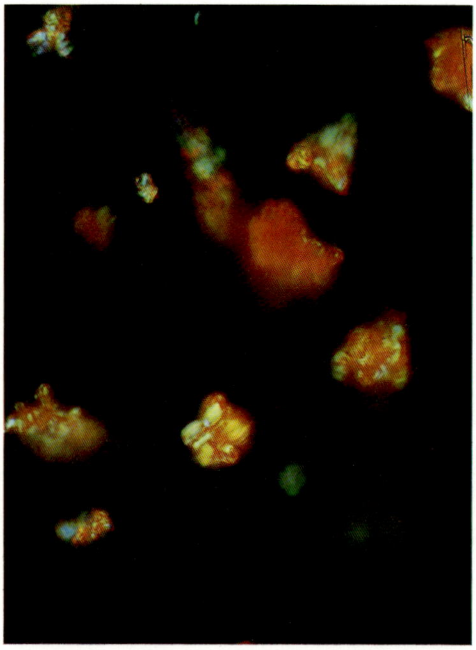

(hydrous iron oxide) (Roberts et al., 1974). The habit of the natural mineral is massive, earthy, or powdery. Microscopically it is cryptocrystalline and is said to be fibrous in character, the optical orientation being parallel to the fibers (Larsen & Berman, 1934). A specimen of the minium used by Larsen and Berman in their publication (USNM 84226, Stone Mine, Leadville, Colorado) is in the Department of Mineral Sciences, Smithsonian Institution. This particular specimen is so finely divided that the optical properties are difficult to determine, even at 1000x; the particles appeared to be birefringent, but no blue-green colors could be seen. No samples of red lead pigment examined to date by the author have been as finely divided as this material.

Artificially prepared pigment occurs in particles made up of tiny, irregularly rounded aggregates, which are red to orange in transmitted light. According to Merwin (1917), particles of red lead produced by oxidation of litharge are "scaly or irregular grains . . . without definite outlines," whereas the pigment prepared from lead carbonate is in the form of "aggregates of extremely small particles with almost no crystalline structure." Dunn (1973) has described the microscopical differentiation between furnace red lead and fumed red lead. The particles of the former are tabular to chunky, with a roughened surface, while those of the latter are flaky and more transparent with a smoother surface (see figs. 4 and 11).

5.2 Chemical Identification

The effect of acids, as noted in 3.5, can give a preliminary indication that an orange pigment is red lead. The formation of dendritic crystals of lead nitrate when the sample is dissolved in dilute nitric acid is perhaps the most characteristic result. The formation of lead chloride has also been described (see 3.5). There are several specific tests for lead, however, that are useful.

5.21 ***Test with Potassium Iodide to Form Lead Iodide.*** The lead nitrate residue from the nitric acid treatment may be taken up with a drop of water or a drop of dilute acetic acid, and a tiny crystal of potassium iodide (KI) added to the solution. PbI_2 precipitates immediately in the vicinity of the crystal in thin, bright yellow scales or hexagonal plates, which appear greenish, brownish, or even gray by transmitted light (fig. 14). By reflected light they scintillate and glisten, sometimes with iridescence, because of

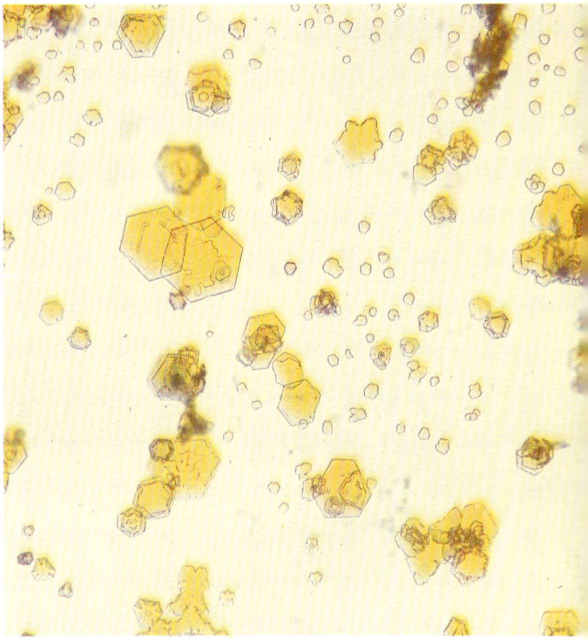

Fig. 14. Lead iodide, by transmitted light. Wet mount with cover glass. 250x. Photo: W. T. Chase. Reprinted by permission from *Studies in Conservation* (Gettens, Kühn & Chase, 1967).

Brownian movement. Two precautions are necessary (Gettens, et al., 1967). The test particle of KI must not be too large because PbI_2 is soluble in an excess of that reagent. Futhermore, too high a concentration of HNO_3 will decompose KI with release of elemental iodine. The limit of identification is 0.0005% lead acetate (5 ppm) (Short, 1940).

It should be noted that if a mixture of vermilion and red lead is tested for lead with potassium iodide, any vermilion, which will not have dissolved in the nitric acid and is still present, will dissolve when the KI is added (Gettens, et al., 1972, footnote, p. 57). Moreover, a mixture of vermilion and lead white will yield a lead iodide test that could lead to the erroneous deduction that the colored paint was based solely on red lead, if the sample is not examined microscopically (Feller, 1971).

5.22 ***Test with Copper Acetate and Potassium Nitrite to Form $K_2PbCu(NO_2)_6$.*** To the residue after treatment with nitric acid a test solution is added, made up of 0.5% copper acetate, and 0.5% sodium acetate, and 1.0%

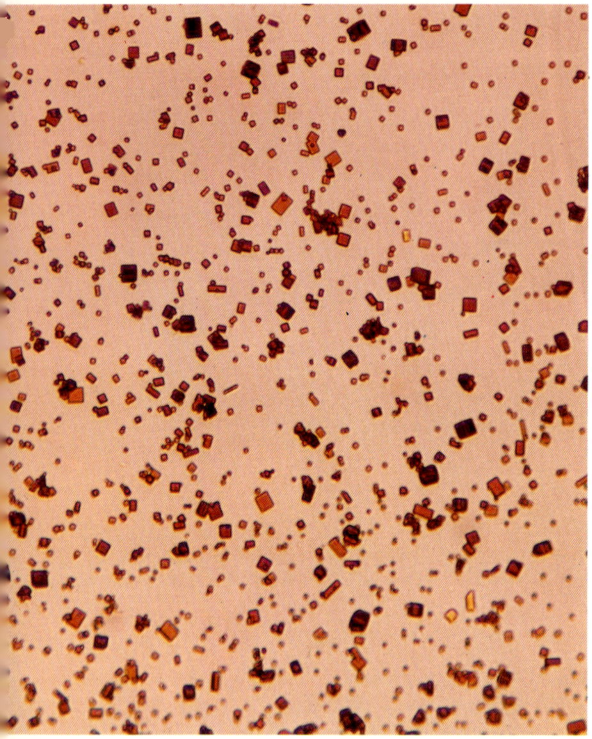

Fig. 15. Triple nitrite test, $K_2PbCu(NO_2)_6$ by transmitted light. 400x. Photo: W. T. Chase. Reprinted by permission from R. J. Gettens and E. W. FitzHugh, "Azurite and Blue Verditer," *Studies in Conservation*, 11 (1966), 54–61.

acetic acid. A single fragment of solid potassium nitrite (KNO_2) is introduced into the solution. When lead is present it immediately precipitates as the triple salt, $K_2PbCu(NO_2)_6$, in the form of cubic crystals. Illustrated in fig. 15, these crystals are black in transmitted light but brownish by reflected light; when very thin they may be brown or orange-colored by transmitted light (Chamot & Mason, 1940). Sensitivity is 0.03 micrograms (Augusti, 1942).

5.23 ***Test with Potassium Chromate to Form $PbCrO_4$.*** A drop of dilute HC1, followed by a drop of K_2CrO_4 solution, is added to the nitric acid solution of red lead to give a heavy, bright yellow precipitate of $PbCrO_4$ (Plesters, 1956).

5.24 ***Test with Sodium Sulfide or Hydrogen Sulfide to Give PbS.*** Red lead turns brown on addition of Na_2S or H_2S owing to the formation of black lead sulfide; the reaction with Na_2S has been used to determine lead microchemically (Augusti, 1946).

5.25 ***Other.*** Many of the tests just described can also be used on cross sections of paint films. Lead may be detected in a cross section with H_2S (Dunn, 1945), or with the lead iodide test (Plesters, 1956). A spot test or drop reaction for lead using KI and stannous chloride, which gives a red-orange precipitate of the complex salt $2PbI_2 \cdot SnI_2$, has also been described (Augusti, 1942). Lead can also be identified by the Weisz ring-oven technique (Locke & Riley, 1970).

5.3 Instrumental Methods of Analysis

5.31 ***X-ray Diffraction.*** The x-ray diffraction data for red lead are shown in table 1. Typical Debye-Scherrer, powder diffraction patterns are reproduced in fig. 16. Because litharge so frequently occurs as a constituent of modern red leads, it is appropriate to include the data for both forms of PbO, yellow painters' litharge (mineral massicot) and the red mineral litharge. PbO_2 (plattnerite) is also included because it may be encountered as a component of darkened red lead (table 2). This is illustrated in table 4 and in fig. 17 in the analysis of a fragment of a wall painting from Kizil, Chinese Turkestan.

Typical patterns of the three lead oxides other than red lead are shown in fig. 18. The three strongest lines for each of the four oxides are given in table 3. Pb_3O_4 also exists in a distorted form that gives anomalous broadening of the x-ray diffraction lines and is called pseudotetragonal Pb_3O_4 (Gillibrand & Halliwell, 1972).

5.32 ***Infrared Spectroscopy.*** The infrared absorption curve of red lead (Pb_3O_4 97% PbO 3%) has been published and the key infrared absorption bands noted (Afremow & Vandeberg, 1966; *Infrared Spectroscopy*, 1969; McDevitt and Baun, 1964). Typical of many oxide pigments, the distinctive absorption bands are all below 600 cm^{-1}, as shown in fig. 19, which also includes the spectrum of litharge (PbO). The location of the distinctive bands can be expected to vary somewhat depending upon the purity of the sample and the calibration of individual instruments. A list of the principal absorption peaks determined from published data is therefore given in table 5, along with the values from fig. 19. The peak data was kindly provided by the Research Center on the Materials of the Artist and Conservator, Carnegie-Mellon Institute of Research,

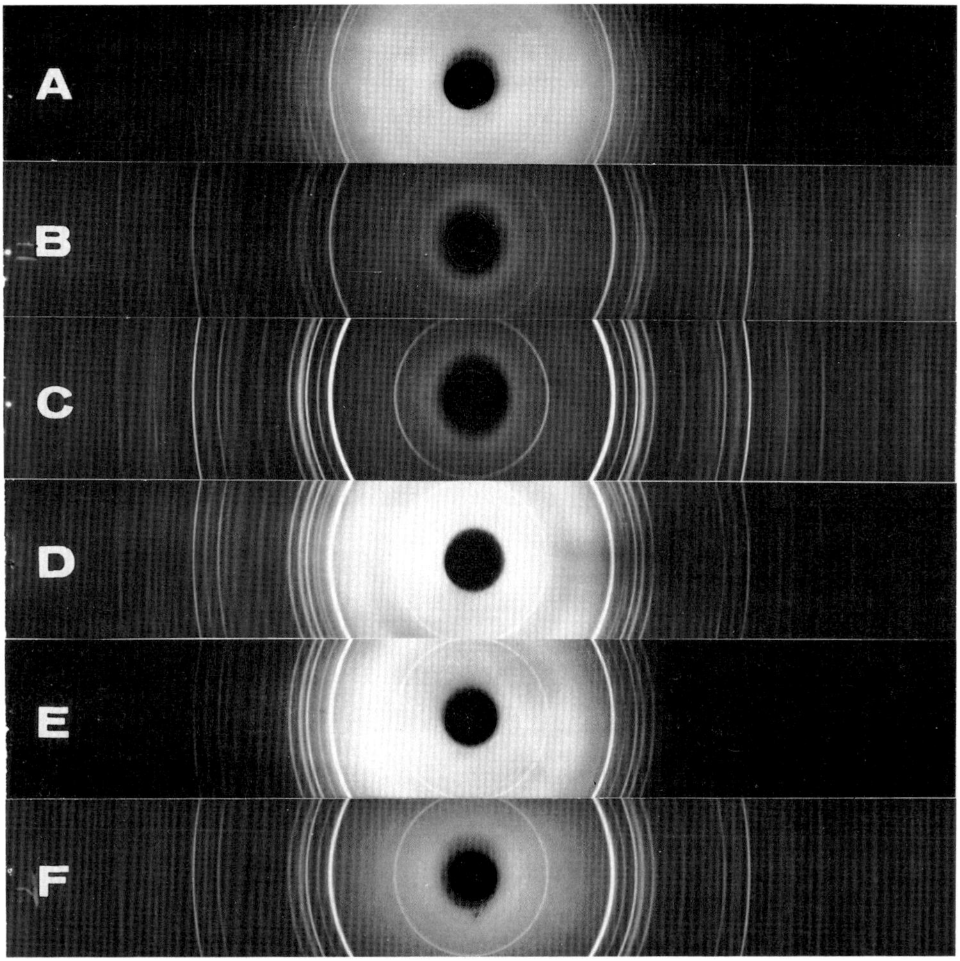

Fig. 16. X-ray powder diffraction patterns of red lead and minium (Pb_3O_4).
A. Minium. Stone Mine, Leadville, Colorado. Department of Mineral Sciences, Smithsonian Institution. USNM 84426. Film F1838.
B. Furnace red lead. National Lead Co., 1936. Film F1674.
C. Red lead, *tan*. Yashiro (Japanese pigment company), 1931, No. 8. Film F1675.
D. Red lead on orange sticks of fan, *A Boy with a Fan*, Ukiyo-e, late seventeenth century. Freer Gallery of Art, Washington, 05.96. Film F1870.
E. Red lead in orange area of carpet, *A Prince with an Attendant*, late Mughal, nineteenth century. Freer Gallery of Art, Washington, 07.621. Film F1835.
F. Red lead. Turner palette, No. 21 (see J. W. Hanson, 1954). Film F1664.
A, D, and E, Gandolfi powder diffraction; B, C, and F, Debye-Scherrer powder diffraction. Camera diameter 114.6 mm. Photographs not to be used for measurement purposes.

Fig. 17. X-ray powder diffraction pattern of darkened red lead (see table 4). Photograph not to be used for measurement purposes.

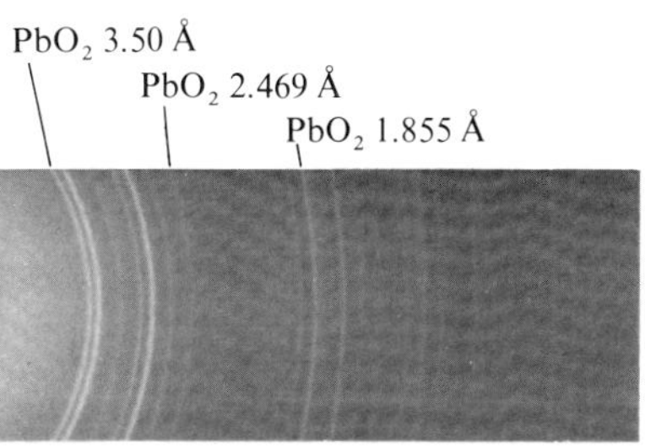

Table 1. **X-RAY DIFFRACTION DATA FOR RED LEAD (Pb_3O_4)[a] (Diffractometric Data, $CuK\alpha_1$ = 1.5405 Å, 25°C)**

d(Å)	*I*
6.23	12
3.66	4
3.38	100
3.28	8
3.113	20
2.903	50
2.787	45
2.632	30
2.444	2
2.289	4
2.260	8
2.205	2
2.076	2
2.032	12
1.970	12
1.903	20
1.887	<1
1.829	20
1.755	30
1.729	2
1.7025	2
1.6897	2
1.6417	8
1.6302	<1
1.5876	12
1.5580	8
1.5292	8
1.5116	2
1.4744	<1
1.4687	4
1.4521	2
1.4144	14
1.3944	<1
1.3728	2
1.3471	4
1.3109	4
1.2830	4
1.2614	4
1.2461	4
Plus 43 lines to 0.7872	

[a] JCPDS Powder Diffraction File 8–19, Courtesy of JCPDS International Centre for Diffraction Data (JCPDS, 1982).

Table 2. **X-RAY DIFFRACTION DATA FOR OTHER LEAD OXIDES WHICH MIGHT BE ASSOCIATED WITH RED LEAD (Diffractometric Data, $CuK\alpha_1$ = 1.5418 Å, 27°C)**

PbO (red)[a] *(mineral litharge)*		*PbO (yellow)*[b] *(mineral massicot)*		*PbO_2 (black)*[c] *(mineral plattnerite)*	
d(Å)	*I*	*d(Å)*	*I*	*d(Å)*	*I*
5.018	5	5.893	6		
3.115	100	3.067	100	3.50	100
2.809	62	2.946	31	2.793	95
2.510	18	2.744	28	2.469	40
2.124	1	2.493	<1	2.438	6
1.988	8	2.377	20	2.224	6
1.872	37	2.278	<1	1.855	80
1.675	24	2.203	<1	1.752	18
1.558	6	2.008	12	1.692	14
1.542	11	1.963	2	1.568	20
1.438	2	1.850	14	1.524	25
1.405	5	1.797	14	1.486	20
1.282	2	1.724	15	1.398	14
1.256	3	1.640	13	1.274	20
1.226	4	1.596	<1	1.240	8
1.219	5	1.534	9	1.218	12
1.1977	<1	1.514	2	1.169	8
1.1462	2	1.474	11	1.151	14
1.1232	2	1.408	<1	1.133	14
1.0768	3	1.372	1	1.109	10
1.0610	2	1.363	<1	1.102	10
1.0386	<1	1.325	1	1.005	12
1.0254	<1	1.297	2	0.972	6
0.9738	1	1.289	3	0.961	14
0.9462	1	1.252	2	0.951	16
0.9365	3	1.244	2	0.934	8
0.9200	3	1.203	4	0.927	10
		1.188	3	0.887	12
		1.174	4	0.877	6
		1.139	2	0.872	10
		1.120	2	0.851	6
		1.102	4	0.842	8
		1.091	2	0.823	20
				0.802	8
				0.793	12
				0.784	10
				0.778	12

a. JCPDS Powder Diffraction File 5-0561.
b. JCPDS Powder Diffraction File 5-0570.
c. JCPDS Powder Diffraction File 25-447.
Courtesy of JCPDS International Centre for Diffraction Data (1982).

Table 3. **THREE STRONGEST LINES IN THE X-RAY DIFFRACTION PATTERNS OF THE FOUR LEAD OXIDES**

	d(Å)	*I*
Pb_3O_4	3.38	100
	2.903	50
	2.787	45
PbO (red)	3.115	100
	2.809	62
	1.872	37
PbO (yellow)	3.067	100
	2.946	31
	2.744	28
PbO_2	3.50	100
	2.793	95
	1.855	80

Table 4. **X-RAY DIFFRACTION DATA FOR SAMPLE OF DARKENED RED LEAD**

Darkened Red Lead[a] *Kizil wall painting fragment, (Freer SC-PA-144)*		Pb_3O_4 *(red)*[b] *(red lead)* (lines *I* = 30 and above)		PbO_2 *(black)*[c] *(plattnerite)* (lines *I* = 30 and above)	
d(Å)	*I*	*d(Å)*	*I*	*d(Å)*	*I*
6.28	30	3.38	100	3.50	100
3.51	90	2.903	50	2.793	95
3.35	100	2.787	45	2.469	40
3.10	20	2.632	30	1.855	80
2.89	60	1.755	30		
2.80	10				
2.63	30				
2.49	30				
2.25	10W				
1.97	10W				
1.91	30				
1.86	80				
1.76	80				
1.70	20				
1.64	30				
1.56	20W				
1.52	30				
1.48	20				
1.41	20				
1.39	20				

a. Gandolfi powder diffraction pattern, Freer Gallery of Art, Film F2386. Intensities determined visually.
b. JCPDS Powder Diffraction File 8-19.
c. JCPDS Powder Diffraction File 25-447.
Courtesy of JCPDS International Centre for Diffraction Data (1982).

Pittsburgh. The spectrum of red lead extending to 45 cm^{-1} has been published by Nyquist and Kagel (1971).

Litharge, which may be present in some samples of Pb_3O_4, usually exhibits only one major and two minor absorption peaks, but again the specific shape of the curve will vary somewhat depending upon purity.

5.33 ***Spectrochemical Analysis.*** Lead is readily detected by emission spectrography. Prominent lines in the visible range are 6001.8 and 4057.8 Å (Peterson & Jaffe, 1953); in the ultraviolet range they are 2873.3, 2833.1, 2663.2, 2614.2 Å, with sensitivity of 0.01% (Waring & Annell, 1953). It has been suggested that determination by emission spectrographic analysis of the copper/silver ratio in lead white from paintings will give some indication as to whether the work originated in Italy or the Netherlands (Kühn, 1966). Identification of the same impurities in red lead might give similar information.

Lead may also be detected by x-ray fluorescence and electron microprobe analysis. The main lead lines are $K\beta_1$ 84.9 kV (0.14590 Å), $K\alpha_2$ 72.8 kV (0.170294 Å), $L\gamma_1$ 14.8 kV (0.83973 Å), $L\beta_2$ 12.6 kV (0.98221 Å), $L\alpha_1$ 10.6 kV (1.17501 Å), and $L\alpha_2$ 10.5 kV (1.18648 Å) (V. F. Hanson, 1973; Bearden, 1972–1973). The presence of red lead in the lead white on a painting has been detected by electron probe microanalysis (Hornblower, 1963), but no analysis of red lead by this method or by x-ray fluorescence has been reported.

5.34 ***X-radiography.*** Because of lead's high absorption coefficient, areas of red lead in paintings are opaque to x-rays. Distinguishing red lead from vermilion by radiography alone would be difficult, however, since the absorption coefficients of lead (101.7 cm^2g^{-1}) and mercury (101.2 cm^2g^{-1}) are similar (de Wild, 1929). Thus, x-radiography does not provide an independent means of identifying red lead; in special cases it might supply confirmatory evidence when used in conjunction with other examination methods.

5.35 ***Neutron Activation Analysis.*** Characteristic impurities in lead white, estimated by neutron activation, gave some indication of the date of a series of Dutch paintings from the sixteenth to the twentieth centuries (Houtman & Turkstra, 1965). Impurities in lead

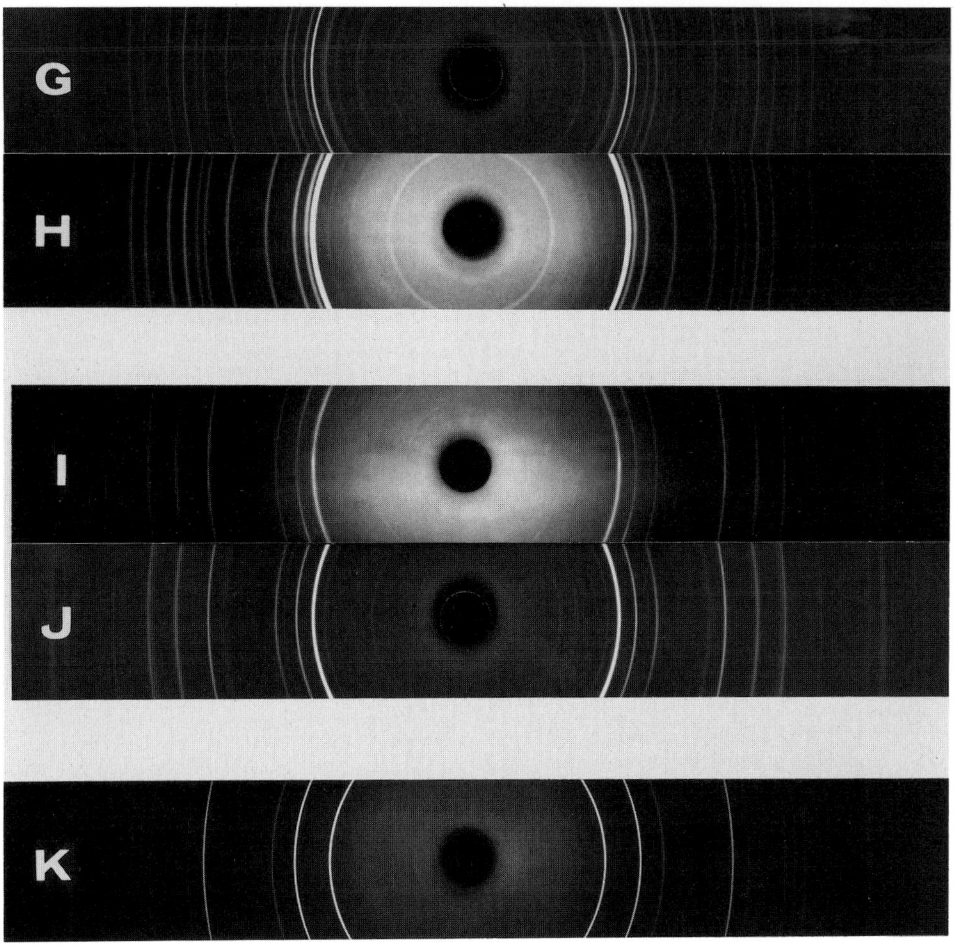

Fig. 18. X-ray diffraction powder patterns of lead oxides other than red lead.
Yellow PbO (orthorhombic), mineral massicot.
G. Sublimed litharge. Eagle Picher. Line marked is d = 2.623, one of six lines sometimes occurring with yellow PbO due to a green phase formed after exposure to air (See *JCPDS Powder Diffraction File* 5-0570). Film F845.
H. Lead monoxide, yellow 99.9995%. Gallard-Schlesinger. Film F1844.
Red PbO (tetragonal), mineral litharge.
I. Litharge. Fundicon Espanol, Quoma Mine, Mizque Province, Cochebamba, Bolivia. Department of Mineral Sciences, Smithsonian Institution, USNM 66580. Film F1869.
J. Litharge. Red corrosion product, lead pad under column, U.S. Capitol, 1959. Film F768.
Black PbO_2, mineral plattnerite.
K. PbO_2, Baker Adamson. Film F1890.
H, I, and K, Gandolfi powder diffraction; G and J, Debye-Scherrer powder diffraction. Camera diameter 114.6 mm. Photographs not to be used for measurement purposes.

white from Norwegian altar frontals have also been determined by this method. (Steinnes, 1976). Presumably the same impurities in red lead on objects or paintings might give similar information.

5.36 ***Mass Spectrography.*** Isotope mass spectrography has been used to identify the source of lead white based on lead isotope ratios, Pb-206/Pb-204 (Keisch, 1970). The lead isotope ratios from the lead on Rubens' *The Gerbier Family* indicate that the red lead and the lead white on this painting are from the same source (Keisch, 1970; Feller, 1973). Isotope analyses have been made of white and red lead from Norwegian medieval art (Plahter, 1976). Spark-source mass spectrography has been used to identify red lead on sixteenth- to seventeenth-century Mughal Indian miniature paintings (Cairns & Johnson, 1972).

Table 5
FREQUENCIES OF PRINCIPAL INFRARED ABSORPTION BANDS IN RED LEAD AND LITHARGE (IN RECIPROCAL CENTIMETERS, cm^{-1})

	Fig. 19	*Nyquist and Kagel (1971) Fig. 327*	*McDevitt and Baun (1964) Table 1*	*Infrared Spectroscopy (1969) Figs. 511, 538*
Red Lead	132	136	—	—
	152	152	—	—
	282	282	—	292
	320	325	320	332
	380	382	380	383
	455	455	445	458
	515	—	—	515
	530	530	525	530
	—	—	650	—
Litharge	295	—	300	300
	355	—	—	355
	375	—	377	—
	—	—	500	478

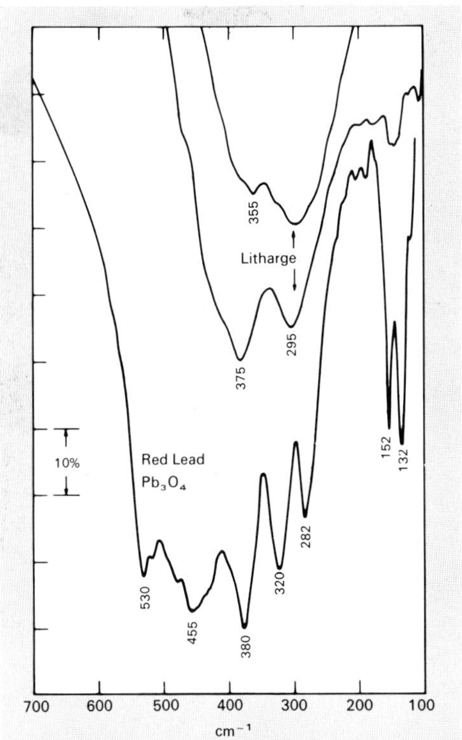

Fig. 19. Infrared absorption spectra of the significant absorption bands in red lead and litharge in the region of 700 to 100 cm^{-1}. Litharge occasionally may show absorption between 355 and 375 cm^{-1} which probably is due to traces of Pb_3O_4. The spectra are arbitrarily displaced to facilitate comparison.

5.37 β-***Ray Backscattering.*** Backscattering of beta rays emitted from radioisotopes has been used to identify pigments on Japanese paintings (Yamasaki, 1957). As with x-radiography, the differentiation between vermilion and red lead is not possible because both give a high intensity of backscattering of beta particles. In special cases confirmatory identification of red lead by this method might be achieved.

5.4 Criteria for Positive Identification

Red lead can usually be recognized with considerable confidence under the polarizing microscope from its color and optical properties; the typical blue-green interference colors between crossed polars are highly distinctive and readily distinguish the pigment from vermilion and hematite. Determination of lead by chemical microscopy, emission spectrographic analysis, or other means provides additional confirmation. Positive identification, of course, is provided by x-ray diffraction analysis.

The blue-green interference color that can often be seen in the finest of particles, the orange color under ordinary light, the high refractive index, the ready solubility in dilute nitric acid, and the subsequent detection of lead — usually as lead iodide — constitute a combination of five attributes that provide practically unequivocal identification, making red lead among the three or four pigments easiest to identify with confidence.

6.0 **NOTABLE OCCURRENCES**

EUROPE

Artist or School *Collection or Location*	*Analyst and Method of Identification*
German Medieval wall painting in the Briefkapelle, Kirche St. Marien, Lübeck	(E. Richter*, 1977), Chem.
Fourteen European illuminated manuscripts, 10th–16th centuries, Bibliothèque Nationale and Archives Nationales, Paris	(Flieder, 1968), Micr.
Crucifix from Lichtenstein castle, c. 1050, Württembergisches Landesmuseum, Stuttgart	(Westhoff, et al. 1979), Chem.
Seven German wall paintings, 12th–16th centuries, Freiberg Cathedral and other locations	(F. Preusser*, 1977), ESA, Micr., Chem.
Virgin and Child, polychrome wood, 12th century, Hovland Stave Church, Eggedal, Norway	(Simonson & Plahter, 1963–64), Micr.
Freudenstädter Lesepult, Romanesque sculpture, c. 1150, Town church of Freudenstädt (Schwarzwald), West Germany	(E. Richter*, 1977), XRD
Virgin and Child, Church of Dal, polychrome wood, 2nd half 13th century, University Museum of National Antiquities, Oslo	(Plahter & Wiik, 1970), Micr.
Painted wooden altar frontals, 1250–1350, Church of Tingelstad, Norway	(Plahter, Skaug, & Plahter, 1974), Cross section, Micr., XRD
Master of Waltensburg, Wall painting, 1330–1350, Waltensburg, Kanton Graubünden	(H. Kühn*, 1976), XRD
Attr. Spinello Aretino, *Madonna and Child Enthroned with Angels*, 14th–15th centuries, Fogg Museum of Art, Harvard University, Cambridge	(Gettens, 1947), Micr.
Part of an altar, 1420–1430, Wallraf-Richartz Museum, Cologne. 60-62-389, 755	(H. Kühn*, 1976), XRD (in a kind of imprimitura)
German medieval playing cards, 1430, Württembergisches Landesmuseum, Stuttgart	(Richter & Härlin, 1976), Micr. ESA, XRD
Master of the Heisterbacher Altars, Altar, 1440–1450, Wallraf-Richartz Museum, Cologne. Inv. No. 73	(H. Kühn*, 1976), XRD (as drying agent in oil under gold leaf)
Paolo Uccello, *Niccolo da Tolentino at the Battle of San Romano*, 1450s, National Gallery, London. No. 583	(J. Plesters*, 1978), Micr., Chem., ESA
German School, *Church of San Sebastian*, 15th–16th centuries, Zuoz	(H. Kühn*, 1976), XRD
Andreani Giusto, *Madonna and Child Enthroned with Four Saints*, First half 15th century, Los Angeles County Museum of Art	(N. Muller*, 1974), Micr.
Presentation of the Virgin in the Temple (Russian icon), 15th century, Walters Art Gallery, Baltimore. 37.2410	(J. Caraher*, 1974), Micr. (Back of panel)
Herlin Altarpiece, 1466, Church of St. James, Rothenberg, West Germany	(Broekman-Bokstijn et al., 1970), Micr., IR
High altar, 1479, Cathedral, Århus, Denmark	(Larsen, 1970), Cross section, Micr.
Blaubeuren High Altarpiece, 1490, Württemberg, West Germany	(E. Richter*, 1977), XRD

XRD = x-ray diffraction; Micr. = microscopic examination; ESA = emission spectrographic analysis; Chem = microchemical tests; IR = infrared; * = private communication; ** Bayerische Staatsgemäldesammlungen.

Artist or School *Title, Date,* *Collection or Location*	*Analyst and Method of Identification*
English, Painted design on lining of cope, Late 15th century, Art Institute of Chicago	(M. Butler*, 1974), Micr.
Leinberger Erhard, and others, Twelve polychromed wood sculptures, 13–18th centuries, Bavarian National Museum, Herz-Jesu-Kirche, and others	(F. Preusser*, 1977), ESA, Micr., Chem.
Master of the Schlutup Altar, Sippenaltar, c. 1500, St. Annen-Museum, Lübeck, West Germany	(E. Richter*, 1977), XRD
Albrecht Dürer, *Madonna and Child with Saint Anne*, 1519, The Metropolitan Museum of Art, New York. 14.40.633	(F. Preusser*, 1977), ESA, Micr., Chem.
Peter Paul Rubens, *The Gerbier Family*, 1577–1640, National Gallery of Art, Washington	(Feller, 1973), Micr.
Kirchenväter-altar, 16th century, Church of Bordesholm (Schleswig-Holstein), West Germany	(E. Richter*, 1977), XRD
David Teniers, *Triptych of Saints Hilduard and Christiane*, 1617, Church of Notre Dame, Termonde, Belgium	(Thissen, 1961), Micr. (ground, mixed with ochre and lead white)
Jan Molenaer, *A Young Man Playing a Theorbo and a Young Woman Playing a Cittern*, 1630s, National Gallery, London. No. 1293	(J. Plesters*, 1978), Micr., Chem., Laser ESA
Gerbrand van der Eeckhout, *Four Officers of the Amsterdam Coopers' and Wine-Rackers' Guild*, 1657, National Gallery, London. No. 1459	(J. Plesters*, 1978) Micr., Chem., Laser ESA (with red ochre in the ground)
El Greco, *The Stripping of Christ*, 1600, BSTG**, Munich. 8573	(F. Preusser*, 1977), ESA, Micr., Chem.
Bulgarian mural paintings, 17th century, St. George Orthodox Church, Weliko Turnowo	(Brochwicz & Walkowa, 1972), Chem., Micr.
Rembrandt van Ryn, *Self Portrait*, 1659, National Gallery of Art, Washington	(F. Preusser*, 1977), ESA, Micr., Chem.
Giovanni Battista Tiepolo, *Adoration of the Holy Trinity of Pope Clement*, 1659, BSTG, Munich. L877	(F. Preusser*, 1977), ESA, Micr., Chem.
Jacopo Tintoretto, *Portrait of a Sculptor*, 16th century, BSTG, Munich. 3721	(F. Preusser*, 1977), ESA, Micr., Chem.
Jan van Huysum, *Fruits, Flowers, and Insects*, 1735, BSTG, Munich. 2077	(F. Preusser*, 1977), ESA, Micr., Chem.
Joseph Mallord William Turner, Collection of painting materials, 1775–1851, Courtauld Institute of Art, London	(Hanson, 1954), ESA
Arnold Böcklin, *Villa by the Sea II*, 1865, Schack-Galerie, Munich. 11,536	(Kühn, 1969), Micr. ESA, XRD
Vincent van Gogh, *View of Arles*, 1889, BSTG, Munich. 8671	(F. Preusser*, 1977), ESA, Micr., Chem.
Vincent van Gogh, *The Plain at Auvers*, 1890, BSTG, Munich. 9584	(F. Preusser*, 1977), ESA, Micr., Chem.
Arnold Böcklin, Collection of pigments, 1827–1901, Institute für Technologie der Malerie, Stuttgart	(Richter & Härlin, 1974b), Micr., ESA, XRD
Auguste Renoir, *Lady at the Piano*, 1876, Art Institute of Chicago	(Butler, 1973), Micr. (mixed with vermilion)
German collection of pigments, 19th century, Darmstadt pharmacy	(Richter & Härlin, 1974a), mixed with red lake and barytes

XRD = x-ray diffraction; Micr. = microscopic examination; ESA = emission spectrographic analysis; Chem = microchemical tests; IR = infrared; * = private communication; ** Bayerische Staatsgemäldesammlungen.

NEAR EAST, INDIA AND FAR EAST

Artist or School Title, Date, Collection or Location	*Analyst and Method of Identification*
Egyptian, Two Fayum portraits, 2nd century A.D., Petrie Museum, University College, London. UC 19611, UC 19612	(Roy, 1979), Micr. Laser ESA
Egyptian, Fayum Portrait of a Woman, 3rd century A.D., Toledo Museum of Art	(Buck & Feller, 1972), XRD
Egyptian, Figure of a woman. Distemper on canvas, 3rd–4th century A.D., Private collection	(Sack et al, 1981), Micr., Chem.
Egyptian, Fayum Portrait of a Man, 4th century A.D., Petrie Museum, University College, London. UC 14768	(Roy, 1979), Micr. Laser ESA
Afghanistan, Buddhist, Wall paintings, late 6th century, Cave chapels, Bāmiyān	(Gettens, 1938a), Micr.
Iraqi, Manuscript, Qazwini, *Wonders of Creation*, 1370, Freer Gallery of Art, Washington, 54.33-54.114, 57.13	(E. FitzHugh), Micr.
Armenian archaic Mark manuscript, 17th century, University of Chicago. 972	(M. Butler*, 1974), Micr.
Nine Persian miniatures, Early 14th–17th centuries, Freer Gallery of Art, Washington	(E. FitzHugh), Micr.
Three Turkish miniatures, Early 16th–17th centuries, Freer Gallery of Art, Washington	(E. FitzHugh), Micr.
Persian miniature, *Shahnama of Firdowsy*, 17th century, Walters Art Gallery, Baltimore, W.603	(J. Caraher*, 1974), Micr.
Nine Indian paintings, Persian style, 15th century; Mughal, 16th–19th centuries; Jaipur, late 18th century; Agra, early 17th century, Freer Gallery of Art, Washington	(E. FitzHugh), Micr.
Indian Miniatures, Mughal, 1600–1640; Rajasthani, 18th century, Los Angeles County Museum of Art	(Johnson, 1972), Mass spectrometry
Indian wall paintings, Late 18th century. Devi Mural at Kulu	(Argrawal, 1969–70), Micr.
Indian Tanjore polychrome wood sculpture, Chauri-bearer, late 18th century, National Museum New Delhi. 62.679/13	(Agrawal, 1971), Micr.
Two Tibetan Thang-ka paintings, Lhasa style, 18th or early 19th century, Private Collection	(Mehra, 1970), Micr. by C.M. Verduyn-Groen
Japanese Buddhist wall paintings, 7th–8th centuries, Horyuji Temple, Nara	(Yamasaki, 1948 & 1967)
Japanese polychromed clay statue, 8th century, Tōdaiji Temple, Nara	(Yamasaki & Nishikawa, 1970), Micr., ESA
Japanese wall paintings, 951, Daigoji Pagoda, Kyoto	(Yamasaki, 1959b)
Japanese polychromed wood sculptures, 7th–13th centuries, Various	(Yamasaki & Nishikawa, 1970), Micr., ESA
Japanese, ten wooden panels, sacred figures, 14th century, Rinnō-ji temple, Nikko	(Nakazato & Tatsuta, 1971), x-ray fluorescence
Japanese, Early Heian, Kamakura Period, Seven paintings on silk or paper, 12th–14th centuries, Freer Gallery of Art, Washington	(R.J. Gettens*), Micr.

XRD = x-ray diffraction; Micr. = microscopic examination; ESA = emission spectrographic analysis; Chem = microchemical tests; IR = infrared; * = private communication; ** Bayerische Staatsgemäldesammlungen.

Artist or School *Title, Date,* *Collection or Location*	*Analyst and Method of Identification*
Japanese, Ashikaga, Pair 6-fold screens, Early 16th century, Freer Gallery of Art, Washington. 71.1	(R.J. Gettens*), Micr.
Japanese, Twenty-five Ukiyo-e prints, Late 17th–mid 18th centuries, Art Institute of Chicago	(Grove, 1971), Micr.
Japanese, Forty-six paintings on silk and paper, Late 17th–mid 19th centuries, Freer Gallery of Art, Washington	(FitzHugh, 1979), Micr.
Early Buddhist wall paintings, *Vision of Kuan Yin*, Kizil, Chinese Turkestan, 5th–8th centuries, A.D. Fogg Art Museum, Harvard University	(Gettens, 1938b), Micr.
Chinese wood sculptures, Tang, 8th–9th centuries, Freer Gallery of Art, Washington. 45.11, 45.12	(R.J. Gettens*), Micr.
Chinese stone sculpture, Northern Chi, 3rd–6th centuries, Freer Gallery of Art, Washington. 11.411	(R.J. Gettens*), Micr.
Chinese Buddhist wall paintings, 9th–10th centuries, Tun-huang, West China	(Warner, 1938), Gettens*, Micr.
Chinese Buddhist paintings on silk from caves of The Thousand Buddhas, Tun-huang, West China, 9th–10th centuries, British Museum, London, M. Aurel Stein Collection	(E. FitzHugh), Micr.
Chinese paintings on silk, Tun-huang, Sung, 10th and 11th centuries, Freer Gallery of Art, Washington. 30.36, 35.11	(R.J. Gettens*), Micr.
Central Asian wall painting fragments, Kara Khoto, 11th–13th centuries, Freer Gallery of Art, Washington, Study Collection, SC-PA-143	(E. FitzHugh), Micr.
Chinese wall painting, 1280–1360, Royal Ontario Museum, Toronto	(Moffat & Adair, 1982)
Ming Dynasty, Central China, wall paintings, 1551, Museum of Fine Arts, Boston	(Gettens, 1938c), Micr.

XRD = x-ray diffraction; Micr. = microscopic examination; ESA = emission spectrographic analysis; Chem = microchemical tests; IR = infrared; * = private communication; ** Bayerische Staatsgemäldesammlungen.

ACKNOWLEDGMENTS

John Winter and W. T. Chase of the Freer Gallery of Art made many helpful comments on early drafts of this paper. Elsewhere in the Smithsonian Institution, Walter R. Brown of the National Museum of Natural History kindly supplied the scanning electron micrographs, and John White of the Department of Mineral Sciences contributed useful samples of the minerals concerned. Hermann Kühn of the Deutsches Museum, Munich, provided a list of occurrences and made several valuable suggestions. Joyce Plesters of the National Gallery, London, Frank Preusser, formerly of the Doerner-Institut, Munich, now of the J. Paul Getty Conservation Institute, Santa Monica, California, and Ernst Ludwig Richter of the Institut für Technologie der Malerei, Stuttgart, also were generous in supplying lists of occurrences. Finally, this paper is dedicated to the memory of R. J. Gettens; it could not have been written without the recollection of his constant example.

REFERENCES

L. C. Afremow and J. T. Vandeberg, "High Resolution Spectra of Inorganic Pigments and Extenders in the Mid-infrared Region from 1500 cm^{-1} to 200 cm^{-1}," *Journal of Paint Technology* 38 (1966), 169–202.

O. P. Agrawal, "A Study on the Techniques of Indian Wall Paintings," *Journal of Indian Museums* 25–26 (1969–1970), 99–118.

O. P. Agrawal, "A Study of Indian Polychrome Wooden Sculpture," *Studies in Conservation* 16 (1971), 56–58.

S. Augusti, "Sul riconoscimento microchimico del catione Pb^{++} nei colori minerali a base di piombo," *Mikrochemie* verbunden mit *Mikrochimica Acta* 30 (1942), 237–240.

S. Augusti, "Sulla sensibilità della reazione con sulfuro sodico, per il reconoscimento microchimico dei cationi piombo ed antimonio," *Annali di Chimica Applicata* 36 (1946), 314–316.

S. Augusti, "Analysis of the Material and Technique of Ancient Mural Paintings," *Application of Science in Examination of Works of Art* (Boston, 1965), 67–70.

S. Augusti, *I colori Pompeiani* (Rome, 1967).

K. C. Bailey, ed. *The Elder Pliny's Chapters on Chemical Subjects*, Part I. Notes and trans. K. C. Bailey (London, 1929).

N. F. Barnes, "Color Characteristics of Artists' Pigments," *Journal of the Optical Society of America* 29 (May 1939), 208–214.

J. A. Bearden, "X-ray Wave-lengths," in *Handbook of Chemistry and Physics*, 53rd ed. (Cleveland, 1972–1973), E131–E169.

J. G. Bearn, *The Chemistry of Paints, Pigments and Varnishes* (London, 1923).

V. Borradaile and R. Borradaile, *The Strasburg Manuscript* (London, 1966).

Z. Brochwicz and D. M. Walkowa, "The Structure of the Seventeenth-century Mural Paintings in the St. George Orthodox Church at Weliko Turnowo (Bulgaria)," *Ochrona Zabytkow* 25, No. 3 (1972), 143–159.

M. Broekman-Bokstijn, J. R. J. van Aspern de Boer, E. H. van't Hul-Ehrneich, and C. M. Verduyn-Groen, "The Scientific Examination of the Polychromed Sculpture in the Herlin Altarpiece," *Studies in Conservation* 15 (1970), 370–400.

O. W. Brown and A. R. Nees, "A Study of the Variations of the Physical and Chemical Properties of Red Lead," *Journal of Industrial and Engineering Chemistry* 4 (1912), 867–876.

R. D. Buck and R. L. Feller, "The Examination and Treatment of a Fayum Portrait," *Conservation of Paintings and the Graphic Arts. Lisbon Congress Preprints, IIC* (London, 1972), 801–807.

M. H. Butler, technical note in *Paintings by Renoir* by John Maxon (Chicago 1973), 208–214.

T. Cairns and B. B. Johnson, "Recent Interfaces between Mass Spectrometry and Art," *Developments in Applied Spectroscopy* 10 (1972), 3–14.

E. R. Caley and J. F. C. Richards, *Theophrastus on Stones* (Columbus, 1956).

R. M. Candee, "Housepaints in Colonial America, Their Materials, Manufacture, and Application", *Color Engineer* 4–5 (1966–67).

E. M. Chamot and C. W. Mason, *Handbook of Chemical Microscopy*, vol. 2, 2nd ed., 7th printing (New York, 1960).

G. Champetier and H. Rabaté, eds., *Chimie de peintures, vernis et pigments*, vol. 2, *Pigments* (Paris, 1956).

M. Chandra, *The Technique of Mughal Painting* (Lucknow, 1949).

Colour Index, vol. 3, 3rd ed. (Bradford, Yorkshire, England, and Research Triangle Park, North Carolina, 1971).

A. M. de Wild, *The Scientific Examination of Pictures* (London, 1929).

M. Doerner, *The Materials of the Artist and Their Use in Painting*, rev. ed., trans. E. Neuhaus (New York, 1949), 75.

E. J. Dunn, "Microscopy in the Paint Industry," *Official Digest of the Federation of Societies for Paint Technology*, No. 247 (1945), 288–300.

E. J. Dunn, "Physical and Chemical Phenomena that Account for Red Lead's Metal Protective Ability," *Official Digest of the Federation of Societies for Paint Technology*, No. 260 (September 1946), 375–383.

E. J. Dunn, "Red Lead," in *Pigment Handbook*, vol. 1, ed. T. C. Patton (New York, 1973), 837–842.

E. J. Dunn, "Lead Pigments," *Treatise on Coatings*, vol. 3, *Pigments*, part I, ed. R. R. Myers and J. S. Long (New York, 1975).

A. J. Eikhoff, "Pigmentation of Coatings," *Pigment Handbook*, vol. 2, ed. T. C. Patton (New York, 1973), 49–55.

Y. Emoto, "On Deterioration of Pigment Layers of Screen and Wall Panel Paintings," *Science for Conservation* 12 (March 1974), 77–82.

R. L. Feller, "Analysis of Pigments," in *American Painting to 1776: A Reappraisal*, ed. I. M. G. Quimby (Charlottesville, Va., 1971), 327–344.

R. L. Feller, "Rubens's *The Gerbier Family*: Technical Examination of the Pigments and Paint Layers," *Studies in the History of Art* (Washington, 1973), 54–74.

E. W. FitzHugh, "A Pigment Census of Japanese Ukiyo-e Paintings in the Freer Gallery of Art," *Ars Orientalis* 11 (1979), 27–38.

F. Fleider, "Mise au point des techniques d'identification des pigments et des liants inclus dans la couche picturale des enluminures de manuscripts," *Studies in Conservation* 13 (1968), 49–86.

R. J. Forbes, *Studies in Ancient Technology*, vol. 3 (Leiden, 1955).

W. R. Fuller, *Federation Series on Coatings Technology. Unit 8. Inorganic Color Pigments*, section III (Philadelphia, 1968), 33–43.

R. J. Gettens, "The Materials in the Wall Paintings of Bāmiyān, Afghanistan," *Technical Studies in the Field of the Fine Arts* 6 (1938a), 186–193.

R. J. Gettens, "The Materials in the Wall Paintings from Kizil in Chinese Turkestan," *Technical Studies in the Field of the Fine Arts* 6 (1938b) 281–294.

R. J. Gettens, "Pigments in a Wall Painting from Central China," *Technical Studies in the Field of the Fine Arts* 7 (1938c), 99–105.

R. J. Gettens, "The Materials in a Painting by Spinello Aretino," *Bulletin of the Fogg Museum of Art* 10, no. 6 (1947), 188–193.

R. J. Gettens, "True and False Vermilion on Early Chinese Ceramics," *Far Eastern Ceramic Bulletin* 6, no. 1 (March 1954), 15–27.

R. J. Gettens, R. L. Feller, and W. T. Chase, "Vermilion and Cinnabar," *Studies in Conservation* 17 (1972), 45–69.

R. J. Gettens and G. L. Stout, *Painting Materials, A Short Encyclopedia* (1942; New York, 1966).

R. J. Gettens, H. Kühn, and W. T. Chase, "Lead White," *Studies in Conservation* 12 (1967), 125–139.

M. I. Gillibrand and B. Halliwell, "The Lead-oxygen System — I. Thermal Decomposition of Lead Dioxide," *Journal of Inorganic and Nuclear Chemistry* 34 (1972), 1143–1159.

N. Grove, appendix to *Ukiyo-e Prints and Paintings; the Primitive Period* by Donald Jenkins (Chicago, 1971), 128.

N. W. Hanson, "Some Painting Materials of J. M. W. Turner," *Studies in Conservation* 1 (1954), 162–173.

V. F. Hanson, "Quantitative Elemental Analysis of Art Objects by Energy-Dispersive X-ray Spectrometry," *Applied Spectroscopy* 27 (1973), 309–334.

R. D. Harley, *Artists' Pigments; c. 1600–1835, a Study in English Documentary Sources*, 2nd ed. (London, 1982).

J. R. Harris, *Lexicographical Studies in Ancient Egyptian Minerals* (Berlin, 1961).

J. G. Hawthorne and C. S. Smith, *On divers arts; the Treatise of Theophilus Translated from the Medieval Latin* (Chicago, 1963).

H. C. Hoover and L. H. Hoover, trans. *De re metallica by Agricola 1556* (New York, 1950).

A. P. Hornblower, "Some Examples of Work with the Electron Probe Microanalyser," *Archaeometry* 6 (1963), 37–42.

J. P. W. Houtman and J. Turkstra, "Neutron Activation Analysis and Its Possible Application for Age Determination of Paintings," *Radiochemical Methods of Analysis*, vol. 1 (Vienna, 1965), 85–103.

Infrared Spectroscopy — Its Use in the Coatings Industry (Philadelphia, 1969), Spectra 511 and 538.

JCPDS Powder Diffraction File, JCPDS International Centre for Diffraction Data (Swarthmore, Pennsylvania, 1982).

B. B. Johnson, "A Preliminary Study of the Technique of Indian Miniature Paintings," *Papers Presented in a Symposium at the Los Angeles County Museum of Art, October 1970*, ed. Pratapaditia Pal (Leiden, 1972), 139–146.

B. Keisch, "On the Use of Isotope Mass Spectrometry in the Identification of Artists' Pigments," *Studies in Conservation* 15 (1970), 1–11.

K. L. Kelly and D. B. Judd, *Color, Universal Language and Dictionary of Names* (Washington, 1976).

T. Kenjo and K. Toishi, "A Basic Study of the Prevention of Decoloration and Discoloration of Pigments. I," *Shikizai Kyokaishi* 37 (1964), 133–137.

The Kline Guide to the Paint Industry, 4th ed. (Fairfield, New Jersey, 1975).

H. Kühn, "Trace Elements in White Lead and Their Determination by Emission Spectrum and Neutron Activation Analysis," *Studies in Conservation* 11 (1966), 163–169.

H. Kühn, *Die Pigmente in den Gemälden der Schack-Galerie* (Munich, 1969).

H. Kühn, "Terminal Dates for Paintings Derived from Pigment Analysis," *Application of Science in the Examination of Works of Art* (Boston, 1973), 199—205.

E. S. Larsen and H. Berman, *The Microscopic Determination of the Nonopaque Minerals* (Washington, 1934).

M. Larsen, "The High Altar in the Cathedral at Åhrus — A Color Examination," *Meddelelser om Konservering* 1, no. 9 (1970), 3–18.

A. P. Laurie, *The Materials of the Painter's Craft* (London, 1910).

R. Levenson, "The Materials and Techniques of the Quebec Artists, 1760–1850," *Journal of Canadian Art History*, 7, no. 1 (1983).

M. Levey, "Medieval Arabic Bookbinding and Its Relation to Early Chemistry and Pharmacology," *Transactions of the American Philosophical Society* n.s. 52, part 4 (1962).

D. C. Locke and O. H. Riley, "Chemical Analysis of Paint Samples Using the Weisz Ring-Oven Technique," *Studies in Conservation* 15 (1970), 94–101.

G. Loumyer, ed. *Un traité de peinture du moyen-age: L'anonymus Bernesis* (Berne, 1908).

N. T. McDevitt and W. L. Baun, "Infrared Absorption Study of Metal Oxides in the Low Frequency Region (700–240 cm^{-1})," *Spectrochimica Acta* 20 (1964), 799–808.

V. A. Malevanny, "The Chemical Study of Archeological Pigments," *Soviet Archaeology*, no. 4 (1968), 224–233.

J. J. Mattiello, ed., *Protective and Decorative Coatings*, vol. 2. *Raw Materials: Pigments, Metallic Powders, and Metallic Soaps* (New York, 1942).

R. Mayer, *The Artists' Handbook of Materials and Techniques*, 3rd ed. (New York, 1970).

V. R. Mehra, "Note on the Technique and Conservation of Some *Thang-ka* Paintings," *Studies in Conservation* 15 (1970), 206–214.

H. E. Merwin, "Optical Properties and Theory of Color of Pigments and Paints," *Proceedings of the American Society for Testing Materials* 17 (1917), 494–520.

E. Moffatt and N. Adair, "The Lord of the Southern Dipper," *C.C.I. Analytical Report*, ARS1900, ARS Project 13D (Ottawa, 1982).

R. P. Multhauf, *The Origins of Chemistry* (New York, 1967).

V. J. Murrell, "The Restoration of Portrait Miniatures," *Conservation and Restoration of Pictorial Art*, eds. N. Brommelle and P. Smith (London, 1976), 129–133.

T. Nakazato and S. Tatsuta, "Note on the Techniques of Tableaux with Polycolored Painting of Sacred Figures at the Rinnō-ji Temple, Nikkō, and Conservation Treatment on Them," *Science for Conservation* 7 (1971), 13–26.

J. Needham, *Science and Civilization in China*, vol. 5. *Chemistry and Chemical Technology*, Part II (Cambridge, 1974).

R. A. Nyquist and R. O. Kagel, *Infrared Spectra of Inorganic Compounds* (New York, 1971), spectrum 327.

H. Oguchi, "Scientific Investigation on Color Materials in Japanese Painting," *Bulletin of the Faculty of Fine Arts* (Tokyo University of the Arts), no. 5 (1969), 27–82.

C. Palache, H. Berman, and C. Frondel, *Dana's System of Mineralogy*, vol. 1, 7th ed. (New York, 1944).

G. D. Parkes, ed., *Mellor's Modern Inorganic Chemistry* (New York, 1961).

P. V. H. Pascal, *Nouveau traité de chimie minérale*, vol. 3, fasc. 3. *Germanium, étain, plomb* (Paris, 1963).

D. Y. Perera and P. M. Heertjes, "Water Transport through Paint Films. Part V: Pigmented Films," *Journal of the Oil and Colour Chemists Association* 54 (1971), 774–794.

M. J. Peterson and H. W. Jaffe, *Visual Arc Spectroscopic Analysis* (Washington, 1953).

W. M. F. Petrie, *Hawara, Biahmu and Arsinoe* (London, 1889), 11.

L. E. Plahter, E. Skaug, and U. Plahter, *Gothic Painted Altar Frontals from the Church of Tingelstad* (Oslo, 1974).

U. Plahter and S. Wiik, "The *Virgin and Child* from the Church of Dal (Norway) — Examination and Restoration," *Studies in Conservation* 15 (1970), 304–315.

U. Plahter, "Isotope Ratios of Lead as Source Identification in Pigments from Norwegian Medieval Art, with Some Comments on Trace Element Composition of White Lead," *Universitetets Oldsaksamlings Årbok, 1972–1974* (1976), 136–138.

J. Plesters, "Cross-sections and Chemical Analysis of Paint Samples," *Studies in Conservation* 2 (1956), 110–157.

V. Radosavljevic, "The Technique of Old Writings and Miniatures," in *Conservation and Restoration of Pictorial Art*, eds. N. Brommelle and P. Smith (London, 1976), 202–206.

Raw Materials Index, Pigment section (Washington, 1975).

K. M. Reese, "Chinese Rediscover Old Means of Protecting Books," *Chemical and Engineering News* 56, no. 31 (July 31, 1978), 44.

E. L. Richter and H. Härlin, "A Nineteenth-century Collection of Pigments and Painting Materials," *Studies in Conservation* 19 (1974a), 76–82.

E. L. Richter and H. Härlin, "The Pigments of the Swiss Nineteenth-century Painter Arnold Böcklin," *Studies in Conservation* 19 (1974b), 83–87.

E. L. Richter and H. Härlin, "The Stuttgarter Kartenspiel — Scientific Examination of the Pigments and Paint Layers of Medieval Playing Cards," *Studies in Conservation* 21 (1976), 18–24.

J. Riederer, "Technik und Farbstoffe der frühmittelalterlichen Wandmalereien Ostturkistans," *Veröffentlichungen des Museums für Indische Kunst* (Berlin) 4 (1977), 353-423.

W. L. Roberts, G. R. Rapp, Jr., and J. Weber, *Encyclopedia of Minerals* (New York, 1974).

A. Roy, in "The Technology, Examination and Conservation of the Fayum Portraits in the Petrie Museum" by B. Ramer, *Studies in Conservation* 24, no. 1 (1979), 1–13.

S. P. Sack, F. C. Tahk and T. Peters, Jr., "A Technical Examination of an Ancient Egyptian Painting on Canvas," *Studies in Conservation* 26 (1981), 15–23.

E. H. Schafer, "The Early History of Lead Pigments and Cosmetics in China," *T'oung Pao* 44 (1955), 413–438.

M. N. Short, *Microscopic Determinations of the Ore Minerals*, 2nd ed. (Washington, 1940).

U. Simonsen and L. E. Plahter, "The Virgin from Hovland Stave Church in Eggedal — Examination and Conservation," *Universitetets Oldsaksamlings Årbok* (1963–1964), 79–97.

C. A. Smith, "Early Protective Coatings, Part 1," *Modern Paint and Coatings* (March, 1978), 59–62.

C. S. Smith and J. G. Hawthorne, "Mappae Clavicula — A Little Key to the World of Medieval Techniques," *Transactions of the American Philosophical Society* n.s. 64, Part 4 (1974).

E. Steinnes, "Trace Elements in the White Lead from Medieval Altar Frontals, Determined by Neutron Activation Analysis," *Universitetets Oldsaksamlings Årbok*, 1972–1974 (1976), 139–140.

T. Takamatsu, *On Japanese Pigments* (Tokyo, 1879–1880).

C. J. A. Taylor and J. Marks, eds., *Pigments, Dyestuffs and Lakes*, Oil and Colour Chemists' Association, Part 6 (New York, 1966).

J. Thissen, "Note de laboratoire sur deux peintures baroques conservée à Tamise et à Termonde," *Institut royal du Patrimoine artistique bulletin* 4 (1961), 207–214.

D. V. Thompson, *The Materials of Medieval Painting* (London, 1936).

D. V. Thompson, trans., *Il libro dell'arte, the Craftsman's Handbook of Cennino d'Andrea Cennini* (New Haven, 1933; New York, 1960).

R. C. Thompson, *On the Chemistry of the Ancient Assyrians* (London, 1925).

C. L. Waring and C. S. Annell, "Semiquantitative Spectrographic Method for Analysis of Minerals, Rocks, and Ores," *Analytical Chemistry* 25 (1953), 1174–1179.

L. Warner, *Buddhist Wall-Paintings: A Study of a Ninth-century Grotto at Wan Fo Hsia* (Cambridge, Massachusetts, 1938).

Weber, F. W., *Artists' Pigments, Their Chemical and Physical Properties* (New York, 1923).

K. Wehlte, *The Materials and Techniques of Painting*, trans. U. Dix (New York, 1975).

C. Welch, *The Houghton Shahnamah* (Cambridge, Massachusetts, 1981), 268.

A. F. Wells, *Structural Inorganic Chemistry*, 4th ed. (Oxford, 1975).

T. A. Wertime, "Pyrotechnology: Man's First Industrial Uses of Fire," *American Scientist* 61 (1973), 670–682.

H. Westhoff, E. L. Richter, B. Becker, and H. Meurer, "Der Kruzifixus von Schloss Lichtenstein," *Jahrbuch der Staatlichen Kunstsammlungen*, Baden-Württemberg (1979).

H. E. Wulff, *The Traditional Crafts of Persia* (Cambridge, Massachusetts, 1966).

K. Yamasaki, "The Chemical Studies on the Pigments Used in the Main Hall and Pagoda of Horyuji Temple," *Proceedings of the Japan Academy* 24, no. 2 (1948), 11–15.

K. Yamasaki, "Review of Conservation of Old Art Objects in Japan," *Studies in Conservation* 3 (1957), 83–88.

K. Yamasaki, "Chemical Studies on the Eighth-century Red Lead Preserved in the Shōsō-in at Nara," *Studies in Conservation* 4 (1959a), 1–4.

K. Yamasaki, "The Pigments Used in the Wall Paintings," *Wall-Paintings in the Daigo-ji Pagoda* (Tokyo, 1959b).

K. Yamasaki, "Pigments Employed in Old Paintings of Japan," *Archeological Chemistry*, ed. M. Levey (Philadelphia, 1967), 347–365.

K. Yamasaki and K. Nishikawa, "Polychromed Sculptures in Japan," *Studies in Conservation* 15 (1970), 278–293.

G. Zerr and R. Rübencamp, *A Treatise on Colour Manufacture*, trans. C. Mayer (London, 1908).

Green Earth

CAROL A. GRISSOM

NOMENCLATURE IN VARIOUS LANGUAGES

English: green earth
German: Grüne Erde
Russian: Земельная зелень
French: terre verte
Italian: terra verde
Spanish: tierra verda,

1.0 INTRODUCTION

Siliceous minerals of dull grayish green color, known generally as green earth, are found widely throughout the world and have long appealed to the artist. The pigment was used by the American Indian as well as by the artists of Pompeii and of the Ajanta caves in India. In the West, perhaps the most familiar use of the pigment was for the green underpainting and shadows in the flesh colors of medieval painting. Lightfast and inexpensive to prepare by crushing and grinding, green earth was ideal for the artist, although the hue was not as high in chroma as was that of malachite and azurite. In the modern day, the pigment is still being sold from the high-quality deposits found on Cyprus.

The present monograph will describe the characterization of a number of closely related green earth minerals for which we now have evidence of use or potential use by artists: celadonite, glauconite, cronstedtite, and chlorite. Because of similarities in these clay minerals, precise and unequivocal identification has largely defied the analyst until just the last few decades. Indeed, significant new information has been produced by mineralogists since the basic research was done for this manuscript in 1973. While results have been incorporated in this text, the reader would be well-advised to search the mineralogical literature for latest developments.

CAROL A. GRISSOM, senior objects conservator, Conservation Analytical Laboratory, Smithsonian Institution, Washington, D.C.

1.1 Brief Definition of the Pigment

Green earth is prepared principally from either of two closely related clay minerals, celadonite and glauconite. Celadonite occurs as a relatively pure substance found in small quantities in vesicular cavities (amygdules) or fractures in volcanic rocks. The mineral glauconite, less pure but more widely distributed, is often found in the form of more widely distributed small greenish pellets (green sand). Other greenish clay minerals may occasionally have been incorporated into pigments labeled as green earth. One reason for this circumstance is that green sands are morphologically rather than chemically defined and may be composed, partly or wholly, of other clay minerals such as montmorillonite, chlorite, and kaolinite (Burst, 1958a, 1958b).

This monograph will consider primarily celadonite and mineral glauconite, but also will include information regarding other greenish clays. Their inclusion is warranted by the author's instrumental analyses confirming a significant percentage of secondary minerals in commercial green earth pigments (see fig. 16).

Green earth is listed in the *Colour Index* (1971) as the historical pigment CI Pigment Green 23,

No. 77009. No present-day supplier is given although the pigment is indeed still available.

1.2 Current Terminology

The most common current designation for the pigment is either the English, green earth, the French *terre verte*, or the Italian *terra verde*. The principal source minerals, glauconite and celadonite, have similar chemical compositions, but the convention has been to retain the separate names, using the term glauconite for a mineral of sedimentary origin and celadonite for a mineral of altered volcanic origin. Only recently has research permitted the unambiguous distinction of the two minerals on the basis of fractional differences in ionic substitution (Buckley, et al., 1978).

The designation green earth or its foreign language equivalent has been applied to both minerals since their first formal study. An early description of the mineral celadonite by DeLisle (1783) was *terra verdi*. Shortly thereafter, Hoffmann (1788) employed the name *Grünerde*. De Brignoli de Brunnhoff (1820), using the term *la terre verte*, further identified the mineral as chlorite, which at that time was the appropriate category. The name celadonite was proposed by Glocker (1847), based on the French word *céladon*, meaning sea green.

One of the earliest descriptions of the mineral glauconite was by von Humboldt's 1823 phrase *Grün Erde* (Schneider, 1927). The designations *terre verte* and green sand or their equivalents were used in other countries (Burst, 1958b). *Glaukonit* from the Greek *glaucos*, meaning bluish green or gray, was adopted by Keferstein (1828).

The usage of glauconite to describe both a specific micaceous mineral as well as a particular morphological form (green sand) is unfortunate. The practice stems from the failure for many years to adequately define components of glauconite pellets. In 1958 Burst (1958a, 1958b), employing a modern analytical technique, x-ray diffraction, divided the materials into four classes, designating the first class as the mineral genus, glauconite, and the other three as rock categories (see 5.31). Burst's system has proved most useful for the study of green earths. In the presentation that follows, care has been taken to specify, as the case may be, "mineral glauconite" and "rock glauconite". Recently mineralogists have proposed that the term glauconite be used only for the mineral species, that mixed mineral pellets be identified by their constituents, e.g. (mineral) glauconite and chlorite, and that glauconitic be used in all other contexts (Buckley, et al., 1978, p. 381).

2.0 HISTORY

2.1 Archaic, Obsolete, and Misrepresentative Names

Green earth has a long history as an artists' pigment. In his chapter on the natural colors, for example, Vitruvius, in the first century B.C. (Granger, 1970), mentioned *creta viridis* (literally green chalk or green clay). The Greeks, he said, called it *Theodotion* because it was first found on the estate of a person named Theodotus. Traditionally, Vitruvius' *creta viridis* has been understood to refer to green earth.

In the first century A.D., Pliny described Appian green as a derivative of green earth, a cheap imitation of malachite. One translator conjectured that the name Appian green might be an emendation of *apiacum* or *apianum*, meaning parsley (Rackham, 1968). Merrifield (1967, p. ccxix) renders such an identification plausible: "Green pigments were also prepared from rue, parsley, columbine, and from the black nightshade. The juice of these plants was incorporated with green earth." In modern times as well, green earth has continued to be used as a substrate for dyes (see 3.5).

In the Middle Ages, green earth was in some instances known by terms derived from the Greek word for leek: *prason*. In the ancient literature, Theophrastus (Caley & Richards, 1956; Eichholz, 1965) and Pliny (Rackham, 1968) had described stones that they respectively called "*prasitis*" and "*prasius*"; these are probably identifiable as the modern mineral prase, a green chalcedony unrelated to green earth. A similar derivative first appears to be connected with green earth in the seventh century; according to Thompson & Hamilton (1933, p. 26), Isidore of Seville (c. 570–636) called green earth *prasina* in his encyclopedic *Originum sive etymologiarum*. Jehan le Begue, in his fifteenth-century *Table of Synonyms*, wrote that according to the *Catholicon* published in 1286, "prasis est creta viridis" (Merrifield, 1967, p. 33).

Unfortunately, a term derived from the root meaning leek did not invariably indicate green earth in the Middle Ages. In the *Mappae Clavicula*, of the seventh-to-eighth centuries (Smith and Hawthorne, 1974), the description of *prasinus* is inappropriate for green earth: "Prasinus terra est viridis, ex quo metallo manat argentum." Theophilus is ambiguous; in his twelfth-century treatise, he mentions two green pigments: *viridis*, a copper-based colorant, and *prasinus*, a green and black colorant that disintegrates in water. In their translation of Theophilus, Hawthorne and Smith (1979, p. 16) retain the word *prasinus*, stating in a footnote:

> *Prasinus* is usually regarded as earth green and Dodwell so translates. However, the easy disintegration of the material in water would be characteristic of a soluble or finely crystalline substance, not an earthy one, hence we have preferred not to translate the word. *Prasinus* is perhaps related to the leek-green gem *prasitis* or *prasius*.

In later treatises, such as the *Bolognese Manuscript* (fifteenth century) and *Paduan Manuscript* (mid-sixteenth century), a distinction between green earth and a leek green pigment is indicated by their separate entries (Merrifield, 1967, pp. 502–505, 648–650).

The *verdeterra* of Cennino Cennini (Thompson, 1960, p. xv) refers to green earth. However, the term has occasionally been translated as *Verditer* (Berger, 1973, pp. 142–143) which is an entirely different pigment, basic copper carbonate.

The most famous historical source of the pigment, that near Verona, Italy, was described as early as 1574 by Mercati in the catalogue of the Vatican mineral collection (de Brignoli de Brunnhoff, 1820, pp. 356–357). The designations *terra verde di Verona*, *terra di Verona*, Verona earth, and Verona green have long been used to denote pigment from this source. Its excellent reputation encouraged the adoption of the name for green earths from other sources.

Twentieth-century trade names sometimes also referred to specific sources: Bohemia, Cyprus, Tirol, and Venice; still other terms have been indiscriminately used to describe green earth as well as other greens: for example, permanent, celadon, and stone green (Beutel, 1907). In the *Colour Index* (1956), the following names are listed: Belgian Earth, Hessian Earth, Rhenish Earth, Saxon Earth, and Green Stone. The term Veronese green has occasionally been used, but generally this refers to pale viridian or, more rarely, to emerald green and chrome green (Mayer, 1970).

2.2 History of Use

The ancient use of green earth as a pigment is well documented. It was identified among the first sets of pigments ever technically examined, in jars from a shop at Pompeii (Chaptal, 1809), in a vase found in the baths of Titus in Rome (Davy, 1815), and in wall paintings at Pompeii (Raehlmann, 1910; Augusti, 1967). Green earth was not the only pigment used in the ancient world to achieve a green color, however. In Cretan fresco paintings, for example, mixed greens composed of Egyptian blue and ochre were found (Neuberger, 1969); in ancient Egypt, copper greens rather than green earth were employed (Lucas & Harris, 1963). Green earth may not even have been used extensively in manuscript illumination, although it is mentioned in *De arte illuminandi* (Thompson & Hamilton, 1933) (an extensive footnote, No. 12, to the latter text is noteworthy for its excellent discussion of early references to green earth and related pigments). In published results of the technical examination of illuminated manuscripts Flieder (1968), for example, reports only copper greens, although she did not analyze any specimens from fourteenth-century Italy.

Perhaps the most well-known use of green earth has been the underpainting of flesh in medieval painting. The tradition of employing greenish mixtures in rendering flesh is recorded in texts from as early as the eleventh century. The eleventh–twelfth century Mt. Athos manuscript (Schäfer, 1855) described a greenish black mixture called *Proplasmus* to be used for the shadows of skin color, and Theophilus (twelfth century) described a mixture containing *prasinus* employed for the same purpose (Hawthorne & Smith, 1979, pp. 17–18). Cennino Cennini gives an account in his fourteenth-century treatise, *Il libro dell'arte* (Thompson, 1960, p. 45):

> Then take a little terre-verte in another dish, well thinned out; and with a bristle brush, half squeezed out between the thumb and forefinger of your left hand, start shading under the chin, and mostly on the side where the face is to be darkest; and go on by shaping up the under side of the mouth; and the sides of the mouth; under the nose, and on the side under the eyebrows, especially in toward the nose; a little in the end of the eye toward the ear; and in this way you pick out the whole of the face and the hands, wherever flesh color is to come.

The purpose of this traditional underpainting technique is not elucidated in the contemporary literature. Cennino stated only that the green must "always show through a little" (Thompson, 1960, p. 94). One modern authority attributes this practice to green earth's transparency (Thompson, 1962); another considers that "a far less muddy effect is achieved than could be got by admixture of black or brown" as "the pure pink of the upper paint is neutralized to a reasonable half tone where the green which is the opposite or complementary of the pink shines through the thinning strokes" (Ruhemann, 1968). Ironically, in one of the best-known examples of green underpaint, the artist altered the traditional technique. In *Madonna and Child, St. John and Angels* at the National Gallery in London, an unfinished work attributed to Michelangelo, little of the green is left showing through the painted flesh (Laurie, 1967; Ruhemann & Plesters, 1964, p. 548).

Green earth was also used for gilding in medieval painting. Its clay properties made it a suitable bole, as noted by Cennino Cennini (Thompson, 1960, p. 30):

> And, if you temper it as I shall show you (for) the bole for gilding, you may gild with this terre verte in the same way. And know that the ancients never used to gild on panel except with this green.

Harley (1982) cites de Mayerne's reference to *bol vert*.

When working in oil, Cennini (Thompson, 1960, pp. 94–95) preferred verdigris to green earth. As oil painting increased in Italy during the High Renaissance, mention of green earth for use in easel painting practically ceased. Vasari (Maclehose, 1960), for example, discussed green earth only in the context of imitative paintings for decoration. Perhaps the use of an oil medium may also account for the absence of green earth from the palette of painters in the North. The pigment has not been found in the early Netherlandish paintings which have been sampled extensively at the Institut royal du Patrimoine artistique in Brussels, nor are there clear references to it in the contemporary literature (Stechow, 1966; Stechow, private communication, 1974).

The use of green earth seems to have declined after the Renaissance but did not cease entirely. Harley (1982) notes that toward the end of the seventeenth century only two English writers mention the pigment. Her investigations of eighteenth-century customs records support the view that the pigment was not then extensively used in England but was growing in popularity during that time. Green earth is the only green listed by Bardwell (1756), although a recent technical study shows that, in actual practice, he most often used a mixture of Prussian blue and Naples yellow (lead antimonate yellow) or yellow ochre to achieve a green color. Green earth was found in only one of fifteen Bardwell paintings studied by Talley and Groen (1975); moreover, the identification of the particular green mineral used, cronstedtite, is singular (see 6.0).

On the continent, in contrast to England, it seems the pigment was used extensively during the seventeenth century. Frequent references to green earth occur in painters' manuals published there throughout the seventeenth, eighteenth, and nineteenth centuries (de Mayerne ms 1620, published Berger, 1973 pp. 100–101; Corneille, 1684; Dupuy du Grez, 1699; Lindenbergh, 1742; Pernety, 1757; Nismes, 1760; DePiles, 1776; Watin, 1778; Field, 1835; Fielding, 1839; Fernbach, 1843, pp. 104, 111; *Handbook of Young Artists . . .*, 1865; *The Painter, Gilder . . .*, 1867).

2.3 Terminal Date

Green earth is still being prepared and sold today, both as dry pigment and in proprietary mixtures. One of the most important modern sources has been Cyprus (Winsor & Newton, 1974, private communication). The site near Verona is said by the local citizenry to have been abandoned after World War II.

3.0 GENERAL PIGMENT PROPERTIES

3.1 Color and Spectral Reflectance

The absorption of light by green earth results from an interaction of the two valency states of iron. A high Fe^{II}/Fe^{III} ratio has been measured by Mössbauer-effect spectroscopy for a bluish pigment (celadonite), while a low ratio has been found for a more yellowish green earth (see 5.37, table 3, and fig. 19) (Grissom, 1974). These ratios may be tentatively correlated with variations in pigment color.

Typical spectral reflectance curves of green earth are shown in fig. 1. The broad central peak with a slight shoulder on the short wavelength side is characteristic, with the percentage reflectance slightly higher at the blue end of the spectrum (400 nm) than at the red (700 nm). The wavelengths at which the maxima and the shoulder occur are not fixed. Note the displacement of the celadonite sample curve toward the blue end of the spectrum; the burnt green earth curve definitely shifts toward the yellow-red region.

The results of spectral reflectance measure-

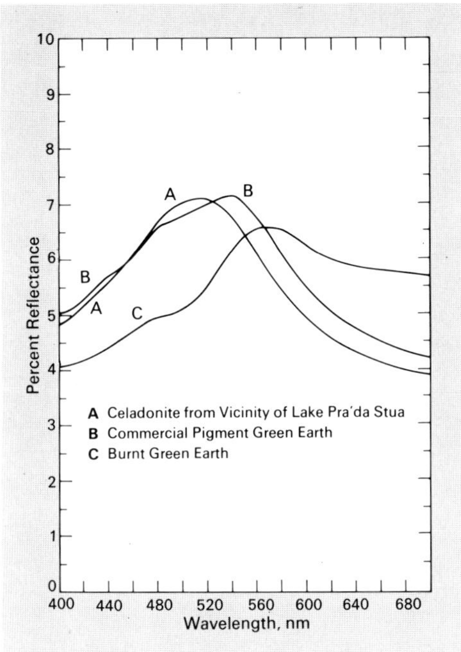

Fig. 1. Spectral reflectance curves for green earth pigments in Acryloid® B-72 acrylic resin (copolymer of methyl acrylate and ethyl methacrylate). (Note expanded vertical scale; the highest reflectance value is only 7% relative to barium sulfate white standard.)

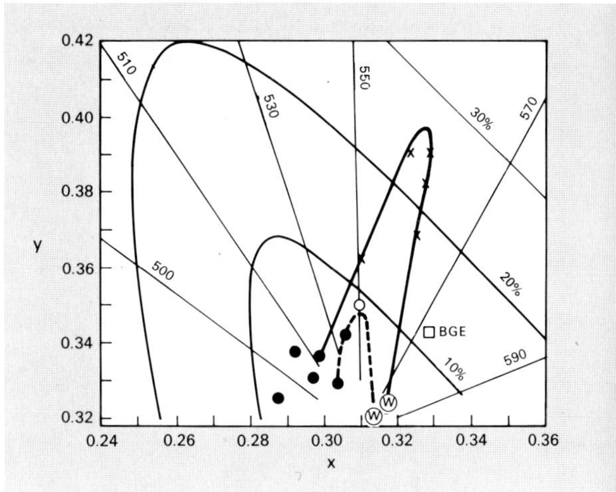

Fig. 2. Chromaticity coordinates for paints made with green earth showing dominant wavelengths in range of 500 to 590 nm and lines of 10, 20 and 30% purity for CIE Illuminant C. ● Masstone for various samples; ○ Barnes (1939) data; - - - Locus of chromaticities for mixtures with lead white; x—x Locus of chromaticities for series at incomplete hiding (glazes over white paint); Ⓦ Chromaticity of the white paint in each series; □ BGE Burnt green earth. All samples were dispersed in Acryloid® B-72 acrylic resin, except the Barnes sample, which was in an animal glue medium.

ments of the masstone color of four samples of green earth ground in acrylic resin are summarized in table 1 and their chromaticity coordinates plotted in fig. 2. These data indicate the range of color that may be encountered and are consistent with the chromaticity reported by Barnes (1939). The dominant wavelength of the masstone samples varies from 497 to 540 nm, with the celadonite sample possessing the shortest (bluest) wavelength; the excitation purity of the masstone colors is low, ranging from 4.2 to 8%. ISCC-NBS standard color names, based on these data, identify the color as dark grayish green or dark greenish gray (Kelly & Judd, 1976). The equivalent Munsell hues range from 2G to 8G, values from 2.7 to 3.0, and chromas from 1.0 to 1.5.

The paints used by the author for these masstone measurements were relatively dark, because they were formulated below the critical pigment-volume concentration (CPVC) in the acrylic medium. Partially for the same reason, green earths ground in an oil medium are usually dark. The effect of pigment concentrations above the CPVC would be to increase the scattering of light, producing a significantly lighter green paint. The relative lightness of the Barnes' green earth, noted in table 1 and fig. 2, may be attributed to the high pigment volume concentration (PVC), so frequently characteristic of a glue medium. The high PVC typifying tempera paints may account for the popularity of green earth in that medium.

Color measurements clearly demonstrate that when green earth is used as a glaze over white — when the paint is applied at incomplete hiding — both the excitation purity and the lightness increase significantly (fig. 2). Indeed, at least one manual for painters in oil recommended using green earth as a glaze (Bardwell, 1756, p. 38). To a lesser extent, mixing green earth with white also increases lightness and excitation purity. When the pigment is mixed with white or glazed over white, the color shifts toward yellow (fig. 2).

The color of burnt green earth is displaced toward yellowish brown; in the example given in table 1 and figs. 1 and 2, the dominant wavelength is shifted to 572 nm. The standard ISCC-

Table 1. **MEASUREMENTS ON THE COLOR OF GREEN EARTH PAINTS**[a]

Sample No.[b]	*Pigment*	*Vehicle*	*CIE Chromaticity Coordinates x*	*y*	*Luminous Reflectance Y*	*Dominant Wavelength (nm)*	*Purity (%)*	*Approximate Munsell Notation (hue/value/chroma)*	ISCC-NBS *Standard Color Names (Kelly and Judd, 1976)*
	Green earth (Barnes, 1939)	Glue	.3092	.3510	29.04	549.8	9.2	/2.9/	
40	Celadonite, Pra' da Stua	B72[c]	.2873	.3259	6.42	497	8.0	4G/3.0/1.5	Dark grayish green
34	Green earth, Fezandie and Sperrle	B72	.2971	.3311	5.84	505	4.2	8G/2.8/1.0	Dark greenish gray
38	Green earth, Winsor & Newton	B72	.2935	.3369	6.28	507.5	5.5	7.5G/2.9/1.5	Dark grayish green
35	Green earth, Fezandie and Sperrle	B72	.3056	.3422	5.51	540	6.0	2G/2.7/1.5	Dark grayish green
37	Burnt green earth	B72	.3276	.3423	5.58	572	11.8	7GY/2.8/2.8	Dark yellowish green

a. Data for first sample from Barnes (1939). For remaining samples, Robert L. Feller and Mary Curran, unpublished data on green earths over white, March 1974, based on spectral reflectance measurements. Samples may not be at complete hiding.
b. The first sample is identified as terre verte by Barnes (1939). Sample 40 is the bluish celadonite from Pra' da Stua, Monte Baldo, and samples 34, 35, 37 and 38 are green earths obtained from shelf stock at the Intermuseum Laboratory, Oberlin.
c. Acrylic resin, Acryloid® B-72, manufactured by the Rohm and Haas Company, Philadelphia.

NBS (Kelly & Judd, 1976) name for the hue is dark yellowish green.

The color of the secondary source mineral, chlorite, is distinctly bluish. Brownish tones are characteristic of montmorillonite, but its hue varies because it is easily colored by impurities and may be altered by absorption of dyes.

The reflectance of green earth in the infrared region is relatively low (Van Asperen de Boer, 1970).

3.2 Hiding Power and Tinting Strength

Because of its low refractive index, about 1.62 (Gettens & Stout, 1966), green earth exhibits little light scattering in oil; the resultant paint is relatively translucent. Tinting strength and the hiding power are poor. Highest saturation (chroma) is obtained when the pigment is glazed over white (see fig. 2) or when it is used in traditional water-dispersed mediums, such as glue or egg tempera, vehicles that normally yield films at high pigment volume concentration.

3.3 Permanence

Green earth is a "permanent" pigment, stable to light and air. If, over a period of time, varnishes or oils gradually penetrate the particles, wetting or filling the submicroscopic interstices, darkening may occur; but this would not be the consequence of instability in the mineral. Green earth was included in the list of permanent pigments published by the German Society for Rational Painting Techniques (Munkert, 1905).

3.4 Compatibility

The pigment is compatible with all media and all other pigments. It was much used in fresco painting where green pigments based on copper could not be employed.

3.5 Chemical Properties

Green earth is partially soluble in acids and alkalis, readily providing sufficient ferrous and ferric ions for microchemical tests (see 5.2). When heated, the pigment turns brownish in color forming what is known as burnt green earth (see table 1 and figs. 1 and 2). In spite of the color change, which may be due to conversion of much of the ferrous iron to ferric, the infrared and x-ray diffraction patterns of burnt green earth samples examined by the author could not be differentiated from those of the ordinary green pigment.

Because of the absorbency of this clay mineral, it has long been a substrate for dyes, particularly in the preparation of lakes with more intense green shades (Bearn, 1923; Doerner, 1949). Basic dyes have been frequently used in this process; for example, the CI Basic Green 1, listed in the *Colour Index* of 1971, yields a colorant known as Lime Green.

3.6 Oil Absorption and Grinding Qualities

The principal minerals from which green earths are derived have a Mohs hardness of 2; when pure, the raw material can be easily ground. The specific gravity for glauconite is 2.79 g/cm^3 (Bentor & Kastner, 1965). Because the pigments are of the clay mineral family, the oil absorption is moderate to high (Grissom, 1974). Munkert (1905) states that Verona earth requires 100% oil; Bohemian earth requires 80% oil; burnt green earth requires about 70%. Cennino Cennini described green earth as a "fat" color, by which he probably meant that it absorbed a large amount of the media (Thompson, 1960, p. 30).

3.7 Toxicity

As with most well-washed clays and earths, green earth poses no particular toxicity problem.

4.0 COMPOSITION

4.1 Chemical Composition

The primary source minerals, glauconite and celadonite, have layer silicate structures similar to that of mica. Fig. 3B illustrates the basic structural unit, composed of a layer of octahedrally coordinated cations (Al, Fe^{II}, Fe^{III}, Mg) sandwiched between two layers of silicate tetrahedra. Interlayer potassium ions bind units together. However, the relative strength of the silicon-oxygen bonds in the xy plane (fig. 3A) results in easy cleavage at the potassium interlayer; hence, the minerals have a platy morphology.

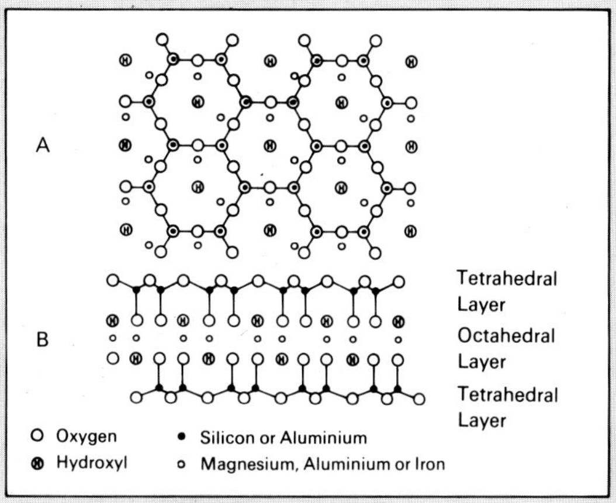

Fig. 3. Schematic representation of the structure of glauconite-celadonite. Diagram after Farmer and Russell (1964). A. Single silicate sheet, parallel to the plane of cleavage (xy plane). B. Tetrahedra of two silicate sheets linked by cations in octahedral coordination, shown normal to the plane of cleavage.

The chemical composition of both minerals varies because of ionic substitutions. This can be expressed by the general formula:
$K[(Al, Fe^{III}), (Fe^{II}, Mg)]\,(AlSi_3, Si_4)\,O_{10}(OH)_2$.
Although the chemical compositions of the minerals are very similar, glauconite has been differentiated from celadonite by its slightly greater number of trivalent ions in the octahedrally coordinated layer and the partial substitution of Al^{III} for Si^{IV} in the tetrahedrally coordinated layer (Hendricks and Ross, 1941; Buckley, et al., 1978). Fig 4, based on the work of Yoder and Eugster (1955, p. 257), illustrates the interrelations of these and a number of closely related minerals. The valency variations of the ions in the octahedral layer are of particular interest, because ferric/ferrous ratios may be correlated with variations in color (see 5.36).

Fig. 5 gives the schematic structures of some primary and secondary minerals that have been employed as green earth. The composition and structure of montmorillonite is similar to that of glauconite and celadonite but the interlayer spaces between the structural units in the first are expanded by water. Chlorites have a structural unit containing a component similar to that of the primary source minerals, but here an additional layer of aluminium or magnesium ions is octahedrally coordinated to hydroxyl groups. The septechlorite mineral, cronstedtite,

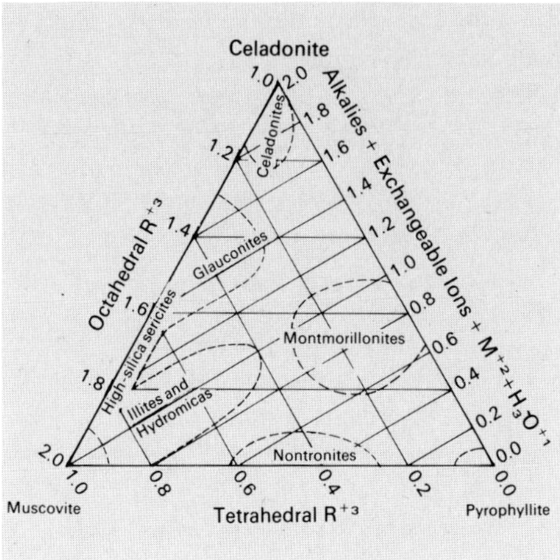

Fig. 4. Relation of tetrahedral and octahedral trivalent ions in dioctahedral micas and minerals (after Yoder and Eugster, 1955, p. 257).

Fig. 5. Schematic representation of the structure of several clay minerals (after Millot 1970). ○ Oxygen; ◎ (OH); ● Silicon or aluminium; ⊘ Aluminium, magnesium, or iron.

is chemically similar to chlorite, but structurally related to the serpentine and kandite minerals. Cronstedtite has a layered structural unit composed of a single tetrahedral component linked to a single tri-octahedral component.

4.2 Sources

The traditional distinction between the primary source minerals of green earth, celadonite and glauconite, arises more from their dissimilar occurrences and formations than from their hardly detectable chemical differences. Glauconite, an authigenic mineral of marine origin, occurs abundantly throughout the world, often as small dark green pellets that readily give rise to the appellation "green sand" (see fig. 7). These pellets may also contain the clay minerals chlorite, montmorillonite, and illite. Celadonite is a less common mineral, usually occurring as a powdery, blue-green substance in vesicular cavities (amygdules) or fractures in volcanic rocks such as basalt (fig. 6, fig. 7).

The abundance of green earth throughout the world was probably clear even in ancient times

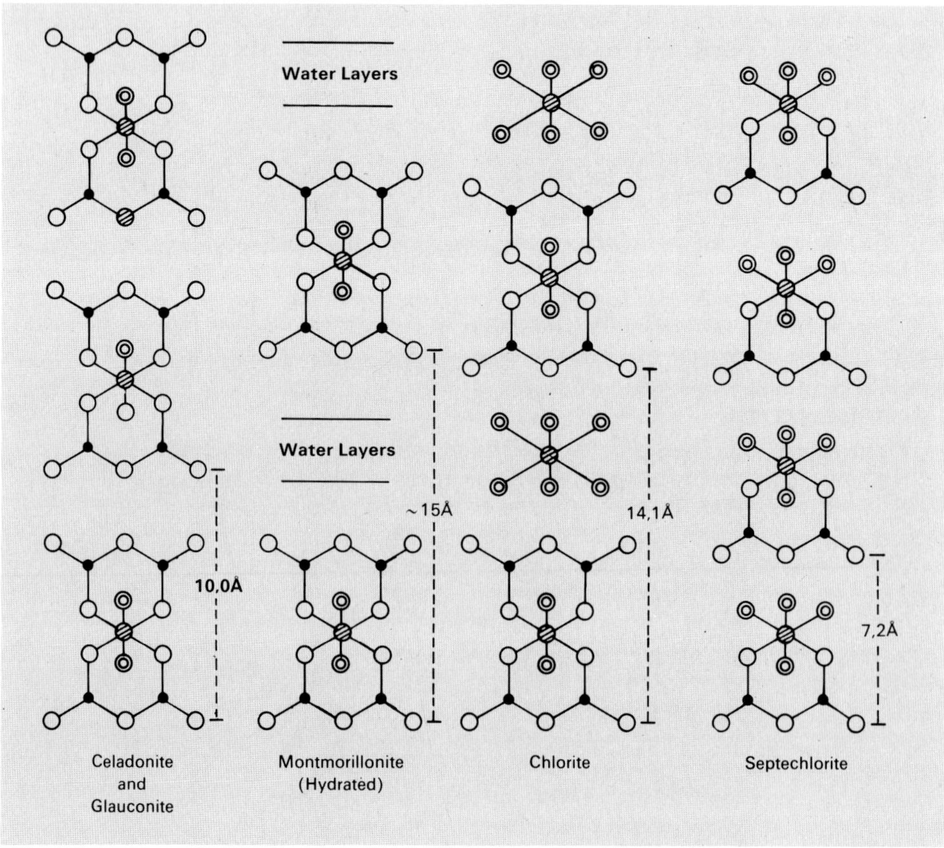

and may account for Pliny's description of a derivative pigment as "cheap" (Rackham, 1968). Qualitative differences between sources have long been recognized as well. The superiority of green earth from Smyrna, a port in Asia Minor, was asserted by Vitruvius (Granger, 1970). Isidore of Seville specified the pigment from Lybia Cyrenensi, i.e., Cyrene, now Ain Shahat-Greena, Libya (Thompson & Hamilton, 1933). Cennino Cennini (Thompson, 1960, p. 27) described veins of green earth, yellow ochre, and sinoper in the mountains of Colle di Val d'Elsa. As noted earlier, the excellent earth from Monte Baldo near Verona was already mentioned in the literature in 1574 (de Brignoli de Brunnhoff, 1820, pp. 356–358); apparently the deposits were mined there until World War II (Grissom, 1975). This green earth, a celadonite, has long been considered the best variety, on the basis of its high chroma and bluish green color (Pernety, 1757; Thompson, 1962, pp. 76, 85).

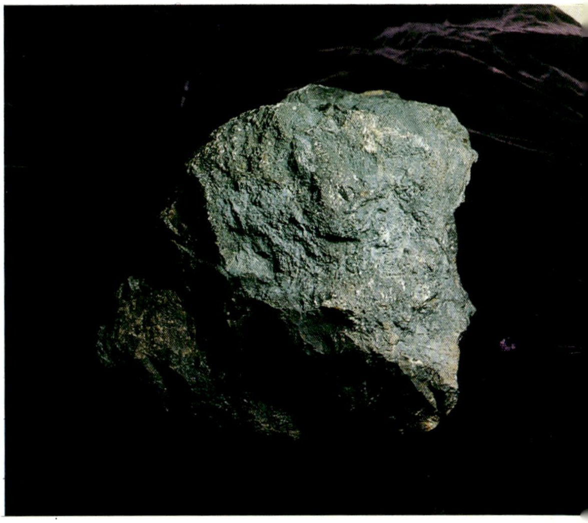

Fig. 6. Celadonite mineral from Pra' da Stua, Monte Baldo. Specimen approximately 8″ in diameter.

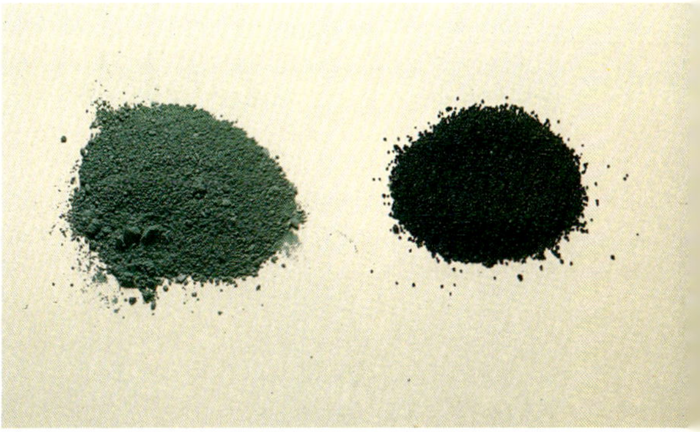

Fig. 7. Celadonite, left, from Pra' da Stua, Monte Baldo, and rock glauconite (green sand), right, from New Jersey. Each pile approximately 8″ in diameter.

In the nineteenth-century literature, other sources of green earth were noted: the Tirol, Bohemia, Saxony, Poland, Hungary, France, Cypress, and the Mendip Hills of England (Leuchs, 1825); Cornwall, Silesia, Moravia, Elbe, and Tunaberg (Lemoine & du Manoir, 1893). The yellow-green Bohemian earth was esteemed by the dye industry because its considerable montmorillonite content allowed it to absorb a high percentage of organic dyes (Remington, 1942). The geological sources of French green earth are reportedly glauconite (Charrin, 1947, 1948, 1955). The Cornwall source is probably the septechlorite mineral, cronstedtite, as shown in the analysis by the Department of Geology and Mineralogy, University Museum, Oxford (JCPDS 17-470). Fernbach (1843, p. 111) referred to the transport of green earth from Cyprus to Europe as ballast in Dutch ships, and Winsor & Newton have long obtained green earth from A. L. Mantovani and Sons in Larnaca, Cyprus (Winsor & Newton, private communication, 1974).

There are references to stilpnomelane, hedenbergite, lievrite, and amphibole which are singular within the literature and should be accepted as sources of green earth with caution; the properties of these minerals do not seem those necessary for the production of good pigments. Lemoine and du Manoir (1893) identify Silesia and Moravia as the source for stilpnomelane; Tunaberg, Sweden, for hedenbergite; and Elbe for lievrite. More recently, Kittel (1960) describes amphibole found at Heiger in Westerwald as a source of green earth.

American Indians presumably obtained the pigment from the large glauconite deposits in the Missouri River basin (R. J. Gettens, private communication, 1974). "Blue earth" from the Blue Earth River in Minnesota has also been suggested as a source of Indian pigment (John C. Ewer, private communication, 1978). Maire (1908) remarked upon sources of green earth in New England in the early twentieth century, stating that "many manufacturers grind it under proprietary names. Manufacturers of mixed paints can also use it to good advantage for certain tints on account of its imparting good wearing qualities and because of its cheapness."

Specimens from specific deposits are often associated with particular pigmentary properties such as color, fineness, or purity, but once the material has been ground into pigment, it may not even be possible to identify the geological source (marine or volcanic) except by analysis with more than one sophisticated instrument. Identification of geologically associated minerals is not helpful for determination of the source, because the same accessory minerals — principally quartz, various clays, calcite, iron oxides, and feldspars — are common to both glauconite and celadonite.

4.3 Preparation

Green earth is prepared simply by grinding. Sometimes levigation and acid treatment are used to eliminate impurities.

4.4 Adulteration and Sophistication

As noted, green earth was often used as a substrate for the absorption of dyes. Hence the investigator should be alert to the possible presence of an absorbed dyestuff, either on the principal green earth minerals or on a mineral such as montmorillonite. A basic dye, CI Basic Green 1, is used in this way to produce a pigment known as Lime Green, which is often employed with lime, sodium silicate, and size media (see 3.5).

5.0 IDENTIFICATION — CHARACTERIZATION

5.1 Characterization by Optical and Scanning Electron Microscopy

For all practical purposes, the optical characteristics of celadonite and mineral glauconite are identical. The author cannot state at present whether a specialist might be able to tell the two apart under the microscope. It may also be difficult to differentiate rock glauconites, but it is possible to distinguish certain other clay minerals.

Fig. 8 illustrates the relation of the crystal faces, crystal axes, and optic axes of celadonite and glauconite. The minerals are monoclinic with the following refractive indices: α = 1.592–1.610, $\beta = \gamma$ = 1.614–1.641; birefringence (γ–α) varies from 0.014–0.030. The minerals are pleochroic: $\alpha = \beta$ = greenish yellow or yellow, and γ = bluish green or yellowish green. The crystals are biaxial negative, and the acute bisectrix (*X*) is nearly perpendicular to the (001) cleavage. 2V ranges from 0–20°. The dispersion is distinct ($\rho < v$) (Deer, et al., 1963; Milner, 1962; Tröger, 1971).

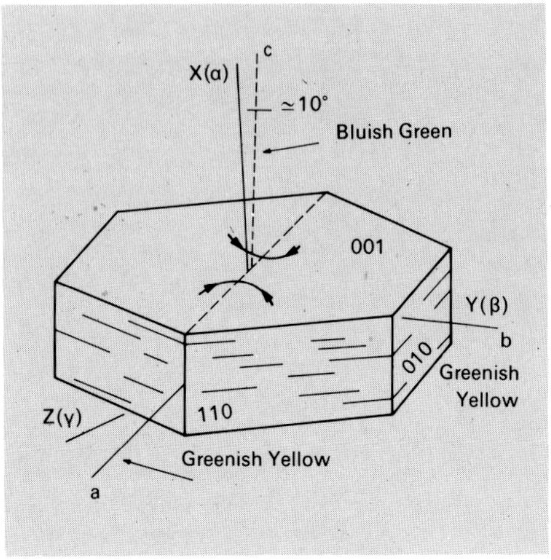

Fig. 8. Schematic representation of celadonite/glauconite, showing the crystal faces; the crystal axes a, b, c; the indices α, β, γ (equivalent to the vibrational axes X, Y, Z), with X the acute bisectrix. Note that $\alpha \neq$ c and $\gamma \neq$ a (after Tröger, 1971).

Green earths usually appear as an extensive mixture of rounded translucent particles of various shades of green intermixed with traces of yellow or brown earths (fig. 9). Some particles appear to have mottled or "grainy" textures. To an experienced microscopist the low mottled birefringence under plane polarized light often presents a characteristic signature for the pigment.

Under the microscope green earth particles are generally viewed in their preferential orientation, lying on the (001) face. In this position they appear as rounded translucent grains of low profile in Canada balsam, usually greenish yellow or yellow. A single particle often occurs as an aggregate of platelets. This structure produces the highly characteristic granular texture, especially when viewed under crossed polars with plane polarized light (fig. 10B) or with a retardation plate in position (fig. 10C). Occasionally a particle has the high degree of optical orientation typical of a single crystal; this phenomenon may also be readily confirmed by the use of a retardation plate, which shows a single color in such a particle.

Less commonly, particles are viewed in the "edge-on" orientation. In this position, they may assume the appearance of a stack of parallel

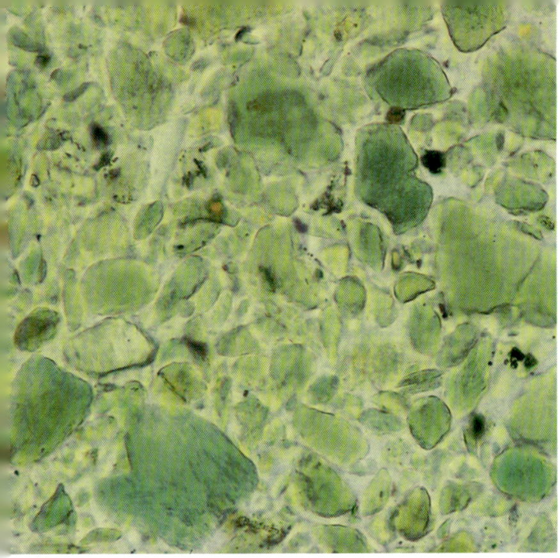

Fig. 9. Typical green earth sample, mounted in medium of refractive index 1.53 with some substage shadowing to bring out shape of the particles. 500x.

Fig. 10. Particles of green earth pigment showing appearance A. under normal transmission, B. under plane polarized light, and C. with retardation plate in position. Examination with the retardation plate shows solid color particles (high degree of orientation) and mottled particles (low degree of orientation). Note also the central particle of fanned or radiated plates. Sample of green earth from Fezandie and Sperrle, New York, in 1950s. Magnification 500x. Sample mounted in medium of refractive index 1.53 without substage shadowing.

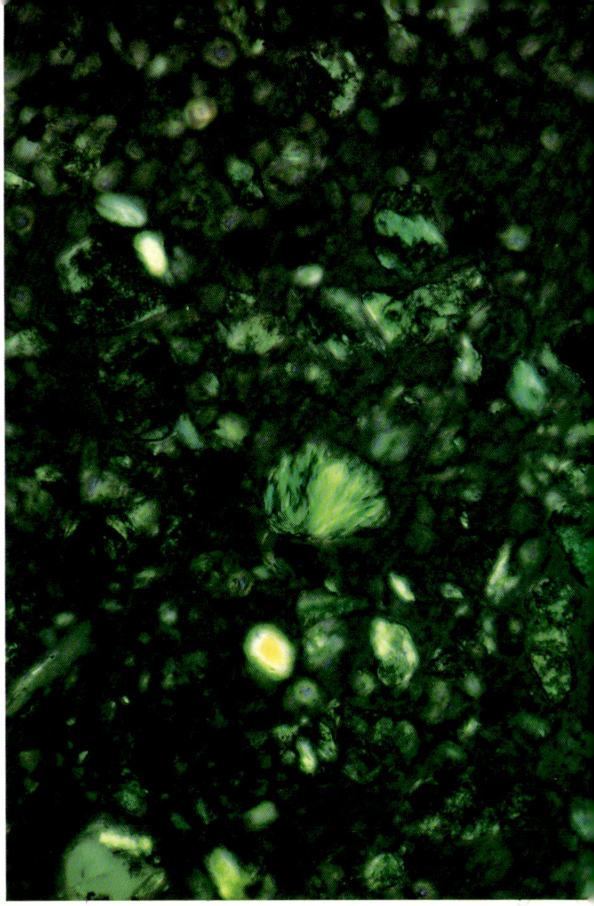

B

A

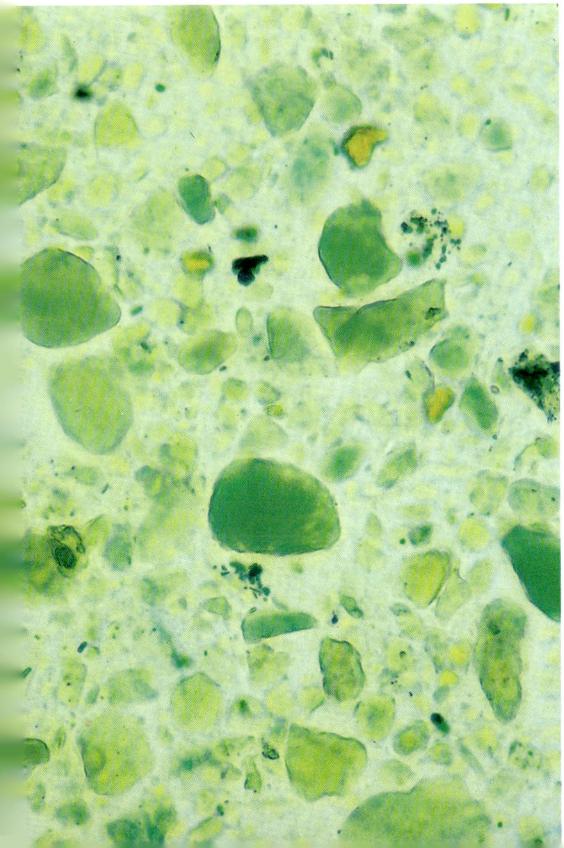

C

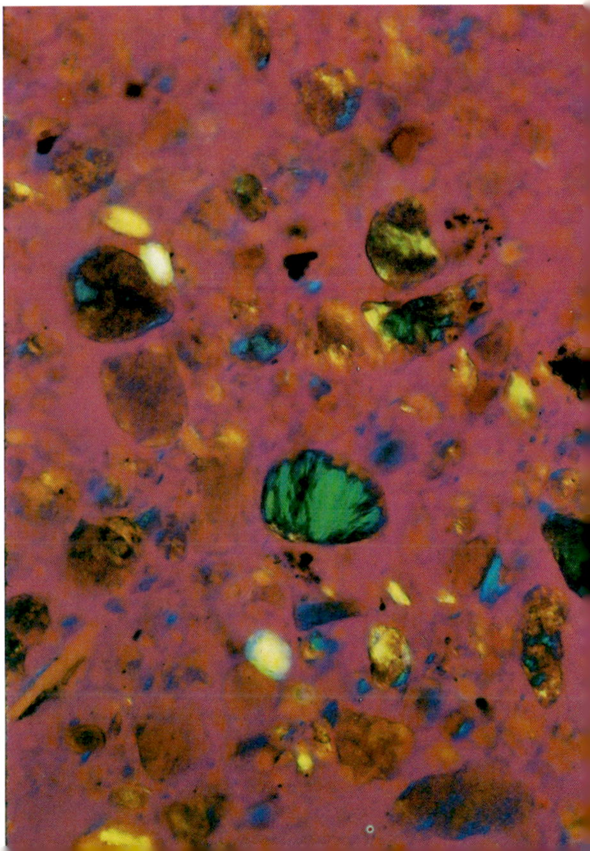

plates or of a bundle of fanned or radiated plates. Because the minerals are pleochroic with γ generally darker than $\alpha = \beta$, the edge-on particle usually appears darker than the preferentially oriented particles. In practice, the higher bluish green chroma (saturation) often enables the viewer to easily spot the edge-on particles in a field of green earth.

Particles range in size from 0.5 to 50 micrometers in diameter in commercial samples, but in samples from paintings they tend to measure less than 10 micrometers. The birefringence colors are usually first-order white to pale yellow for pigment-sized particles. With a "first-order red" retardation plate, the addition color is usually a second-order greenish blue and the subtraction color, a first-order bright yellow (see fig. 10C). In other words, the particles have low to moderate birefringence when viewed under crossed polars. Oriented preferentially, the superimposed lamellae generally extinguish in a randomly undulating fashion as the microscope stage is rotated. An occasional high degree of optical orientation produces simultaneous extinction throughout an entire particle. On the other hand, when lamellar particles are viewed edge-on, they exhibit parallel (straight) extinction. Extinction of radiated plates tends to proceed from spoke to spoke as the microscope stage is rotated. With a retardation plate inserted, the particles viewed edge-on appear length slow (positive elongation).

The color of burnt green earth particles is brownish orange by ordinary or plane-polarized illumination. Under crossed polars, they appear optically identical to unburnt particles.

Montmorillonite and chlorite, secondary mineral constituents of green earth, are also monoclinic and possess platy morphologies. Montmorillonite usually occurs as scales and plates too small to detect with an optical microscope. Aggregates in the shape of shards, pseudomorphous after volcanic glass, are common, as are vermicular clusters. The birefringence of montmorillonite is similar to that of celadonite and glauconite, but its refractive indices are generally lower. Iron-bearing types are pleochroic in shades of yellow, brown and green. Chloritic green earths tend to have large platy particles, often with a conchoidal fractured appearance. Chlorites are distinguished by their low birefringence and refractive indices; however, it should be noted that the latter increase, as does the color, with an increase in ferric or ferrous ion (Heinrich, 1965). Fig. 11 illustrates the particle morphology of chlorite, which is noticeably different from that of the usual glauconite/celadonite green earth.

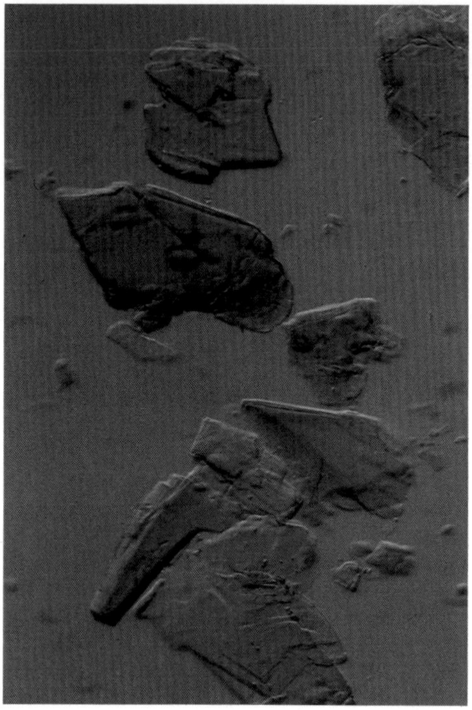

Fig. 11. Photomicrograph of a chlorite mineral. Sample labeled "Terra Verde di Verona, 'molto buono,'" Forbes Collection 9.01.24, obtained by Edward W. Forbes in Venice, 1923; Grissom sample 24 (1974). Verified by infrared and x-ray diffraction analysis as a chlorite. Particles have little noticeable birefringence but often have small inclusions that are highly birefringent. Sample mounted in medium of refractive index 1.53 with substage shadowing. Magnification 300x.

Associated minerals, such as feldspars, clays, calcite and especially iron oxides, are usually observed mixed with the green particles in varying degree.

Scanning electron micrographs have been published by a number of authors. Buckley et al. (1978) characterize a vesicular celadonite as radiating clusters of needles, distinct from the appearance of glauconite pellets as either platy crystals, often with swirling foliation, or crystals with more ordered appearance. Earlier Beutelspacher and Marel (1968) found that particles of celadonite from the Verona area consisted of well-developed blades, while glau-

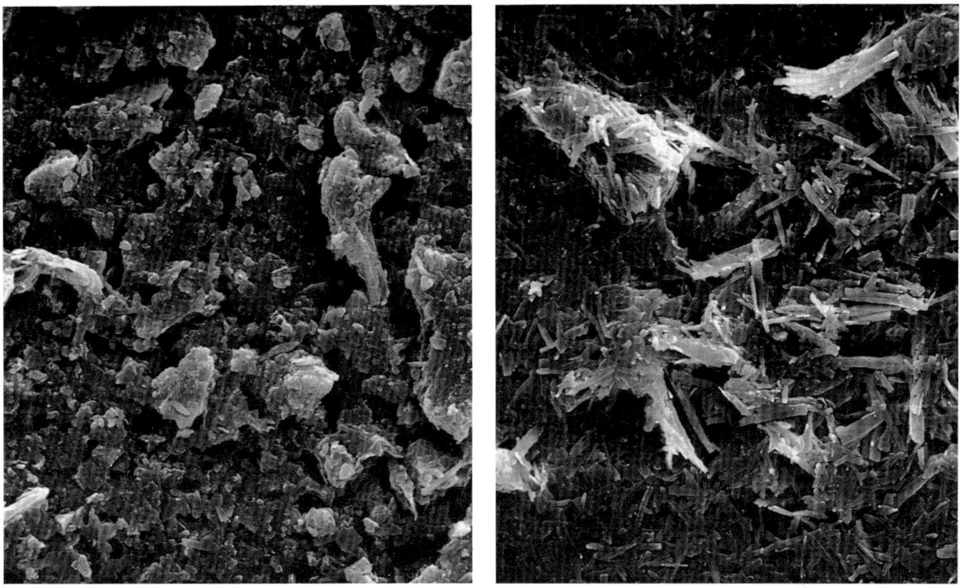

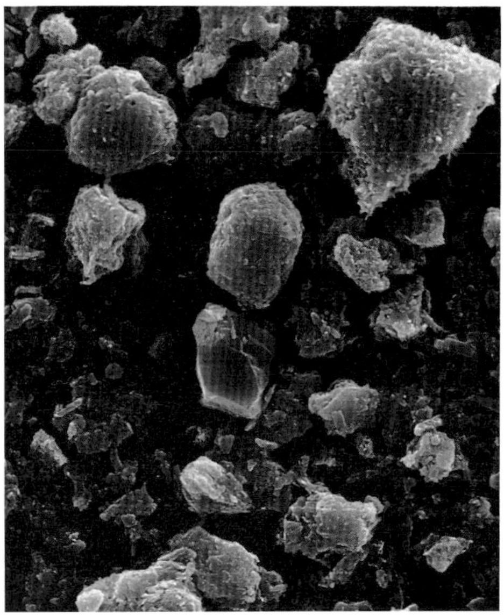

Fig. 12. Scanning electron micrographs of green earth pigments. A. "Terra di Verona," purchased in Venice, 1928, Forbes Collection 9.01.06. Grissom sample 6 (1974). B. Pulverised celadonite sample, Terra di Verona, obtained by the author from Pra' da Stua, Monte Baldo, 1973; Grissom sample 40 (1974). C. Sample of green earth obtained from the Fezandie and Sperrle Company, New York, in 1950s. Magnification 2,000x. Courtesy of John Winter, Freer Gallery of Art, and Walter R. Brown, National Museum of Natural History, Smithsonian Institution, Washington.

conite particles tended to be aggregated, striated lengthwise by cleavage with ends broken in an irregular way, and mostly opaque because of high iron content. For the present study micrographs of Verona and other green earths were made, revealing bundles of micaceous flakes (fig. 12). It should be noted that if the instrument is equipped for elemental analysis, one can readily detect iron, aluminum, and silicon.

5.2 Chemical Identification

Microchemical methods may be employed to identify both the ferric and ferrous ion in green earth, provided the ions are put into solution by treatment with acid. Detection of the presence of both ferric and ferrous ions would distinguish green earth from other green pigments with reasonable assurance. Chrome green contains iron but this mixture of blue and yellow pigments is usually recognizable upon microscopic examination at 100 to 1000x. Furthermore, the disclosure of the Prussian blue upon dissolution of the chrome yellow by acid treatment precludes chrome green being confounded with green earth. In contrast, the Prussian blue component can be dissolved by using alkali.

It has been reported that the Fe^{2+}/Fe^{3+} ratio of green earth minerals can be determined using a modification of the ferrous 2,2′-dipyridyl iron complex. This is said to be useful for distinguishing celadonite from mineral glauconite, if accompanied by determination of total iron, as by microprobe (Buckley, et al., 1978, p. 380).

5.21 ***Decomposition of Specimen with Acid.*** Green earth is only partially soluble in acids. Treatment with acid may result in a certain amount of effervescence from carbonate impurities in the specimen, such as chalk, lead white, and copper carbonate pigments. For the purposes of the following tests for iron, it has been the author's experience and that of Plesters (1956) that the use of concentrated hydrochloric acid is preferable. The solutions need not be evaporated to dryness.

5.22 ***Alpha, alpha′-dipyridyl Test for Ferrous Iron.*** Because iron oxide pigments as well as green earth may yield the ferric ion, a test for ferrous ion is recommended. The organic base, alpha, alpha′-dipyridyl, reacts with ferrous salts in acid solutions to give a stable, deep red complex cation. This reagent provides perhaps the simplest chemical test for green earth; no other common pigment readily yields ferrous iron. The test, based on that of Feigl and Anger (1972, pp. 263–264), is highly sensitive, and other metallic ions do not interfere.

The procedure is initiated by adding a drop of 2% alpha, alpha′-dipyridyl in alcohol to an acidified test solution on a microscope slide. A positive reaction produces a red solution. A spot test may also be made on dried filter paper impregnated with the reagent. The limit of identification is 0.03 micrograms of iron, and the limit of dilution is 1:1,666,000. Less costly reagents such as alpha, alpha′-phenanthroline may be substituted for the alpha, alpha′-dipyridyl.

Feigl and Anger state that large amounts of ferrous ion can be detected in the presence of ferric ion because the pink color of the ferrous dipyridyl complex is easily visible, even in a yellow solution of ferric salts. If there is a danger that the yellow color of a concentrated ferric solution might mask the slight pink coloration of small amounts of ferrous ions, the ferric ion can be converted to the colorless hexafluoride ion $(FeF_6)^{-3}$ by the addition of potassium fluoride. This is done by placing a drop of the acid test solution in a porcelain microcrucible coated with paraffin and decolorizing by the addition of a few crystals of potassium fluoride. The coating of paraffin prevents the silicate crucible from being attacked by hydrofluoric acid. A drop of alpha, alpha′-dipyridyl produces a red or pink coloration in the presence of the ferrous ion.

5.23 ***Other Tests.*** Plesters (1956) mentions a second test for the ferrous ion that can be made with chemicals that are usually more readily at hand than the above: the green precipitate of $Fe(OH)_2$, formed upon treatment of green earth with NaOH. This precipitate can be further oxidized subsequently to brown ferric hydroxide by hydrogen peroxide.

5.24 ***Potassium Ferrocyanide Test for Ferric Iron.*** Potassium ferrocyanide, which forms Prussian blue, $Fe_4[Fe(CN)_6]_3$, with ferric salts in acid solution, usually gives a satisfactory test for iron in a sample of green earth that has been treated with concentrated hydrochloric acid. However, the reaction may be masked in the presence of large proportions of other metal salts that also form insoluble ferrocyanides; azurite, for example, would yield brown copper ferrocyanide.

The test can be carried out by adding an equal amount of ferrocyanide solution to the acid test solution on a microscope slide. The ferric ion gives a dark blue precipitate or, when only a small amount of iron is present, simply a light blue coloration. A blank is essential if only a slight coloration is obtained. The test may also be conducted on paper. According to Feigl and Anger (1972, p. 269), the limit of identification is 0.1 microgram of iron on paper and 0.05 microgram of iron on a spot plate; the limit of dilution is 1:5,000,000 on paper and 1:1,000,000 on a spot plate.

5.25 ***Potassium Thiocyanate Test for Ferric Iron.*** Potassium thiocyanate, which reacts with ferric ions in acid solution to yield a red color, provides a somewhat less satisfactory test. The sensitivity of the procedure is reduced by certain metallic salts, particularly copper, which could be contributed by the presence of azurite. Nonetheless, Feigl and Anger (1972, p. 271) report that 0.025 microgram of iron can be detected in the presence of 640 times that amount of copper and 1.25 microgram of iron in the presence of 320 times that amount of cobalt or chromium. The test is carried out by mixing a drop of the test solution on a spot plate with a drop of 1% potassium thiocyanate solution. A more or less deep red color appears in the presence of ferric iron.

5.3 Instrumental Methods of Analysis

Instrumentally, green earths are best identified by x-ray diffraction and infrared spectrophotometry, preferably both. Furthermore, because pigment specimens are usually heterogeneous and the minerals themselves variable in composition, these two analytical methods may provide means of "fingerprinting" specific samples. Several supplementary methods can provide additional information. It should be remarked here that because research in this area has produced much new information in recent years, the serious analyst should search the mineralogical literature for the latest developments.

5.31 *X-ray Diffraction.*

X-ray diffraction analysis can be used to identify celadonite/glauconite and is highly effective in distinguishing rock glauconites. Recent studies also indicate that the diffractometer patterns of celadonite and mineral glauconite can be differentiated, but it appears unlikely that the minerals can be distinguished using a powder camera.

Table 2 gives the spacings, intensities, and reflections reported for glauconite and celadonite. Strong basal spacings (near 10.0 and 3.33 Å) are characteristic of micaceous structure with good cleavage of the (001) plane (see fig. 5). Reflections at 3.09 and 3.64 Å indicate that the layers are superimposed in an ordered manner.

Recently it has been reported that celadonite patterns exhibit sharp basal and *hkl* reflections, in contrast to mineral glauconite patterns which have broader basal and reduced *hkl* reflections. In addition, the *d* 060 spacings are said to be useful for differentiating the minerals, celadonites ranging from 1.507–1.509 Å and glauconites, 1.512–1.517 Å, particularly when they are plotted against the amount of Fe^{3+} ion (Buckley, et al., 1978).

In practice, x-ray diffractometer traces of large specimens of primary green earth minerals displayed nearly all the peaks listed for these two minerals in the *JCPDS Powder Diffraction File* (see fig. 13A). In comparison, Debye-Scherrer (powder camera) films tended to show fewer d-spacings, primarily those at 10.0, 4.53, 3.64, 3.09 and 2.58 Å; sometimes those at 4.35 and 4.14 Å (see fig. 14A). Burnt green earth displayed essentially the same diffractometer patterns as did the uncalcined samples.

X-ray diffraction provides the best means of identifying the heterogeneous components of glauconite pellets as Burst (1958a, b) showed. He employed x-ray diffraction analysis to differentiate the degree of interlayering with expanded clay minerals, such as montmorillonite, or intermixing with clay minerals, such as chlorite, montmorillonite, and kaolinite. He then divided the glauconites into four groups with the following criteria: 1) *Ordered (1M) glauconites*, having the diffraction properties usually attributed to the mineral glauconite, especially sharp and symmetrical basal diffraction peaks. Recently the consensus has been that mineral glauconite be limited to those specimens showing less than 5% interlayering (Buckley, et al., 1978). 2) *Disordered (1Md) glauconites*, having subdued peaks displaying broad bases and asymmetric sides. Another worker prefers an additional criterion for differentiation between groups 1 and 2: the presence or absence, respectively, of the 112 and $11\bar{2}$ reflections (Warshaw, 1957). 3) *Interlayered clay/glauconite pellets*,

Table 2. **X-RAY DIFFRACTION DATA FOR GREEN EARTH (Diffractometer Data, Cu$K\alpha_1$ = 1.5418 Å)**

Glauconite[a]		*Celadonite*[b]	
d (Å)	*I*	*d (Å)*	*I*
10.1	100	9.97	45
4.53	80	4.53	85
4.35	20	4.35	40
4.12	10	4.14	35
3.63	40	3.64	80
3.33	60	3.35	60
3.09	40	3.09	80
2.89	5	2.90	10
2.67	10	2.67	75
2.537	100	2.580	100
2.396	60	2.402	75
2.263	20	2.264	20
2.210	10	2.209	25
2.154	20	2.148	30
1.994	20	2.092	10
1.917	5	—	—
1.715	10	—	—
1.66	30	1.65	15
1.511	60	1.509	60
1.495	10	—	—
1.307	30	—	—
1.258	10	—	—

a. JCPDS Powder Diffraction File, 9-439.
b. JCPDS Powder Diffraction File, 17-521.
Courtesy of JCPDS International Centre for Diffraction Data (1982).

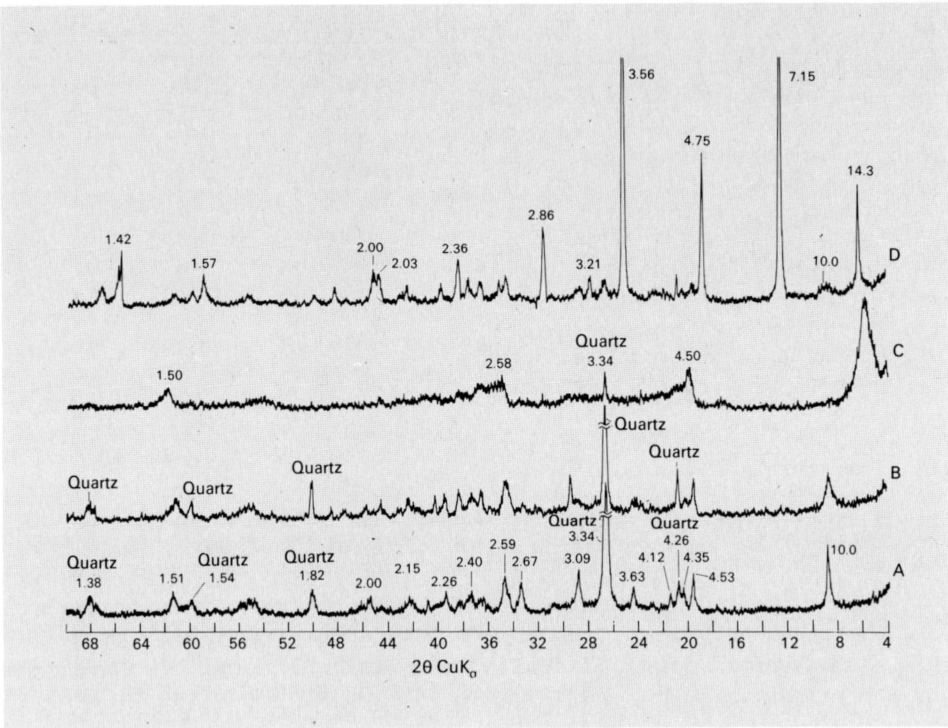

Fig. 13. X-ray diffractometer patterns of green earth and related minerals. A. Typical celadonite or mineral glauconite pattern; quartz impurities also present. Grissom sample 40 (1974). Celadonite. Pra' da Stua, Monte Baldo. Collected by the author, 1973. B. Disordered glauconite; quartz impurities also present. Green earth from the collection of Edward Waldo Forbes, bought in Lucerne in 1914. Forbes Collection 9.01.14; Grissom sample 14 (1974). C. Bentonite, National Gallery of Art Pigment Bank, Pittsburgh. D. Chlorite, "Terra verde di Verona" from the collection of Edward Waldo Forbes, bought in Verona in 1928. Forbes Collection 9.01.2; Grissom sample 2 (1974). Data courtesy Dr. Sidney Pollack, Carnegie-Mellon Institute of Research, Pittsburgh.

possessing swelling clay properties, apparent in first-order basal spacings greater than 10 Å. 4) *Mixed-mineral pellets*, exhibiting the diffraction patterns of such clay minerals as montmorillonite, chlorite, and kaolinite. The diffractometer patterns corresponding to these groups are shown in fig. 15. Minor variants of Burst's system have been used by other investigators (Warshaw, 1957; Hower, 1961).

Forty samples of green earth, primarily from the Forbes Collection, were analyzed with x-ray diffraction. Diffractometer patterns and to a lesser degree, Debye-Scherrer patterns, revealed examples of all four Burst groups (figs. 13 and 14) (Grissom, 1974). Distribution of samples between the groups is shown in fig. 16. Celadonite and mineral glauconite were not distinguished, but differentiation might be feasible at the present time. The small size of the sample that can be taken from paintings, however, may preclude both distinction of the Burst groups and the mineral species.

Minor impurities, frequently present, sometimes produce more intense lines than does the green earth itself. Especially noteworthy is quartz; its most intense *d*-spacing (3.34 Å) is coincident with the 003 spacing for celadonite and glauconite. However, potential misinterpretation can be checked by a search for the less intense 4.26 Å quartz spacing (see fig. 13). Feldspars have a strong line at about 3.20 Å; calcite at 3.03 Å. Iron oxide impurities, though common in green earth, are rarely observed in diffraction patterns. Acid treatment to eliminate carbonate impurities does not appear to affect the diffraction pattern of green earth.

Electron diffraction of powder samples of celadonite have been reported (Zviagin and Shakhova, 1957).

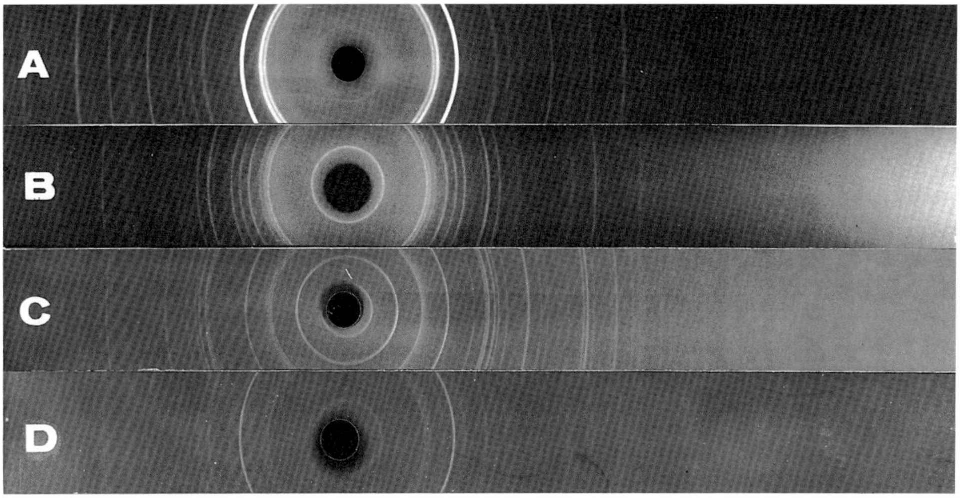

Fig. 14. Debye-Scherrer x-ray powder diffraction patterns of green earth and related minerals. Patterns courtesy of Dr. Sidney Pollack.
A. Typical celadonite pattern; quartz impurities also present. Pra' da Stua, Monte Baldo, Sample collected by Grissom (1974). Grissom sample 40.
B. Green earth from Cyprus, Winsor & Newton, Sample B-626. Grissom (1974) sample 39.
C. Chlorite, Wards Scientific Company, Sample B-733.
D. Disordered glauconite, green earth from the Collection of Edward Waldo Forbes, bought in Lucerne, 1914. Forbes Collection 9.01.14. Grissom (1974) sample 14.
(Photographs are not to be used for measurement purposes.)

5.32 ***Infrared Spectroscopy.*** There is considerable evidence in the literature that celadonite and glauconite can be differentiated by infrared spectrophotometry, and recently two authors have so stated (van der Marel and Beutelspacher, 1976, p. 120). Celadonite spectra are characteristically sharp and distinct, suggesting good ordering of the crystals. On the other hand, glauconite spectra tend to be broad, which can be attributed to the substitution of aluminum in the silicate tetrahedra and larger divergence from the optimum $R^{3+} R^{2+}$ ratio of 1:1 in the octahedral layer. Because of the sensitivity of infrared spectra to variations in ionic substitutions, the spectrograms of both minerals exhibit some variation in band locations; these would seem to have potential for fingerprinting individual specimens.

The principal absorption bands for celadonite were analyzed in detail by Farmer and Russell (1964), as follows: 3601, 3557, and 3534 cm^{-1} (hydroxyl stretching); 1640 cm^{-1} (hydroxyl bending); 1105, 1075, and 975 cm^{-1} (Si-O stretching); 845, 800, 746, and 681 cm^{-1} (R-O-H bending, where R is the octahedral ion); 494, 457, and 442 cm^{-1} (Si-O-R and R-OH absorption). With slight variations other published spectra (van der Marel & Beutelspacher, 1976, p. 162) as well as those obtained by the author (figs. 17A and 18) conform to this absorption pattern. In addition, van der Marel and Beutelspacher (1976, p. 120) specify certain peaks which may be used to differentiate celadonite from glauconite: 3672, 3650, 970, 956, 799, and 743 cm^{-1}. Buckley, et al. (1978, p. 377), emphasize the presence of two to four narrow OH stretching bands in the 3610–3530 cm^{-1} region for celadonite, as well as the vibration lying at 800 cm^{-1}. For the author, one of the most distinctive areas of the spectra is the Si-O stretching region (1100–900 cm^{-1}), characterized by two double peaks in the case of celadonite.

Much less detailed work has been published regarding mineral glauconite's spectra, perhaps because the absorption bands are broad and indistinct. Moreover, the frequent confusion of mineral and rock glauconite brings into question the results of some investigators. In any case, a number of authors (Lyon & Tuddenham, 1960; Milkey, 1960; Stubican & Roy, 1961; Farmer, 1974; Moenke, 1962; Manghnani & Hower, 1964) postulate that substitution of aluminum in the silicate tetrahedra results in broadening of the spectra and especially in shifting of the wavelengths in the Si-O stretching

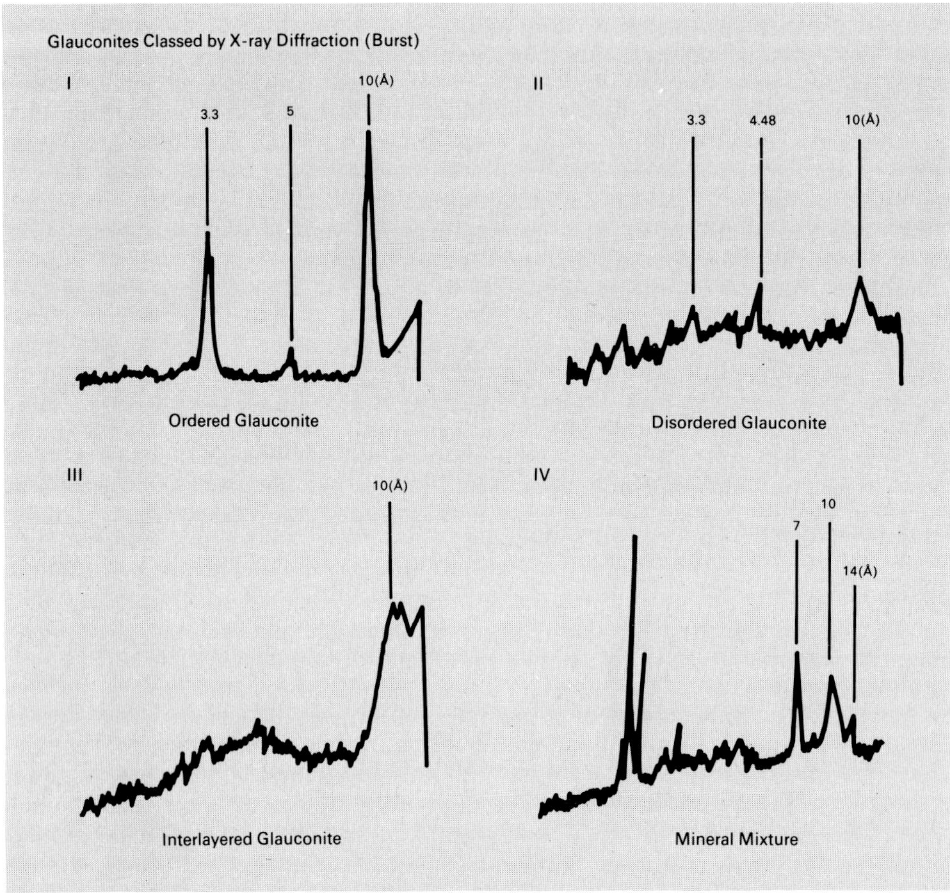

Fig. 15. Method of classifying glauconites according to the characteristics of x-ray diffractometer patterns (Burst, 1958b, pp. 486–488).

region to intermediate wavelengths. Indeed the five spectra of glauconite published by van der Marel and Beutelspacher (1976, pp. 160–161) are distinguished by absorption in a much narrower range, e.g. 1075 and 1010 cm^{-1}; moreover, the two double absorption bands found in celadonite are replaced by broad double bands or even a single one in some cases. Van der Marel and Beutelspacher (1976, p. 120) also point out that glauconite shows absorption at 872, 813, and 566 cm^{-1} which can be used to distinguish the mineral from celadonite. Buckley, et al. (1978, p. 377), emphasize the shifting of the 800 cm^{-1} band of celadonite to 815 cm^{-1} or a doublet for mineral glauconite.

Infrared analysis does not seem useful for

Fig. 16. Relative frequency of various mineral types classified according to the four Burst groups (Burst, 1958b). Samples include thirty-three examples of green earth from the Edward W. Forbes Collection, Oberlin; six samples of ICA shelf stock, Intermuseum Conservation Association, Oberlin; and two related minerals (Grissom, 1974).

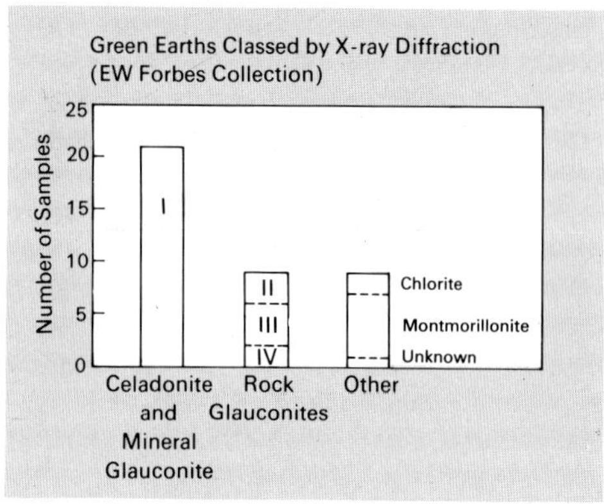

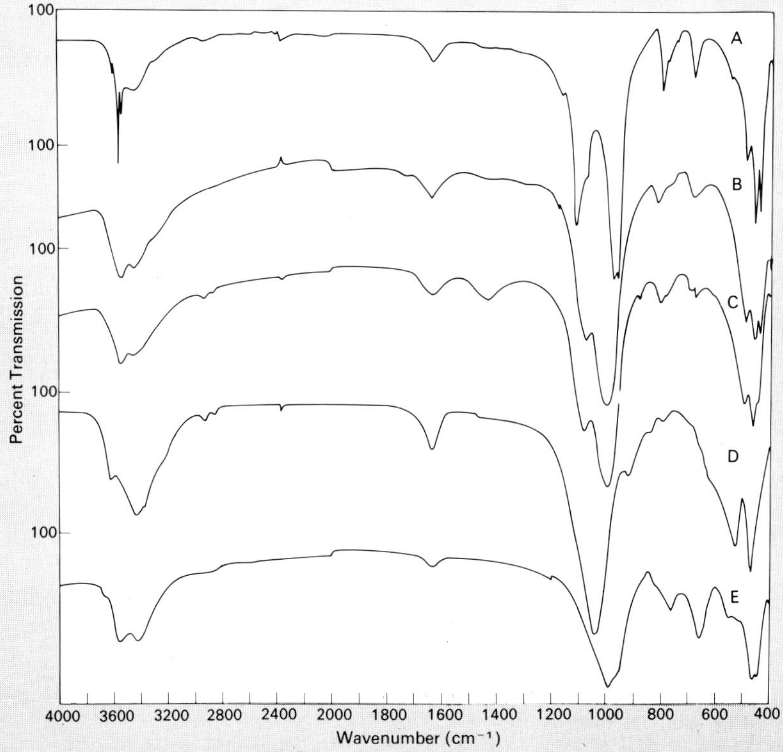

Fig. 17. Infrared absorption spectra of green earth and associated minerals. Beckman IR9 spectrophotometer. Pigment was dispersed in potassium bromide by grinding. Courtesy Research Center on the Materials of the Artist and Conservator, Carnegie-Mellon University, Pittsburgh. A. Celadonite. Pra' da Stua, Monte Baldo. Collected by Grissom sample 40 (1974). B. Green sand from USNM 91934. C. Disordered glauconite. Green earth from the collection of Edward Waldo Forbes, bought in Lucerne in 1914. Grissom sample 14 (1974). Forbes Collection 9.01.14. D. Montmorillonite (bentonite), National Gallery of Art Pigment Bank No. 516, Polkville, Mississippi, Ward's Natural Science Establishment. E. Chlorite. Wards Scientific Company.

distinguishing the Burst categories of rock glauconites. Green earth samples previously identified (by x-ray diffraction analysis) as disordered glauconites and interlayered glauconite/clay showed broadened and shifted bands in the Si-O stretching region and generally less distinct patterns which could not be readily differentiated (fig. 17C).

The spectra of secondary green earth minerals often will be distinctive. For example, the spectrum of chlorite is sufficiently unique to be readily differentiated from that of celadonite (fig. 17E). On the other hand, the principal bands of bentonite (montmorillonite) contain so few features that the infrared spectrum of this mineral is of limited use for identification purposes (fig. 17D).

Calcite and other carbonate impurities were easily detected, because their absorption peaks in the vicinity of 1400 cm^{-1} are sharp and the spectra do not interfere with the celadonite or glauconite patterns. Other common impurities, such as quartz and plagioclase minerals, are less easily distinguished; they exhibit broad absorption in the region around 1100 cm^{-1}, and other bands coincident with those of celadonite and mineral glauconite. Quartz, however, has a pair

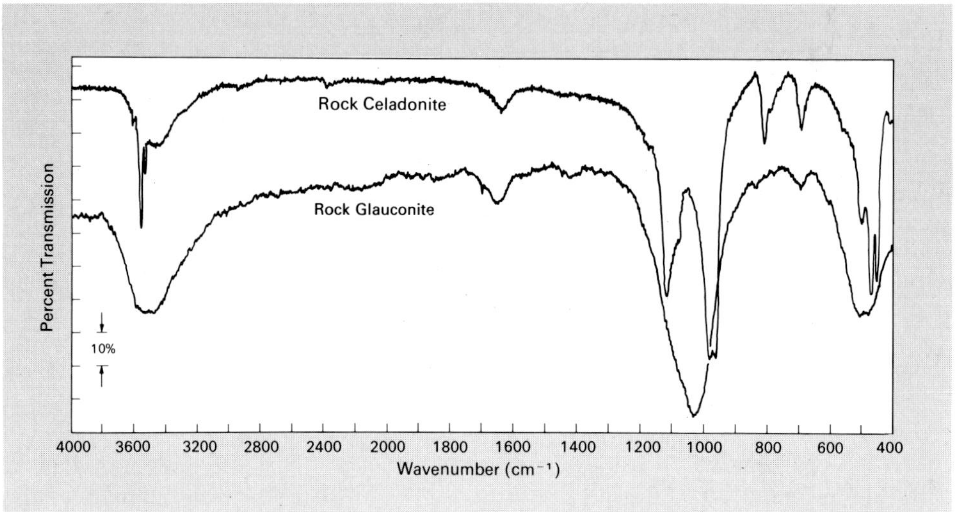

Fig. 18. Infrared spectra of celadonite [Pra' da Stua, Grissom sample 40 (1974)] and rock glauconite (green sand from Birmingham, New Jersey), illustrating the reasonably distinctive differences that one may expect to observe in the 1,200 to 900 cm^{-1} region.

of distinctive bands at 798–778 cm^{-1} that often are noticeable even in extensive mixtures. The absorption bands of iron oxides were not identifiable with assurance in any of the infrared spectra examined.

As the author found in the case of x-ray diffraction analysis, treatment of the paint samples with mild acid to eliminate carbonate impurities did not seem to affect the infrared pattern of green earth.

5.33 ***Electron Probe Micro-Analysis.*** Electron probe micro-analysis (EPMA) may be useful for distinguishing green earth from copper greens by indicating the presence of iron. Quantification of the amount of iron present in conjunction with determination of the Fe^{2+}/Fe^{3+} ratio might also be used to differentiate celadonite and mineral glauconite (Buckley, et al., 1978, pp. 377, 380). Finally, fingerprinting samples may be possible with elemental quantification.

5.34 ***Spectrochemical Analysis.*** Emission spectrographic analysis (ESA) may prove useful for distinguishing green earth from copper greens by the presence of iron. Detection of the other significant elements in celadonite and glauconite may also be used to confirm the identification of green earth, but the method shows no potential for distinguishing between the classes of the pigment.

In a dozen samples of green earth examined from the Forbes Collection, the main elements of celadonite and glauconite were found (iron, silicon, aluminium, magnesium) as were minor amounts of potassium-replacement ions (calcium, manganese and sodium) and a trace element (titanium). Not present were arsenic, boron, beryllium, bismuth, cadmium, cobalt, mercury, molybdenum, nickel, tin, strontium, vanadium, zinc, and zirconium (Dr. Edwin Hodge, Mellon Institute, Pittsburgh, 1974; private communication).

5.35 ***X-Radiography.*** Like many other silicate minerals, green earth exhibits negligible density to typical x-radiography.

5.36 ***Neutron Activation Analysis.*** Adon Gordus of the University of Michigan (private communication) has carried out neutron activation analyses on the pigments in the Forbes Collection, but the results have not been published.

5.37 ***Mössbauer-Effect Spectroscopy.*** Because Mössbauer-effect spectroscopy quantitatively measures the oxidation states of iron, it is particularly relevant for the study of green earth, as the pigment's color depends on the interaction of the two valency states of iron (see fig. 19 and table 3). The method may also have potential for distinguishing celadonite from glauconite; the former is reported to have a slightly higher Fe^2/Fe^3 ratio than that of min-

Table 3. **DATA FROM MÖSSBAUER-EFFECT SPECTRA OF THREE SAMPLES OF GREEN EARTH**[a]

Peak Characteristic	*Sample No.*[b]	*Peaks*		
		1 (FeIII)	*2 (FeIII)*	*3 (FeII)*
Position (velocity in mm/sec)	40	.2486	.6227	2.0940
	12	.1451	.6214	2.1022
	14	.1621	.6290	2.3228
Amplitude (Relative Absorption)	40	.0798	.0604	.0192
	12	.0729	.0650	.0055
	14	.0847	.0727	.0063
Area/relative area	40	.0620/50%	.0469/38%	.0149/12%
	12	.0724/51%	.0645/45%	.0054/4%
	14	.0881/52%	.0756/44%	.0065/4%

a. Bernard Keisch, unpublished Mössbauer-effect spectra of green earth, 1974.
b. Sample 40 is the bluish celadonite from the vicinity of Pra' da Stua; samples 12 and 14 are yellowish-green commercial samples from the Forbes Collection (Grissom, 1974, 1975).

eral glauconite. This characteristic is seen in fig. 19 in the greater height of the Fe^2 peak at about 2.1 mm/sec for the celadonite. Burnt green earth is easily distinguished, owing to the absence of ferrous iron. To a certain extent the technique could be used to fingerprint samples (Dr. Bernard Keisch, private communication). However, Mössbauer results should always be interpreted in conjunction with additional analyses because other clays, such as chlorite, might produce similar results.

5.4 Criteria for Positive Identification

In consideration of the microscopically small samples that the investigator may have to work with in examining paintings, green earth may be one of the pigments that is most difficult to identify unequivocally. However, the appearance of celadonite/glauconite under the microscope is almost unmistakable: varied shades of green particles of low refractive index and a characteristic low birefringence — greenish gray, mottled, and striated. The additional detection of ferric and ferrous ions, especially the latter, by treatment with concentrated acid solution would make identification practically certain.

Provided a large enough sample can be obtained, an x-ray diffraction pattern and infrared spectrogram should be effective in identifying the pigment. Preferably both of these analyses should be made for positive identification. However, the principal bands in the infrared spectrum may be masked by the patterns of other minerals present, and the paucity of sharp

Fig. 19. Mössbauer-effect spectra of celadonite, Pra' da Stua, green earth, Grissom sample No. 40 (green earth group I) and Forbes Collection 9.01.12, of disordered glauconite, Grissom sample 12 (green earth group II) (Grissom, 1974). Courtesy Dr. B. Keisch, Carnegie-Mellon Institute of Research, Pittsburgh.

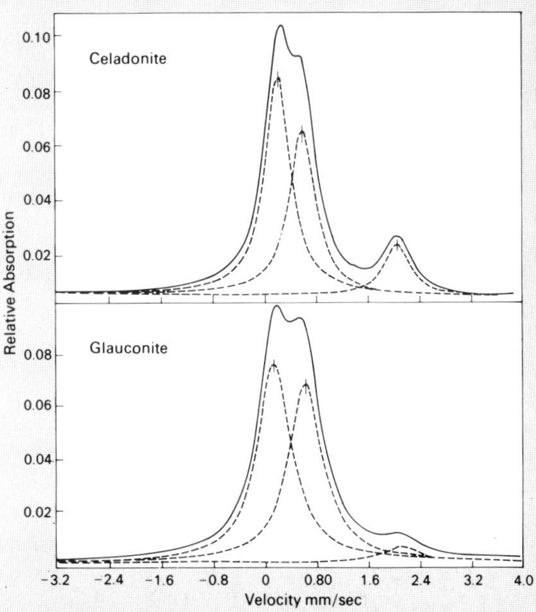

lines in the usual Debye-Scherrer pattern may rule out finer distinctions, such as assignment of the pigment to a Burst classification or differentiation of celadonite and mineral glauconite.

The following additional analytical results would corroborate the identification. Electron probe micro-analysis and emission spectrographic analysis, showing the presence of iron, aluminum, and silicon, as well as the absence of copper, would constitute supporting evidence. Confirmation of the absence of copper can also be made on the basis of a microchemical test. Verdigris, perhaps the principal pigment that would need to be differentiated from very finely divided particles of green earth, is readily dissolved by dilute acids and yields a positive test for copper. Even if the Debye-Scherrer pattern is unclear, it might at least show clay mineral lines and the absence of a verdigris pattern; similarly, a poor infrared spectra would at least show a broad band between 1200 and 900 cm^{-1}.

6.0 **NOTABLE OCCURRENCES**

Green earth, once a popular pigment in the arts, has now almost passed into oblivion. Rarely present in the modern artist's palette, the source minerals have been found in objects around the globe; the most famous sites of the Western world are those in south central Europe. In the East, the pigment has been identified in paintings from India and Nepal (Kahn, 1949; Agrawal, 1971), as well as in early Japanese wall paintings (Yamasaki and Emoto, 1979).

Green earth has not been found in studies of Cretan wall paintings (Neuberger, 1930), medieval manuscript illumination (Flieder, 1968), or Chinese and Japanese scroll paintings (FitzHugh, 1979).

In view of the range of clay-related minerals, the author has been intrigued with the possibility of identifying specific types of green earths. An outstanding example of such classification has been reported only recently; on the basis of x-ray diffraction analysis, Mosk (1975) was able to characterize the green earth from the Bardwell portrait of William Crowe as the septechlorite mineral, cronstedtite. This identification is unique. The author has found bulk samples of green earths whose sources are chlorite minerals, similar in composition to the septechlorites, but the Bardwell cronstedtite is the first sample taken from a work of art so identified. Cronstedite is known to occur in Cornwall; a source of green earth in the Mendip Hills near Bristol, England was also known and reported in the German literature of the nineteenth century (Leuchs, 1825).

Artist or School *Title, Date* *Collection or Location*	*Analyst and Method* *of Identification*
Contents of a container from the Baths of Titus, 1st century A.D., Rome	Davy (1815), chem.
Walls of Roman villa at Bignor, Sussex, 1–4th century, England	Davy (Wall, n.d.)
Various Pompeian paintings, 1st century, Museo Nazionale, Naples	Raehlmann (1910), micr., chem. Augusti (1967), micr., chem.
Fragments of paintings from ruins near Rome, Fogg Art Museum, Harvard University, Cambridge. 25.6.36	Grissom (unpublished data, 1974), micr.
Fragments of Roman wall paintings from Dura-Europos, Syria, 1–3rd century, Fogg Art Museum, Harvard University, Cambridge. 1940.97.	Grissom (unpublished data, 1974), micr., XRD
Wall painting, mid-6th century, Ōtsuka Tomb, Fukuoka prefecture, Japan	Yamasaki and Emoto (1979), (glauconite), XRD
Istra, Wall painting, Church of St. Agata, 11th–12th centuries, Kanfenar, Yugoslavia	Kojić-Prodić, et al. (1976), XRD
Byzantine wall paintings, Kariye Camii, 11th–14th centuries, Istanbul	Gettens & Stout (1958), micr.

Chem. = microchemical tests; micr. = microscopic examination; XRD = x-ray diffraction; IR = infrared; ESA = emmission spectrographic analysis.

Artist or School *Title, Date* *Collection or Location*	*Analyst and Method* *of Identification*
Wall paintings at the Abbey at St. Bavo, c. 1175, Ghent	Bontinck (1940), chem.
Thirteen medieval paintings in churches, 11th–14th centuries, Bulgaria	Prashkov & Zhelninskaya (1971)
Apse frescoes from monk's chapel at Berzé la Ville near Cluny, 12th century, Fogg Art Museum, Harvard University, Cambridge. 1940.97	Grissom (unpublished data, 1974), micr.
Byzantine wall paintings, Santa Sophia, 13th–14th centuries, Trebizond, Turkey	Plesters (1963), micr., chem.
Vincent of Kastav, Wall painting, Church of Saint Maria, 1472, Beram, Yugoslavia	Kojić-Prodić, et al. (1976), XRD
Five Italian paintings, 14th–15th centuries, Louvre, Paris	Gay (1976), micr.
Herlin Altarpiece, 1466, Rothenburg, West Germany	Broekman-Bokstijn, et al. (1970), chem, IR, micr.
Spinello Aretino (att.), *Madonna and Child Enthroned with Angels*, 14th century, Fogg Art Museum, Harvard University, Cambridge. 1905.1	Gettens (1947), micr.
Michelangelo (att.), *Madonna and Child with St. John and Angels*, 16th century, National Gallery, London	Ruhemann & Plesters (1964), micr., chem.
Saloman van Ruysdael, *The Bridge*, 1630, Mauritshuis, The Hague	de Wild (1929), micr., chem.
Jan Willemsz Lapp, *Italian Landscape*, 1640, Mauritshuis, The Hague	de Wild (1929), micr., chem.
Jan Victors, *The Grocer's Shop "de Buyskool,"* 1654, Rijksmuseum, Amsterdam	de Wild (1929), micr., chem.
Master Blaž, Wall painting, Church of the Holy Spirit, 16th century, near Sušnjevica, Yugoslavia	Kojić-Prodić, et al. (1976), XRD
Jan Vermeer, *The Concert, Allegory of the New Testament, A Lady Receiving a Letter*, 17th century, Isabella Stewart Gardner Museum, Boston; Metropolitan Museum, New York; Gemäldegalerie, Dresden	Kühn (1968), micr., ESA
Thomas Bardwell, *Portrait of William Crowe*, c. 1746, Castle Museum, Norwich	Mosk (1975) (cronstedtite), XRD
Various Schools, Approximately 239 paintings, 14th–19th centuries, Doerner-Institute, Munich	Kühn (1973), various methods
Seventeen German paintings, 19th century, Schack-Galerie, Munich	Kühn, (1969), micr.
B. J. van Hoven, *The Heerengracht at the Hague*, 1840, Municipal Museum, The Hague	de Wild (1929), micr., chem.
Joseph Mallord William Turner, Materials from his studio, 19th century, Tate Gallery, London	Hanson (1954), ESA
Mexican potsherds, Casas Grandes, Freer Gallery of Art, Washington	Gettens (private communication, 1974), micr.

Chem. = microchemical tests; micr. = microscopic examination; XRD = x-ray diffraction; IR = infrared; ESA = emmission spectrographic analysis.

ACKNOWLEDGMENTS

The author wishes to express her appreciation to the Intermuseum Conservation Association, Oberlin, Ohio, for allowing her to analyze the green earths in the Edward W. Forbes collection of pigments. While affiliated with the Mellon Institute, Pittsburgh, Dr. Sidney Pollack, now at the U.S. Department of Energy, Pittsburgh, provided assistance with x-ray diffractometry, and Dr. Bernard Keisch, now with Brookhaven National Laboratories, Upton, New York, provided data on Mössbauer spectra. The encouragement of the late Richard D. Buck, then director of the Intermuseum Conservation Association, was essential, and the late Rutherford John Gettens provided inspiration by his example and enthusiasm. Most of all, gratitude is due to Dr. Robert L. Feller, director of the Research Center on the Materials of the Artist and Conservator, Carnegie-Mellon Institute of Research, Pittsburgh. He was a valued adviser and mentor, initially suggesting the topic of this project, providing judicious criticism, editing with authority, and offering generous and patient support throughout.

REFERENCES

O. P. Agrawal, "A Study of Indian Polychrome Wooden Sculpture," *Studies in Conservation* 16 (1971), 56–68.

S. Augusti, *I colori Pompeiani* (Rome, 1967), 100.

T. Bardwell, *The Practice of Painting and Perspective Made Easy* (London, 1756).

N. F. Barnes, "A Spectrophotometric Study of Artists' Pigments," *Technical Studies in the Field of the Fine Arts* 7 (1939), 120–138.

J. G. Bearn, *The Chemistry of Paints, Pigments and Varnishes* (London, 1923), 94–95.

Y. K. Bentor and M. Kastner, "Notes on the Mineralogy and Origin of Glauconite," *Journal of Sedimentary Petrology* 35, no. 1 (1965), 155–166.

E. Berger, *Böcklin's Technik* (Munich, 1906), 151.

E. Berger, *Quellen für Maltechnik während der Renaissance und deren Folgezeit* (Munich, 1901; Wiesbaden, 1973).

E. Beutel, *Die Materialien des Kunst und Dekorationsmalers* (Vienna, 1907), 32.

H. Beutelspacher and H. W. van der Marel, *Atlas of Electron Microscopy of Clay Minerals and their Admixtures* (Amsterdam, 1968), 121–122, 134–135.

E. Bontinck, "Microchemical Analysis of the Wall Paintings at St. Baafsabtei in Ghent (about 1175)," *Mededeelingen van de Koninklijke Vlaamse Academie voor Wetenschappen, Letteren en Schone Kunsten van Belgie, Klasse der Wetenschappen*, no. 13 (1940), 3–12.

Giovani de Brignoli de Brunnhoff, "Dissertation sur la chlorite ou terre verte de Vérone," *Journal de physique, de chimie et d'histoire* 90 (1820), 355–361, 423–442.

M. Broekman-Bokstijn, J. R. J. Van Asperen de Boer, E. H. Van't Hul-Ehrnreich, and C. M. Verduyn-Groen, "The Scientific Examination of the Polychromed Sculpture in the Herlin Altarpiece," *Studies in Conservation* 15 (1970), 370–400.

H. A. Buckley, J. C. Bevan, K. M. Brown, L. R. Johnson, and V. C. Farmer, "Glauconite and Celadonite: Two Separate Mineral Species," *Mineralogical Magazine* 42, (1978), 373–382.

J. F. Burst, "'Glauconite' Pellets: Their Mineral Nature and Applications to Stratigraphic Interpretations," *Bulletin of the American Association of Petroleum Geologists* 42, nc. 2 (1958a), 310–327.

J. F. Burst, "Mineral Heterogeneity in 'Glauconite' Pellets," *American Mineralogist* 43 (1958b), 481–497.

E. R. Caley and J. F. C. Richards, *Theophrastus on Stones* (Columbus, 1956), 137–138.

M. Chaptal, "Sur quelques couleurs trouvées à Pompeia," *Annales de chimie* 52 (1809), 22.

V. Charrin, "Les terres vertes," *Travaux de peinture* 3 (1948), 172.

V. Charrin, "Les terres vertes — glauconie et serpentine," *Peintures, pigments, vernis* 31, no. 4 (1955), 334–336.

V. Charrin, "Les terres vertes, leur gisements en France," *Peintures, pigments, vernis*, 23 (1947), 307.

Colour Index vol. 2, 2nd ed. (Bradford, Yorkshire, England, and Lowell, Massachusetts, 1956).

Colour Index vol. 3, 3rd ed. (Bradford, Yorkshire, England, and Research Triangle Park, North Carolina, 1971).

J. B. Corneille, *Les Premiers Elémens de la peinture pratique* (Paris, 1684), 40, 72, 82.

H. Davy, "Some Experiments and Observations on the Colours used in Painting by the Ancients," *Royal Society of London, Philosophical Transactions* 105 (1815), 110.

W. A. Deer, R. A. Howie, and J. Zussman, *Rock-Forming Minerals*, vol. 3 (New York, 1963), 35.

R. DeLisle, *Cristallographie, ou Description des Formes Propres a Tous les Corps du Regne Mineral*, vol. 2 (Paris, 1783), 502.

DePiles, *Elémens de peinture pratique* (Amsterdam 1776), 222.

M. Doerner, *The Materials of the Artist and Their Use in Painting*, rev. ed. trans. E. Neuhaus (New York, 1949), 87.

B. Dupuy du Grez, *Traité, sur la peinture* (Toulouse, 1699), 235.

D. E. Eichholz, trans. of Theophrastus, *De lapidibus* (London, 1965), 70–71.

V. C. Farmer, *The Infrared Spectra of Minerals* (London, 1974), 339.

V. C. Farmer and J. D. Russell, "The Infrared Spectra of Layer Silicates," *Spectrochimica Acta* 20 (1964), 1,149–1,173.

F. Feigl and V. Anger, *Spot Tests in Inorganic Analysis* 6th rev. ed. trans. R. E. Oesper (Amsterdam, 1972),263-264.

F. X. Fernbach, *Die Oelmalerei* (Munich, 1843), 104, 111.

G. Field, *Chromatography; or, A Treatise on Colours and Pigments, and Their Powers in Painting, & c.* (London, 1835), 128.

T. H. Fielding, *On Painting in Oil and Water Colours* (London, 1839), 124.

E. W. FitzHugh, "A Pigment Census of Ukiyo-e Paintings in the Freer Gallery of Art," *Ars Orientalis* 11 (1979), 27–38.

F. Flieder, "Mise au point des techniques d'identification des pigments et des liants inclus dans la couche picturale des enluminures de manuscrits," *Studies in Conservation* 13 (1968), 60.

M. C. Gay, "Application of the Staining Method to Cross-sections in the Study of the Media of Various Italian Paintings of the Fourteenth and Fifteenth Centuries," in *Conservation and Restoration of Pictorial Art*, eds. Bromelle and Smith (London, 1976), 78–83.

R. J. Gettens, "The Materials in a Painting by Spinello Aretino," *Bulletin of the Fogg Museum of Art* 10, no. 6 (1947), 188–193.

R. J. Gettens and G. L. Stout, "A Monument of Byzantine Wall Painting — the Method of Construction," *Studies in Conservation* 3 (1958), 107–118.

R. J. Gettens and G. L. Stout, *Painting Materials, A Short Encyclopedia* (1942; New York, 1966).

E. F. Glocker, *Generum et specierum mineralium secundum ordines naturales digestorum synopsis* (Halle, Germany, 1847), 193.

F. Granger, trans. of Vitruvius, *On Architecture*, VII, vii, 4 (Cambridge, Massachusetts, 1970), 112–113.

C. A. Grissom, "A Literature Search for a Pigment Study," *International Council of Museums, Committee for Conservation, Preprints*, 75/21/5, (Paris, 1975).

C. A. Grissom, "A Study of Green Earth" (M.A. thesis, Oberlin College, 1974).

Handbook of Young Artists and Amateurs in Oil Painting (New York, 1865), 50.

N. W. Hanson, "Some Painting Materials of J. M. W. Turner," *Studies in Conservation* 1 (1954), 170.

R. D. Harley, *Artists' Pigments; c. 1600–1835, A Study in English Documentary Sources*, 2nd ed. (London, 1982), 76, 77.

J. G. Hawthorne and C. S. Smith, *Theophilus On Divers Arts: The Foremost Medieval Treatise on Painting, Glassmaking and Metalwork* (Chicago, 1963; New York, 1979).

E. W. Heinrich, *Microscopic Identification of Minerals* (New York 1965), 299, 321–323.

S. B. Hendricks and C. S. Ross, "Chemical Composition and Genesis of Glauconite and Celadonite," *The American Mineralogist* 26 (1941) 683–708.

C. A. S. Hoffmann, "Versuch einer Oriktographie von Kursachsen," *Bergmannisches Journal* 1 (1788) 519.

J. Hower, "Some Factors Concerning the Nature and Origin of Glauconite," *The American Mineralogist* 46 (1961), 313–334.

JCPDS Powder Diffraction File. JCPDS International Centre for Diffraction Data, Swarthmore, Pennsylvania, 1982.

R. B. Kahn, "Fresco Paintings of Ajanta," *Journal of the Oil and Colour Chemists' Association* 32 (1949), 23–31.

C. Keferstein, *Deutschland geognostisch geologisch dargestellt* 5, III (1828), 508–512.

K. L. Kelly and D. B. Judd, *Color, Universal Language and Dictionary of Names* (Washington, 1976).

H. Kittel, ed., *Pigmente. Herstellung, Eigenschaften, Anwendung* (Stuttgart, 1960), 333.

B. Kojić-Prodić, A. Deanović, and N. Brnicević, "X-ray Identification of Some Pigments on the Istra (Yugoslavia) Frescoes," *Journal of Applied Crystallography* 9 (1976), 485–490.

H. Kühn, *Die Pigmente in den Gemälden der Schack-Galerie* (Munich, 1969), 103.

H. Kühn, "A Study of the Pigments and the Grounds Used by Jan Vermeer," *Report and Studies in the History of Art*, National Gallery of Art, Washington (1968), 161, 181.

H. Kühn, "Terminal Dates for Paintings Derived from Pigment Analysis," in *Application of Science in the Examination of Works of Art* ed. W. J. Young (Boston, 1973), 201.

A. P. Laurie, *The Painters' Methods and Materials* (London, 1926; New York, 1967), 146–148.

R. Lemoine and C. du Manoir, *Les matières premières employées en peinture* (Rouen, 1893), 317–319.

J. C. Leuchs, *Anleitung zur Bereitung aller Farben und Farbflüssigkeiten* (Nuremberg, 1825), 362.

J. F. Lindenbergh, *Die nieuwe verligter; verwen; fernissen* (Rotterdam, 1742), 42.

A. Lucas and J. R. Harris, *Ancient Egyptian Materials and Industries* (London, 1963), 344–345.

R. J. P. Lyon and W. M. Tuddenham, "Determination of Tetrahedral Aluminum in Mica by Infra-red Absorption Analysis," *Nature* 185 (1960), 374–375.

L. S. Maclehose, trans. of *Vasari on Technique* (1907; New York, 1960), 242.

F. Maire, *Modern Pigments and Their Vehicles* (New York, 1908), 140.

M. H. Manghnani and J. Hower, "Glauconites: Cation Exchange Capacities and Infrared Spectra," *The American Mineralogist* 49 nos. 5, 6 (1964), 586–598, and nos. 11, 12, 1631–1642.

H. W. van der Marel and H. Beutelspacher, *Atlas of Infrared Spectroscopy of Clay Minerals and Their Admixtures* (Amsterdam, 1976).

R. Mayer, *The Artist's Handbook of Materials and Techniques*, 3rd ed. (New York, 1970), 73.

M. Merrifield, *Original Treatises on the Arts of Painting*, vol. 1 (London, 1849; New York, 1967).

R. G. Milkey, "Infrared Spectra of Some Tectosilicates," *The American Mineralogist* 45 (September–October, 1960), 990–1007.

G. Millot, *Geology of Clays*, trans. W. R. Farrand and H. Paquet (New York and Paris, 1970), 2.

H. B. Milner, *Sedimentary Petrography* (New York, 1962), 111.

H. Moenke, *Mineralspektren* (Berlin, 1962), no. 6.92.

J. A. Mosk, "Analytical Methods Applied in the Investigation of the Bardwell Samples," appendix to "Thomas Bardwell and His Practice of Painting," *Studies in Conservation* 20 (1975), 103–107.

A. Munkert, *Die Normalfarben* (Stuttgart, 1905), 154–156.

A. Neuberger, *The Technical Arts and Sciences of the Ancients*, trans. H. L. Brose (1930; New York, 1969), 197.

G. di Nismes, *L'arte di acquerellare* (Lucca, 1760), 38, 58, 71.

The Painter, Gilder, and Varnisher's Companion (Philadelphia, 1867), 38.

A.-J. Pernety, *Dictionnaire portatif de peinture, sculpture et gravure*, vol. 40 (Paris, 1757), 532.

J. Plesters, "Cross-sections and Chemical Analysis of Paint Samples," *Studies in Conservation* 2, (1956), 138.

J. Plesters, "Sancta Sophia, Trebizond — A Note on the Materials and Technique," *Studies in Conservation* 8 (1963), 133.

L. Prashkov and Z. M. Zhelninskaya, "Nature of Pigments Used in Some Bulgarian Murals from the Middle Ages," *Soobshcheniia vsesoiuznoi tsentralnoi nauchno-issledovztelskoi laboratorii po konservatsii i restauratsii muzeinykh khudozhestvennykh tsennostei*, no. 26 (1971), 141–154.

H. Rackham, trans. of C. Plinius Secundus, *Naturalis historia*, book XXXV, p. XXIX (Cambridge, Massachusetts, 1968), 296–297.

E. Raehlmann, *Uber die Maltechnik der Alten* (Berlin, 1910), 14.

J. S. Remington, *The Paint Laboratory Note Book* (London, 1942), 53.

H. Ruhemann, *The Cleaning of Paintings* (New York, 1968), 351.

H. Ruhemann and J. Plesters, "The Technique of Painting in a *Madonna* Attributed to Michelangelo," *The Burlington Magazine* 106 (1964), 546–554.

G. Schäfer, *Das Handbuch der Malerei vom Berge Athos*, trans. Godeh (Trier, 1855).

H. Schneider, "A Study of Glauconite," *Journal of Geology* 35, no. 4 (1927), 290.

C. S. Smith and J. G. Hawthorne, "*Mappae Clavicula*: A Little Key to the World of Medieval Techniques," *Transactions of the American Philosophical Society*, n.s. 64, pt. 4 (1974).

W. Stechow, *Northern Renaissance Art, 1400–1600* (Englewood Cliffs, New Jersey, 1966), 25.

V. Stubican and R. Roy, "Isomorphous Substitution and Infra-red Spectra of the Layer Lattice Silicates," *American Mineralogist* 46 (1961), 32–51.

M. K. Talley, Jr., and K. Groen, "Thomas Bardwell and His Practice of Painting: A Comparative Investigation between Described and Actual Painting Technique," *Studies in Conservation* 20 (1975), 44–108.

D. V. Thompson, Jr., *The Practice of Tempera Paintings* (New Haven, 1936; New York, 1962), 85.

D. V. Thompson, Jr., trans., *Il libro dell'arte, the Craftsman's Handbook of Cennino d'Andrea Cennini* (New Haven, 1933; New York, 1960).

D. V. Thompson, Jr. and G. H. Hamilton, trans. of *De arte illuminandi* (New Haven, 1933), 26.

W. E. Tröger, *Optische Bestimmung der gesteinsbildenden Minerale* (Stuttgart, 1971), 105, 108.

J. R. J. Van Asperen de Boer, *Infrared Reflectography, A Contribution to the Examination of Earlier European Paintings* (Amsterdam, 1970), 32.

J. C. Wall, *Medieval Wall Paintings* (London, n.d.), 30.

C. M. Warshaw, "The Mineralogy of Glauconite," (Ph.D. diss., Pennsylvania State University, 1957).

Watin, *L'art du peinture, doreur, vernisseur* (Liege, 1778), 30.

A. M. de Wild, *The Scientific Examination of Pictures*, trans. L. C. Jackson (London, 1929), 74–75.

K. Yamasaki and Y. Emoto, "Pigments Used on Japanese Paintings from the Protohistoric Period through the 19th Century," *Ars Orientalis* 11 (1979) 1–14.

H. S. Yoder and H. P. Eugster, "Synthetic and Natural Muscovites," *Geochimica et cosmochimica acta* 8 (1955), 225–280.

B. B. Zviagin, "Determination of the Structure of Celadonite by Electron Diffraction," *Soviet Physics, Crystallography* 2, no. 3 (1957), 388–394.

B. B. Zviagin and R. A. Shakhova, "Electron Diffraction Study by Reflection of Powder Samples of Celadonite," *Soviet Physics, Crystallography* 2, no. 1 (1957), 173–175.

Zinc White

HERMANN KÜHN

NOMENCLATURE IN VARIOUS LANGUAGES

English: zinc white
German: Zinkweiss (Zinkoxid)
Russian: цинковые белина
French: blanc de zinc
Italian: bianco di zinco
Spanish: blanco de cinc

1.0 INTRODUCTION

Zinc oxide, CI Pigment White 4, No. 77947 (*Colour Index*, 1971), represents one of the three white pigments of good hiding power that are of principal interest in the commercial world of artistic and decorative paints — lead, titanium, and zinc white. Although known since ancient times, zinc white apparently was not seriously considered as an artists' pigment until the end of the eighteenth century. The widest application of the pure pigment has been in watercolor, long sold under the name of Chinese white. Today, many manufacturers incorporate it as a supplementary pigment in titanium and lead white oil paints, to improve the performance particularly for use out-of-doors.

1.1 Brief Definition of Pigment

Zinc white, ZnO, a substance readily available in high purity, has been widely used in water-dispersed mediums. In oil, its hiding power is not as great as that of titanium or lead white, and today it is generally employed in combination with these pigments.

HERMANN KÜHN, curator of the Department of Handicraft and Old Technology at Deutsches Museum, Munich, and lecturer on art history, Munich University.

Zinc oxide does not seem to have been suggested as a painters' white until considerable concern about lead poisoning arose during the early stages of the Industrial Revolution. Tests for possible substitute whites were reported by Guyton de Morveau about 1780. It was fifty years later before an attempt was made, by Leclaire, to produce the pigment commercially (Petit, 1907).

The virtues of zinc white in oil paint — clear white color, good suspension properties, and absorbance of ultraviolet radiation — have long been recognized and appreciated. A book issued by the New Jersey Zinc Company (Brown, 1957) provides an excellent overview of a remarkably wide range of industrially useful properties.

1.2 Current Terminology

One regularly encounters the designations "French process" zinc oxide (made from the metal) and "direct" or "American process" zinc oxide (made from the ore and, hence, traditionally somewhat less pure, see 4.3). The terms nodular and acicular are used to describe the particle shapes. Before the increased concern in recent years for environmental and health hazards, "leaded" zinc oxides were also extensively manufactured (see 4.1). In the art world, the pigment is commonly referred to as zinc white.

2.0 HISTORY

2.1 Archaic, Obsolete, and Misrepresentative Names

Names found in the ancient world of alchemy include *pómpholyx* (*Hüttenrauch*), *spodós*, *spodium* (*Ofenbruch*), *lana philosophica*, *nix alba*, *nihilum album* (*nil album*, *nihili*), *spiritus volatilis cadmiae*, and *flores zinci* (flowers of zinc, *Zink-*

blüte, Blütenweiss, fleurs de zinc). All of these, as well as the term *tutia* or *tucia* (derived from the Persian *dûd* for smoke) introduced by Arabian physicians during the Middle Ages, can be found in the *Lexicon alchemiae sive dictionarium alchemisticum* published by Martin Ruland in 1612 in Frankfurt. Some more recent terms, principally of the nineteenth century, are *Schneeweiss* (snow white, *blanc de neige*), *blanc de trémie, blanc léger*, constant white, Hubbocks white and Chinese white. Perhaps the latter is most familiar to the reader.

2.2 History of Use

As mentioned, oxide of zinc in the form of a fine white powder has been known since antiquity, but its use in painting appears to date only from the end of the eighteenth century. In the ancient world the material was familiar as a by-product of brass production by the cementation process, in which copper was heated with zinc spar. Dioscorides describes zinc oxide's production: *cadmia* (*kadmeia*) was mixed with coal and burned in a hearth chamber which was connected to another chamber where the zinc oxide that formed was able to settle (Kopp, 1845). This material often resembled wads of wool, to which the medieval term *lana philosophica* can probably be traced. During that time, the fine white *pómpholyx* (smelter smoke), which settled in the upper part of the brass furnaces, was differentiated from the *spódos* (*spodium, tutty*), stained dark by the metallic zinc (Lippmann, 1919). Both products served medicinal purposes; the writings of Hippocrates (about 460–377 B.C.), for example, mention *spódos*, and the *Materia medica* of Dioscorides (mid-first century) refers to *pómpholyx* (Dierbach, 1969; Wellmann, 1958). Zinc oxide can also form fine light dust or flakes, and for this reason the alchemists called it *nihilum album* (*nil album, nihili*) and *nix alba*. The alchemist Libavius (Andreas Libau) in his 1613–1615 *Opera omnia medicochymica* designated zinc oxide as "spiritus volatilis cadmiae" because of its derivation from calamine (*cadmia*). Evidence suggests that a clear distinction between *cadmia* and *pómpholyx* has not always been made, and as a consequence, the two have often been confused in the literature.

The key to zinc oxide's use in medicine was its effectiveness in preventing inflammation; since antiquity the compound has had special significance in the field of ophthalmology. The pharmacopoeia of the sixteenth century employed the terms *nihilum album* and *tutia* (sometimes *tutia alexandrina*). During the period 1670–1750 zinc oxide as a rule was called *flores zinci* (zinc bloom); thereafter *zincum oxydatum* became popular (Schneider, 1972).

The analysis by the author of two samples from apothecary jars of the sixteenth to seventeenth century, labeled "tutia, nix album" (Apothecary Biberach) and "tutia, nihil album" (Braunschweigisches Landesmuseum), proved the contents to be pumice powder and gypsum, respectively. Samples from about 1860 labeled "flores zinci" and "zincum oxydulatum" were in fact zinc oxide, whereas a "Zinc White No. 1" from the same source was a mixture of zinc oxide and zinc carbonate. The samples were generously placed at the author's disposal by Prof. Dr. W. Schneider, Pharmacy Historical Seminar, Technical University, Brunswick. In all probability a very bright white substance such as zinc oxide, uncontaminated by metallic zinc or other matter, was rather rare and expensive before the eighteenth to nineteenth century. For this reason the material was not infrequently "replaced" by other compounds medically less effective.

The first attempt to introduce zinc white as a pigment for paint dates to 1780. Some attribute this effort to Courtois, a demonstrator for the laboratory of the Academy of Dijon; others to Guyton de Morveau, magistrate at the court of Dijon, with whom Courtois might have collaborated. Whatever the truth, these are the names associated with the introduction of zinc oxide as a pigment, an effort that had a philanthropic objective: to save the lead white workers from the scourge of lead poisoning caused by the primitive methods of manufacture in that epoch (Petit, 1907). However, the enormous difference in cost between zinc oxide, which sold at 8 francs per pound, and lead white, which only cost one quarter of that sum, caused the former's failure to gain wide acceptance. In *Chemical News* (1783), page 282, D. Lorenz Crell, Leipzig, reports in "The Latest Discoveries in Chemistry (Part 8)": "M. Courtois of Dijon prepares a white paint for artists from zinc, which is lasting, mixes easily with all paints, and has no adverse effect on health, all properties which justify the effort to conduct thorough tests with it. If subsequently proven to be satisfactory, it should be preferred to the leaded paints used heretofore." Vandemaele (1965) mentions a German, Cramer, who apparently in 1739 made an earlier attempt to prepare paints with the pigment.

References for the method of making zinc white in the eighteenth century are to be found in Johann Karl Gottfried Jacobsson's *Technologischës Wörterbuch* (1795). An English patent was issued to Atkinson in 1796 (Leuchs, 1825; Harley, 1982). Zinc vitriol, known in the Middle Ages as white vitriol (*Galitzenstein*), usually served as base material. The *Chemisches Wörterbuch*, published by Klaproth and Wolff (1810), reports: "White zinc oxide is used in painting. For very superior results the zinc has to be carefully purified of foreign admixtures."

Despite the mentioned production of zinc white in the 1780s (Harley, 1982; *Practical Treatise*, 1795), Lemoine and Manoir (1893) state that the popularity of the pigment declined. It was not until Rouquette in 1842, Mathieu in 1844, and Leclaire and Barruel in 1849 again took up production that interest revived in zinc oxide's use in paints. Mierzinski (1881) cites Mollerat as an early manufacturer. Principal early manufacturers in Germany were Göttling, Höpfner, and Lampadius.

Analytical investigations of paintings also indicate zinc white was not used extensively as an artists' pigment prior to the second quarter of the nineteenth century. Presumably, this failure resulted from the absence of economical procedures for the pigment's production. The longstanding and still-practiced method of manufacture using metallic zinc (see fig. 1) would not have come into prominence until economic mining and manufacturing methods produced the metal in significant quantity. Improvements in the production of zinc in fact resulted from the advent of the muffle furnace, described in the first quarter of the nineteenth century, evidence of potentially increased availability of the metal. Thus, it is only in relatively recent history that we should expect to find zinc white used significantly as a pigment for commercial paints and artists' colors.

Early trials with zinc white in oil revealed certain disadvantages. In particular, the paints did not dry well and possessed poor hiding power. For this reason the pigment was popular initially in watercolor, where unlike lead white it did not darken under the influence of hydrogen sulfide (see 3.3). Winsor & Newton marketed zinc white as "Chinese White" for the first time in 1834 (Winsor & Newton, 1930). Saint-Victor (1835) mentions a *blanc léger* (light white) in his book *Aquarelles miniatures*, along with other pigments used or recommended both for watercolor and miniature painting. *Blanc léger*

Fig. 1. Cross section of early indirect process for burning molten zinc to form zinc oxide. A. Retort. B. Furnace. C. Flue. D. Collecting hoppers. F, G. Ventilators for cooling. H. Air main. J. Carbon monoxide flue to retort. K. Molten zinc (Zerr and Rübencamp, 1908).

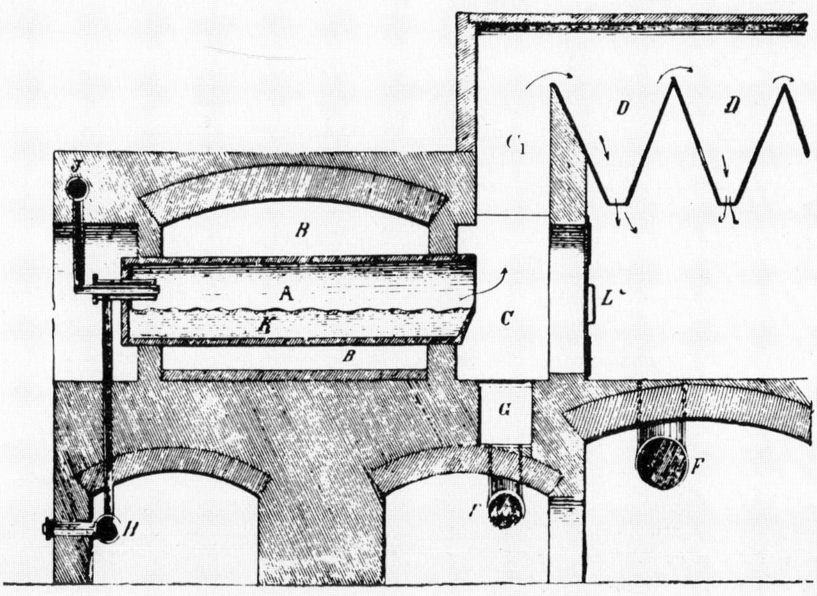

was probably zinc white, because Saint-Victor states, "This color must be pure; one must take care not to buy lead white instead, which tends to darken."

During the years 1835–1844 Leclaire worked in France to introduce zinc white for oil painting. He finally succeeded in producing a pigment which had excellent hiding power in oil and, by using siccatives, obtained better drying. As a consequence, he was able to start zinc white production on an industrial scale in 1845 in the vicinity of Paris (Gettens and Stout, 1966; Leclaire, 1849). In other European countries, as well as in the U.S.A., major industrial manufacture started about 1850 or shortly thereafter (Gentele, 1860).

According to the *Handwörterbuch der reinen und angewandten Chemie* (1864) the following types of zinc white were distinguished in the nineteenth century:

1) Snow white, covers as well as slate white (lead white).
2) Zinc White No. 1, replaces the best lead white.
3) Stone gray, serves as primer to coat stone and metal.
4) Gray zinc oxide, used for lesser paint jobs.

By adding lead white the hiding power of the paint was increased.

In the course of the systematic pigment analyses carried out at the Doerner Institute, zinc white was found in paintings created at the very end of the eighteenth century (see 6.0). However, as mentioned above, the pigment is more frequently encountered, and at a steadily increasing rate, after 1850. It is uncertain in which country zinc white was first used. Early examples have been found in paintings from Britain (G. Morland) and Germany (J. G. von Dillis) as well as Italy (J. Rebell) and Austria (J. A. v. Rhomberg) (see 6.0). Investigations of paintings in Munich's Schack-Galerie (Kühn, 1969) and more recent, still-unpublished findings of F. Preusser, formerly at the Doerner-Institute, Munich, indicate that zinc white was used both for priming and in the paint proper. In white paints it occurs more frequently mixed with lead white than by itself. X-ray powder diffraction reveals no lead-sulfate bearing "leaded zinc oxide" in the paintings in the Schack-Galerie. In the majority of the works investigated the whites do not contain zinc white, whereas the colored paints often do. This suggests that zinc white was added to various pigments by the manufacturer as a lightening agent. In that way painters probably used zinc white without realizing it.

To a large extent the greater hiding power of titanium white has resulted in its replacing zinc white on the artists' palette. Nonetheless, zinc oxide continues to be found as an additive in titanium white pigments as well as in mixtures of toners of high tinting strength. Like lead white, zinc white is frequently found blended with barytes.

2.3 Terminal Date

The pigment is available today, manufactured by numerous companies. In preparing this manuscript, the editors found that the manufacture of leaded zinc oxide had been completely discontinued in the United States for a number of years.

3.0 GENERAL PIGMENT PROPERTIES

3.1 Color and Spectral Reflectance

Zinc white produces a cool, clear white, while leaded varieties usually exhibit a yellowish or grayish tinge. The spectral reflectance curves, shown in fig. 2, indicate that the pigment tends to absorb more in the blue range than does titanium white, but absorbs less than lead white. Fig. 3 illustrates in further detail the differences in the abosrption of zinc and titanium whites in the short wavelength region of the spectrum.

3.2 Hiding Power and Tinting Strength

The refractive indices ($\epsilon = 2.02$, $\omega = 2.00$) of zinc white correspond closely to those of lead white ($\epsilon = 1.94$, $\omega = 2.09$), but in contrast to the usual commercial lead white products, zinc white exhibits very weak birefringence under the polarizing microscope (Merwin, 1917). Under ultraviolet radiation zinc oxide frequently fluoresces; impurities or admixtures can inhibit the phenomenon as well as affect the color of the fluorescence.

The excellent ultraviolet-absorbing qualities of the pigment are utilized for light-screening purposes (Stutz, 1926). In the early years of commercial paints, the addition of zinc white to lead white was shown to provide protection to the paint vehicle during outdoor exposure. Because the pigment reflects little ultraviolet, using zinc white paint on ceilings and walls results in a considerable lowering of the ultraviolet component of natural illumination indoors or of fluorescent lamp light when indirectly reflected from such surfaces (Thomson, 1961; Feller, 1964).

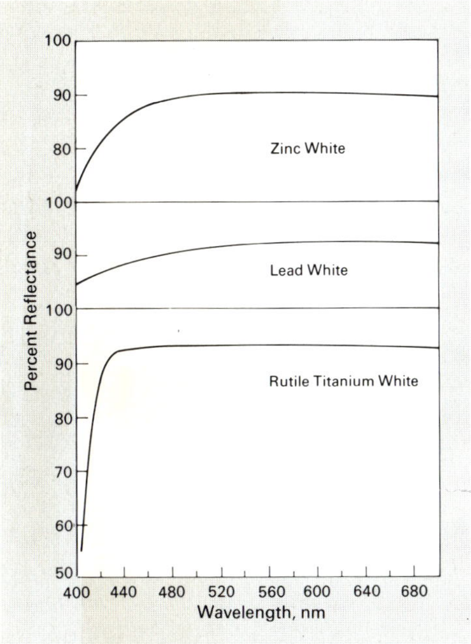

Fig. 2. Spectral reflectance of zinc white, with those of lead and titanium whites for comparison.

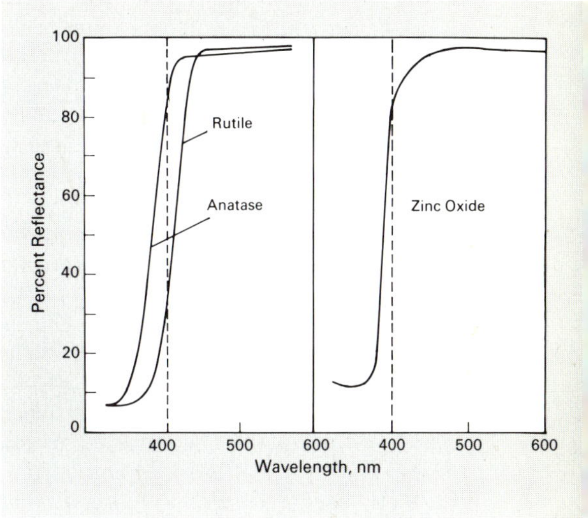

Fig. 3. Spectral reflectance curves of zinc and titanium white (rutile and anatase) pigments illustrating differences in the reflectivity in the short wavelength region of the spectrum.

Fig. 4. Scanning electron micrographs of nodular grade (A) and acicular (B) zinc oxide pigments. Magnification 4000 x. Micrographs courtesy L. E. Carlson, St. Joe Zinc Company, Monaca, Pa.

Although zinc white possesses nearly the same refractive index as lead white, it has less hiding power in oil. This characteristic is attributed mainly to the higher oil requirement of zinc white, which means a lower pigment volume concentration in the ready-to-spread paint. The hiding power is also influenced by the particle size; the slightly coarser types such as Green Seal and Red Seal hide better than the very finely divided White Seal type, although the latter possesses the greatest degree of whiteness.

With the help of the electron microscope various shapes of zinc white pigments can be distinguished (fig. 4); Vandemaele (1965) published more detailed electron micrographs of crystal structures. The subject of particle morphology is discussed in 5.1.

3.3 Permanence

Zinc white paints tend to retain their original whiteness during aging, not only in watercolor but to a great extent in oil paint as well. Compared with other white pigments ground with oil, zinc white frequently exhibits the least tendency to yellow; paints formulated with zinc white and poppy oil are believed to yellow the least. Levison (1976), for example, has clearly demonstrated that the addition of zinc white can be a major factor in reducing the yellowing of titanium white-barium sulfate paints formulated with both linseed and safflower oil. Judicious additions of lead white or lead sulfate in leaded zinc oxides further improve the weatherproof properties of zinc white in oil as well as such characteristics as adherence, resiliency, and rust-protection.

The pigment tends to react with acidic components in vehicles, leading to the formation of zinc soaps. A soap is the metal salt of an organic acid; in this case zinc tends to react with the fatty-acid components of the drying oils. The reactivity of zinc oxide with vehicles is affected considerably by the particle size, being most rapid in the colloidal (finely dispersed) grades (Payne, 1961; Bell, 1970; Morley-Smith, 1958). The highly birefringent clusters of soap crystals that result from such reactions can frequently be detected upon microscopic examination of the aged paints. In the case of linseed oil–zinc oxide paints this action is considered to harden the films. However, when zinc white is added as a secondary component in the titanium-white-based paints, the effect has been found to be beneficial. Thus, zinc white has long been added to increase the weather resistance of titanium white and lithopone coatings (Kittel, 1960, p. 185). The pigment has also been found to stabilize paints based on cellulose nitrate because the oxide is able to neutralize the nitric acid that forms during decomposition of the vehicle resulting from age.

Although the pigment may be converted into the sulfide upon exposure to industrial pollutants, this change has negligible influence on the color because zinc sulfide is also white and it hides well. As also in the case of lead white, this conversion occurs primarily in paints with aqueous vehicles (nonwashable distemper, aquarelle paints, etc.).

The finding of many examples of faded retouchings in blue segments of paintings induced the Doerner Institute some years ago to initiate exposure experiments to test different mixtures of Prussian blue with lead white, zinc white, and rutile (titanium dioxide) in oil, resin, and synthetic-resin bonding agents. The results were interesting: all mixtures faded, the ones containing lead white most, followed by the ones with titanium white, while the Prussian blue–zinc white blends showed relatively little change. According to a private communication from Dr. F. Preusser, these experiments, begun by the author, at the Doerner Institute, were continued and showed that with oil-base binding media the course and speed of the drying process influenced fading. Furthermore, consideration must be given to certain partially conflicting results which indicate that crystal habit and lattice imperfections, as well as trace impurities, can influence the photochemical behavior of zinc white. It has been known for some time that zinc white addition may often provide protection against light (Müller, 1962).

Research in recent years has revealed much about the tendency of zinc oxides to catalyze the formation of hydrogen peroxide when irradiated by the near ultraviolet (Rubin et al., 1953; Hoffmann and Savacz, 1969). This reaction has long been known as a cause of the rapid loss of strength in zinc-oxide-treated textiles (Egerton, 1948). Moisture is an essential ingredient for the production of hydrogen peroxide, the presence of which in turn can cause chalking of paints or tendering (embrittlement) of textiles. Feller (1966) has reported an interesting interlayer chalking effect on retouches in dammar varnish which may be due to chemical, perhaps soap, formation, as well as photochemical activity between zinc white and components of the resin.

3.4 Compatibility

Zinc white is compatible with all inorganic pigments. It is stable toward light in the sense that it does not discolor. However, owing primarily to the photochemical activity just mentioned, it has been known for many years that zinc oxide will affect the fading process of various organic pigments to a large degree. Following extensive testing, Eibner (1911) concluded: "Zinc white indeed accelerates light-fading of artists' paints. This manifests itself when coats of the corresponding mixtures with gum arabic, i.e., in aquarelle colors, are exposed to light, this being most pronounced when under glass." According to Eibner and Gerstacker (1913) Prussian blue–zinc white colors that have been faded by light will regain their original intensities, but not their original hues, in the dark. The rapid fading of mixtures of Prussian blue and zinc white has also been reviewed in more recent literature (Wulf, 1974). Haug (1951) discovered that moisture is the determining factor in the fading process of nonwashable distemper coats.

3.5 **Chemical Properties**

The pH value of zinc oxide is 7.1–7.5; that of leaded zinc oxide is 6.0–6.8. Because of its basic character, the oxide tends to form soaps with resin acids and with the free acids in oil; it also induces saponification of the triglycerides that are the major constituents of drying oils. The pigment, insoluble in water, dissolves readily in acids as well as in alkalis, even weak ones such as acetic acid or ammonia. The solubility in dilute ammonium hydroxide, owing to the formation of the complex ion of zinc and ammonia, is specific for this white. Therefore the addition of ammonium hydroxide is a useful test to indicate the presence of the pigment in bulk samples or in a minute specimen on a microscope slide.

In humid air the compound may change into granular zinc carbonate, which is not suitable as a pigment because of its minimal hiding power. In the presence of moist sulfur dioxide, conversion to zinc sulfate occurs. With hydrogen sulfide, the pigment forms white zinc sulfide, although a dark color may result if the paint contains lead compounds as well. Because zinc white is equally sensitive to acid and alkalis, it is not suitable for vehicles with a high proportion of free fatty or resinic acids nor for lime-based vehicles.

Owing to its mild antiseptic properties, zinc oxide has been used to great advantage in preventing the growth of fungi and bacteria in paints. Its pharmaceutical use in zinc ointment has long been known.

The pigment turns yellow upon heating but becomes white again upon cooling.

For paints based on sodium silicate, zinc white has been used as a setting catalyst. The formation of soaps between zinc oxide and fatty acids and certain colloidochemical reactions, such as adsorption of the bonding agent on the pigment, are considered to be causes of thickening in oil-based paints. Therefore, leveling agents, usually benzoic acid or α-crotonic acid, are often added to facilitate flow, particularly when the oil content is low.

In oil, the pigment tends to exhibit a lower rate of drying than does lead white and to form hard, often brittle films. The latter behavior frquently causes zinc white paints to crack easily when placed over a more resilient, softer undercoat or primer (Müller, 1962). As a rule bodied oil (stand oil) is added today to counteract embrittlement. It was found as early as the nineteenth century that the properties of pure zinc white paints were improved through additions of lead white; increased hiding power, resiliency, and drying rate were achieved. This explains in part the advantage of leaded zinc oxide.

3.6 **Oil Absorption and Grinding Qualities**

There are physical varieties of the pigment on the market (see remarks in 5.1 on morphology), and as a consequence, its oil absorption varies considerably, at least from 20–25% (*Raw Materials Index*, 1975). Certain manufacturing procedures make it possible to produce pigment grades that demand little oil. These will contain 14–15% oil in the paste or 24–30% oil in the ready-to-spread paints, as opposed to the normal oil contents of 20–25% or 40–50%, respectively (Kittel, 1960, p. 185). The low-oil types are termed "zinc white double" or "zinc oxide double." The oil requirement depends in part on the degree of fineness of the zinc white, increasing with decreasing particle size. Hiding power also diminishes as particle size decreases below about 0.25 μ diameter (Stephenson, 1973). As stated earlier, although the refractive index is similar to that of lead white, the hiding power of zinc white oil paints is usually lower, owing in part to their higher oil requirement.

The presence of zinc white often assists in the dispersion of more inert pigments (Mattiello, 1942).

3.7 **Toxicity**

Zinc white is not particularly toxic. The use of zinc ointment as a pharmaceutical preparation is evidence of its beneficial rather than harmful effects. However, in manufacturing plants "when freshly formed ZnO is inhaled, a disorder known as the *oxide shakes* or *zinc chills* sometimes occurs." It is recommended that where large quantities of zinc oxide dust are encountered good ventilation be provided "to avoid concentrations exceeding 15 mg of zinc oxide per cubic meter over an eight-hour exposure" (*Handbook of Chemistry and Physics*, 1971–1972).

4.0 COMPOSITION

4.1 **Chemical Composition**

In Germany only those products with a minimum content of 99% zinc oxide are termed "zinc white" or "metal zinc white"; the maximum lead content is specified as 0.4% (Kittel, 1960).

In the United States the term "lead free" implies products that are 99% ZnO if made by the French Process, 98% ZnO if manufactured by the American Process (Stephenson, 1973). Customarily the pigment is marketed under various Seal brands, White Seal being the finest and purest type (0.5–0.7-μm medium-sized granules), followed by the slightly coarser, but equally pure Green Seal, and finally the coarsest, less pure Red Seal.

"Zinc oxide" or "leaded zinc oxide" are terms used in Germany for a product with at least 90% zinc oxide and a maximum of 10% lead (calculated as lead oxide). In contrast to traditional white pigments, this material may possess a yellow or gray cast. "Plumbic zinc oxide" may contain more lead but must contain at least 75% zinc oxide. "Leaded zinc oxide," now no longer available in the United States, commonly contained 12–55% basic lead sulfate, but there were also grades containing up to 70% lead compounds. Leaded zinc oxides produced from zinciferrous and plumbiferrous ores are termed "co-fumed" (in Germany "Harzsiegel" brands), in contrast to the "blended" types obtained by subsequent mixing of zinc oxide with lead sulfate. Basic zinc carbonate has been suggested as a substitute for zinc white, but it has not acquired significance as a pigment.

Commercial pigments may contain small amounts of zinc sulfate, sulfite, or chloride as well as traces of lead, iron, cadmium, and sulfur (Stephenson, 1973).

4.2 Sources

The natural mineral was never used for pigment, and the manufacturing sources today are so varied that it is not profitable to discuss them as a special topic. From the practical point of view, one can distinguish between pigments made by the direct (American) and indirect (French) processes and the leaded or co-fumed types. The first two may be considered to represent distinctly different sources of pigment (see 4.3).

The companies producing both American and French process zinc oxide are ASARCO Incorporated, New Jersey Zinc Company, and St. Joe Minerals Corporation. Fig. 5 shows a picture of the St. Joe Balmat mine, located in St. Lawrence County, New York. Eagle-Picher and Sherwin-Williams Chemicals were listed as manufacturers of American zinc oxide (*Raw Materials Index*, 1975).

4.3 Preparation

Zinc oxide occurs in nature as the red-colored zinc ore, zincite, useless as a pigment because of impurities. According to Dana (1952), manganese is often present and is probably responsible for coloring the mineral; the traditional borax bead test in an oxidizing flame often produces a reddish violet color with native zincite, indicating manganese. At the present time zinc oxide is produced by three methods. The French or indirect process starts with metallic zinc. The metal is simply distilled in crucibles, or retorts, and becomes inflamed upon contact with the air, producing the oxide (fig. 1). It may be added that it is almost impossible to melt the metal without producing the oxide. Most treatises on commercial pigments describe this mode of manufacture in considerable detail (Petit, 1907, p. 68). The American or direct process involves the heating of zinc ores, usually various combinations of franklinite, zincite, and willemite.

According to Faloon (1925), the means of producing zinc oxide from the ore was said to have been discovered about 1854 by a workman named Burrows who was employed by the Passaic Chemical Company at Newark, New Jersey. There was a zinc plant adjoining the property, and Burrows, while on night duty, found a badly leaking fire flue in one of the furnaces he was tending. He mended this hole by throwing an old piece of fire grate over the opening and covering it with some of the zinc company's mixture of ore and coal, which he found handy. He was surprised a short time later to see a cloud of zinc oxide arising from the burning patch. He later related his experience to Wetherill and Jones of the zinc company. His discovery led to manufacturing developments along this line, patented by Wetherill between 1855 and 1868. Jones perfected a method of collecting the zinc oxide in fabric bags. The processes developed by these two men have been used extensively for the manufacture of American, or direct, process zinc oxide. Until 1892 zinc white was made in the U.S.A. exclusively by the direct method; the purer grade was imported (Ullman, 1932).

American process pigment traditionally was a less pure product of higher lead content, which in Germany is called "zinc oxide" (see 4.1). However, today pigment of high quality can be recovered by purification of the zinc vapors generated as an intermediate product in the

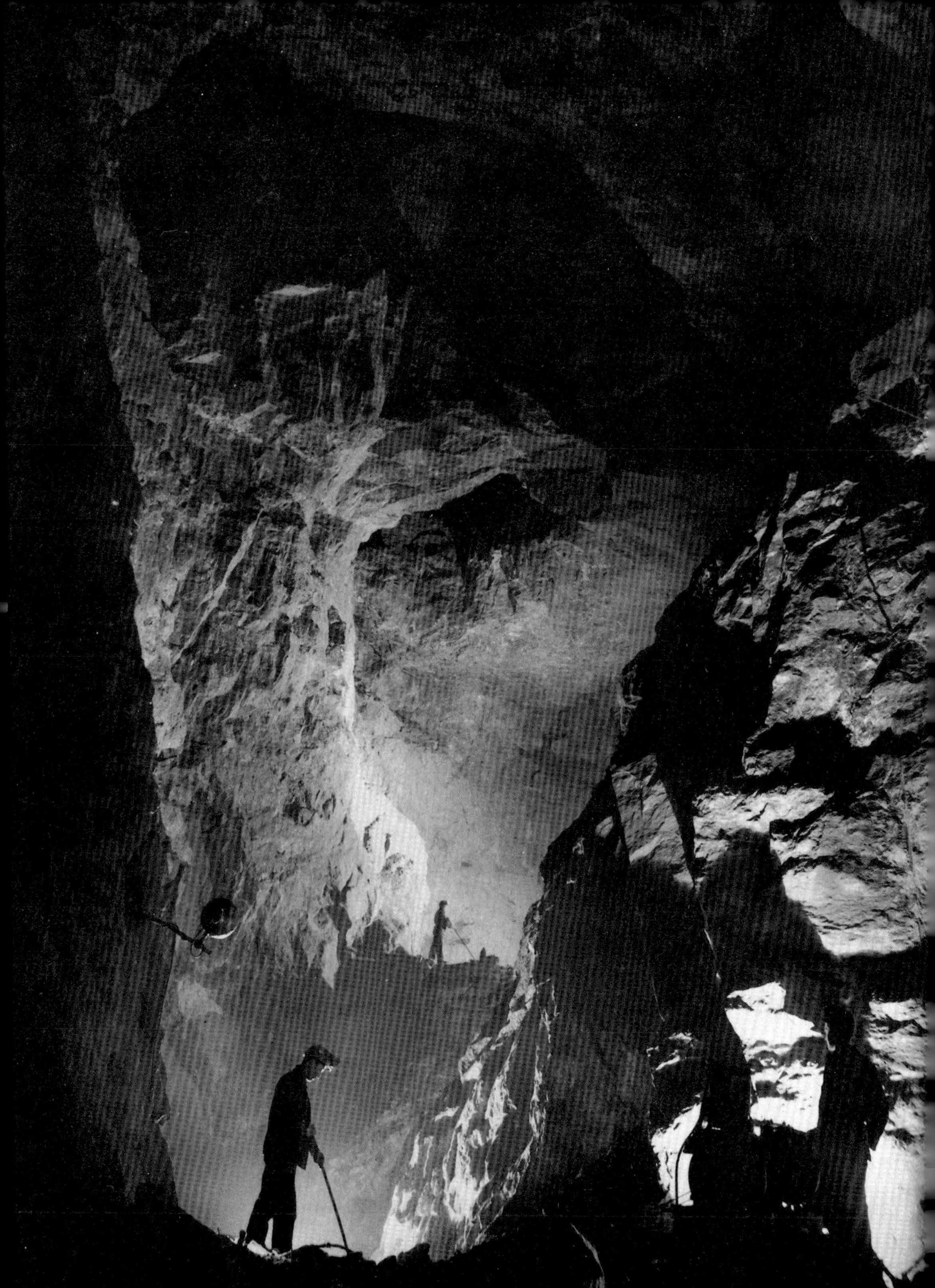

Fig. 5. St. Joe Minerals' Balmat mine, St. Lawrence County, New York, where veins of sphalterite occur in limestone. Photo: St. Joe Minerals Corporation.

indirect process. Such pigment can also be manufactured directly from the ores or even from the refuse products of the smelter. A diagram of a process for making zinc oxide of very high purity, essentially through removal of cadmium and lead by fractional distillation, is shown in fig. 6.

Sulfidic ores such as zinc blende must be first roasted to yield the oxide. By-products of smelting and metal residues also can be processed into zinc oxide. From the commonly occurring mixed ores based on zinc blende and lead glance, leaded zinc oxides are produced by the direct process ("co-fumed leaded zinc oxide," "Harzsiegel" by Unterharzer-Berg- und Hüttenwerke, Goslar/Oker). This third major type of pigment obviously contained significant amounts of lead sulfate.

There also exists a group of processes based on the precipitation or hydrolysis of solutions of zinc salts, the resulting products being called "precipitated zinc white." When ores or waste products from smelting are leached with sulfuric acid, zinc carbonate may be precipitated from the purified zinc sulfate solution by introducing carbon dioxide, and the carbonate is then converted into zinc white by roasting. The leaching of zinciferrous materials with ammonia or sodium hydroxide solution produces ammoniates or zincates which yield zinc hydroxide and zinc oxide respectively by subsequent hydrolysis. Zinc white is also produced on a small scale electrolytically (Kittel, 1960, pp. 175–182).

4.4 Adulteration and Sophistication

In most industrial paints zinc oxide usually was accompanied by judiciously chosen percentages of lead and titanium white, if not also by selected inert substances. As mentioned in 3.3, these additions were customarily made to improve the properties of the resultant paints and therefore should not be regarded as unwanted or unnecessary. Even in artists' paints, Kühn (1969) has found that zinc white in oil generally occurs in mixtures with lead white, barium sulfate, and titanium whites.

Griffith's zinc white (Bersch, 1901), an old product prepared by igniting zinc sulfide, obviously was not pure zinc oxide.

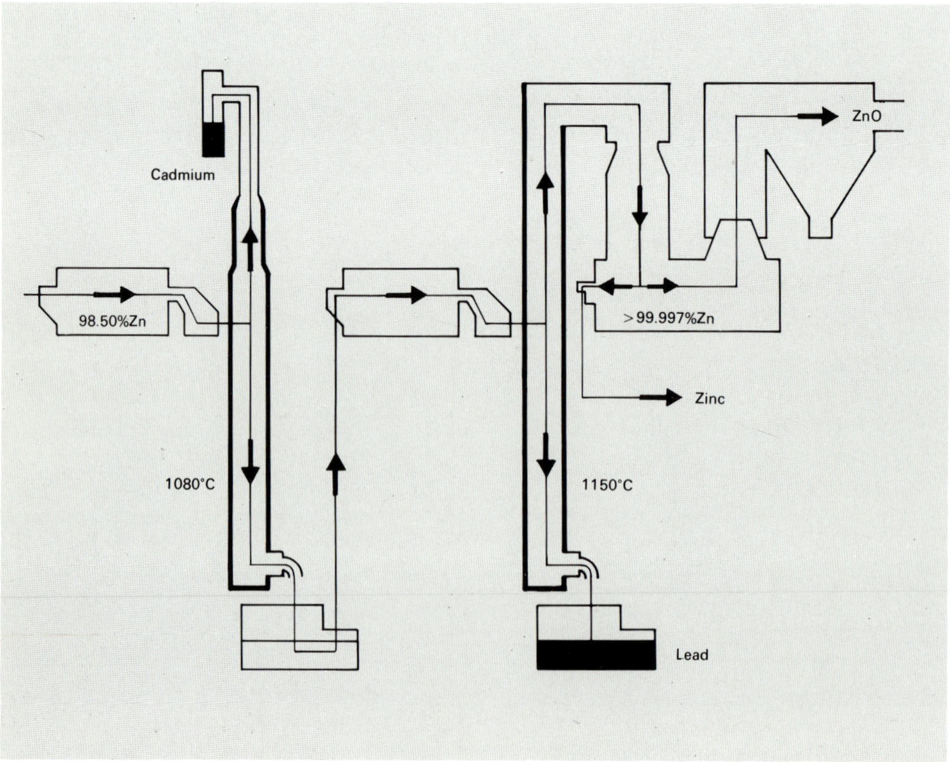

Fig. 6. Flow diagram of the New Jersey Zinc Company procedure for making zinc oxide of high purity by the French process (after Vandemaele, 1965).

5.0 IDENTIFICATION — CHARACTERIZATION

5.1 Characterization by Optical Microscopy

Zinc oxide belongs to the hexagonal crystal system. Morphologically we can distinguish three shapes: planar, isometric, and linear. The first two are common technical products; the linear variety is the basis for acicular zinc oxide. The various types are most clearly distinguished under the electron microscope, as illustrated in fig. 4. Fig. 7 illustrates the boule shape of typical acicular-shaped crystals. Vandemaele (1965) has published transmission electron photomicrographs of a number of acicular types. The needles and twins of acicular types can be detected even under the optical microscope (fig. 8).

The particle size of nodular zinc white as a rule lies between 0.25 and 1.0 μm, but there are also finer specialty grades. The particle size of the less pure types may vary much more than this.

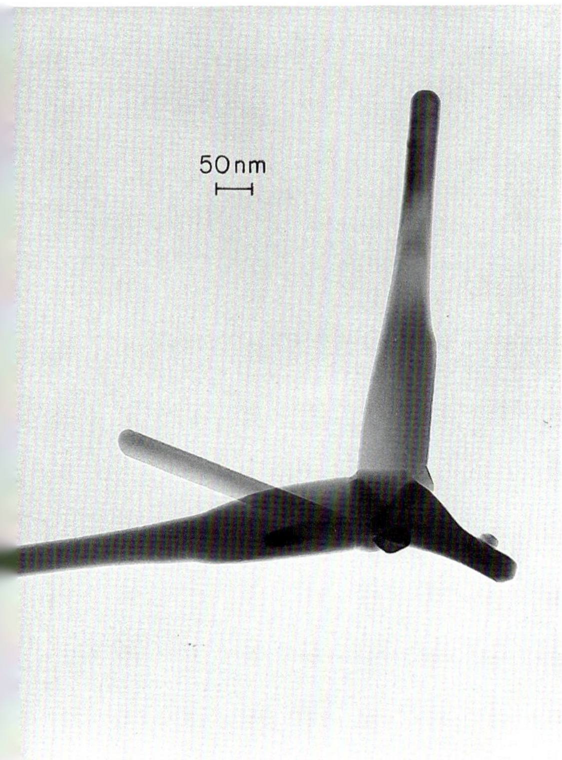

Fig. 7. Acicular crystals of zinc oxide showing boule shape of individual needles. Magnification 100,000x. Micrograph: St. Joe Minerals Corporation.

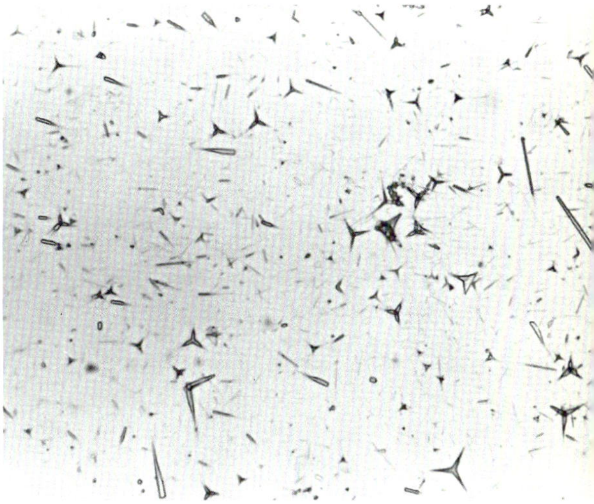

Fig. 8. Photomicrograph of acicular grade zinc oxide at 550x magnification.

Fig. 9. Photomicrograph of nodular-shaped pigment grade zinc oxide at 550x magnification.

Under the optical microscope, most pigment varieties display very fine, rounded particles, which are not distinctive (fig. 9). An exception is the acicular type, which at 100–500x can be seen as individual, twinned, or quadruply joined needles (fig. 8). The latter distinctive shape forms with slow burning of the zinc vapor, the primary particles grouping together into needles of variable length. Photomicrographs of the nodular and acicular pigments at 1,285x have been published by Nelson (1940).

5.2 Chemical Identification

5.21 *Mixed Crystals with Thiocyanate.*

For the detection of zinc, the coprecipitation of zinc and cobalt ions with mercury thiocyanate has proved to be an effective micro-chemical reaction (Feigl and Anger, 1972, p. 514). The sample is heated on the slide with a drop of dilute nitric acid (1:1) and evaporated to dryness. To the residue one drop of distilled water is added, the slide heated slightly, and one drop each of cobaltic salt solution and mercuric thiocyanate solution added. The sample on the slide is then rubbed with the glass rod for about 15 seconds. If the solution contains zinc, a blue precipitate of mixed crystals of zinc and cobalt mercury thiocyanate soon appears. Under the microscope the crystals usually show forked growth structures of striated stars. In the absence of zinc, blue crystals only appear after two or three minutes, first mostly in the shape of dark blue prisms (cobalt-mercury thiocyanate) along the scratch lines of the rod upon the slide. A blank test is recommended. Detection limit: 0.1–0.2 microgram zinc. Composition of test solutions:

a) Cobalt solution: 0.05 g cobalt-II-chloride in 100 ml hydrochloric acid (dilution: 1 part conc. hydrochloric acid to 19 parts water).

b) Mercury thiocyanate solution: 8 g mercury-II-chloride plus 9 g ammonium thiocyanate in 100 ml water.

5.22 *Dithizone Test.*

The spot test with dithizone for the determination of zinc in alkaline solution, although less sensitive, is specific (Feigl and Anger, 1972, p. 512).

5.3 Instrumental Methods of Analysis

5.31 *X-ray Diffraction.*

Provided the pigment is in sufficient concentration, a Debye-Scherrer powder diffraction pattern provides unequivocal identification of zinc white in a typically small sample from a painting or other object (König, 1956). This affords a positive distinction between zinc white and lithopone, besides distinguishing it from the sulfide or carbonate, both also white. The latter has occasionally been substituted for zinc white (see 2.2). To obtain a Debye-Scherrer powder pattern, a grain of material weighing only a few micrograms is sufficient (Kühn, 1974). Fig. 10 shows powder diffraction patterns of American process zinc white and various leaded zinc whites.

Values for the *d*-spacings and intensities of the lines are given in table 1. Table 2 gives the diffraction data for lead sulfate, which is present in leaded types and occasionally in the American process type. Nelson (1940) published powder diffraction patterns of normal zinc oxide as well as both co-fumed and blended leaded types, pointing out that the weaker-appearing lines of lead sulfate appeared the same in the patterns of the latter two.

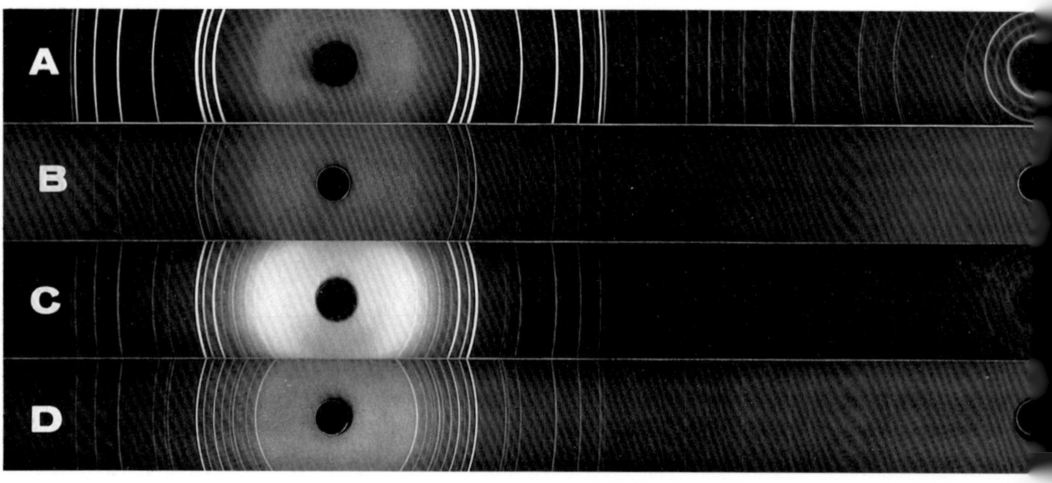

Fig. 10. X-ray powder diffraction patterns of various zinc oxides. A. American process, sample B-706. B. Leaded zinc oxide, 88% ZnO, sample B-672. C. Leaded zinc oxide, 65% ZnO, sample B-725. D. Leaded zinc oxide, sample B-671. Photographs not to be used for measurement purposes. (Patterns courtesy of The Freer Gallery of Art, Washington. Sample numbers from the National Gallery of Art Pigment Bank, Carnegie-Mellon Institute of Research, Pittsburgh.)

Table 1. **X-RAY DIFFRACTION DATA FOR ZINC OXIDE (ZnO)[a] (Diffractometric Data, Cu$K\alpha_1$ = 1.5405 Å)**

d(Å)	*I*	*d(Å)*	*I*
2.816	71	1.0422	10
2.602	56	1.0158	5
2.476	100	0.9848	4
1.911	29	.9764	7
1.626	40	.9555	1
1.477	35	.9382	4
1.407	6	.9069	12
1.379	28	.8826	6
1.359	14	.8675	1
1.301	3	.8369	6
1.238	5	.8290	2
1.1812	3	.8237	2
1.0929	10	.8125	5
1.0639	4		

a *JCPDS Powder Diffraction File* 5-0664. Courtesy of JCPDS International Centre for Diffraction Data (1982).

Table 2. **X-RAY DIFFRACTION DATA FOR LEAD SULFATE (ANGLESITE) POSSIBLY PRESENT IN ZINC OXIDE PIGMENTS[a] (Diffractometric Data, Cu$K\alpha_1$ = 1.5405 Å)**

d(Å)	*I*	*d(Å)*	*I*
5.381	3	2.028	48
4.26	87	1.973	21
3.813	57	1.905	3
3.622	23	1.879	6
3.479	33	1.793	15
3.333	86	1.741	8
3.220	71	1.716	3
3.001	100	1.703	16
2.773	35	1.656	7
2.699	46	1.648	3
2.618	8	1.621	19
2.406	17	1.611	10
2.355	<1	1.571	6
2.276	20	1.542±	2
2.235	5	1.525±	1
2.193	7	1.493	15
2.164	26	1.467	7
2.133	5	1.441	8
2.067	76	1.429	4
2.031	34	1.406	3

a *JCPDS Powder Diffraction File* 5-0577. Courtesy of JCPDS International Centre for Diffraction Data (1982).

Fig. 11. Two variations that may be encountered in the shape of the principal infrared absorption band of zinc oxide. Spectra arbitrarily displaced on vertical scale for comparison.

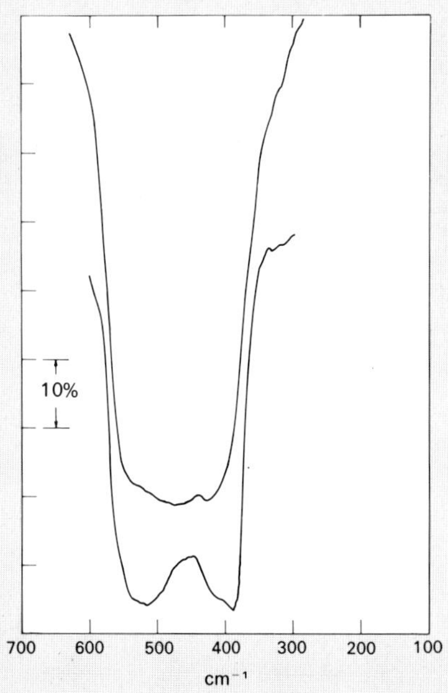

5.32 ***Infrared and Ultraviolet Spectroscopy.*** The infrared spectrum of zinc oxide has practically no distinctive bands of wavelength shorter than 17 microns (590 cm^{-1}); hence, many of the older infrared instruments could not be used to characterize the pigment (*Infrared Spectroscopy*, 1980). Fig. 11 shows two shapes of principal absorption bands that may be observed in the region between 400 and 500 cm^{-1} (Nyquist and Kagel, 1971). This type of variation, which can be encountered in various samples, is also seen in published spectra in the book *Infrared Spectroscopy* (1980).

The curves in fig. 12 indicate that sulfate bands usually can be detected in leaded zinc oxides. As noted above, the presence of lead sulfate is usually also detected in x-ray diffraction measurements. Thus, specimens A and B in fig. 12 (infrared) correspond respectively to samples C and D in fig. 10.

Zinc white absorbs radiation in the ultraviolet region. The absorption edge of the band occurring at the beginning of the visible region

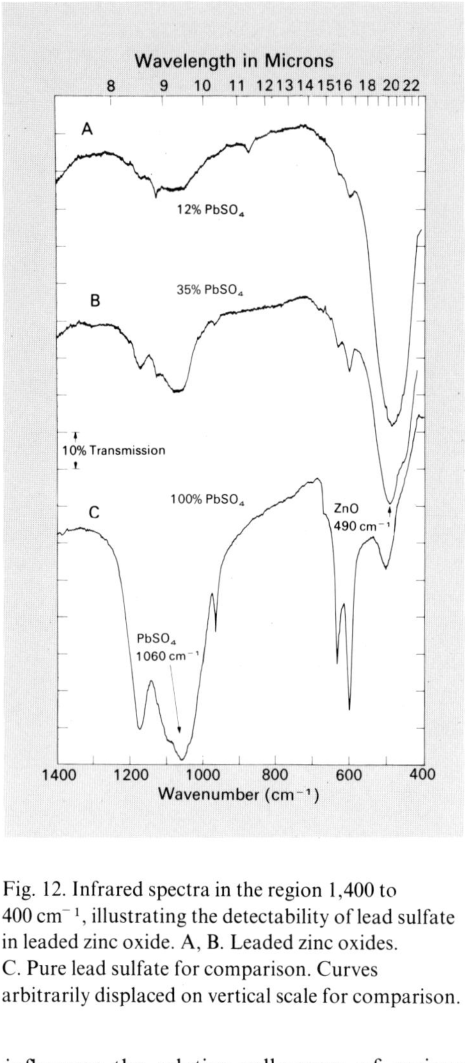

Fig. 12. Infrared spectra in the region 1,400 to 400 cm^{-1}, illustrating the detectability of lead sulfate in leaded zinc oxide. A, B. Leaded zinc oxides. C. Pure lead sulfate for comparison. Curves arbitrarily displaced on vertical scale for comparison.

influences the relative yellowness of various preparations of the oxide (note fig. 3).

5.33 ***Spectrochemical Analysis.*** Zinc is easily detected by emission spectrographic analysis (ESA), the lower limits of detection being between 0.01% (100 ppm) and 0.0003% (3 ppm). Certain lines fall in the ultraviolet at 3282.33, 3302.59, and 3345.02 Å (limit of detection 0.01%). The highest sensitivity, about 0.0003%, is in the 2138.56 Å emission; however, ordinary photographic emulsions are insufficiently sensitive in this region (Ahrens and Taylor, 1961). Analysis by ESA has an advantage over microchemical analyses, as it is unaffected by cations, anions, or the organic binding media. At the same time a single determination will provide a permanent record of the other metallic elements that are present (Kühn, 1974). Cadmium is a common impurity (Stephenson, 1973), although it has not yet been detected by emission spectrographic analysis of paint samples examined thus far by the author.

5.34 ***X-radiography.*** At a wavelength of 0.708 Å the mass absorption coefficient of zinc white is 47.6 cm^2g^{-1} (37.9 cm^2g^{-1} in a paste containing 15% by weight of oil). This value is less than half that of lead white (109.2, or 97.3 cm^2g^{-1} in a paste containing 11% oil). As a consequence zinc white films are darker in the usual radiographs than lead white films of the same thickness.

5.35 ***Mass Spectrography.*** Mass spectral analysis can provide even higher sensitivity than emission spectrographic analysis. The limits of detection are reported to be about 0.00001% (0.1 ppm) (Kienitz, 1968).

5.36 ***Energy-Dispersive X-ray Fluorescence.*** The principal lines in the energy-dispersive x-ray fluorescence spectrum are $ZnK(1)\alpha_1$ 1.432 Å, $ZnK(1)\alpha_2$ 1.436 Å, $ZnL(1)\beta_1$ 11.985 Å, and $ZnL(1)\alpha$ 12.257 Å (Sagel, 1959). Parenthetical numerals refer to spectral orders.

5.4 Criteria for Positive Identification.

Positive identification of zinc white is achieved by x-ray diffraction. A Debye-Scherrer pattern will enable one to detect pigment that is associated with blanc fixe (barium sulfate), or to distinguish the pigment from lithopone (zinc sulfide/barium sulfate). The usual infrared spectra in the region of 3 to 15 μm (3,330 to 666 cm^{-1}) will aid little to identify zinc white; however, in the region between 15 and 100 μm (666 to 100 cm^{-1}) one can differentiate zinc white, leaded zinc white, blanc fixe, and lithopone.

The pigment should be completely soluble in dilute nitric acid without effervescence. Solubility of a white pigment in dilute ammonium hydroxide is highly indicative of zinc white.

The detection of a major amount of zinc in a white pigment either by emission spectrographic analysis, energy dispersive x-ray fluorescence analysis, or a microchemical test is a fair indicator of the presence of either zinc oxide or zinc sulfide (Kühn, 1966). Varying amounts of zinc white, of course, occur in modern varieties of lead and titanium white.

6.0 NOTABLE OCCURRENCES

At the end of the nineteenth century Russell and Abney (1888) sent questionnaires to renowned English artists who worked in watercolors; of the forty-six who reported the pigments they used, twelve reported Chinese White. In 1923 Moreau-Vauthier mentioned twelve artists who had zinc white on their palettes. But in a more recent publication listing artists' palettes Birren (1965) notes only two who use zinc white.

Up to 1979, at the time this manuscript was prepared, the Doerner Institute in Munich had determined the presence of zinc white in 204 paintings in the Bavarian State painting collections (Bayerische Staatsgemäldesammlungen, BSTG). The accompanying table of notable occurrences is derived largely from these determinations (Kühn, 1969).

Zinc White — Notable Occurrences in Works from the Bayerische Staatsgemäldesammlungen, Munich.

Artist or School *Title, Date* *Inventory Number*
Johann Georg von Dillis, *Triva Castle*, 1797, 9392
George Morland, *Beach Scene*, 1798, 3862
Johann Georg von Dillis, *Landscape with Waterfall*, about 1800, 10 243
Joseph Rebell, *Roman Wine Shop*, about 1800, 5452
J. A. Koch, *Landscape with St. George*, 1807, 4907
Caspar David Friedrich, *Summer*, 1808, 9702
Karl Friedrich Schinkel, *Cathedral above a City*, 1813, 19,422
M. J. Wagenbauer, *Morning in a Village*, 1816, WAF 1173
J. A. Rhomberg, *Brother Graycoat with Female Pilgrim*, 1820, 2629
Ferdinand von Olivier, *Italian Mountain Landscape with Hermits*, 1820–1830, 10,852
J. P. von Langer, *Madonna with Child and John the Baptist*, 1823, 7614
B. F. Bauer, *Madonna*, 1824, 4902
Lorenzo Quaglio II, *Peasant Couple before a Tavern*, 1826, 14,187
Ludwig Schnorr von Carolsfeld, *St. George* — high altar of parish church in Wein-Kahlenbergdorf, 1826, 2745
Johann Georg von Dillis, *View of the Quirinal*, 1817/1818, 9106
J. C. Zeigler, *Landscape*, 1829, 9544
Heinrich Bürkel, *Sabine Hills*, 1831–1833, 11,086
Ludwig Richter, *Camping People in the Mountains*, 1833, 12,031
Heinrich Bürkel, *Rainstorm in Partenkirchen*, 1838, WAF 133
Johann Adam Klein, *Cattle with Herdsman*, 1844, 14,186
Moritz von Schwind, *The Dream of Erwin von Steinbach*, about 1845, 11,558 x
Leo von Klenze, *The Acropolis, Athens*, 1846, 9463
Heinrich Bürkel, *Shearing Sheep in the Campagna*, about 1850, 13,046
Heinrich Bürkel, *Before an Alpine Hut*, about 1850, 9195
Carl Spitzweg, *The Farewell*, about 1855, 11,561 x
Franz von Lenbach, *Village Street in Aresing*, 1856, 8375
Carl Millner, *View of Upper Lake near Berchtesgaden*, about 1856, 11,601 x
Leo von Klenze, *Campo Santo in Pisa*, 1858, 13,078
Bonaventure Genelli, *Rape of Europa*, 1859, 11,547 x
Arnold Böcklin, *Shepherdess*, about 1860, 11,529 x
Moritz von Schwind, *Hermit in a Grotto*, about 1860, 11,759 x
Arnold Böcklin, *The Bagpipe Player*, 1861, 9471
Arnold Böcklin, *Villa at the Sea, I*, 1864, 11,528 x
Arnold Böcklin, *Dragon in a Gorge*, 1870, 11,531 x
Arnold Böcklin, *Ideal Spring Landscape*, 1870/1871, 11,527 x
Hans Makart, *Diana with Attendants*, sketch, about 1870, 13,294
Arnold Böcklin, *Triton and Nereid*, 1874, 11,534 x
Paul Cézanne, *Self-Portrait*, about 1877, 8648

Most of these findings are based on either emission spectrographic analyses or x-ray diffraction by the Debye-Scherrer technique.

x, see Kühn (1969).

Artist or School *Title, Date* *Inventory Number*
Arnold Böcklin, *Veritas*, 1881, 10,444
Paul Cézanne, *Still Life*, about 1885, 8647
Vincent van Gogh, *Vase with Sunflowers*, 1888, 8672
Vincent van Gogh, *View of Arles*, 1889, 8671
Vincent van Gogh, *The Plain of Auvers*, 1890, 9584
Gustav Klimt, *Music*, 1895, 8195
Max Slevogt, *Karl Voll*, 1898, 10,014
Paul von Ravenstein, *View into the Valley*, 1905, 8461–8136 a
Ferdinand Hodler, *Landscape, Lake of Geneva*, 1907, 8715
Max Slevogt, *Prince Regent Luitpold*, 1908/1910, 9088
Max Slevogt, *Prince Regent Luitpold*, 1909, 9091
Odilon Redon, *Rose Window in a Cathedral*, about 1912, 13,080
Max Slevogt, *Still Life, Fruit*, 1925, 9948

Most of these findings are based on either emission spectrographic analyses or x-ray diffraction by the Debye-Scherrer technique.

x, see Kühn (1969).

ACKNOWLEDGMENTS

The author thanks Dr. H. von Sonnenburg, of the Doerner Institute, Munich, and Dr. F. Preusser of the J. Paul Getty Conservation Institute, formerly of the Doerner Institute, for the list of notable occurrences; Dr. W. Schneider of the Pharmacy Historical Seminar, Technical University, Brunswick, for the opportunity to examine the contents of several apothecary jars of the sixteenth to seventeenth centuries (see 2.2); and Mary Curran and Robert L. Feller for references to a number of twentieth-century publications.

REFERENCES

L. H. Ahrens and S. R. Taylor, *Spectrochemical Analysis*, 2nd ed. (Reading, Massachusetts, 1961), 75, 422.

S. H. Bell, "Controlling Zinc Oxide Hazing in Gloss Alkyds," *Paint and Varnish Production* 60, no. 4 (April 1970), 55–60.

J. Bersch, *The Manufacture of Mineral and Lake Pigments*, 3rd ed. rev., trans. A. C. Wright (London, 1901), 128.

F. Birren, *History of Color in Painting* (New York, 1965), 42–74.

H. E. Brown, *Zinc Oxide Rediscovered* (New York, 1957).

Colour Index, vol. 3, 3rd ed. (Bradford, Yorkshire, England, and Research Triangle Park, North Carolina, 1971).

E. S. Dana and W. E. Ford, *Dana's Manual of Mineralogy*, 16th ed. rev. (New York, 1952).

J. H. Dierbach, *Die Arzneimittel des Hippokrates* (Heidelberg, 1824; Hildesheim, 1969), 253.

G. S. Egerton, "The Role of Hydrogen Peroxide in the Photochemical Degradation of Cotton Sensitized by Vat Dyes and Some Metallic Oxides," *The Journal of the Textile Institute* 39 (1948), 305–318.

A. Eibner, "Uber die Lichtwirkungen auf Malerei-Farbstoffe," *Chemiker Zeitung* 35 (1911), 753–755, 774–776, 786–788.

A. Eibner and L. Gerstacker, "Über Lichtreaktionen auf Malerfarbstoffe II. Die Zusammensetzung des Turnbullsblaues," *Chemiker Zeitung* 37 (1913), 137–139.

D. B. Faloon, *Zinc Oxide History, Manufacture and Properties as a Pigment*, 1st ed. (New York, 1925).

F. Feigl and V. Anger, *Spot Tests in Inorganic Analysis*, 6th ed. rev., trans. R. E. Oesper (Amsterdam, 1972).

R. L. Feller, "Control of Deteriorating Effects of Light upon Museum Objects," *Museum* 17, no. 2 (1964), 57–98.

R. L. Feller, "Problems of Retouching: Chalking of Intermediate Layers," *Bulletin of the American Group — IIC* 7, no. 1 (1966), 32–34.

J. G. Gentele, *Lehrbuch der Farbenfabrikation* (Braunschweig, 1860), 54–55.

R. G. Gettens and G. L. Stout, *Painting Materials, A Short Encyclopedia* (1942; New York, 1966).

Handbook of Chemistry and Physics, ed. R. C. Weast (Cleveland, 1971–1972), B-38.

Handwörterbuch der reinen und angewandten Chemie, vol. 9 (Braunschweig, 1864), 962–968.

R. D. Harley, *Artist's Pigments; c. 1600–1835, A Study in English Documentary Sources*, 2nd ed. (London, 1982), 178.

R. Haug, "Uber die photochemische Ausbleichung von Mischungen aus Berlinerblau und Zinkweiss," *Farbe und Lack* 57 (1951), 49–54.

E. Hoffmann and A. Savacz, "Weathering of Paint Films: 1. Chalking Caused by Zinc Oxide in Latex Paints," *Journal of the Oil and Colour Chemists' Association* 52 (1969), 113–132.

Infrared Spectroscopy-Atlas for the Coatings Industry (Philadelphia, 1980), Spectra 811, 835, 836.

J. K. G. Jacobsson, *Technologisches Wörterbuch* (Berlin, 1795), 262, 418.

JCPDS Powder Diffraction File, JCPDS International Centre for Diffraction Data (Swarthmore, Pennsylvania, 1982).

H. Kienitz, *Massenspektrometrie* (Weinheim, 1968), 377.

H. Kittel, ed., *Pigmente. Herstellung, Eigenschaften, Anwendung* (Stuttgart, 1960).

M. H. Klaproth and F. Wolff, *Chemisches Wörterbuch*, vol. 5 (Berlin, 1810), 703.

R. König, *Anorganische Pigmente und Röntgenstrahlen* (Stuttgart, 1956), 56.

H. Kopp, *Geschichte der Chemie*, vol. 4 (Braunschweig, 1845), 121.

H. Kühn, "Trace Elements in White Lead and Their Determination by Emission Spectrum and Neutron Activation Analysis," *Studies in Conservation* 11 (1966), 163–169.

H. Kühn, *Die Pigmente in den Gemälden der Schack-Galerie* (Munich, 1969).

H. Kühn, "Möglichkeiten und Grenzen der Untersuchung von Gemälden mit Hilfe von naturwissenschaftlichen Methoden," *Maltechnik/Restauro*, no. 3 (1974), 149–162.

E. C. Leclaire, *Peinture au blanc de zinc et aux couleurs à base de zinc* (Paris, 1849).

R. Lemoine and C. du Manoir, *Les matières premières employées en peinture* (Rouen, 1893).

J. C. Leuchs, *Anleitung zur Bereitung aller Farben und Flüssigkeiten* (Nuremberg, 1825), 65.

H. W. Levison, *Artists' Pigments* (Hallandale, Florida, 1976), 30.

E. O. von Lippmann, *Entstehung und Ausbreitung der Alchemie* (Berlin, 1919), 592.

J. J. Mattiello, ed., *Protective and Decorative Coatings*. Vol. 2. *Raw Materials. Pigments, Metallic Powders and Metallic Soaps* (New York, 1942), 373.

H. E. Merwin, "Optical Properties and Theory of Color of Pigments and Paints," *Proceedings of the American Society for Testing Materials* 17 (1917), 494–530.

S. Mierzinski, *Die Erd-, Mineral- und Lack-farben*, 4th ed. (Weimar, 1881), 473.

C. Moreau-Vauthier, *Comment on peint*, ed. H. Floury (Paris, 1923), 23–34.

C. T. Morley-Smith, "Zinc Oxide — A Reactive Pigment," *Journal of the Oil and Colour Chemists' Association* 41 (1958), 85–97.

H. G. Müller, *Schoenfeld's Malerfibel* (Düsseldorf, 1962), 70.

H. A. Nelson, *The Versatile Paint-Making Properties of Zinc Oxide* (New York, 1940).

R. A. Nyquist and R. O. Kagel, *Infrared Spectra of Inorganic Compounds* (New York, 1971), 221.

H. F. Payne, *Organic Coating Technology*, vol. 2 (New York, 1961), 763.

G. Petit, *The Manufacture and Comparative Merits of White Lead and Zinc White Paints*, trans. D. Grant (London, 1907), 67–79.

Practical Treatise on Painting in Oil-Colours (London, 1795), 218–246.

Raw Materials Index, Pigment section (Washington, 1975), 2–3.

T. R. Rubin, J. G. Calvert, G. T. Rankin, and W. MacNevin, "Photochemical Synthesis of Hydrogen Peroxide at Zinc Oxide Surfaces," *Journal of the American Chemical Society* 75 (1953), 2850–2853.

W. J. Russell and W. de W. Abney, *Report on the Action of Light on Water Colours* (London, 1888), 73–77.

K. Sagel, *Tabellen zur Röntgen- Emissions- und Absorptions-Analyse* (Berlin, 1959), 128.

C. Saint-Victor, *Aquarelles miniatures* (Milan, 1835), 38.

W. Schneider, *Geschichte der pharmazeutischen Chemie* (Weinheim, 1972), 26 and 148.

H. B. Stephenson, "Zinc Oxide and Leaded Zinc Oxide," in *Pigment Handbook*, vol. 1, ed. T. C. Patton (New York, 1973), 37–51.

G. F. A. Stutz, "The Testing of Paint Pigments for Transparency to Ultraviolet Radiation in ZnO," *Journal Franklin Institute* 202 (1926), 89–99.

G. Thomson, "A New Look at Colour Rendering, Level of Illumination, and Protection from Ultraviolet Radiation in Museum Lighting," *Studies in Conservation* 6 (1961), 49–70.

F. Ullmann, *Enzyklopädie der technischen Chemie*, vol. 10 (Berlin, 1932), 708.

J. Vandemaele, "Les oxydes de zinc," *Double-Liaison*, no. 119 (July 1965), 47–63.

M. Wellmann, *Pedanii Dioscuridis anazarbei de materia medica libri quinque* (Berlin, 1958), 40–42.

Winsor & Newton, *Descriptive Handbook of Modern Water Colour Pigments* (London, 1930), 15.

H. Wulf, *Grosse Farbwarenkunde* (Cologne, 1974), 98.

G. Zerr and R. Rübencamp, *A Treatise on Colour Manufacture*, trans. C. Mayer (London, 1908).

BIBLIOGRAPHY

C. Coffignier, *Couleurs et peintures* (Paris, 1924).

P. Fleury, *The Preparation and Uses of White Zinc Paints*, trans. D. Grant (London, 1912).

C. D. Holley, *The Lead and Zinc Pigments* (New York, 1909).

C. J. Lancucki and E. Hoffmann, "A Note of the Chalking of Zinc Oxide Latex Paints," *Journal of the Oil and Colour Chemists' Association* 53 (1970), 218–219.

C. H. Mathewson, *Zinc: The Science and Technology of the Metal, Its Alloys and Compounds* (New York, 1959), 334–385.

Chrome Yellow and Other Chromate Pigments

HERMANN KÜHN and MARY CURRAN

A. Chrome Yellow

NOMENCLATURE IN VARIOUS LANGUAGES

English: chrome yellow
German: Chromgelb
Russian: желтый крон
French: jaune de chrôme
Italian: giallo de cromo
Spanish: amarillo de cromo, chromato de plomo

1.0 INTRODUCTION

The chromate ion, $CrO_4^{=}$, combines with several heavy metals to form yellow-to-red compounds suitable for use as pigments. By far the most important historically have been the lead chromates: chrome yellow ($PbCrO_4$ and $PbCrO_4 \cdot PbSO_4$) and chrome orange ($PbCrO_4 \cdot PbO$), although today we find chrome orange slowly disappearing from the market place. Next in importance is molybdate orange ($PbCrO_4 \cdot PbMoO_4 \cdot PbSO_4$), which is not discussed in this chapter. Other chromate pigments are zinc yellow (zinc potassium chromate) (see section B), strontium yellow ($SrCrO_4$), and barium yellow ($BaCrO_4$) (see section C).

HERMANN KÜHN, Curator of the Department of Handicraft and Old Technology at Deutsches Museum, Munich, and lecturer on art history, Munich University.

MARY CURRAN, Administrative assistant, Research Center on the Materials of the Artist and Conservator, Mellon Institute, Carnegie-Mellon University, Pittsburgh.

1.1 Brief Definition of Pigment and Current Terminology

The name chrome yellow refers to a group of pigments ranging from medium yellow to primrose in color. The medium shade is nearly pure lead chromate while the primrose and light or lemon types are solid solutions of lead chromate and lead sulfate. According to ASTM Standard D211-67 (ASTM, 1977), there are three types of chrome yellow: Type I, primrose, has a minimum of 50% $PbCrO_4$; Type II, lemon, has a minimum of 65% $PbCrO_4$; and Type III, medium, has a minimum of 87% $PbCrO_4$. The remaining component is most often but not necessarily lead sulfate. The "lemon" variety is also frequently called "light." The designation of the pigment is CI Pigment Yellow 34, No. 77600 when pure; No. 77603 with coprecipitated lead sulfate (*Colour Index*, 1971).

2.0 HISTORY

2.1 Archaic, Obsolete, and Misrepresentative Names

Citronengelb, Zitronengelb; lemon yellow
Königsgelb; king's yellow
Neugelb; new yellow
Neuchromgelb; new chrome yellow
Leipzigergelb; Leipzig yellow
Wienergelb; Vienna yellow
Parisgelb; Paris yellow
Kölnergelb, Cölnergelb; Cologne yellow
Amerikanisches Chromgelb; American chrome yellow
Baltimorechromgelb; Baltimore chrome yellow
Französisches Chromgelb; French chrome yellow

jaune d'or; golden chrome yellow
jaune de chrôme jonquille; jonquil chrome yellow
jaune de chrôme Spooner; Spooner's chrome yellow

The above names are largely fanciful and arbitrary. Although sometimes indictive of certain additives or extenders (calcium carbonate, calcium sulfate, lead sulfate, barium sulfate) (Mierzinski, 1881, p. 132), particular names have not always been rigorously applied to pigments containing a specific extender. In the last century, for example, Baltimore chrome yellow probably included lead sulfate or calcium sulfate as an additive; today the name is used for a pale, voluminous chrome yellow that is precipitated on gypsum or asbestine (fibrous talc) and pressed while damp into cubes (Kittel, 1960, p. 284). Different shades have also been named, but again the terms cannot be associated invariably with a given chemical composition: extra bright, bright, medium bright, sulfur yellow, canary yellow, and bright, medium, and dark lemon yellow.

Cologne yellow was the name originally given to pigment containing lead chromate and sulfate; American chrome yellow contained alum in place of Glauber's salt (Hurst, 1906, p. 133). The French, Spooner, and Baltimore yellows of yesteryear were also made with alum (Mierzinski, 1881, p. 134). Reduced chromes were traditionally mixed with terra alba (gypsum) (Bearn, 1923, p. 71). A Cologne yellow formula containing gypsum is given by Schützenberger (1867). *Gothaergelb* and *Zwickauergelb* are mentioned by Tschelnitz (1857). For those interested in these matters, a list of more than a dozen obsolete terms in French is available (Fiche Technique, 1966).

2.2 History of Use

Lead chromate occurs in nature as the rare mineral crocoite, a red lead-bearing ore first described by J. G. Lehmann (fig. 1). According to Parkes (1967), Lehmann (1766) mentioned the mineral in a letter of 1762 addressed to the naturalist G. L. L. de Buffon (Kopp, 1845). The composition of the mineral, named "Siberian red lead" after its original place of discovery, was, of course, not understood until the end of the eighteenth century. At that time Vauquelin (1797) reported his discovery of a new metallic substance in the mineral and suggested that the metal be named chrome (Greek for *color*) owing to its ability to impart color. Although the red ore yields an orange-red to reddish yellow powder upon pulverization, there is no indication in the mineral's early descriptions of its use as a pigment.

In the period shortly before 1800 a number of alkali, alkaline-earth, and heavy-metal chromates were produced in the laboratory, but the recognition of lead chromate as a pigment did not come until 1803 or 1804 (Berthollet & Vauquelin, 1804, p. 225). Vauquelin (1809) conducted an extensive investigation concerning the influence of the precipitating conditions (acid, neutral, or alkaline) on the color and came to the conclusion that the dark lemon yellow, which painters valued most highly and which was also generally the most durable, was the result of precipitation from a weakly acidic solution. Basic lead chromate (chrome orange, chrome red), prepared under alkaline conditions, was also described.

A prerequisite for increased use of chrome compounds in arts and industry was the availability of an abundant chrome mineral, namely, chromite ($FeO{\cdot}Cr_2O_3$). The mineral became popular in the latter years of the nineteenth century, after the mining deposits in the region of Var, France, had become known in about 1818 and the deposits in the Shetland Islands recognized in 1820 (Guignet, 1888, p. 98; Harley, 1982, p. 102).

A note by Peclet states that the mines in Var appeared to be exhausted in 1829 (Leuchs, 1829). However, in the period from 1808 to 1902, mines in the Pennsylvania-Maryland region of the United States were internationally important (Pearre & Heyl, 1959). Harley (1982) tells the story of Dr. Bollmann, who became acquainted with the source of chrome in the United States and later began to manufacture chrome pigments in England sometime between 1814 and 1816.

The recognized usefulness of chrome yellow as a pigment is shown by the following entry in the *Nouveau dictionnaire d'histoire naturelle appliquée aux arts*, vol. 27 (1818): "The lead chromate is used in painting by Russian artists — who pay quite dearly for it — in the preparation of a particular sort of color that increases the value of the devotional tableaux with which every room is decorated." In the diary of the Danish painter Eckersberg (1783–1853) an entry has been found stating that he bought one ounce of "jaune de crome" in Rome on 18

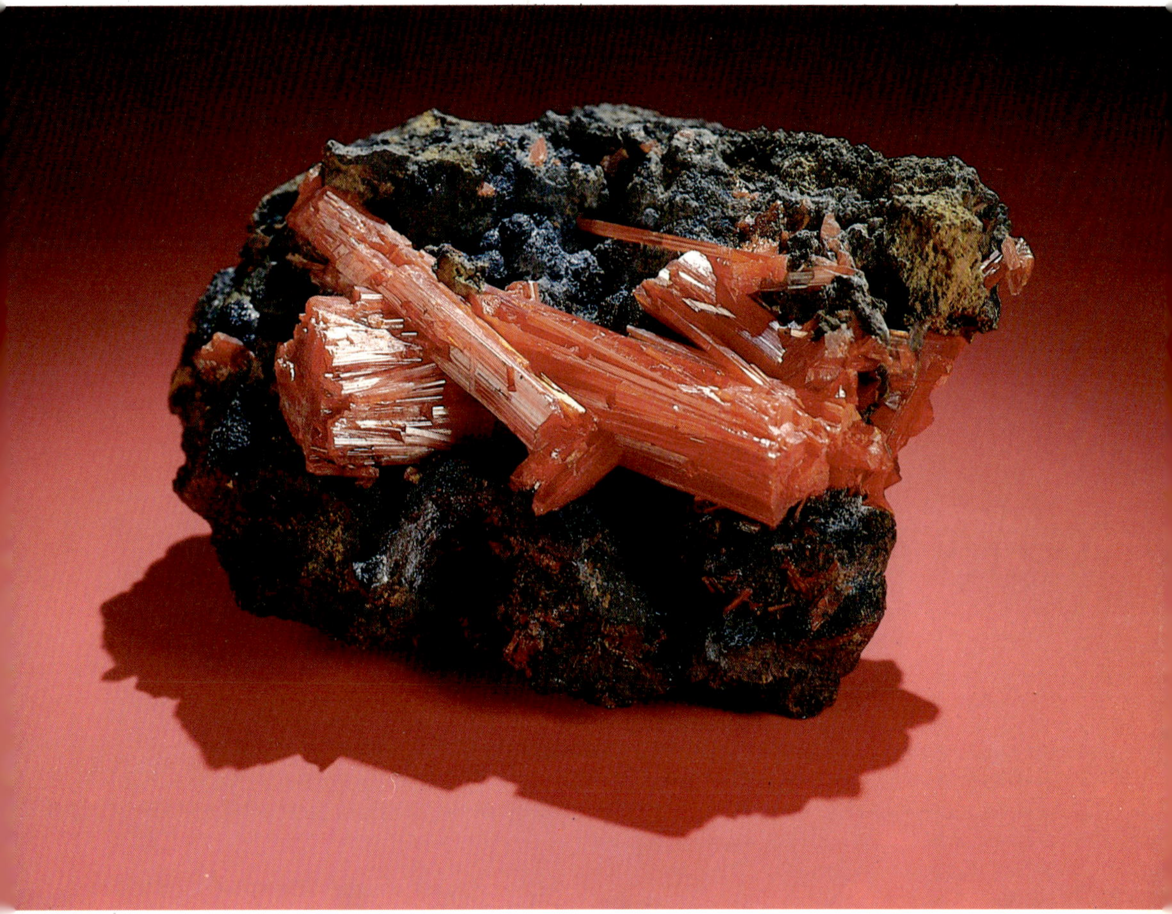

Fig. 1. Prismatic red crystals of crocoite (lead chromate) with goethite. National Museum of Natural History, Smithsonian Institution, Washington, sample no. 139743 from Dundas, Tasmania, Australia. Photograph by Dane Penland.

September 1815 (Raft, 1973). Thus far, this appears to be the earliest known mention of the pigment in the literature on artists' materials. Investigations at the Doerner Institute, Munich, have uncovered three examples of the use of chrome yellow in the first quarter of the nineteenth century (Kühn, 1969), the earliest in a painting by Sir Thomas Lawrence dated before 1810 (see 6.0). The introduction of lead chromate as a pigment must have occurred some time between 1804 and 1809, judging from the remarks of Vauquelin (1809).

Chrome yellow began to be used widely as a pigment in watercolors, coach painting, and general-purpose oil paints in the second quarter of the nineteenth century. The *Handwörterbuch der reinen und angewandten Chemie* (Liebig, et al., 1842) mentions chrome yellow as a very expensive color, which was extended greatly with barite, gypsum, alumina, and lead sulfate and sold under the commercial name "Cologne yellow." In modern times one of the major commercial uses of the pigment has been in traffic paints on highways (*Kirk-Othmer*, 1968). The famous yellow of taxi cabs is principally chrome.

Shortly after the arrival of the pigment in the painter's palette, it also appeared in greens, mostly prepared with Prussian blue. A mixture of chrome yellow and Prussian blue was once known as "green cinnabar" (Liebig, et al., 1842) and later as "chrome green." Because of the enhancement of pigment performance properties by phthalocyanine blue, today mixtures of chrome yellow and phthalocyanine blue are available, sold under tradenames such as Monarchrome Green® (Ciba-Geigy).

The Doerner Institute has found chrome yellow in several hundred paintings of the nineteenth and early twentieth centuries. It occurred in green paints as often as in yellow and, in the latter, was often mixed with other yellow pigments. When chrome yellow's lack of durability became known, it was replaced — primarily in expensive painting sets — by the more durable cadmium yellow. In the 1950s and 1960s special light-stabilized chrome-yellow shades came to play an important role as artists' colors (H. G. Müller, Schoenfeld & Co., Dusseldorf, private communication), but all types of chrome-containing pigments may someday pass into obsolescence with the increased concern for the health hazard they pose.

2.3 Terminal Date

At the present time chrome yellows are still being widely manufactured but their use will probably diminish.

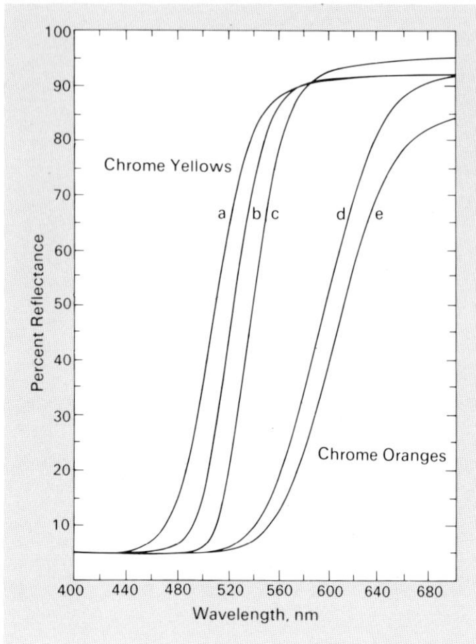

Fig. 2. Typical spectral reflectance curves of chrome yellow, chrome orange and chrome red pigments in masstone. a. primrose; b. lemon; c. medium; d. chrome orange medium; e. chrome orange extra deep. Courtesy of Dr. R. C. Schiek, Ciba-Geigy Corporation.

3.0 GENERAL PIGMENT PROPERTIES

3.1 Color and Spectral Reflectance

The approximate masstone hues of chrome yellows range from the reddish yellow of the medium type to the greenish yellow of the primrose. Spectral reflectance curves are shown in fig. 2. Tristimulus values for masstones and tints in mixture with titanium white are given by Schiek (1973).

An extensive discussion of various investigations on the influence of preparational procedure upon color can be found in Mellor (1931, p. 295). Loeffler and Burns (1967) state that the color of crocoite arises from an oxygen-to-metal charge-transfer transition in the $CrO_4^{=}$ group.

3.2 Hiding Power and Tinting Strength

Owing to its high refractive index, lead chromate possesses good hiding power. The pigment's high tinting strength allows the pure lead chromate to be diluted by extenders such as barium sulfate, calcium sulfate, china clay, and diatomaceous earth. However, with the darker varieties, addition of an extender usually results in duller colors.

The medium chrome yellows, which usually contain the minimum amount of lead sulfate or extenders, tend to have the greatest opacity, tinting strength, and — at least among the older types — fastness to light. Light and primrose yellows possess the next highest levels of these qualities. The high hiding power of pure lead chromate can be appreciated by reference to its mean refractive index of 2.3 to 2.7, markedly higher than that of lead sulfate (about 1.88–1.89). Special "shading yellows," usually based on the primrose type, are sold for mixing with Prussian blue or phthalocyanine blue to make chrome greens (Robinson, 1973). Wagner (1931) considered the rhombic type particularly important in the formation of such mixtures.

3.3 Permanence

The greatest drawback of the yellows, especially the older types, has long been a lack of lightfastness. On the other hand, the basic chromate (chrome red) is practically lightfast. Chrome yellows, particularly the rhombic crystalline varieties, tend to darken with age and become brown (Haug, 1951); a change toward green is sometimes observed. The darkening or browning is a photochemically induced reaction in which an important part is played by visible as well as ultra-violet radiation. Darkening can take place in the absence of oxygen and moisture. Research into the problem has been reported by Haug (1951) and by Watson and Clay (1955). The lead chromes also tend to darken upon exposure to hydrogen sulfide (Church, 1915).

Today there are well-known light-fast types of the monoclinic lead chromate (see 5.1). Lightfastness is achieved by the addition of photochemical stabilizers such as zinc salts, or else by coating the lead chromate with lead sulfate during the precipitation. Patents have been granted for the addition of other stabilizers such as antimony salts, aluminum hydroxide, titanium dioxide, and rare earth oxides (Schofield, 1948). The resulting products are further protected by cleaning with hydrophobic materials before hydrolysis (Kittel, 1960, p. 283). Treating or encapsulating with amorphous silica is also used to promote permanence (Feller, 1972).

3.4 Compatibility

Common types of chrome yellow, neutral lead chromate for example, may not be combined with alkali binders such as lime and water glass because they are thereby transformed to the red basic chromate (chrome red). These varieties also tend to exhibit poor resistance to acids (Schiek, 1973).

Lead chromates can be considered compatible with all common pigments (Wulf, 1974, p. 111), although older references sometimes warned against mixing with chalk because of the latter's basicity. If acid residues should remain after manufacture, these can affect acid-sensitive pigments such as ultramarine blue. When made suitably stable, the pigments have long been used in mixtures with blue to produce a green, traditionally known as chrome green.

3.5 Chemical Properties

Lead chromates are insoluble in water and in weak acids such as acetic acid, but they are dissolved by alkalis and mineral acids. Neutral lead chromates can be transformed by lime and dilute alkalis into redder shades. Under alkaline conditions, the chromate ion, $CrO_4^{=}$, is formed and the solution becomes straw-yellow. Acidification changes the hue to orange, the color imparted by the dichromate ion, $Cr_2O_7^{=}$. Pigment samples treated with hydrochloric acid yield a green solution of chromic chloride. When bulk samples are available, the latter reaction constitutes a reasonably fast and simple test for chromate. The text by Koch (1953, tables 3 and 4) contains an interesting color plate that shows the effect of an acid, alkali, or hydrogen sulfide attack on five of the yellow chrome pigments, two of the oranges, and several greens based on mixtures with chrome yellow. Using such tests on a house paint in situ, the painter could easily identify many of the types of chrome pigment that had been present.

Older texts state that the lead chromates dry moderately well. It is known that they enhance the film-forming qualities of oil paints. Modern texts confirm that they dry well in oil, especially in the case of chrome red, and the texts further state that the pigments do not form lead soaps in the process.

Lead chromates react with hydrogen sulfide or soluble metal sulfides, becoming dark owing to the formation of black lead sulfide. The process occurs primarily in aqueous-based vehicles. To a considerable degree, oil and resin binders protect the pigment from such influences.

3.6 Oil Absorption and Grinding Qualities

Preuss (1974, p. 85) states that the pigment has excellent dispersibility; he gives the oil requirement for chrome yellow paste as 20 to 25%. The oil absorption demand ranges between 16 to 27% (*Raw Materials Index*, 1975).

3.7 Toxicity

In recent years concern has mounted regarding the potential hazards to health that chromates represent. There still does not seem to be great alarm over the risks in using the pigment in small quantities in paints, but exposure to high concentration of soluble chromate salts in industrial plants has been known for decades to cause serious lesions in the respiratory passages and, ultimately, cancer. The problem is currently under active investigation (Langård & Norseth, 1975). Clay and Watson (1944) discuss the lead poisoning aspects of the pigment. A recent publication (Davies, 1979) on the history of workers in three English manufacturing plants stressed the significant danger of cancer for those involved with zinc chromate as opposed to lead chromate. Because of its very low solubility in body fluids, the use of lead chromate in paints represents little potential hazard (Imperial Chemicals Industries, 1981). Of course, ingestion of and prolonged breathing of pigment dust should be avoided.

4.0 COMPOSITION

4.1 Chemical Composition and Crystal Structure

Lead chromates provide a particularly striking example of how composition and crystal structure are related to major properties of materials — in this case, pigmentary properties. The mineral crocoite, $PbCrO_4$, is monoclinic, but a second low-temperature variety can exist as orthorhombic crystals, normally referred to as the metastable form. Existence of a high-temperature tetragonal structure, attributed to remarks by Jaeger and Germs (1921), has never been demonstrated (R. Schiek, private communication, 1978). Lead chromate can form two series of solid solutions with lead sulfate, one monoclinic in structure and the other orthorhombic. When the lead chromate content exceeds 30–35%, stable monoclinic crystals tend to form, seen in the transmission electron photomicrograph in fig. 3A; below 10% lead chromate, crystals are orthorhombic. If not stabilized, however, the latter will revert to the monoclinic form. When the lead chromate content is between 10% and 30%, both monoclinic and orthorhombic structures can form (Watson and Clay, 1955).

Commercial chrome yellows can be divided into three types:

1) Medium chrome yellows: monoclinic, practically pure lead chromate.

2) The lighter yellows: monoclinic solid solutions of lead chromate containing from 10% to 40% lead sulfate; and

3) Primrose yellows: orthorhombic solid solutions, up to 10% lead chromate according to Watson and Clay (1955), stabilized by some material such as lead tartrate, metal oxide, or hydroxide (Wormald, 1975, p. 320). Schiek (private communication, 1979) informed the authors that orthorhombic types are now made that can contain up to 85% lead chromate. Tables 1 and 2 give some typical compositions for pigments in the United States. Fig. 3B shows a transmission electron photomicrograph of a primrose yellow.

According to Schiek (private communication, 1979), the designation C.P. indicates the pigments are commercially pure, not chemically pure. Silicates, aluminum phosphate, titanium, aluminum, and zirconium compounds are frequently present, introduced to enhance pigment properties and working characteristics such as lightfastness, oil absorption, gloss, etc.

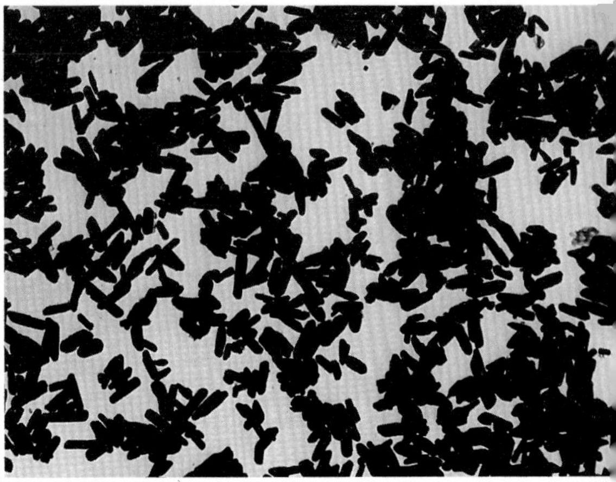

Fig. 3.A. Transmission electron micrograph of light chrome yellow (monoclinic) at 10,000x.
B. Transmission electron micrograph of primrose chrome yellow pigment at 10,000x. Courtesy E. I. du Pont de Nemours & Company.

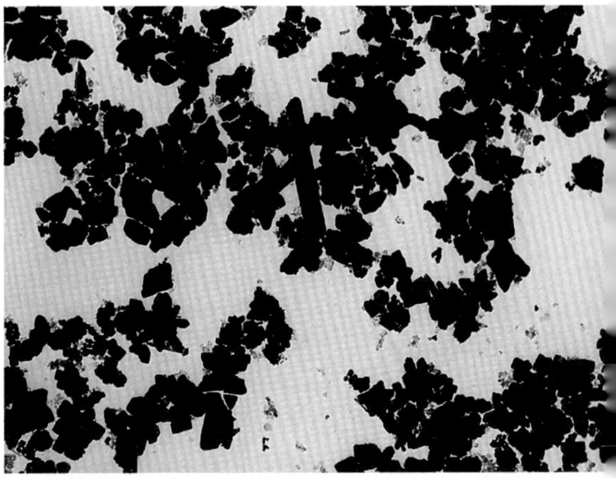

4.2 Sources

In the period 1828–1882, the Wood Mine in Lancaster County, Pennsylvania, supplied much of the world's needs, and considerable pigment was produced at the Baltimore Chrome Works, but by 1902 this mineral source was exhausted (Socolow, 1976; Pearre & Heyl, 1959). Mention has been made of the early, important source of chromite ore in the region of Var, France, from about 1818 to perhaps 1829. Through the 1970s, American manufacturers of chrome yellows were American Cyanamid, DuPont, Hercules (now Ciba-Geigy), Harshaw, Mineral Pigments, and Reichhold (R. Schiek, private communication, 1979).

Table 1. **COMPOSITIONAL CLASSIFICATION OF LEAD CHROMATE PIGMENTS ACCORDING TO SCHIEK (1973)**

		Minimum $PbCrO_4$				
		A	*B*	*C*	*$PbCrO_4$*	*PbO*
Type I	Primrose chrome yellow	58	50	~50		
Type II	Lemon or light chrome yellow	65	65	~50		
Type III	Medium chrome yellow	87	87	~82		
Light chrome orange					~60	~40
Dark chrome					~60	~40

A = Minimum amounts, U.S. Government Specification TT-P-346b.
B = ASTM D211-67, reapproved 1973 (American Society for Testing and Materials, 1977).
C = Classification by Schiek (1973), p. 358.

Table 2. **CLASSIFICATION OF LEAD CHROMATE PIGMENTS ACCORDING TO SCHIEK (private communication, 1979)**

Shade of chrome	*Components*	*Crystal system*
Pale Primrose	Lead chromate with 5–40% lead sulfate, stabilized	Orthorhombic
Pale Lemon	Lead chromate with 45–55% lead sulfate	Monoclinic
Lemon	Lead chromate with 15–40% lead sulfate	Monoclinic
Middle	Lead chromate	Monoclinic
Orange	Basic lead chromate	Monoclinic
Scarlet	Lead chromate, lead sulfate, and lead molybdate	(Commercial) Modified Monoclinic

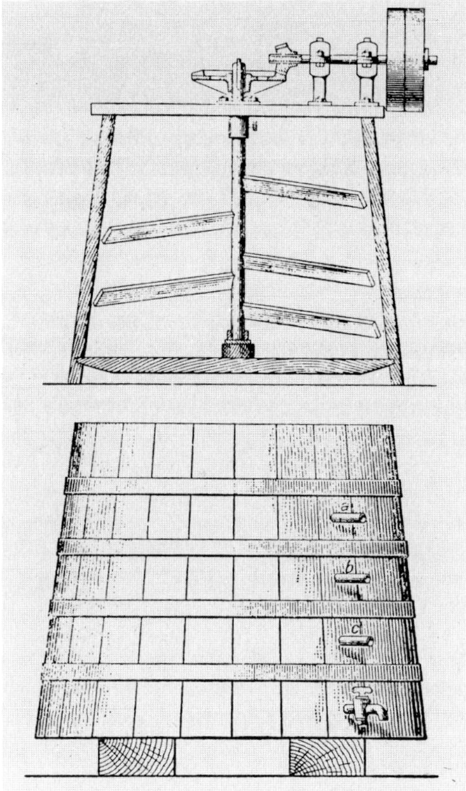

Fig. 4. Cross section of nineteenth-century precipitating vat with stirring blades. Also general view of settling tank with taps to draw off suspended pigment having particle sizes of varying fineness (from Bearn, 1923).

4.3 Preparation

Details of the manufacturing process, starting materials, and blenders used in the production of chrome yellow during the nineteenth century can be found in Gentele (1860, pp. 183, 198, and 273), Mierzinski (1881, pp. 132, 147, and 471), Bearn (1923), and Hurst (1906, pp. 131–137). The procedures have remained largely the same to this day: chromate or bichromate solutions are mixed with neutral solutions or suspensions of lead salts (acetate, nitrate, chloride, etc.), the latter often produced from lead monoxide. Hurst (1906, p. 132) shows pictures of the traditional wooden precipitating and settling vats used in a rather straightforward manufacturing process employing double decomposition reactions. A typical vat, shown in fig. 4, had taps at the side to allow the top liquor to be drawn off as the pigment settled (Bearn, 1923).

Stoichiometric relationships are regulated during precipitation so that either pure $PbCrO_4$ or a solid solution of lead sulfate and lead chromate can form. The sulfate ions can be supplied by sulfuric acid, alkali metal sulfate, or aluminum sulfate, which is added to the chromate solution. Lead sulfate, in solid solution with lead chromate, need not act as a diluent but can contribute to the scattering power, somewhat in the manner of the coprecipitate pigment of titanium white and calcium sulfate that is now no longer manufactured. Chrome yellow can also be obtained through the direct addition of lead monoxide to chromates in paste form (Kittel, 1960, p. 279).

The color of chrome yellow is dependent upon the interactions of three factors: (1) chemical composition, (2) crystal structure, and (3) particle size or shape.

4.4 Adulteration and Sophistication

Barytes, china clay, gypsum, asbestine, and whiting were widely used to tone lead chromates (Brown, 1944, p. 279; Hurst, 1906, p. 135; Mierzinski, 1881, pp. 132–144; Smith, 1915). However, Hurst (1906, p. 127) points out that the colorant was only rarely adulterated in the usual sense of the term, for there were no cheaper yellows in the early days of manufacture.

4.5 Modified Varieties

In modern times silica-treated and silica-encapsulated chrome pigments exhibiting outstanding lightfastness have been produced. The presence of silica can be detected by emission spectrographic analysis, and by infrared spectrophotemtry, as Feller (1972) has shown (see fig. 10). Johnston and Feller (1965) have drawn attention to the excellent lightfastness of Ciba-Geigy "Regal"® chrome yellow. A silica-cored type of yellow and a chrome orange containing about 50% SiO_2 by weight (Dunn, 1973) were first marketed by the National Lead Company in 1963.

5.0 IDENTIFICATION — CHARACTERIZATION

5.1 Characterization by Optical and Chemical Microscopy

Modern chrome yellow pigments often consist of 1 micrometer equant particles or slightly larger prisms. Because of the fineness of the particles, the crystal habit is not always readily discernible by optical microscopy. Moreover, the crystal morphology is not sufficiently distinctive to permit ready differentiation of the various chromates under the microscope. Manufacturing techniques can modify the crystal morphology practically at will. Clay and Watson (1944, 1948), and Wagner, et al. (1933) have published photomicrographs of various prismatic and rhombic forms. Fig. 3 shows transmission electron micrographs of typical modern pigments.

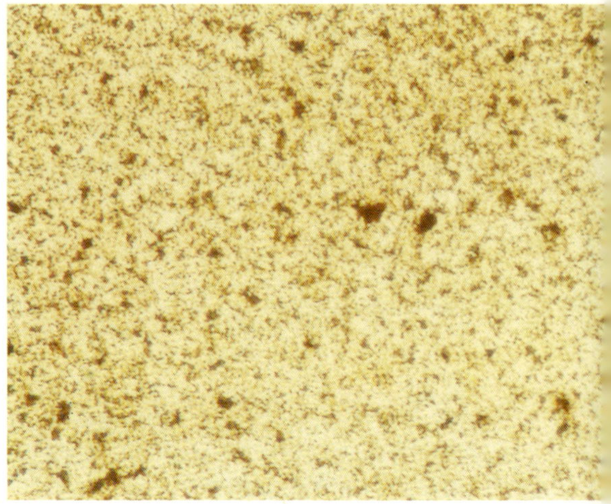

A

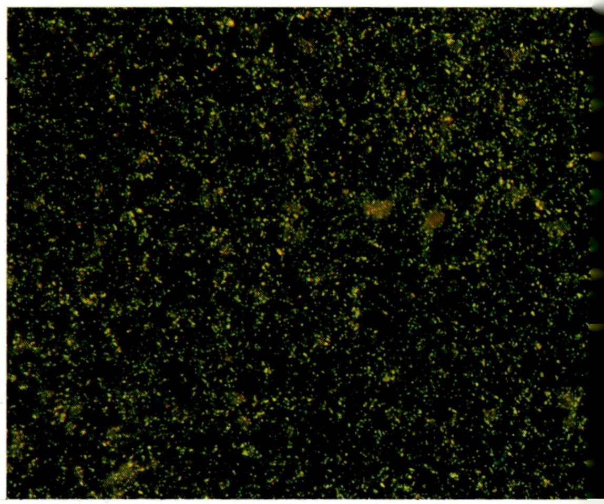

B

Fig. 5. Photomicrographs of chromate pigments under plane-polarized light and under crossed polars. 400x. Refractive index of mounting medium, 1.53. Courtesy of Research Center on the Materials of the Artist and Conservator, Carnegie-Mellon University, Pittsburgh. A and B. Chrome yellow, 95.5% $PbCrO_4$; C and D. Calcium chromate; E and F. Zinc chromate; and G and H. Chrome orange, medium.

Chrome yellow mounted in Canada balsam appears as fine, high-profile particles that usually exhibit their birefringence only under very high magnification. Figs. 5A and B illustrate the appearance and birefringence of lead chromate. Pure lead chromate and varieties containing up to 50% lead sulfate as a rule appear as strongly

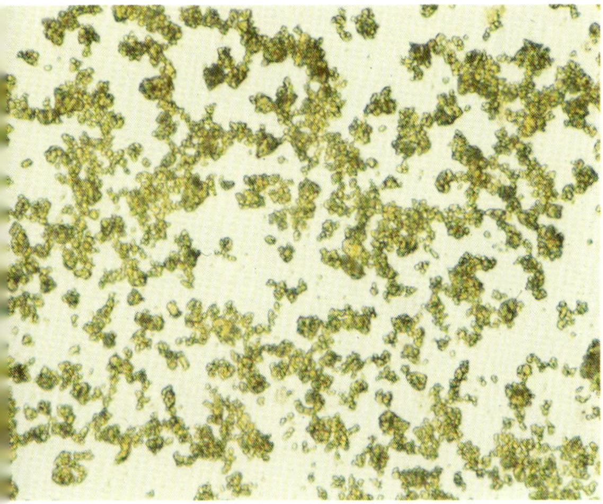

C

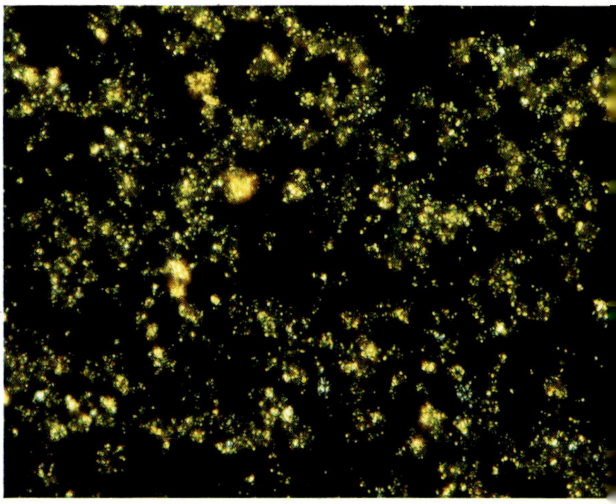

D

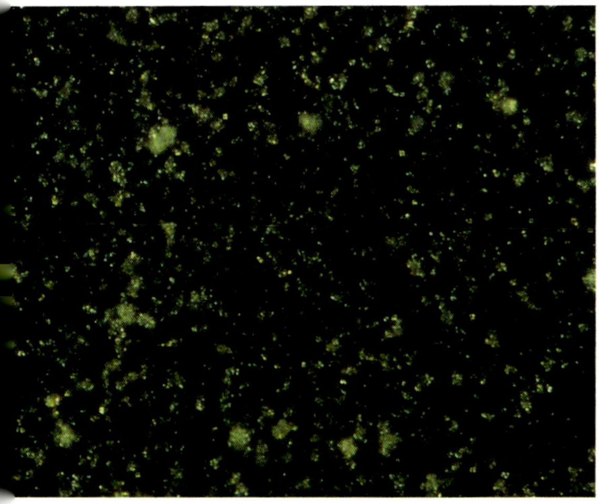

E

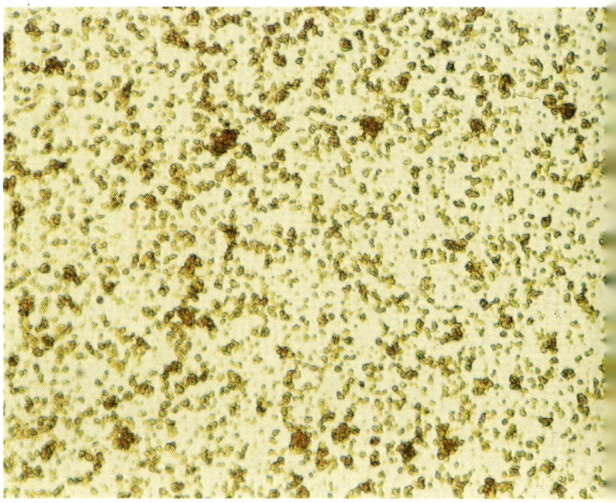

F

G

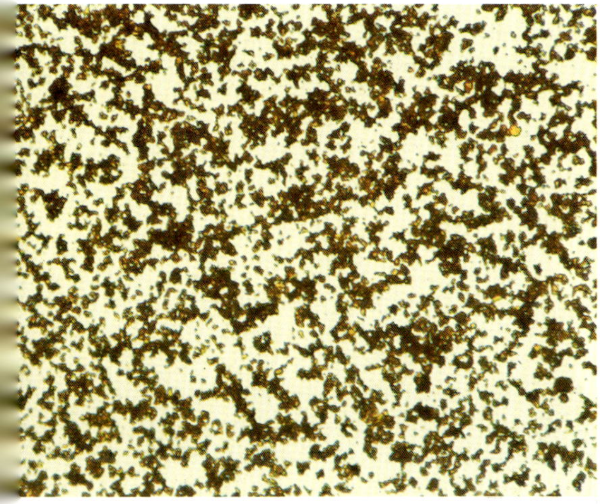

H

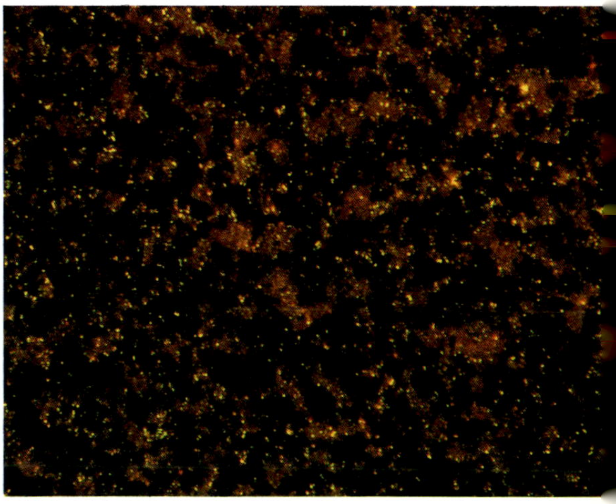

refractive particles. The refractive indices of crocoite crystals with the lithium lamp are $\alpha = 2.29$, $\beta = 2.36$, and $\gamma = 2.66$ (Merwin, 1917). Mellor (1931, p. 296) states that crocoite is optically positive; optic axial angle $2V = 54°$. In the mineral, cleavage is distinct on (110), poor on (001) and (100). Although the material is pleochroic [orange red (α, β) to blood red (γ)], the change in color is difficult to notice in the fine, pigment-grade particles. Colors range from gray to bright yellow under crossed polars (McCrone & Delly, 1973, pl. 439, p. 494).

Small additions of chrome yellow in ochre and synthetic iron oxide colors are used as "beautifiers," and the bright yellow particles can often be observed microscopically under high magnification.

A number of greens have been based on mixtures of chrome yellows and various blue pigments, especially Prussian blue. Under the optical microscope, mixtures of yellow and blue pigments usually are immediately apparent, and in a preliminary examination, each pigment's separate behavior toward acids and bases can usually be observed.

5.2 Chemical Identification

5.21 ***Test with Acid and Base.*** A simple identification test for lead chromate is provided by its behavior with boiling dilute hydrochloric acid and boiling caustic soda. With the former, pure lead chromate dissolves completely to give an orange-colored solution from which needle-shaped crystals of colorless lead chloride are produced on cooling. With the latter reagent, a pure lead chromate dissolves to give a yellow solution (Taylor & Marks, 1966). House painters can make these convenient tests with acid and base directly on patches of old paint (Koch, 1953).

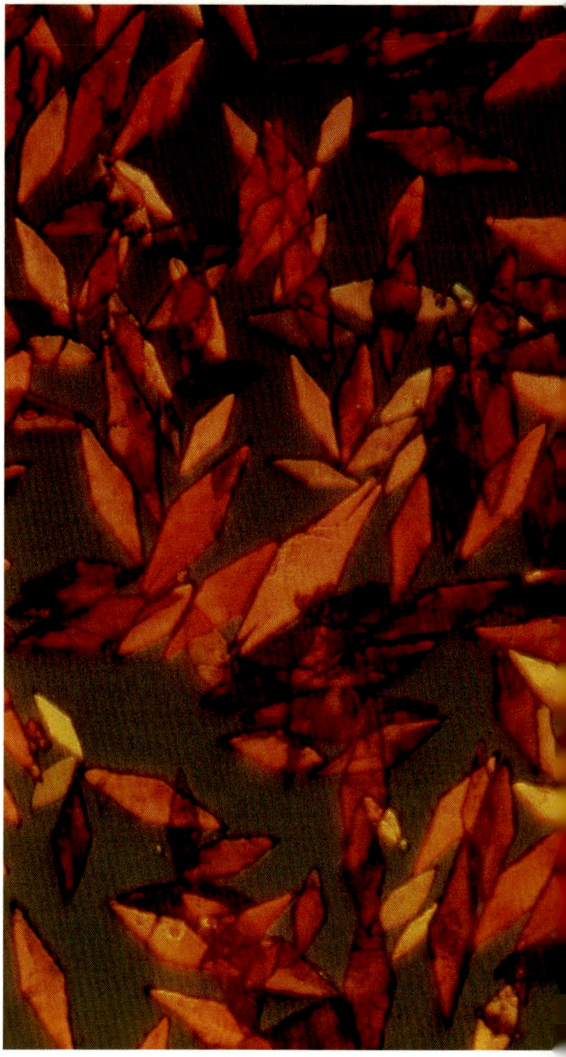

Fig. 6. Crystals of silver chromate between partially crossed polars. Refractive index of mounting medium, 1.52. The precipitation of silver chromate is a convenient test for chromate. 650x.

5.22 ***Spot Test with Diphenylcarbazide.*** The identification of chromium in pigments in the laboratory is frequently carried out as a spot test (Plesters, 1956, p. 140). The specimen is agitated for a short time in a micro-drop of 5–10% solution of potassium or sodium hydroxide on a white porcelain spot plate. A drop of 1% alcohol solution of diphenylcarbazide is then added, which produces an intense red color. The addition of a few drops of 2N sulfuric acid results in a blue-violet color if the unknown contains the chromate ion. If no chromate ion is present, the red color of the alkaline solution disappears quickly following the addition of the sulfuric acid. As usual, it is recommended that the tests be accompanied by a blank (Hetterich, 1931). The limits of detection lie between 0.5 and 2.5 micrograms (dilution: 1,000,000 or 1:2,500,000). Mercury salts or molybdates, which cause similar color reactions, interfere with the reaction. This can be prevented by the addition of hydrochloric acid or sodium chloride in the presence of mercuric salts, or oxalic acid in the presence of molybdate ion (Feigl & Anger, 1972).

5.23 ***Formation of Silver Chromate.*** A second microchemical test comprises dissolving the sample with a drop of potassium hydroxide solution, acidifying with nitric acid, and adding a drop of silver nitrate solution. In the presence of the chromate ion, dark red, triclinic, flat, rectangular, or rhombic crystals of silver chromate appear (fig. 6), occasionally deformed into hexagons.

5.3 Instrumental Methods of Analysis

5.31 ***X-ray Diffraction.*** The crystalline compounds chrome yellow, zinc yellow, and other chromates, if they are present in a pigment sample in sufficient concentration, can be identified by the Debye-Scherrer powder diffraction method (König, 1956). Because of its smaller particle size, chrome yellow produces a pattern with sharp lines, while chrome red, if not ground, may yield spotty patterns. The ease of detection of chrome or zinc yellow depends on the quantity present and the compounds with which they are mixed. In the presence of lead white, barytes, or calcite, small amounts of chrome yellow are difficult to detect because its strongest lines overlap major lines of the other compounds. If several milligrams of the unknown are available, the greater resolution afforded by diffractometry makes this method preferable to the Debye-Scherrer technique. With a few particles of the colored substance, weighing less than 10 micrograms, a Debye-Scherrer powder pattern can easily be obtained (Kühn, 1974). Lattice *d*-values and associated intensities of chrome yellow (crocoite) are given in table 3. In the case of primrose chrome yellows, the orthorhombic character of the pattern is evident, but the *d*-spacings gradually change as the concentration of chromate ion is increased. Diffraction data on lemon and primrose type pigments are given in table 4. The *d*-values and associated intensities for strontium, barium, and calcium chromates are given in table 6.

Powder diffraction patterns of four chrome yellow pigments and two zinc chromates are shown in fig. 7. Fig. 8 shows the powder patterns for lead chromate and a silica-treated pigment, Regal® chrome yellow. Both the silica-treated and encapsulated types give essentially the same pattern as does lead chromate itself.

Table 3. **X-RAY DIFFRACTION DATA FOR CHROME YELLOW, CROCOITE ($PbCrO_4$)* (Guinier camera, Cu$K\alpha_1$ = 1.5405 Å)**

d(Å)	*I*	*d(Å)*	*I*
5.43	10	2.252	25
5.10	6	2.243	10
4.96	25	2.214	4
4.38	25	2.154	4
4.37	14	2.130	2
3.76	12	2.120	2
3.72	8	2.090	25
3.48	55	2.080	4
3.32	4	2.056	4
3.28	100	2.046	4
3.24	6	2.002	8
3.15	12	1.988	4
3.09	6	1.978	20
3.03	65	1.967	14
3.00	30	1.900	8
2.710	16	1.866	4
2.653	2	1.857	4
2.597	14	1.847	25
2.549	18	1.829	2
2.510	2	1.814	6
2.460	4	1.806	4
2.351	4	1.796	6
2.320	14	1.760	2
		1.740	4
		1.734	2
		1.710	4
		Plus numerous lines of smaller *d* values	

**JCPDS Powder Diffraction File* 8-209, 8-209a.
Courtesy of JCPDS International Centre for Diffraction Data (1982).

5.32 ***Infrared Spectroscopy.*** Campbell (1965) states that it is possible to identify most of the chromates unambiguously by measuring their infrared spectra between 600 and 1,000 cm^{-1}. He notes that there are two types of spectra: I, the chromate type; and II, the dichromate type, represented by the zinc, cadmium, iron, and bismuth salts. The more common type I has no significant absorption between 700 and 800 cm^{-1}. Instead, a principal peak in the region of 847 cm^{-1} is assigned to the V_1 symmetrical stretch and one at 884 cm^{-1} to the V_3 asymmetrical stretch. Although bending frequencies (450 to 250 cm^{-1}) are especially sensitive to changes in crystal structure, the accompanying curves (fig. 9) show only the principal stretching peaks, because this is the region covered by the usual infrared spectrometer. The most readily available reference to spectra from 1,500 to 200

Table 4. **COMPARISON OF X-RAY DIFFRACTION DATA: LIGHT CHROME YELLOW VS. PRIMROSE CHROME YELLOW OF CLOSE CHEMICAL COMPOSITION (Commercial Products)***

Lead Chromate Light or Lemon Type 71% $PbCrO_4$ 22% $PbSO_4$		*Lead Chromate Primrose Type 69% $PbCrO_4$ 21% $PbSO_4$*	
d(Å)	*I*	*d(Å)*	*I*
5.40	9		
4.95	34		
4.35	42 ⎤[1]	4.37	(54)
4.26	S ⎦		
		3.90	(42)
3.75	14 ⎤		
3.70	11 ⎦		
		3.67	(19)
		3.62	S ⎤
		3.55	(29) ⎦
3.46	56		
		3.40	(83)
3.32	S ⎤		
3.26	100 ⎦	3.28	(60)
3.14	11		
3.07	(3) ⎤	3.07	(100)
3.03	(66) ⎥		
2.99	S ⎦		
		2.83	(44) ⎤ D[2]
		2.78	(44) ⎦
2.70	20		
		2,65	(10)
2.59	15		
2.53	23		
2.45	6	2.47	(15)
2.31	18	2.31	S
2.24	32		
		2.21	(20) ⎤
		2.19	S ⎦
		2.11	(51)
2.08	26	2.07	S ⎤
		2.06	(38) ⎦
		2.00	(11)
1.967	28 ⎤		
1.958	S ⎦		
1.894	10		
1.838	30		
		1.821	(17)
		1.733	(11)
1.685	15		
		1.651	(24)

1. S = Shoulder. Brackets indicate diffractometer feature tied together visually.
2. D = Doublet.
*Data courtesy of Ciba-Geigy Corporation, Pigments Department, Research and Development, Glens Falls, New York.

cm^{-1} is that of Afremow and Vandeberg (1966). Most of the salts gave good spectra at a concentration of two microformula weights of chromate per 200 mg tablet of KBr (Campbell, 1965, p. 1,336).

The characteristic absorptions between 1,000 and 800 cm^{-1}, shown in fig. 9, correspond very well to published data by Campbell (1965), Doyle and Eddy (1967), Afremow and Vandeberg (1966), the Federation of Societies for Paint Technology (*Infrared Spectroscopy* 1980), Nyquist and Kagel (1971), Newman (1979) and Harkins, et al. (1959). Certain commercial products, usually highly resistant to darkening, are silica treated, and the extra silica band, at about 1,100 cm^{-1}, can be detected in such products by infrared spectroscopy as shown in fig. 10 (Feller, 1972). Here Krolor® (DuPont) KY788D is silica encapsulated, whereas Regal® (Ciba-Geigy) chrome yellow contains silicate as part of the after-treatment to enhance pigment performance properties.

Pigments coprecipitated with lead sulfate can be detected by the appearance of a sulfate band in the region of 1,100 cm^{-1}. However, some care must be exercised in distinguishing between low concentrations of lead sulfate and of barium sulfate by means of infrared spectroscopy in the region of wave numbers higher than 400 cm^{-1}.

5.33 ***Spectrochemical Analysis.*** The proof of the presence of a chromium-containing pigment requires evidence of chromium either with the aid of a spot test or one of the numerous instrumental techniques for determining elemental composition. The methods selected will depend on the sample size, information desired, and availability of equipment.

Difficulties arise in the differentiation of individual chromate pigments, because the sample may contain several different compounds and some can crystallize in various habits. For example, the detection of the zinc, lead, and chromium in a yellow pigment can imply, among others, the following combinations: chrome yellow + zinc white, zinc yellow + lead white, chrome yellow + zinc yellow.

Conventional x-ray fluorescence or energy-dispersive x-ray analysis can be used for samples of milligram size. Smaller samples can be analyzed with a scanning electron microscope equipped with an energy-dispersive x-ray analyzer, by emission spectroscopy, or by neutron acti-

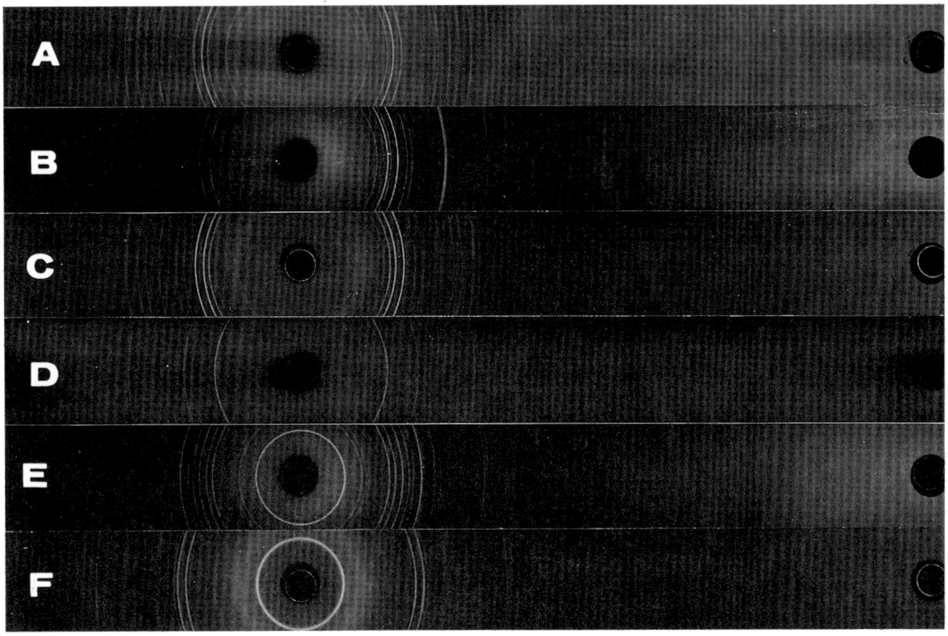

Fig. 7. X-ray powder diffraction patterns for yellow chromate pigments. A. Lead chromate, Certified A.C.S. (Fisher Scientific Company); B. Barium chromate (Fisher Scientific Company); C. Strontium chromate X-2865 pigment (from Ciba-Geigy Corporation); D. Calcium chromate J-1376 (Frank D. Davis Company); E. C. P. zinc chromate (J. T. Baker Chemical Company); F. Basic zinc chromate 505-WJ (Frank D. Davis Company). Patterns courtesy of Freer Gallery of Art, Washington. (Photographs not to be used for measurement purposes.)

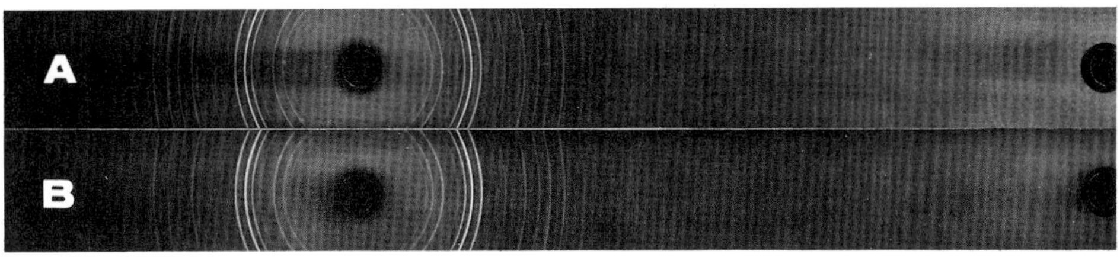

Fig. 8. Debye-Scherrer x-ray powder diffraction patterns for A. Lead chromate (Fisher Scientific Company); B. Silica-treated Regal® chrome yellow X-3218 (Ciba-Geigy Corporation). Patterns courtesy of Freer Gallery of Art, Washington. (Photographs not to be used for measurement purposes.)

vation. The last two techniques are especially useful when the associated trace elements are of interest.

In x-ray spectrometric analysis, the wavelengths of the principal emissions are Cr$K(1)\alpha_1$ 2.285 Å, Cr$K(1)\alpha_2$ 2.289 Å, Cr$K(1)\beta_1$ 2.081 Å, Cr$K(2)\alpha_1$ 4.570 Å, Cr$K(2)\alpha_2$ 4.578 Å (Sagel, 1959). Parenthetical numbers denote spectral orders.

5.34 ***X-radiography.*** At a wavelength of 0.708 Å the mass absorption coefficient of chrome yellow is 92.4 cm^2g^{-1} and that of a paste of 20% oil content is 74.4 cm^2g^{-1} (Hodgman, 1960). These values are lower than those of lead white as pigment (109.2 cm^2g^{-1}) or as a paste of 11% oil content (97.3 cm^2g^{-1}).

5.4 Criteria for Positive Identification

The simplest and most direct way to identify specific crystalline compounds is by x-ray diffraction. Because each chromate pigment has a unique pattern, it is often possible to identify the specific compounds present in limited mixtures of pigments.

Difficulties may arise in distinguishing vari-

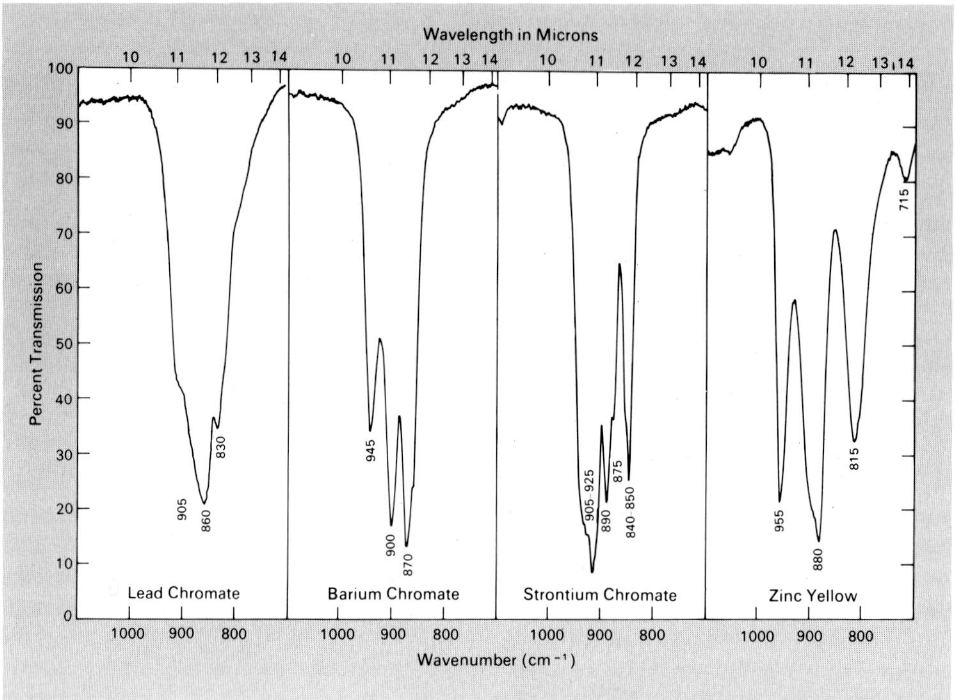

Fig. 9. Spectra showing absorption bands in the infrared region between 1,100 and 700 cm^{-1} for the principal chrome yellow pigments.

ous primrose and light yellow types because the *d*-spacings of the patterns differ only slightly. Table 4 gives the *d*-spacings and intensities for two commercial products having about 70% lead chromate and 21% lead sulfate. Because the light yellows are solid solutions, their patterns are not listed in the *JCPDS Powder Diffraction File*; the investigator must recognize the similarity of a given pattern to that of either monoclinic or orthorhombic lead chromate. The presence of the lead sulfate in the primrose pigment can be seen in fig. 7C. Under these conditions, knowledge of the presence of sulfate, often detectable in the infrared spectrum, would help in providing a more detailed characterization of the pigment.

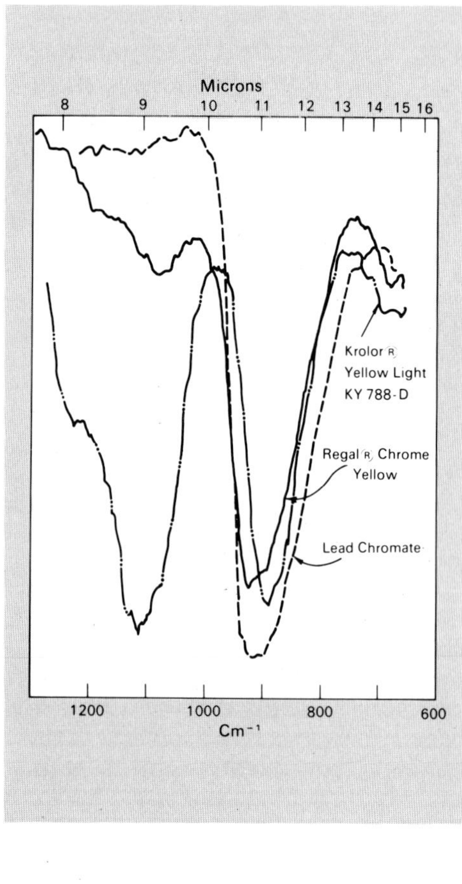

Fig. 10. Principal infrared absorption bands in ordinary lead chromate and modern types of pigments that have been silica treated or encapsulated to improve light stability, showing silicon-oxygen absorption at about 1100 cm^{-1}. Krolor® is a registered trade name of the DuPont Company; Regal®, a registered trade name of the Imperial Colors Division, Ciba-Geigy Corporation.

B. Zinc Yellow

NOMENCLATURE IN VARIOUS LANGUAGES

English: zinc yellow
German: Zinkgelb
Russian: Цинковый крон
French: jaune de zinc
Italian: giallo di zinco
Spanish: amarillo de zinc
Japanese: aenki

1.0 INTRODUCTION

1.1 Brief Definition of Pigment

Unlike the simple salts of the alkaline earth metals, this pigment is a complex zinc potassium chromate with the approximate composition $K_2O{\cdot}4ZnCrO_4{\cdot}3H_2O$. CI Pigment Yellow 36, No. 77955.

1.2 Current Terminology

As well as the yellow described above, also known as zinc chromate, there is a much less water-soluble basic zinc yellow or zinc tetroxychromate of composition $ZnCrO_4{\cdot}4Zn(OH)_2$. This pigment is designated as CI Pigment Yellow, 36:1, No. 77956 (*Colour Index*, 1971).

2.0 HISTORY

2.1 Archaic, Obsolete, and Misrepresentative Names

Some of the names applied to this pigment in the past were: ultramarine yellow (a term also used for barium and strontium chromates), yellow button of gold (Fiche Technique, 1966) *jaune de chrôme malterve*, and permanent yellow.

2.2 History of Use

Zinc yellow was first synthesized about 1800 during a systematic study of the chromates (Guignet, 1888, p. 228), but it found no immediate use as a pigment and is not mentioned as such in chemical dictionaries and technological encyclopedias before 1850 (Gentele, 1860, p. 183; Trillich, 1923). According to Remington (1944), the pigment was discovered in 1847 by a Scottish chemist, James Murdock, who produced three shades: marigold, lemon, and pale primrose. However, Brown (1944, p. 380) and Mellor (1931, p. 278) cite discovery by earlier entrepreneurs such as Thomson in 1825 and Lampidus in 1829. They credit Leclaire as the pioneer in the pigment's introduction. Zinc yellow has been identified in four artworks painted after 1855, currently in the Schack-Galerie, Munich (Kühn, 1969), as well as in a collection of pigments belonging to the Swiss painter Arnold Böcklin (1827–1901) (Richter & Härlin, 1974). In the latter collection, zinc chrome green, a mixture of zinc yellow and Prussian blue, was also found. Zinc yellow was never as popular as chrome yellow, either in commercial paints or in artists' colors; most of it was and is now used in green pigment mixtures (zinc chrome green and "Victoria green," the latter a mixture of zinc chrome green and hydrated chrome oxide). Nonetheless, Moreau-Vauthier (1923) lists four artists' palettes that contained zinc yellow.

According to Toch (1925), industrial interest in the pigment's rust-preventing qualities was stimulated primarily in the decade before 1925. Kastens and Prigotsky (1949) note that zinc yellow's general use depended upon the introduction of phenolic vehicles in the 1930s; its applications for military purposes skyrocketed during World War II. The basic type of zinc

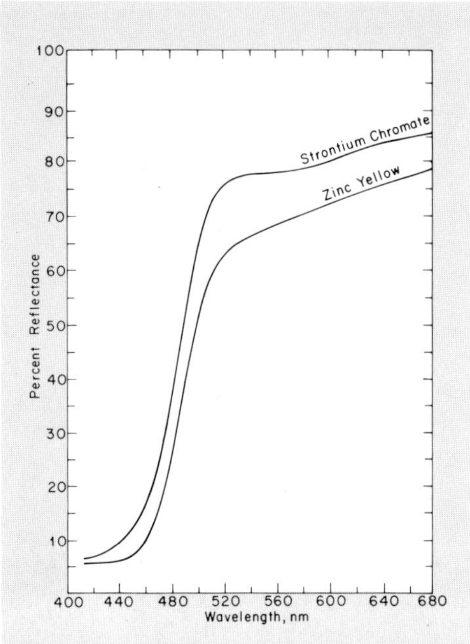

Fig. 11. Spectral reflectance curves of the masstone characteristics of zinc and strontium chromate pigments. Courtesy of Dr. R. C. Schiek, Ciba-Geigy Corporation.

yellow (Leisy, 1941), the zinc tetroxychromate, used in rust-preventative coatings was introduced in 1942 (Kittel, 1960, p. 195).

2.3 Terminal Date

Zinc yellow is still being manufactured.

3.0 GENERAL PIGMENT PROPERTIES

3.1 Color and Spectral Reflectance

Zinc yellow is a greenish yellow pigment similar in hue to primrose chrome yellow. The reflectance spectrum is shown in fig. 11. The zinc tetroxychromate used in rust-preventative coatings displays a reddish tone.

3.2 Hiding Power and Tinting Strength

The pigment lacks the hiding power and tinting strength of the lead chromate yellows. Its refractive index is given as 1.84–1.90 (Gettens & Stout, 1966).

3.3 Permanance

Zinc yellow is not especially lightfast and is apt to turn gray-green owing to the formation of the chromic ion (Gettens & Stout, 1966). Greens made by mixing various blues with zinc yellow are brilliant in color but are said to darken appreciably on exposure to light. Hydrogen sulfide and soluble metal sulfides tend to produce no darkening, because any zinc sulfide that may form as the result of such action is white.

3.4 Compatibility

The lemon-colored basic zinc yellow is compatible with the pigments and binders in current use, as long as they are not alkaline (Wulf, 1974, p. 114).

3.5 Chemical Properties

The pigment dissolves easily in mineral acids, ammonia, and other alkalis (Zerr & Rübenkamp, 1908). Unlike chrome yellow, it is readily soluble in acetic acid. A simple test for distinguishing the two, therefore, is to dissolve the sample in hot dilute sodium hydroxide solution and add acetic acid; in contrast to the lead chromate, the zinc pigment is not reprecipitated (Remington, 1942).

The zinc yellows of the nineteenth century were slightly soluble in cold water. Because of its water-solubility, Toch (1946) recommended that the pigment not be used by artists.

Modern basic zinc chromates are much less soluble, and therefore the basic type is preferred for coatings that are to be immersed in water or exposed to moisture-condensing atmospheres. The slight solubility of zinc yellow provides chromate ions, which inhibit corrosion of ferrous and nonferrous metals. The mildly basic nature of the pigment also neutralizes acid components of the paint film.

The formation of soaps and the pigment's reduced solubility in water may be further reasons for the passivating effect of zinc chromate on iron and steel (rust protection). Wagner (1931, 1937) has published photomicrographs showing the presence of zinc soaps in paints made with this pigment.

3.6 Oil Absorption and Grinding Qualities

Zinc yellow dries well as an oil color, forming zinc soaps. According to Preuss (1974, p. 93) it grinds easily. The oil absorption is from 19 to 34% for zinc chromate and 40 to 49% for basic zinc chromate (*Raw Materials Index*, 1975, p. 80). Uebele (1913) states that the pigment requires 28 parts by weight of poppyseed oil to 72 parts of the dry powder to make 100 parts of paste in oil.

3.7 Toxicity

As mentioned in the remarks on lead chromate in a recent publication, zinc chromate seems to pose a far more positive threat of a carcinogenic effect than does lead chromate (Davies, 1979).

4.0 COMPOSITION

4.1 Chemical Composition

The composition of modern zinc yellow is $K_2O{\cdot}4ZnCrO_4{\cdot}3H_2O$. Older pigments may have been considerably different in composition, because yellows with widely varying $ZnO{:}Cr_2O_3$ ratios can be manufactured (Kittel, 1960, p. 195). Some commercial zinc yellow pigments contain up to 3% sulfate, presumably in solid solution. Mellor (1931, p. 278) gives a curve of the equilibrium conditions in the system $ZnO{\cdot}CrO_3{\cdot}H_2O$ at 25°C. As explained in 1.2, basic zinc chromate has the composition $4Zn(OH)_2{\cdot}ZnCrO_4$.

4.2 Sources

DuPont, Mineral Pigments, and Davis are current manufacturers and suppliers of zinc yellow in America.

4.3 Preparation

Several processes are used for making zinc yellow. All are based on reacting aqueous solutions of zinc and chromate compounds with dissolved potassium salts, in accordance with the following equation: $4ZnO + H_2SO_4 + 2K_2Cr_2O_7 + 2H_2O = K_2O{\cdot}4ZnO{\cdot}4CrO_3{\cdot}3H_2O + K_2SO_4$

5.0 IDENTIFICATION — CHARACTERIZATION

5.1 Characterization by Optical and Chemical Microscopy

Although x-ray diffraction analysis has indicated that zinc yellow crystals are monoclinic, detailed optical studies have not been made and the three principal refractive indices have not been identified. Two of the indices are probably 1.84 and 1.90, known limits for these crystals.

Under high magnification, zinc yellow appears as clumps of 0.5 to 1.0 micrometer-sized spherical particles with high birefringence. The small particle size makes the crystals difficult to study under the optical microscope (see fig. 5E, F). The shape of the particles at 20,000x by transmission electron microscopy is seen in fig. 12. Under plane-polarized light the spheres are pale yellow and the agglomerates greenish yellow. Between crossed polars, particles range from pale gray and blue to bright white and yellow. Agglomerates are pale lemon yellow in reflected light (McCrone & Delly, 1973, pl. 452, p. 499). Schultze (1895) described the *lemon yellow* pigment as prismatic crystals, easily soluble in acids.

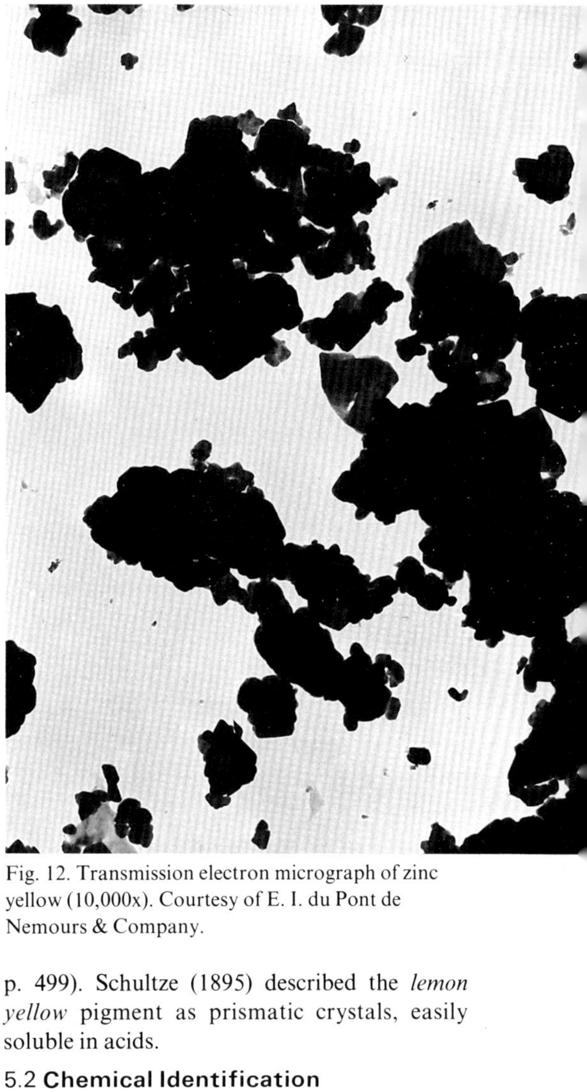

Fig. 12. Transmission electron micrograph of zinc yellow (10,000x). Courtesy of E. I. du Pont de Nemours & Company.

5.2 Chemical Identification

Zinc. The mutual precipitation of zinc and cobalt with mercury thiocyanate is an effective test for zinc. For details the reader is referred to the chapter on Zinc White, section 5.2. A spot test for zinc with dithizone in alkaline solution is less sensitive, although specific (Feigl & Anger, 1972).

Chromium. The chemical tests for the chromate iron are the same as those used for lead chromate.

5.3 Instrumental Methods of Analysis

5.31 ***X-ray Diffraction.*** The crystalline nature of zinc yellow permits positive identification by means of x-ray diffraction. In mixtures

Table 5. **X-RAY DIFFRACTION DATA FOR ZINC YELLOW ($4ZnCrO_4 \cdot K_2O \cdot 3H_2O$)***
(Guinier camera, Cu*K*α = 15405 Å)

d(Å)	*I*	*d(Å)*	*I*
7.02	85	2.212	4
5.07	40	2.102	6
4.73	30	2.030	2
4.55	12	1.992	12
4.20	4	1.945	12
3.69	18	1.922	2
3.52	75	1.885	12
3.31	90	1.863	8
3.18	6	1.845	8
3.08	85	1.754	35B
2.896	70	1.726	30
2.780	40	1.653	14
2.609	100	1.590	16
2.594	30	1.575	14
2.554	16	1.537	18
2.537	16	1.517	4
2.358	45	1.500	6
2.296	4	1.469	6
2.264	20	1.450	4

**JCPDS Powder Diffraction File* 8-202.
Courtesy of JCPDS International Centre for Diffraction Data (1982).

with large amounts of lead white, barite, or calcite, small amounts of zinc yellow can still be detected because it has strong lines that do not overlap those of the white pigments. A Debye-Scherrer pattern for zinc yellow is shown in fig. 7E and that for basic zinc yellow in 7F.

The *JCPDS Powder Diffraction File* gives diffraction patterns for zinc chromate ($ZnCrO_4$) 19-1456; three basic zinc chromate hydrates ($2ZnO \cdot CrO_3 \cdot H_2O$), 11-276 ($4ZnO \cdot CrO_3 \cdot 3H_2O$), 11-275 ($5ZnO \cdot CrO_3 \cdot 4H_2O$), 11-277; and zinc, chromate hydrate (potassium zinc yellow, $4ZnCrO_4 \cdot K_2O \cdot 3H_2O$), 8-202 (see table 5). The lines in the latter correspond to those found in a sample of zinc yellow kindly supplied by the pigment department of Ciba-Geigy (formerly Hercules) and analyzed by Dr. Sidney Pollack of Carnegie-Mellon Institute of Research (Sample No. B456 in the National Gallery of Art Pigment Bank collection at the Research Center on the Materials of the Artist and Conservator, Carnegie-Mellon Institute of Research, Pittsburgh).

5.32 ***Infrared Spectroscopy.*** Infrared can also be used for the identification of zinc yellow in the pure state, but it might be difficult to detect the pigment in a mixture with other chromates. Fig. 9 gives the infrared curves for zinc yellow and other yellow chromates. In contrast to the others, zinc yellow has significant absorption in the region of 3280 and 3460 cm^{-1}.

5.4 Criteria for Positive Identification

Because of the typically fine particle size and the complex chemical nature of both zinc chromate and basic zinc chromate, the only reliable method of identifying and differentiating the two is x-ray diffraction.

C. Strontium, Barium and Calcium Chromates

1.0 STRONTIUM CHROMATE, $SrCrO_4$

Addition of alkali chromate or dichromate to a solution of strontium chloride results in the precipitation of CI Pigment Yellow 32, No. 77839 (*Colour Index*, 1971). Gentele (1860) refers to strontium chromate only under "Additives to the yellow mineral colors," calling it a "very pale yellow, similar to the barium chromate." Bouvier (1910, p. 63) states that a pigment mixture with Prussian blue went under the name of "green cinnabar," whereas a mixture of chrome oxide with cadmium or strontium yellow was called "Behrendt green." The pigment has also appeared under the name "ultramarine yellow" (Bouvier, 1910, p. 48). A product in France was known as citron yellow (Bennett, 1963). In Birren (1965, pp. 67–68) the pigment appears as strontaine yellow. In modern times strontium yellow is used as a corrosion inhibitor, particularly in mixtures with aluminum flake, where, owing to its low tinting strength, it hardly influences the color (*Kirk-Othmer*, 1968).

The bright yellow strontium salt surpasses barium chromate in chromaticity and hiding power, the latter quality making it unsatisfactory for use in watercolors (Winsor & Newton, 1910). The spectral reflectance of the pigment at complete hiding is shown in fig. 11. According to Wagner, et al. (1933), the pigment tends to be more sensitive to light than barium yellow and, when so exposed, gradually acquires a green shade (Eibner, 1909).

Strontium chromate can be decomposed in alkali with formation of $CrO_4^{=}$ and insoluble strontium hydrate. It is dissolved in dilute mineral acids and acetic acid. It is more soluble in water than is barium chromate. Under the microscope, strontium chromate can appear as rhombic flakes or, more usually, as monoclinic needles of high birefringence and positive elongation (fig. 13). Fig. 14 shows a transmission electron micrograph of a modern strontium yellow. Many pigments sold as barium-strontium yellow show these distinctive needles. X-ray diffraction data for strontium chromate will be found in table 6 and a Debye-Scherrer pattern in fig. 7.

Birren (1965), writing about artists' palettes, notes that strontaine yellow was used by Pierre Bonnard and Henri Matisse, while Moreau-Vauthier (1923) lists the following artists as employing this pigment: Cormon, Chabas, D'Espagnat, Domergue, Levy-Dhurmer, Maillart, Point (Armand), and Doigneaw.

2.0 BARIUM CHROMATE, $BaCrO_4$

Barium Chromate, CI Pigment Yellow 31, No. 77103 (*Colour Index*, 1971), marketed under the name barium yellow, ultramarine yellow, lemon yellow, baryta yellow, or Steinbühl yellow, is produced by precipitation with potassium chromate or dichromate from barium chloride solution. Neutral solutions result in a pale lemon shade, and acid solutions yield a bright orange pigment (Meschezerski, 1882). Although barium chromate was early mentioned as a yellow-green pigment for porcelain (Klaproth & Wolff, 1807, p. 605), apparently little attention was given to its use in paints before the middle of the nineteenth century. For example, Gentele (1860, p. 198) states: "Barium chromate has a very pale sulfur-yellow color. It has not been introduced

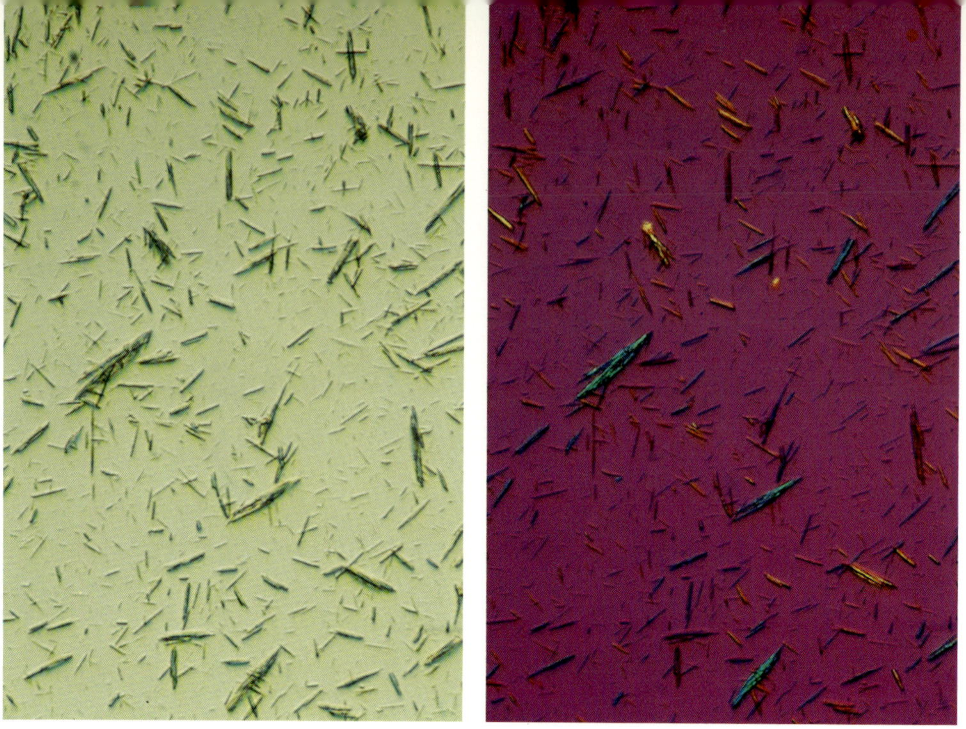

Fig. 13. Typical appearance of strontium chromate pigment under the optical microscope. 500x:A. in plane-polarized light; B. with a full-wave gypsum retardation plate. Note the blue color at 45° (SW-NE direction) in B: positive elongation. Refractive index of mounting medium, 1.53; with substage shadowing.

Fig. 14. Transmission electron micrograph of strontium chromate pigment at 10,000x. Courtesy E. I. du Pont de Nemours & Company.

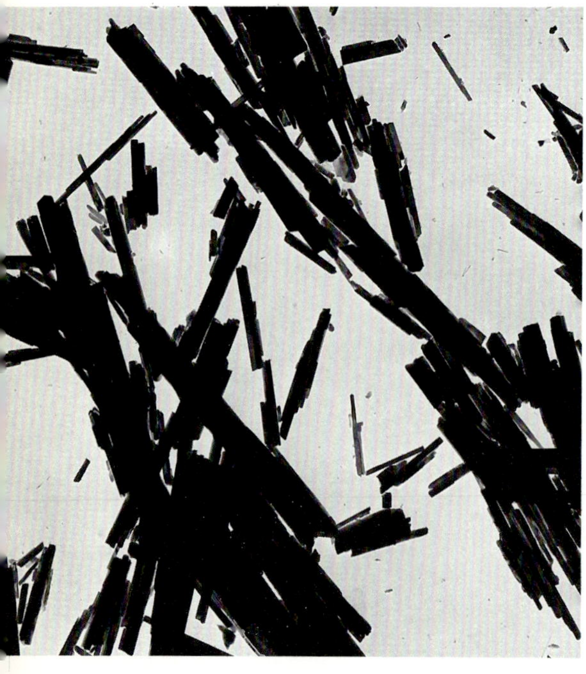

yet commercially as yellow paint and is not to be preferred to the lead chromates."

Apart from the modern, highly stable types of chrome yellow, barium chromate is said to possess the best long-term stability of the chromate pigments, although, like the others, it is also said to turn green slowly in light. Unfortunately its tinting strength and hiding power are low. In contrast to strontium chromate, barium chromate is only very slightly soluble in water; however, it can be decomposed in bases and dilute mineral acids. A barium-potassium chromate was developed as an anti-corrosive pigment by the National Lead Company (Kastens & Prigotsky, 1949). It hydrolyzes slowly, releasing the potassium chromate and leaving behind the barium chromate as a residual coating , CI Pigment Yellow 31.1, No. 77106 (*Colour Index*, 1971).

Some types of barium chromate can be recognized under the microscope as nearly colorless, birefringent rhombic flakes, while other varieties are so fine-grained that their optical properties cannot be readily observed. X-ray diffraction data for the pigment are given in table 6 and a Debye-Scherrer pattern is shown in fig. 7.

Birren (1965) lists only André Derain as having baryta yellow on his palette.

3.0 CALCIUM CHROMATE, $CaCrO_4$

Although calcium chromate, CI Pigment Yellow 33, No. 77223 (*Colour Index*, 1971) was

Table 6. **X-RAY DIFFRACTION DATA FOR STRONTIUM, BARIUM, AND CALCIUM CHROMATES**

$SrCrO_4$[a] Debye-Scherrer camera, MoKα = 0.709 Å		$BaCrO_4$[b] Debye-Scherrer camera, MoKα = 0.709 Å		$CaCrO_4$[c] Diffractometer, $CuK\alpha_1$ = 1.5405 Å	
d(Å)	*I*	*d(Å)*	*I*	*d(Å)*	*I*
3.70	17	4.50	11	4.75	10
3.46	34	4.00	27	3.62	100
3.26	100	3.54	40	2.880	16
3.01	100	3.19	83	2.679	55
2.72	17	2.90	33	2.562	12
2.55	23	2.78	27	2.375	16
2.36	6	2.53	8	2.254	8
2.32	6	2.37	7	2.013	6
2.25	17	2.25	7	1.913	6
2.14	6	2.16	100	1.8510	45
2.06	29	1.97	1	1.8100	16
1.97	40	1.91	17	1.6926	2
1.84	23	1.80	10	1.6195	10
1.79	6	1.71	33	1.5722	6
1.69	6	1.66	11	1.4999	14
1.65	6	1.62	11	1.4499	6
1.61	6	1.56	23	1.4423	6
1.53	6	1.50	5	1.3397	8
1.41	17	1.45	13	1.3146	4
1.30	11	1.41	3	1.2946	10
		Plus additional lines to 1.06 Å		Plus additional lines to 0.9051 Å	

a. *JCPDS Powder Diffraction File* 1-0698.
b. *JCPDS Powder Diffraction File* 1-1221.
c. *JCPDS Powder Diffraction File* 8-458.
Courtesy of JCPDS International Centre for Diffraction Data (1982).

produced in the laboratory at the beginning of the nineteenth century, Gentele (p. 289) mentions it as late as 1860, only under "Additives to the yellow mineral colors." Even in later literature, calcium chromate is frequently reported solely as an extender for other yellow pigments, especially chrome yellow. The color was said to be pale but to have good permanence. When examined under the microscope, samples prepared in the laboratory were observed to be very pale yellow indeed (see fig. 5C, D). Obsolete terms sometimes used in Germany for this pigment were *Gelbin* and *Steinbühler Gelb*. In the United States, one manufacturing source remains, the Frank D. Davis Company of Los Angeles.

Mellor (1931, p. 267) cites the research of Herrmann and coworkers (1926) that describes the crystals as rhombic bipyramids, containing 4 molecules in the unit cell with a = 7.75, b = 10.3, and c = 5.85 Å. X-ray diffraction data for calcium chromate are given in table 6; a Debye-Scherrer pattern in fig. 7.

Calcium chromate is difficult to characterize with confidence by infrared spectroscopy because the curve has few significant features. Its one principal absorption peak has been assigned the wave number 880 cm^{-1} by Doyle and Eddy (1967). Caution will be needed in differentiating the curves of pale yellow calcium chromate and those of lead and strontium molybdates. All have primarily one strong absorption band, but the maxima occur respectively at 880, 860 and 830 cm^{-1}. The curves also differ in detail between 400 and 200 cm^{-1} (Afremow & Vandeberg, 1966).

D. Chrome Orange — Chrome Red

NOMENCLATURE IN VARIOUS LANGUAGES

English: Chrome orange
German: Chromorange, Chromrot
Russian: Красный крон
French: chromorange
Italian: rossetto cromo
Spanish: Rojo de Cromo

1.0 INTRODUCTION

1.1 Brief Definition of Pigment

These pigments, although based on lead chromate, are placed under a separate heading because hue is a key property that the analyst usually employs in initiating the process of identification. Both chrome orange and chrome red are basic lead chromates having the chemical composition, $PbO{\cdot}PbCrO_4$. Different shades are primarily achieved by variations in the particle size, although many authors in the past attributed variation in color to the proportion of PbO present. Chrome orange is designated as CI Pigment Orange 21 (yellowish-orange) or CI Pigment Orange 45 (reddish orange), No. 77601; CI Pigment Orange 21:1 is a similar composition produced on a silica base. Chrome red is CI Pigment Red 103, No. 77601 (*Colour Index*, 1971).

2.0 HISTORY

2.1 Archaic, Obsolete, and Misrepresentative Names

Some of the older terms for basic lead chromate are Derby red, Persian red, Vienna red, Chinese red, Victoria red, American vermilion, chrome cinnabar, Australian cinnabar, garnet chrome, ruby red chrome, golden orange yellow, orange paste, Austrian red, and *rouge de Perse*. Midnight sun is another name listed for chrome scarlet by Kelly and Judd (1976).

2.2 History of Use

Records of the early use of chrome orange and red are sketchy. Most nineteenth-century authors describe the preparation of the pigment, rarely its use in commerce or by artists. Chrome orange was one of the substances prepared by Vauquelin in 1809 during his original investigations of chrome compounds. Its use in dyeing is discussed by Parnell (1846). In modern times a specific bright red-orange shade of basic lead chromate or a light shade of molybdate orange, designated as International Orange, has been used as a standard color for airport markings (*Kirk-Othmer*, 1968).

2.3 Terminal Date

The pigment is still in use in rust-inhibitive primers and paints for metals (Kirk-Othmer, 1968). However, the demand for chrome orange is waning because improved properties and lower costs are attainable through the blending of molybdate orange and medium yellows.

3.0 GENERAL PIGMENT PROPERTIES

3.1 Color and Spectral Reflectance

The pigment can range from light to deep orange. A spectral reflectance curve for dark chrome orange in masstone is shown in fig. 2, along with those of the more yellowish chromate pigments. Chemical analyses indicate that particle size is the key factor determining color of the chrome oranges and reds; the reds change to orange when the pigment is finely ground

(R. Schiek, private communication, 1979). The spectral reflectance curves of various lead pigments in the ultraviolet range and their appearance in ultraviolet, visible, and infrared photographs have been reported by Dunn (1975).

3.2 Hiding Power and Tinting Strength

High refractive indices between 2.42 and 2.7 give chrome orange and red their excellent hiding power in oil paints. Their tinting strength and hiding power are said to be somewhat lower than those of the chrome yellows (Brown, 1944, pp. 13–15, 277–282).

3.3 Permanence

The basic chromate tends to darken on exterior exposure but some grades are less reactive than others (Preuss, 1974, p. 74). The oranges are stated to be more light stable than are the yellows (Brown, 1944). Both are nonbleeding in practically all organic vehicles.

3.4 Compatibility

Chrome orange pigments mix well with oil but are generally sensitive to alkali; as noted in 1.1, the color becomes redder as the basicity increases. In contact with dilute acids, the pigments are satisfactory. They also show little or no change in color at baking temperatures up to 350°F (Preuss, 1974).

3.5 Chemical Properties

The basic character of chrome oranges makes them more resistant to alkalis than are chrome yellows and less resistant to acids (*Kirk-Othmer*, 1968). When mixed with oil, the paint dries quickly. Dunn (1975) considers that the presence of chromium tends to retard formation of lead soaps as compared to their very rapid generation with red lead, and intermediate rate of formation with lead white. This observation has also been made by Wormald (1975, p. 403).

3.6 Oil Absorption and Grinding Qualities

The oil absorption is 10–13% (*Raw Materials Index*, 1975, p. 77). Overgrinding reduces the particle size, resulting in the masstone becoming more yellow (Preuss, 1974, p. 74).

3.7 Toxicity

Because of both its lead and chromate components and its relative chemical activity, the compound should be considered among the most hazardous of the chromate pigments.

4.0 COMPOSITION

4.1 Chemical Composition

Although the shades vary, the chemical composition of all chrome orange/red pigments is fundamentally the same, $PbO \cdot PbCrO_4$. Some workers have described the composition as $Pb(OH)_2 \cdot PbCrO_4$, but a recent investigation has noted that the infrared pattern contains no sharp hydroxyl band, revealing instead only the presence of hygroscopic water (Pollack & Feller, 1976). Brown (1944, p. 277) states that the lighter shades contain the equivalent of 85% lead chromate and 15% lead oxide, whereas the deeper shades may contain as little as 60% lead chromate with 40% lead oxide. These figures are similar to those given by Preuss (1974, p. 74). An equilibrium diagram for the system $PbO\text{-}CrO_3\text{-}H_2O$ at 25°C and freezing-point curves for the binary system $PbO\text{-}PbCrO_4$ can be found in Mellor (1931, p. 301). Dunn (1975) quotes the phase diagrams of Levin, et al. (1956) that establish four lead chromates (1) normal, (2) monobasic, (3) 2½ basic, and (4) tetrabasic.

The orange and scarlet chromes are much easier to make than the yellows, as the conditions of formation are more favorable to the production of orange shades than of yellow shades. This is partly owing to the formation of these hues at a high temperature, while it is necessary to keep the mass as cold as possible if good yellows are to be obtained (Hurst, 1906).

According to Hurst (1906, p. 137) the more nearly pure types still contained some lead sulfate, while common products generally contained barytes, gypsum, or china clay. Bearn (1923) described a "Genuine orange chrome" as being prepared by adding red lead to the pure, orange chrome. Adulterated varieties may contain terra alba, barites, and other inerts.

4.2 Sources

In America the commercial sources of chrome orange and chrome red in 1973 were Hercules, Harshaw, American Cyanamid, DuPont, and Reichhold.

4.3 Preparation

Chrome orange can be produced by precipitation from alkaline solutions or by treatment of neutral lead chromate with alkalis. It can be made also by boiling lead white with chromate solutions or by mixing lead oxychloride with soluble chromates. Shades varying from light orange to orange-red are produced by reactions in which the basic components predominate.

The orange pigments are usually made from more concentrated solutions than are the middle (yellow) shades, and they are prepared at the boil. Unlike the middle and pale shades of yellow, which should be finished with a slight excess of lead in solution, the redder shades are finished with all the lead precipitated and usually with a slight excess of sodium chromate in solution. In the case of the orange and red chromes the order of addition may be reversed, i. e., the lead salt may be added to the bichromate (Harrison, 1930).

5.0 IDENTIFICATION — CHARACTERIZATION

5.1 Characterization by Optical and Chemical Microscopy

Chrome red appears as tabular crystals, 1 micrometer and larger in size, having strong birefringence. Refractive indices are $\alpha = 2.42$, $\beta = 2.7$, $\gamma = 2.7$. Wagner (1931) states that the particles are ordinarily so fine that little can be said about their characteristics under the microscope (see fig. 5G, H). Schiek (1973) has published electron photomicrographs of pigments at 18,000x.

5.2 Chemical Identification

The chemical tests for lead and for the chromate ion are the same as those used for lead chromate.

Table 7. **X-RAY DIFFRACTION DATA FOR CHROME ORANGE-RED[a] (Guinier camera, Cu$K\alpha_1$ = 1.5405 Å)**

d(Å)	*I*	*d(Å)*	*I*
6.45	14	2.315	8
6.32	16	2.269	16 (broad)
5.96	10	2.187	6
5.17	2	2.132	6
4.44	14	2.114	10
3.78	8	2.058	18
3.73	4	2.015	2
3.54	6	1.992	6
3.39	100	1.950	4
3.23	12	1.874	4
2.986	80	1.867	18
2.882	20	1.859	4
2.840	35	1.829	2
2.650	2	1.818	2
2.590	4	1.794	2
2.565	4	1.778	8
2.511	12	1.770	8
2.479	14	1.753	4
2.460	6	1.726	12
2.372	8	Plus 17 lines to 1.419 Å	

Lead chromium oxide (Pb_2CrO_5).
a *JCPDS, Powder Diffraction File* 26-832.
Courtesy of JCPDS International Centre for Diffraction Data (1982).

5.3 Instrumental Methods of Analysis

5.31 ***X-ray Diffraction.*** Although most references characterize chrome orange as tetragonal, it is monoclinic. The original data, which were published by Wagner and Schirmer in 1935, were interpreted incorrectly for lack of a high-quality Debye-Scherrer camera and because of x-ray absorption of the low-angle lines (Pollack & Feller, 1976). More recent and accurate x-ray diffraction data for chrome orange are given in table 7, and a Debye-Scherrer pattern is shown in fig. 15. On most Debye-Scherrer patterns the feeble low-angle lines would be very weak or absent. Dunn (1975) has published a drawing showing the principal lines and intensities of normal, monobasic, and tetrabasic lead chromates.

5.32 ***Infrared Spectroscopy.*** Fig. 16 gives the slight variations that may be found in the major absorption region in the infrared spectrum of chrome orange. Its principal absorption band resembles that of chrome yellow, but five of six samples of the orange examined exhibited distinct bands at 440 and 500 cm^{-1}. The band at 1090 cm^{-1} in the c.p. chrome orange curve is not

Fig. 15. X-ray powder diffraction patterns for chrome orange pigments: A. Extra deep chrome orange, C.P. A-4337 (Ciba-Geigy Corporation), B. Chrome orange medium 2213, 58% $PbCrO_4$ (Harshaw Chemical Company). Patterns courtesy of Freer Gallery of Art. (Photographs not to be used for measurement purposes.)

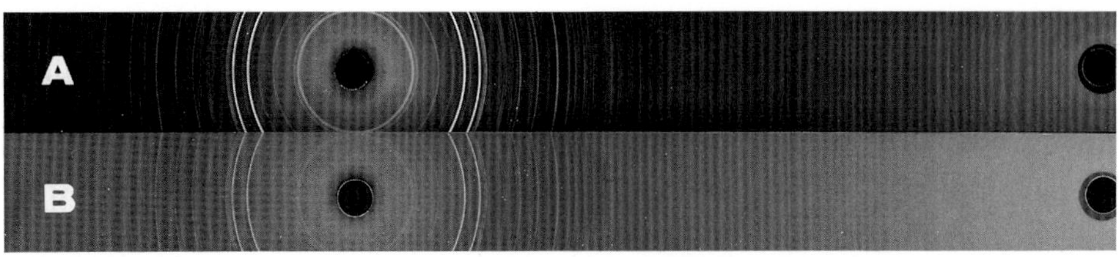

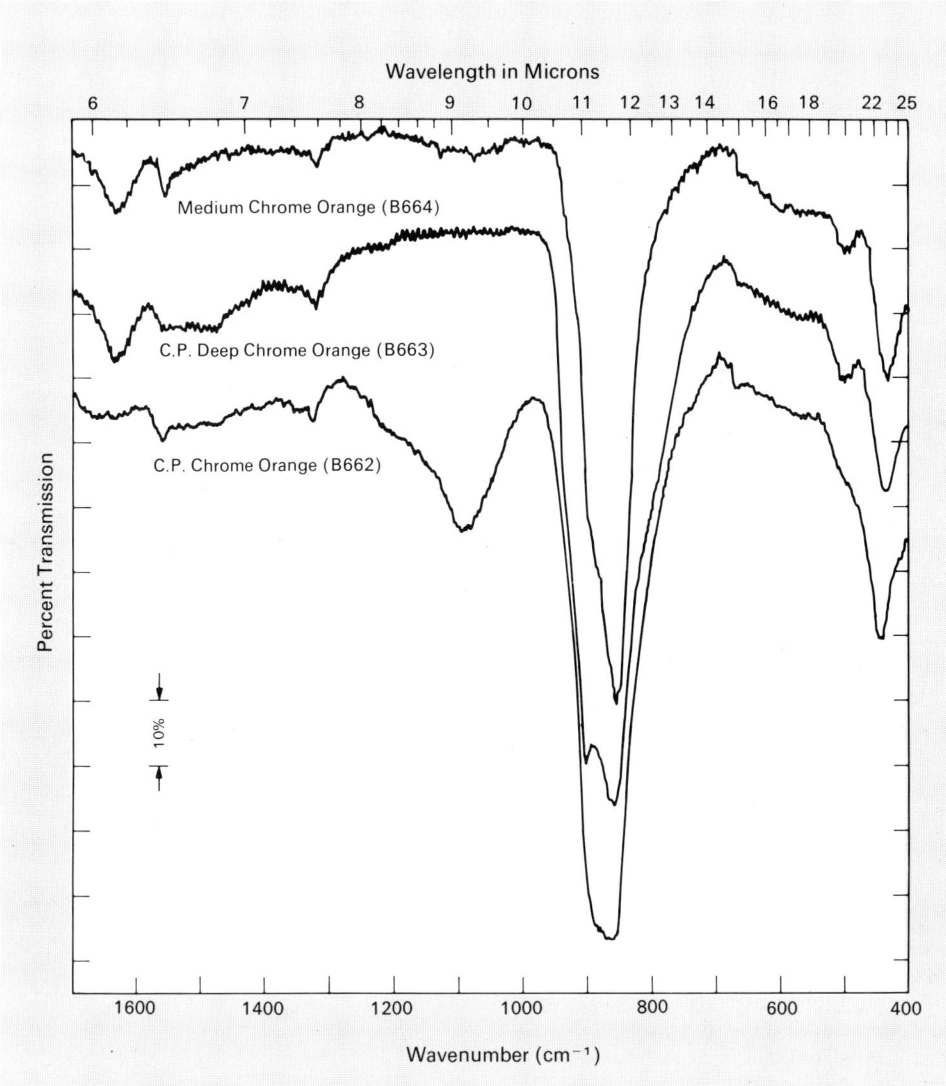

Fig. 16. Infrared spectrum showing principal absorption bands of representative chrome oranges.

characteristic of the pigment but only of the particular commercial product involved.

5.4 **Criteria for Positive Identification**

The only positive method for identification of microscopic-sized samples of chrome orange and chrome red is x-ray diffraction. Rather strong indication, of course, can be obtained by the complete solubility of an orange or red pigment in dilute hydrochloric acid, yielding both lead and chromate ions.

E. Rarely Encountered Chromates

1.0 YELLOWS

1.1 Basic Cadmium Chromate

This bright yellow powder, $CdCrO_4 \cdot Cd(OH)_2$, is made by the precipitation of cadmium salts with potassium chromate. Cadmium chromate was probably first introduced in the last quarter of the nineteenth or in the beginning of the twentieth century. It is expensive and rarely used. Eibner has published information concerning the pigment (1909).

1.2 Bismuth Chromate

In 1804 Godon referred to the combination of bismuth and chromic acid and its possible use as pigment (Berthollet & Vauquelin, 1804, p. 228). Gentele (1860, p. 290) mentions bismuth chromate under "Additives to the yellow mineral colors," but notes that the lemon yellow pigment has no advantage over chrome yellow and, besides, is expensive.

1.3 Basic Iron Chromate

For manufacture, solutions of ferric chloride and potassium chromate are mixed and heated for a long time, causing the iron chromate to separate as sediment. Impurities present in the iron oxide will cause the resulting pigment to be dull orange-yellow rather than bright yellow. Iron chromate was first manufactured by V. Kletzinsky (1872) and is mentioned by Mierzinski (1881, p. 418), who also describes the pigment's manufacture.

Siderin yellow, as it has been called, is compatible with all binders, dries quickly in oil, and with water glass forms a waterproof coating as hard as cement. In porcelain glazes it produces brown shades.

Basic iron chromate results from the precipitation of neutral potassium chromate with iron sulfate. This pigment, of dark brown color, is said to have been used as both an oil and watercolor (Ullmann, 1929).

2.0 REDS

2.1 Mercuric Chromate

When divalent mercury salts are precipitated with potassium chromate, a cinnabar-hue red powder is produced. Precipitation from the nitrate yields a beautiful pigment. Mercuric chromate, Hg_2CrO_4, was mentioned in Klaproth and Wolff's *Chemisches Wörterbuch* (1807, p. 609) and was discussed by Hartmann (1840) as a commercial pigment. Gentele (1860, p. 272) indicated that mercuric chromate had no significance as a paint colorant because it is not permanent and, above all, darkens greatly in oil.

6.0 NOTABLE OCCURRENCES

Moreau-Vauthier (1923) in his list of artists' palettes, notes seven artists who used chrome yellow; Birren (1965) lists four. Of the forty-six watercolor artists who reported their pigments to Russell and Abney (1888), twenty-two stated that they used lemon or permanent yellow, while four reported their pigment as chrome yellow; only one mentioned the use of the orange.

Chrome orange ($PbO \cdot PbCrO_4$), identified by x-ray diffraction, was found in two modern Japanese glass matrix pigments in the collection of the Technical Laboratory, Freer Gallery of Art, Washington. The two samples are labeled *cha seki* (Ishida Pigment Set No. 16, E. W. Forbes Collection) and stone red, *iwa aka* (Randall Pigment Set No. 40). Glass matrix pigments of various colors have been made in Japan since the early twentieth century, evidently in imitation of some of the coarse, naturally occurring mineral pigments. They are said to be ceramic glazes in which pigments are suspended or dispersed (E. W. FitzHugh, private communication).

Chrome Yellow — Notable Occurrences from the Bayerische Staatsgemäldesammlungen.

Artist or School *Title, Date,* *Inventory Number*
Sir Thomas Lawrence, *Portrait of a Gentleman*, 1800/1810, 13035
Johann Georg von Dillis, *Landscape with Cascade*, after 1800, 10243
Lorenz Adolf Schönberger, *Mountainous Landscape*, around 1810, 4901
Josef Anton Koch, *Landscape with Rainbow*, 1812, WAF 447
Josef Anton Koch, *Italian Vineyard Workers' Festival*, 1812, WAF 448
Max Joseph Wagenbauer, *Morning in a Village*, 1816, WAF 1173
Domenico Quaglio, *A Forest Chapel*, 1817, 1615
Josef Anton Koch, *View of Olevano*, c. 1817, 10026
Joseph Karl Stieler, *Portrait of Goethe*, 1818, WAF 1048
Franz Ludwig Catel, *The Theater of Taormina*, c. 1818, 11466 x
Franz Ludwig Catel, *Italian Folk Festival at Pozzuoli*, before 1823, 3798
Carl Rottmann, *The Kochel Lake*, c. 1823, 11476 x
Henrich Maria von Hess, *Apollo and the Muses*, 1826, WAF 347
Friedrich Overbeck, *Italia and Germania*, 1827, WAF 755
Friedrich Wilhelm Schadow, *Fanni Ebers*, 1827, 9671
Domenico Quaglio, *Villa Malta in Rome*, 1830, WAF 784
Caspar David Friedrich, *Church Ruin in the Forest*, shortly before 1831, 9872
Friedrich Loos, *Salzburg-Moenchberg*, 1833, 12017
Karl Blechen, *View of Assisi*, 1830–1835, 10338
Eugen Napoleon Neureuther, *View of Rome*, 1837, 9121
Carl Rottmann, *Landscape with Mountains*, c. 1838, 12032
Heinrich Bürkel, *Brawl in Front of the Tavern*, c. 1840, 11970
Jules Coignet, *Temple of Poseidon at Paestum*, 1844, WAF 158
Moritz von Schwind, *The Dream of Erwin von Steinbach*, c. 1845, 11558 x
Mortiz von Schwind, *Street player at the Einsiedler (Hermit)*, c. 1846, 13029
Moritz von Schwind, *In the Forest (Des Knaben Wunderhorn)*, c. 1848, 11576 x
Johann Christian Claus Dahl, *Norwegian River Landscape*, c. 1850, 12521
Arnold Böcklin, *Nymph at the Spring*, c. 1855, 11538 x
Arnold Böcklin, *Shepherdess*, c. 1860, 11529 x
Anselm Feuerbach, *The Garden of Ariost*, 1862, 11523 x
Anselm Feuerbach, *Madonna*, 1863, 11524 x
Arnold Böcklin, *Villa on the Sea*, 1864, 10811
Anselm Feuerbach, *Nymph Listening to Children Playing Music*, 1864, 11520 x
Carl Spitzweg, *A Hypochondriac*, c. 1865, 11563 x
Anselm Feuerbach, *Hafis at the Well*, 1866, 11511 x
Franz von Lenbach, *Portrait of a Franciscan Monk*, 1866, 11455 x

Chrome yellow has been detected in 205 nineteenth- and twentieth-century paintings at the Doerner-Institute, Munich. A representative selection is given here. In every case chromium was detected spectrochemically. x = see Kühn (1969).

Artist or School *Title, Date,* *Inventory Number*
Franz von Lenbach, *Salome with Head of John the Baptist*, 1867/68, 11662 x
Franz von Lenbach, *Charles V on Horseback* (after Titian), 1868, 11686 x
Franz von Lenbach, *The Vega of Granada*, 1868, 11458 x
James Marshall, *Tartini's Dream*, 1868, 11636 x
N.V. Diaz de la Peña, *Clearing*, 1868, 12523
August Wolf, *Madonna with St. John, Paul, Magdalena and Hieronymus* (after Titian), 1868, 11491 x
Paul Cézanne, *Railroad Cutting*, c. 1870, 8646
August Wolf, *St. Barbara* (after Palma Vecchio), 1870/71, 11668 x
Arnold Böcklin, *Italian Villa*, c. 1871, 11532 x
August Wolf, *St. Agnes Bringing the Son of the Prefect Back to Life* (after Tintoretto), 1873, 11676 x
Hugo von Habermann, *Mother of the Artist*, 1877, 9393
Wilhelm Ferdinand Xylander, *Dutch Landscape*, 1877, 11650 x
August Wolf, *The Child Murder at Bethlehem* (after Bonifazio Veronese), 1878, 11658 x
Mihaly Munkacsy, *Visit to a Woman in Childbed*, 1879, 7933
Friedrich August Kaulbach, *Two Young Women in the Forest*, 1880, 10810
August Wolf, *Madonna of Throne with Saints* (after Paolo Veronese), 1880, 11695 x
Paul Gauguin, *Peasant Women from Bretonne*, 1886, 8701
Paul Sérusier, *Savoy Landscape*, 1890, 12222
Maurice Denis, *Gallic Goddess of the Herds*, c. 1890, 8654
Walter Crane, *The Horses of Neptune*, 1892, 13419
Paul Gauguin, *The Birth of Christ*, 1896, 8652
Ferdinand Hodler, *Jena Student on the Move*, 1908, 8643
Max Slevogt, *Prince Regent Luitpold at the Altar*, 1909, 9090
Lovis Corinth, *Larkspur*, 1924, 9219

Chrome yellow has been detected in 205 nineteenth- and twentieth-century paintings at the Doerner-Institute, Munich. A representative selection is given here. In every case chromium was detected spectrochemically. x = see Kühn (1969).

ACKNOWLEDGMENTS

The authors thank Dr. H. von Sonnenburg of the Doerner-Institute, Munich, and Dr. F. Preusser, formerly of the Doerner-Institute, now of the J. Paul Getty Conservation Institute, for the list of notable occurrences. The transmission electron micrographs were kindly supplied by J. H. Lowell (retired) and B. H. Perkins of E. I. du Pont de Nemours & Company. The authors are particularly grateful for information on crystal structure, chemical composition, and spectral reflectance kindly provided by Dr. Robert C. Schiek of the Ciba-Geigy Corporation, Glens Falls, New York, and for his reading and criticism of technical aspects of the manuscript.

REFERENCES

L. C. Afremow and J. T. Vandeberg, "High Resolution Spectra of Inorganic Pigments and Extenders in the Mid-infrared Region from 1500 cm^{-1} to 200 cm^{-1}," *Journal of Paint Technology* 38 (1966), 169–202.

American Society for Testing and Materials, Annual Book of ASTM Standards Part 28, No. D211-67. Standard specification for chrome yellow and chrome orange (Philadelphia, 1977).

J. G. Bearn, *The Chemistry of Paints, Pigments, and Varnishes* (London, 1923), 65–79.

R. F. Bennett, "Chromate de strontium — Un pigment anticorrosion," *Peintures, pigments, vernis* 39 (1963), 379–383.

M. Berthollet and L. N. Vauquelin, "Rapport fait à la classes des sciences physique et mathématiques de l'institut, sur une mémoire de M. Godon," *Annales de chimie* 53 (1804), 222–229.

F. Birren, *History of Color in Painting* (New York, 1965), 67–68.

P. L. Bouvier, *Handbuch der Ölmalerei*, 8th ed. (Leipzig, 1910).

A. F. Brown, "Yellow and Orange Pigments," in *Protective and Decorative Coatings*, 2nd ed., vol. 2: *Raw Materials: Pigments, Metallic Powders, and Metallic Soaps*, ed. J. J. Mattiello (New York, 1944), 279–282.

J. A. Campbell, "Spectral Evidence for Interionic Forces in Crystals–Chromates and Dichromates," *Spectrochimica Acta* 21 (1965), 1333–1343.

A. H. Church, *The Chemistry of Paints and Painting*, 4th ed. (London, 1915).

H. F. Clay and V. Watson, "The Crystal Structure of Lead Chrome," *Journal of the Oil and Colour Chemists' Association* 31, no. 9 (1948), 418–422.

H. F. Clay and V. Watson, "Some Important Properties of Chromes," *Journal of the Oil and Colour Chemists' Association* 27, no. 283 (1944), 3–18.

Colour Index, vol. 3, 3rd ed. (Bradford, Yorkshire, England, Research Triangle Park, North Carolina, 1971).

J. M. Davies, "Lung Cancer Mortality of Workers in Chromate Pigment Manufacture: An Epidemiological Survey," *Journal of the Oil and Colour Chemists' Association* 62 (1979), 157–163.

W. P. Doyle and P. Eddy, "Infrared Spectra of Chromates (V) of Calcium, Strontium and Barium," *Spectrochimica Acta* 23A (1967), 1903–1907.

E. J. Dunn, Jr., "Lead Pigments," in *Treatise on Coatings*, vol. 3: *Pigments Part I*, ed. R. R. Myers and J. S. Long (New York, 1975).

E. J. Dunn, Jr., "Normal Lead Silico Chromate," in *Pigment Handbook*, vol. 1, ed. T. C. Patton (New York, 1973), 371–373.

A. Eibner, *Malmaterialienkunde als Grundlage der Maltechnik* (Berlin, 1909), 165.

F. Feigl and V. Anger, *Spot Tests in Inorganic Analysis*, 6th ed. rev., trans. R. E. Oesper (Amsterdam, 1972), 190.

R. L. Feller, "Scientific Examination of Artistic and Decorative Colorants," *Journal of Paint Technology* 44, no. 566 (1972), 51–58.

Fiche Technique P.P.V. No. 100 Férrier, *Peintures, pigments, vernis*, 42, no. 2 (1966), 89–92.

J. G. Gentele, *Lehrbuch der Farbenfabrikation* (Braunschweig, 1860).

R. J. Gettens and G. L. Stout, *Painting Materials, A Short Encyclopedia* (1942; New York, 1966).

C.-E. Guignet, *Encyclopédie chimique* (Paris, 1888), 98.

T. R. Harkins, J. T. Harris, and O. D. Shrere, "Identification of Pigments in Paint Products by Infrared Spectroscopy," *Analytical Chemistry* 31 (1959), 541–545.

R. D. Harley, *Artists' Pigments; c. 1600–1835, a Study in English Documentary Sources*, 2nd ed. (London, 1982).

A. W. C. Harrison, *The Manufacture of Lakes and Precipitated Pigments* (London, 1930).

C. Hartmann, *Enzyklopädisches Wörterbuch der Technologie*, vol. 1, 2nd ed. (Augsburg, 1840), 761.

R. Haug, "Über Belichtungsversuche mit Chromgelb," *Deutsche Farben-Zeitschrift* 5 (1951), 343–348.

K. Hermann, M. Mosenfeld, and N. Schönfeldt, *Wissenschaftliche Veröffentlichungen aus dem Siemens-Kanzern* 5 (1926), 189.

H. Hetterich, "Zum Stand und zur künftigen Entwicklung der mikrochemischen Bilduntersuchung" (Ph.D. diss., Technische Hochschule, Munich, 1931), 40.

C. D. Hodgman, in *Handbook of Chemistry and Physics* (Cleveland, 1960), 2664.

G. H. Hurst, *Painters' Colours, Oils, and Varnishes: A Practical Manual*, 4th ed. (London, 1906).

Imperial Chemicals Industries, *Lead Chrome Paints, the Facts* (1981), 12 pp.

Infrared Spectroscopy Atlas — Its Use in the Coatings Industry (Philadelphia, 1969), Spectra 519–529.

F. M. Jaeger and H. C. Germs, "Uber die binären Systeme der Sulfat, Chromat, Molybdat und Wolframat des Bleies," *Zeitschrift für anorganische und allgemeine Chemie* 119 (1921), 145–173.

JCPDS Powder Diffraction File. JCPDS International Centre for Diffraction Data, Swarthmore, Pennsylvania, 1982.

R. M. Johnston and R. L. Feller, "Optics of Paint Films: Glazes and Chalking," in *Application of Science in the Examination of Works of Art*, Proceedings of the Seminar, 7–16 September 1965, Museum of Fine Arts (Boston, 1967), 86–95.

M. L. Kastens and M. J. Prigotsky, "Barium Potassium Chromate Pigment," *Industrial and Engineering Chemistry* 41, pt. 2 (1949), 2376–2382.

K. L. Kelly and D. B. Judd, *Color, Universal Language and Dictionary of Names* (Washington, 1976).

Kirk-Othmer Encyclopedia of Chemical Technology, 2nd ed. (New York, 1968), 530–555.

H. Kittel, ed. *Pigmente. Herstellung, Eigenschaften, Anwendung* (Stuttgart, 1960).

M. H. Klaproth and F. Wolff, *Chemisches Wörterbuch*, vol. 1 (Berlin, 1807).

V. Kletzinsky, "Das basisch-chromsaure Eisenoxyd oder das Sideringelb," *Dinglers polytechnisches Journal* 207 (1872), 83–84.

C. Koch, *Grosses Malerhandbuch* (Giessen, 1953), Tafel 3.

R. König, *Anorganische Pigmente und Röntgenstrahlen* (Stuttgart, 1956), 66 and 77.

H. Kopp, *Geschichte der Chemie*, vol. 4 (Braunschweig, 1845), 81.

H. Kühn, "Möglichkeiten und Grenzen der Untersuchung von Gemälden mit Hilfe von naturwissenschaftlichen Methoden," *Maltechnik/Restauro*, no. 3 (1974), 149–162.

H. Kühn, *Die Pigmente in den Gemälden der Schack Galerie* (Munich, 1969).

S. Langård and T. Norseth, "A Cohort Study of Bronchial Carcinomas in Workers Producing Chromate Pigments," *British Journal of Industrial Medicine* 32 (1975), 62–65.

J. G. Lehmann, "De nova mineral plumbi specine crystallina rubra," *Nov. Comm. Acad. Petropoli* 12 (1766), 356.

R. W. Leisy, A process for the manufacture of basic zinc chromate. United States Patent 2,251, 846. British Patent 547, 859 (August 5, 1941).

J. C. Leuchs, *Traité complet des propriétés, de la préparation et de l'emploi des matières tinctoriales et des couleurs. Deuxième partie. Fabrication des couleurs* (Paris, 1829), 72.

E. M. Levin, H. F. McMurdie, and F. P. Hall, *Phase Diagrams for Ceramists*, ed. The American Ceramic Society, Inc. (Columbus, 1956), p. 58, fig. 98.

J. Liebig, I. C. Poggendorff, and F. Wöhler, *Handwörterbuch der reinen und angewandten Chemie*, vol. 2 (Braunschweig, 1842), 279.

B. M. Loeffler and R. G. Burns, "Shedding Light on the Color of Gems and Minerals," *American Scientist* 64 (1967), 636–647.

W. C. McCrone and J. G. Delly, *The Particle Atlas*, vol. 2, 2nd ed. (Ann Arbor, 1973).

J. W. Mellor, *A Comprehensive Treatise on Inorganic and Theoretical Chemistry*, vol. 11 (New York, 1931).

H. E. Merwin, "Optical Properties and Theory of Color of Pigments and Paints," *Proceedings of the American Society for Testing Materials* 17 (1917), 494–530.

J. Meschezerski, "Trennung des Baryums von Strontium und Calcium durch neutrales chromsaures Kali," *Zeitschrift für analytische Chemie* 21 (1882), 399–400.

S. Mierzinski, *Die Erd-, Mineral- und Lack-Farben*, 4th ed. (Weimar, 1881).

C. Moreau-Vauthier, *Comment on peint* (Paris, 1923), 23–34.

R. Newman, "Some Applications of Infrared Spectroscopy in the Examination of Painting Materials," *Journal American Institute for Conservation* 19, 1 (1979), 42–62.

R. A. Nyquist and R. O. Kagel, *Infrared Spectra of Inorganic Compounds* (New York, 1971), 327–329.

G. D. Parkes, ed. *Mellor's Modern Inorganic Chemistry* (New York, 1967), 866.

E. A. Parnell, *A Practical Treatise on Dyeing and Calico-Printing* (New York, 1846), 142.

N. C. Pearre and A. V. Heyl, Jr., *The History of Chromite Mining in Pennsylvania and Maryland* (Harrisburg, Pennsylvania, 1959).

J. Plesters, "Cross-sections and Chemical Analysis of Paint Samples," *Studies in Conservation* 2 (1956), 110–157.

S. Pollack and R. L. Feller, "On the Crystallography of Chrome Orange," *Journal of Coatings Technology* 48, no. 618 (1976), 68.

H. P. Preuss, *Pigments in Paint* (Park Ridge, New Jersey, 1974), 85.

A. Raft, "Eckersberg's farbekob i Rom," *Meddelelser fra Thorvaldsen's Museum* (1973), 15–17, 152–154.

Raw Materials Index, Pigment section (Washington, 1975).

J. S. Remington, *The Paint Laboratory Note Book* (London, 1942), 42.

J. S. Remington, *Pigments, Their Manufacture and Properties* (London, 1944), 116.

E.-L. Richter and H. Härlin, "The Pigments of the Swiss Nineteenth-century Painter Arnold Böcklin," *Studies in Conservation* 19 (1974), 83–87.

D. J. Robinson, "Chrome Green Pigments," in *Pigment Handbook*, vol. 1, ed. T. C. Patton (New York, 1973), 385–388.

W. J. Russell and W. De W. Abney, *Report on the Action of Light on Water Colours* (London, 1888), 73–77.

K. Sagel, *Tabellen zur Röntgen- Emissions- und Absorptions-Analyse* (Berlin, 1959), 115.

R. C. Schiek, "Lead Chromate Pigments," in *Pigment Handbook*, vol. 1, ed. T. C. Patton (New York, 1973), 357–370.

M. Schofield, "Chromium Pigments," *Paint Manufacturer* 18 (1948), 394–396.

J. Schultze, "Uber einige Salze der Chrom- und Bichromsäure mit den Schwermetallen," *Zeitschrift für anorganische Chemie* 10 (1895), 148.

M. P. Schützenberger, *Traitée des matières colorantes*, vol. 1 (Paris, 1867), 254.

J. C. Smith, *The Manufacture of Paint*, 2nd ed. (London, 1915), 148–151.

A. A. Socolow, "Pennsylvania Chrome-Yellow Heritage," *Pennsylvania Geology* (Harrisburg, June 1976).

C. J. A. Taylor and S. Marks, eds., *Pigments, Dyestuffs and Lakes*, Oil and Colour Chemists' Association, Paint Technology Manuals, pt. 6 (New York, 1966).

M. Toch, *The Chemistry and Technology of Paints*, 3rd ed. (New York, 1925).

M. Toch, *Paint, Paintings and Restoration*, 2nd ed. (New York, 1946), 103.

H. Trillich, *Das Deutsche Farbenbuch*, vol. 3 (Munich, 1923), 55.

S. Tschelnitz, *Farben-Chemie* (Vienna, 1857), 44.

C. L. Uebele, *Paint Making and Color Grinding* (London, 1913), 256.

F. Ullmann, *Enzyklopädie der technischen Chemie*, vol. 3 (Berlin, 1929), 394.

L. N. Vauquelin, "Mémoire sur le chrôme. Nouvelle substance métallique contenue dans le plomb rouge de Sibérie," *Annales de Chimie* 25 (1797), 21–31.

L. N. Vauquelin, "Mémoire sur la meilleure méthode pour décomposer le chrômate de fer, obtenir l'oxide de chrôme, préparer l'acide chrômique, et sur quelques combinaisons de ce dernier," *Annales de Chimie* 70 (1809), 70–94.

H. Wagner, "Dispersion Problems and Pigment Manufacture," *Paint, Varnish, Lacquer, Enamel, and Colour Manufacture* (October 1931), 186–195.

H. Wagner, "Einfluss der Werkstoffumstellung aud die Pigmentfarben," *Angewandte Chemie* 50 (1937), 824–827.

H. Wagner and H. Schirmer, "Basisches Bleichromat und seine Umwandlung," *Zeitschrift für anorganische und allgemeine Chemie* 222 (1935), 245–248.

H. Wagner, M. Zipfel, G. Heintz, and R. Haug, "Ursachen und Behebung der Cromgelb-Lichtunechtheit," *Farben Zeitung* 38 (1933), 932–934.

V. Watson and H. F. Clay, "The Lightfastness of Lead Chrome Pigments," *Journal of the Oil and Colour Chemists' Association* 38 (1955), 167–175.

Winsor & Newton, *Descriptive Handbook of Modern Water Colour Pigments* (London, 1910).

G. Wormald, "Chromate Pigments," in *Treatise on Coatings*, vol. 3, *Pigments* (Part 1), eds. R. R. Myers and J. S. Long (New York, 1975), chap. 6, 317–332.

H. Wulf, *Grosse Farbenwarenkunde* (Cologne-Braunsfeld, 1974).

G. Zerr and R. Rübencamp, *A Treatise on Colour Manufacture*, trans. C. Mayer (London, 1908), 129.

BIBLIOGRAPHY

M. R. Alderson, N. S. Rattan, and L. Bidstrup, "Health of Workmen in the Chromate-producing Industry in Britain," *British Journal of Industrial Medicine* 38, no. 2 (1981), 117–124.

E. E. Christofano, G. L. Kennedy, Jr., D. E. Gordon, M. L. Keplinger, and J. C. Calandra, "Toxicity of Lead Chromate Pigments and Tissue Residues Following 90 Days Feeding to Beagle Dogs. Hercules, Inc., Wilmington, Delaware; and Industrial Bio-Test Laboratories, Inc., Northbrook, Illinois," *Toxicology and Applied Pharmacology* 37 (1976), 161.

N. A. Dalager, T. J. Mason, J. F. Fraumeni, Jr., R. N. Hoover, and W. W. Payne, "Cancer Mortality among Workers Exposed to Zinc Chromate Paints," *Journal of Occupational Medicine* 22, no. 1 (1980), 25–29.

G. L. Kennedy, Jr., M. L. Keplinger, R. J. Wigender, E. E. Christofano, and J. C. Calandra, "Toxicity of Lead Chromate Pigments and Tissue Residues Following 90 Days Feeding to Albino Rats. Industrial Laboratories, Inc., Northbrook, Illinois; and Hercules, Inc., Wilmington, Delaware," *Toxicology and Applied Pharmacology* 37 (1976), 160.

M. L. Somme-Dubru, M. Genet, A. Mathieux, P. G. Rouxhet, and L. Rodrique, "Evaluation by Photoelectron Spectroscopy and Electron Microscopy of the Stabilization of Chrome-Yellow Pigments," *Journal of Coatings Technology* 53, no. 683 (1981), 51–56.

Lead Antimonate Yellow

IAN N. M. WAINWRIGHT, JOHN M. TAYLOR and ROSAMOND D. HARLEY

NOMENCLATURE IN VARIOUS LANGUAGES

English: lead antimonate yellow, Naples yellow, antimony yellow
German: Bleiantimoniat, Neapelgelb, Antimongelb
Russian: неаполитанская желтая
French: jaune d'antimoine jaune de Naples
Italian: giallo di Napoli
Spanish: amarillo de Nápoles
Dutch: Napels geel

1.0 INTRODUCTION

Lead antimonate yellow is an artificial pigment which has been manufactured in various periods and as early as the XVIIIth Dynasty (sixteenth–fourteenth century B.C.) at Thebes. From about the fifteenth century B.C. on, it was the only yellow colorant and opacifier in general use in ancient Egyptian and Mesopotamian glass and glazes; then it was gradually replaced by lead stannate yellows, a change which appears to have been completed by the fourth century A.D. As a pigment, lead antimonate yellow has been lost and rediscovered a number of times throughout history. From studies reported by Kühn (1973) it appears lead antimonate yellow enjoyed its highest popularity in European art between approximately 1750 and 1850 after which it was gradually replaced by lead chromate and cadmium sulfide (see 2.32). The *Colour Index* (1971) lists lead antimonate yellow (Naples yellow) as CI Pigment Yellow 41, No. 77588. The theoretical formula is $Pb_2Sb_2O_7$. The *Colour Index* also lists a modification, No. 77589, that might be made by fusion with zinc oxide or bismuth oxide, as discussed by Josef Bersch (*The Manufacture of Mineral and Lake Pigments* [London, 1901], 149, 151).

1.1 Brief Definition of Pigment

The earliest recorded recipes for the pigment are those of Cipriano Piccolpasso (1934) who, in his *Li tre libri dell'arte del Vasaio* written between 1556 and 1559, gives seven variations in its composition for use in *maiolica*. Vannoccio Biringuccio (1966) also refers to the pigment's use in pottery, enamels, and glass in *Pirotechnia*, written in 1540. On the basis of analyses, lead antimonate yellow seems to appear in European painting only from the second quarter of the seventeenth century when it began to replace lead-tin yellow, a pigment employed until then. It came to be called Naples yellow, a name that first appears in print as *luteolum Napolitanum* in a treatise by Andrea Pozzo published in Rome in 1693–1700 (Merrifield, 1849, 1967). This rather misleading term has survived to the present day to describe a shade of color; production of the genuine lead antimonate was phased out by most manufacturers around the middle of the nineteenth century. Naples yellow is now made

IAN N. M. WAINWRIGHT, Senior conservation scientist, Analytical Research Services, Canadian Conservation Institute, Ottawa.

JOHN M. TAYLOR, Chief, Analytical Research Services, Canadian Conservation Institute, Ottawa.

ROSAMOND D. HARLEY, Senior lecturer, conservation of paintings, Gateshead Technical College, United Kingdom.

from mixtures of other pigments, notably cadmium yellow.

Lead antimonate yellow is synthesized by roasting mixtures of lead and antimony oxides or salts. Bindheimite, a mineral with identical chemistry, occurs in nature but there is no evidence to suggest that it was ever used by artists. The belief that Naples yellow was a volcanic product was possibly based on the occurrence at Naples of a yellow tuff which has been used as a building stone.

1.2 Current Terminology

The name lead antimonate yellow, which is used in the archeological literature and in studies of glass pigments, has much to recommend it over the term Naples yellow, which is, however, firmly entrenched in the art-historical literature of the last three centuries. For over one hundred years, Naples yellow has been used to describe both the artificial lead antimonate pigment and any of a number of other mixtures which have imitated its color. Indeed, unless a composition is specified, references to Naples yellow after about 1850 may well concern substitutes rather than the genuine lead antimonate.

In order to avoid confusion, we will use the term lead antimonate yellow. Kühn (1968) has called two types of related lead stannate yellows "lead-tin yellow I" and "lead-tin yellow II"; the names lead-tin oxide or cubic lead-tin oxide are also encountered. We have adopted Kühn's terminology. In related mineralogical and other studies, the expressions bindheimite, synthetic bindheimite, and lead pyroantimonate (as well as orthi-, di-, and meta-antimonate) are found, all referring to the same pigment.

2.0 HISTORY

2.1 Archaic, Obsolete, and Misrepresentative Names

Giallolino, or variants of this word (*giallorino, zalulino, giallulinum, gialdolino*), are obscure Italian names considered synonymous with Naples yellow in the early literature (de Bondaroy, 1766, 1769; de Massoul, 1797). This diminutive of *giallo*, meaning pale yellow, has also been associated with two other lead-based pigments. Merrifield (1849, vol. 1, pp. clvi–clxiii) made a detailed study of the ambiguity and concluded that there were three kinds of *giallolino*:

1. an artificial yellow lead oxide analogue of massicot (see also Kühn, 1968).

2. an artificial yellow whose manufacture is described in Recipes 272 and 273 of the fifteenth-century Bolognese manuscript "Segreti per colori" (Merrifield, 1849, vol. 2, p. 528) and which Kühn (1968) has subsequently identified as lead-tin yellow.

3. a natural yellow mineral pigment found in the neighborhood of volcanoes (Thompson & Hamilton, 1933; Thompson, 1956).

To these we can now add:

4. an artificial lead antimonate yellow whose manufacture is described by Piccolpasso (see 1.1). (Piccolpasso, 1934).

Unless *giallolino* is qualified by the phrase *di Napoli*, it is unwise to translate it as Naples or lead antimonate yellow. Andrea Pozzo ("the Jesuit") is credited with the earliest (1693–1700) reference to Naples yellow which he calls *luteolum Napolitanum* (Merrifield, 1967, vol. 1, p. clix) (*luteolus*, a diminutive of *luteus*: of the color of the plant *lutum*). Other obsolete names which have been associated with the pigment are *Mineralgelbe* (Dingler's *Polytechnisches Journal*, 1828, p. 224) and *jaune minéral* (Ure, 1843, p. 878); however, Field (1835, pp. 78–79) states that this term refers to lead chromate, and Rabaté (1947, p. 262) that it refers to lead oxychloride.

Naples yellow was frequently confused with either Mérimée's yellow in which the lead antimonate is accompanied by lead oxychloride (Rabaté, 1947, p. 262; Reclus, 1956, p. 377) or with Turner's yellow (Cassel yellow, patent yellow, Paris yellow, *jaune de Vérone*, *jaune Véronèse*) which are exclusively lead oxychloride. The terms brilliant yellow, *jaune brillant*, and Nankeen yellow are also used as synonyms for Naples yellow. Brilliant yellow usually refers to an imitation Naples yellow composed chiefly of cadmium sulfide (see chapter on Cadmium Yellows, Oranges and Reds).

2.2 History of Use

2.21 *Early History* Lead antimonate yellow, like Egyptian blue, is one of the oldest synthetically produced pigments known. Its use as a colorant in glass dates back at least to the XVIIIth Egyptian dynasty. Turner and Rooksby (1959, p. VIII/19; VIII/22) found it in a glass fragment decorated with white, blue, and yellow threads from the site of the palace of Amenhotep II

(1450–1422 B.C.) at Thebes. Farnsworth and Ritchie (1938, pp. 165–166) had earlier discovered that of 73 specimens, mostly from the palace of Amenhotep III at Thebes but also including some of c. 1370 B.C. from Tell el-Amarna, 67 contained lead, 62 tin, and 62 antimony, but they concluded that an explanation of their yellow color lay elsewhere. Sayre (1965, pp. 147–148) has discussed the source of this error, concluding that the tin was introduced accidentally in association with copper (see also Turner and Rooksby, 1959, p. VIII/25).

Lead antimonate yellow was found in opaque yellow glass (Brill, et al., 1973, p. 81) and thin rods (Turner and Rooksby, 1959, pp. VIII/19–22) of yellow opaque glass from Tell el-Amarna in Egypt. Lucas (1962, p. 190) found a lead and antimony compound in a XIXth dynasty yellow glass and cites references to the determination of these elements in the research of others. Contemporaneous artifacts from Mesopotamia indicate that the pigment was also in use there at Tell al-Rimah and Nuzi (Brill, 1968, p. 59; Brill, et al., 1973, p. 79) as well as at Nippur, Hasanlu, and Nimrud in later periods (Brill, 1968, p. 59; Charleston, 1960).

Lead antimonate yellow was also used as a glaze in Mesopotamia, Assyria, and Babylonia. Hedges (1976, p. 120) found it in 1300–1000 B.C. glazes from Nippur (late Kassite, Post Kassite period), and Hedges and Moorey (1975, p. 26, 31) have confirmed its presence in badly weathered glazed bricks from Nineveh where it was used together with calcium antimonate white. Partington (1975, pp. 282–284) cites references to the discovery of the pigment in Assyrian glazed bricks from the period 900–600 B.C. at Borsippa and in the glaze of the tiles of the processional street in Babylon. Contenau found cakes of lead antimonate pigment in the palace of Sargon II at Khorsabad (722–705 B.C.), where it may have been used as the yellow glaze of the doorway arch (Partington, 1975, p. 292).

Antimony has been associated with a glass or glasslike substance, *anzaḫḫu*. This word occurs in prescriptions for making colored glasses found in cuneiform texts from the library of King Assurbanipal (668–627 B.C.) at Nineveh. Oppenheim has suggested that *anzaḫḫu* was a raw material obtained by the glassmakers from specialized craftsmen or through trade; and Brill writes that this theory is consistent with his suspicion that lead antimonate yellow was a commodity traded over considerable distances (Oppenheim, et al., 1970, pp. 36, 78, 84, 116–119).

In the earlier centuries of the first millenium A.D., lead antimonate yellow and its counterpart calcium antimonate white were gradually replaced by lead-tin oxide and tin oxide respectively. Lead antimonate has been found in glass bangles from this era in Britain, two of which, one from Traprain Law and the other from Dod Law, were very likely made by remelting scraps of Roman glass (Newton, 1971; Stevenson, 1954, p. 216). An early Celtic armlet, British Museum Reg. No. 1946 4-2,2 (Brailsford, 1975, p. 81) and the glass of surface finds at Caesarea (Bimson and Werner, 1969, p. 264) provide further evidence of the wide use of the pigment in late Hellenistic and Roman times.

It is apparent from these studies that lead antimonate yellow was the only yellow pigment in general use during the first seventeen centuries of glass making (Brill, 1968, p. 58; Turner and Rooksby, 1963; Rooksby, 1962). The change to the tin-based yellow and white pigments is discussed further in section 2.3; however, it should not be concluded too readily that lead antimonate yellow disappears entirely until after the Renaissance. Turner and Rooksby (1959, pp. VIII/22–25) have found lead antimonate as a solid solution with lead-tin oxide ($Pb_2Sb_2O_7 - PbSnO_3$) in an eleventh-century glass mosaic tessera from a Byzantine church at Novgorod and in the fragment of the neck of a fourteenth-century enameled bottle from Syria (Victoria and Albert Museum, ref. no. C.825-1935). Future studies may reveal that the pigment has been used continuously at various times and in various places from distant antiquity to the present.

2.22 *Fourteenth to Twentieth Centuries*

We know from early literary sources that lead antimonate yellow was used as a yellow enamel pigment in Italian tin-glazed earthenware (*maiolica*) from the beginning of the sixteenth century onward. Pending further chemical analyses of the actual pigments, the year 1500 can be considered, for convenience, to be the date at which the pigment regained popularity, at least in European ceramics.

Giambattista della Porta of Naples (c. 1535–1615) writes of *jalloline* and *jaletto* in the English edition of his *Magia naturalis* (Porta, 1957, p. 185) and his *giallolino* was considered by Wiegleb to be lead antimonate yellow

(Partington, 1961, vol. 2, p. 20). A more specific reference to the use of antimony in making yellow colors appears in the *Pirotechnia* of Vannoccio Biringuccio, first published in 1540, where its use in painting earthenware vases and tinting enamels and glasses is cited (Biringuccio, 1966, p. 92).

As mentioned, the earliest written recipes for lead antimonate yellow are to be found in the sixteenth-century treatise by Piccolpasso (Piccolpasso, 1934; Partington, 1961, vol. 2, pp. 77–80). Previously it was noted (Dingler's *Polytechnisches Journal*, 1828; Mérimée, 1839; Rose, 1916, p. 307; de Wild, 1929, p. 56; Harley, 1982, p. 99; Fleming, 1976, p. 30) that the earliest recipes appeared in an article, "Istoria delle pitture in majolica fatte in Pesaro e ne luoghi," written in 1758 by Giambattista Passeri, but we now know that Passeri's seven recipes for *giallolino* are based either on Piccolpasso's recipes for *zalulino* or on a source very close to Piccolpasso. In table 1, the recipes for yellow and light yellow glazes of the two authors are compared; it can be seen how close the correlation is, not only in the identical number of recipes given, but also in the ratios of the principal ingredients.

Piccolpasso's treatise, which borrows to some extent from the work of Biringuccio, describes the recipes used in *maiolica* glazes in different locales — namely Urbino, neighboring Castel Durante, Castello, and the March (which may mean the March of Ancona). The recipes call for various proportions of lead, antimony, lees, and salt (see table 2, note b), and their methods of preparation and application are discussed (see also Rackham, 1952; Rackham, 1955).

Maiolica was produced primarily in what are now the Italian regions of Tuscany, Umbria, and Marche. Urbino and Castel Durante were important centers of production from about 1500, and the painters Nicola Pellipario, Giovanni Maria and Francesco Xanto Avelli di Rovigio were prominent (*Encyclopaedia*

Table 1. **A COMPARISON OF THE COMPOSITIONS OF LIGHT YELLOW (*ZALULINO, GIALLOLINO*) AND YELLOW (*ZALLO, GIALLO*) as described by Cipriano Piccolpasso in the *Arte del Vasaio*, written 1556–1559 (Piccolpasso, 1934, pp. 33, 37, 38), and by the Abbé Giambattista Passeri in *Istoria delle Pitture in Majolica fatte in Pesaro*, written in 1758 (Passeri, 1758, pp. 103–104). All weights are in lbs.**[a]

Pb	*Sb*	*Lees*	*Salt*		*Pb*	*Sb*	*Iron Rust*	
Piccolpasso's *zalulino*					Piccolpasso's *zallo*			
$1\frac{1}{2}$	1	$\frac{1}{16}$	$\frac{1}{16}$	b	$1\frac{1}{2}$	1	$\frac{1}{2}$	b
3	2	$\frac{1}{16}$	$\frac{1}{32}$	b	5	3	2	b
					2	2	$1\frac{1}{2}$	b
6	4	$\frac{1}{2}$	—	c	6	5	$3\frac{1}{2}$	c
6	4	1	0	d	7	5	3	d
3	2	1	$\frac{1}{2}$	d	2	$1\frac{8}{16}$	1	d
$1\frac{1}{2}$	1	$\frac{1}{16}$	$\frac{1}{16}$	e	5	8	3	e
$3\frac{1}{2}$	2	$\frac{1}{16}$	$\frac{0}{16}$	e	3	2	1	e
Passeri's *giallolino*					Passeri's *giallo*			
$1\frac{1}{2}$	1	$\frac{1}{16}$	$\frac{1}{16}$	*	5	3	2	*
6	4	1	—	*	7	5	3	*
					2	1	1	
3	2	1	$\frac{6}{16}$	**	2	$\frac{8}{16}$	1	
5	4	$\frac{6}{16}$	—		6	5	$3\frac{1}{2}$	*
4	2	$\frac{6}{16}$	—		6	$4\frac{1}{2}$	3	**
$1\frac{1}{2}$	1	1	1	**	5	8	3	*
$3\frac{1}{2}$	2	1	—	**	3	2	1	*

a. Weights given in 16th's are given in ounces in the original. An entry of O means that O was specified in the original
b. Recipes used in Castel Durante. c. "Colours of the March," note 111: presumably the March of Ancona. d. "Marzacotto in the Urbino Manner." e. "Castello Colours." Recipes from Passeri marked * have an exact counterpart in Piccolpasso. Recipes marked ** have a counterpart with the same lead and antimony amounts but differences in the amounts of lees, salt, and iron rust.

Britannica, 1974). It can be inferred that the major use of lead antimonate yellow for *maiolica* occurred during and after the period of Piccolpasso's work. To test this hypothesis we obtained an early ceramic fragment with yellow pigment (c. 363/68-1920) from J. V. G. Mallet, Keeper of the Department of Ceramics, Victoria and Albert Museum. The piece was probably made in Faenza in about 1510; the pigment was identified as lead antimonate yellow by x-ray diffraction analysis.

Antimony in yellow glazes, often in conjunction with tin, also appears in English yellow-glazed earthenware. Miller (1974, pp. 5–6; pp. 119–120) cites references to several such recipes, and Laurie (1914, p. 12) writes of the pigment's use for glazes.

Difficulty in interpretation of the term *giallolino* as used by later Italian writers such as Lomazzo (1584), Armenini (1586), Ferrante (1599), and Bisagno (1642) makes it unclear if lead antimonate is being referred to. However, as several authors mention more than one type of *giallolino* — Lomazzo specifically mentioning three varieties — antimonate may conceivably be among the pigments discussed and thus employed in both fresco and oil painting in Italy during the last quarter of the sixteenth century.

The scarcity of references to identification of lead antimonate yellow is not altogether surprising since in northern Europe in the seventeenth century Naples yellow was reportedly rare. This rarity may have led to recommendations for the pigment's use in watercolor painting at that time rather than in oils, for which it was considered more suitable during the following century. To paint yellow drapery in miniatures, a French author recommended a mixture of opaque Naples yellow and a transparent yellow lake as an alternative to a mixture of massicot and gamboge (C. B. [Claude Boutet], 1684). The English author Elsum (1704) stated that Naples yellow was a middle yellow, more suited for miniatures than for oil painting. Laurie (1910, p. 127) wrote that "The artificial pigment is not to be recommended for use in fresco."

In point of fact, there has been considerable confusion regarding the occurrences of lead antimonate and other yellow pigments in fourteenth- to nineteenth-century European paintings (Jedrzejewska, 1959). Until the rediscovery of lead stannate yellows by Jacobi (1941), and their subsequent investigation by Kühn (1968), it was not recognized that the stannate had been used in paintings from 1300 to 1750, most frequently in those from 1400 to 1700 (Kühn, 1968, p. 9; Kühn, 1973, p. 199). Consequently, there appears to have been a tendency to assume that in paintings from this period a yellow pigment was lead antimonate (Naples yellow) and also a tendency to accept a positive microchemical test for lead, without testing for antimony or tin, as confirmation. This has led, understandably, to a number of questionable and contradictory reports of the early use of lead antimonate yellow. Raehlmann's (1914) determination of Naples yellow in Renaissance paintings has been disputed on the grounds that his analysis of lead was not definitive and that he did not test for antimony (de Wild, 1929; Kühn, 1968). Lemaire (1941) and Doerner (1962, p. 61) have also suggested that lead antimonate yellow was used by "the old masters," although Field (1835) had earlier rejected this idea. The pigment is said to have been used by Rubens (Doerner, 1962, p. 62; Eastlake, 1960, vol. 1, p. 439) but it is lead-tin yellow I that has been identified on seven of his paintings by x-ray diffraction and on a further six by emission spectroscopy (Kühn, 1968, pp. 22–23).

The earliest occurrence in Western painting was detected by the authors in the course of a technical examination of Matthias Stomer's (1630–1632) *Arrest of Christ* (National Gallery of Canada, 4094; see fig. 1). The pigment has also been found in the works of the English portrait painter Thomas Bardwell (1704–1767) (Talley and Groen, 1975). Lead antimonate yellow was found among a collection of nineteenth-century pigments in an old pharmacy near Darmstadt where it was labeled "Neapelgelb Neopolitanisches Gelb Verbindung des Spiessglanz, Bleies" (Richter & Härlin, 1974a). It has also been discovered in a pigment collection of the Swiss painter Arnold Böcklin (1827–1901) labeled "Neapel Gelb, hellst"; this sample also contained zinc white. Another box of pigment labeled "Neapel Gelb, dunkel" contained chrome yellow, lead white, and barium sulfate (Richter & Härlin, 1974b). During the examination of J. M. W. Turner's (1775–1851) work boxes, Naples yellow was identified by Hanson (1954, p. 170) by means of emission spectrography.

In view of the many philological, art-historical, and chemical difficulties surrounding occurrences of lead antimonate yellow, it can be

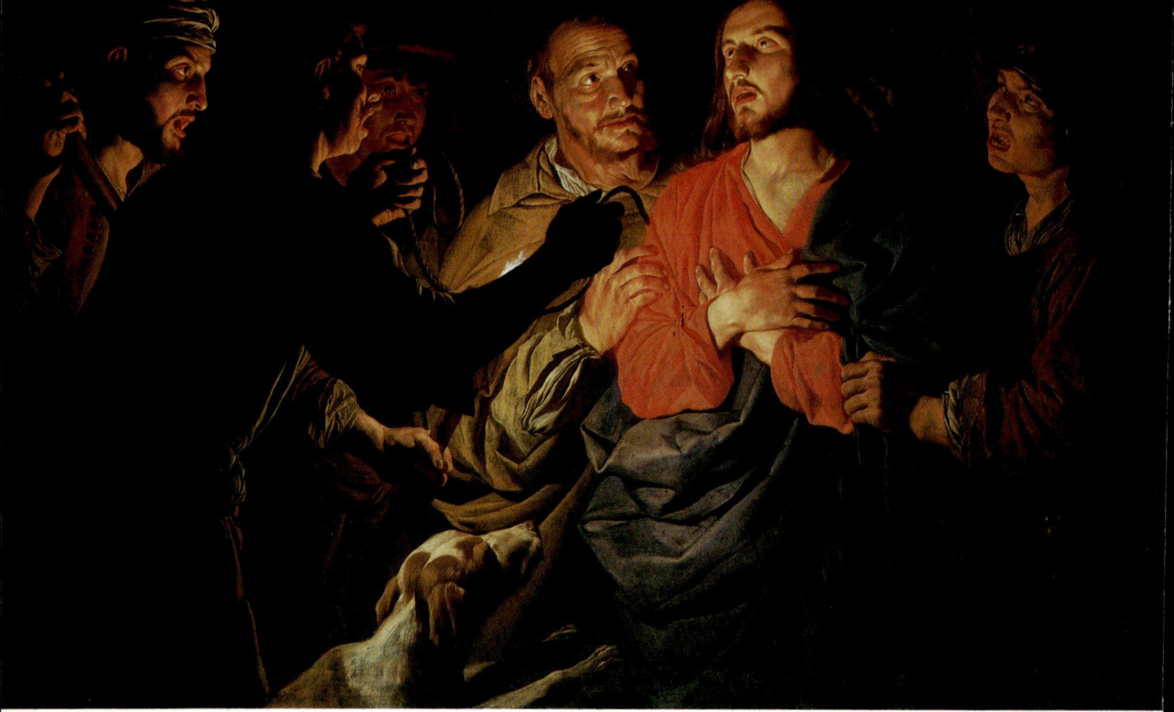

Fig. 1. Matthias Stomer, *Arrest of Christ* (1630–1632), National Gallery of Canada, Ottawa. 4094.

concluded that only careful analyses of samples from paintings of known provenance will reveal the chronologies of this pigment and the lead-tin yellows. In fig. 2 data are presented regarding the distribution of occurrences obtained by studies conducted at the Doerner-Institute (F. Preusser, private communication). Kühn earlier noted that these results are skewed for a number of reasons, not least of which is, obviously, the variation in the number of specimens representing each quarter of a century (Kühn, 1973, p. 205). If such difficulties can be ignored, one can be convinced from this evidence that the use of lead antimonate yellow grew from 1600 to 1800 and then fell off with the invention of safer, cheaper yellows.

2.3 Terminal Date

2.31 *Terminal Dates in the Ancient Period.* The earliest published occurrences of lead antimonate yellow indicate that its use as a colorant and opacifier was established in Egyptian and Mesopotamian artifacts by c. 1500 B.C. Research at the Corning Museum of Glass, Brookhaven National Laboratory, the British Museum, and The General Electric Company Limited has provided a substantial body of knowledge of the pigment's use in glass and as a glaze in the second millenium B.C. (see 2.21).

A definition of the period in which lead antimonate yellow was replaced by lead-tin oxide is of considerable interest. From the literature, it appears that the change occurred in various parts of Europe from c. 50 to 350 A.D. Lead antimonate from this period has been found in the British Isles, Turkey, and Afghanistan (see 2.1 and *Notable Occurrences, Antiquity*).

It was concluded by Turner and Rooksby (1963) and Brill (1968) that the change from lead antimonate yellow ($Pb_2Sb_2O_7$) to lead-tin oxide ($PbSnO_3$) and the coincident change from calcium antimonate white ($Ca_2Sb_2O_7$) to tin oxide (SnO_2) had been accomplished by the middle of the fourth century A.D. Some studies have indicated that lead-tin oxide yellow was in use prior to this time: it has been found in early

Celtic armlets from 50–150 A.D., British Museum Reg. No. 38 7-14 3a and b (Brailsford, 1975, p. 81), a 100–300 A.D. mosaic tessera from Antioch-on-Orontes, a 200–300 A.D. tessera from the mosaic of the *Drinking Contest* in Seleucia Pieria, the port of Antioch, and a first-century glass bead analyzed by Werner and Bimson and reported by Brill (1968, pp. 58–59).

Particularly intriguing are two discoveries of the simultaneous occurrence of calcium antimonate with lead stannate yellow: a first-century A.D. gaming piece from Welwyn Garden City (Bimson & Werner, 1969, pp. 263–264) and the second to fourth centuries A.D. *opus sectile* panels from Kenchreai, the eastern port of Corinth, studied by Brill (Ibrahim, et al., 1976, p. 239). As further analyses are reported, it may be possible to determine the terminal dates for various regions with greater accuracy and to discover whether the antimony and tin-based compounds were produced simultaneously by artisans or whether their admixture is a result of glass having been traded and remelted to make new artifacts (Newton, 1971). Brill has suggested (Ibrahim, et al., 1976, p. 246) that the use of antimony compounds, particularly the lead antimonate yellow, may have been discontinued because of their greater toxicity or, if antimony had become scarce, because of their expense. It is curious, therefore, that lead antimonate yellow reappears as a pigment 1,200 years later only to be again replaced within a relatively short period of time.

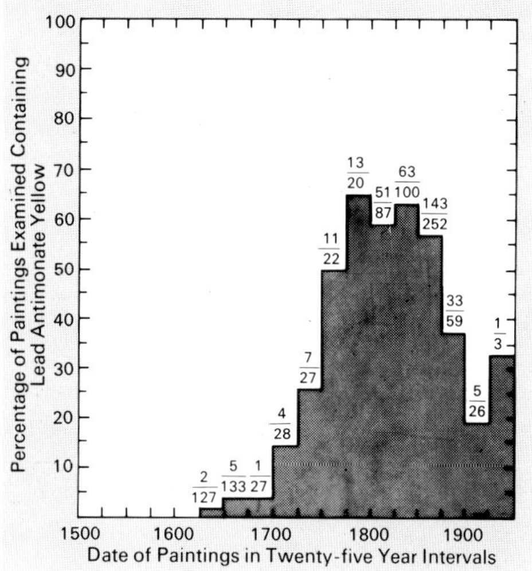

Fig. 2. Frequency distribution of lead antimonate yellow in European easel paintings as determined by emission spectroscopy and light microscopy. Fractions at head of each bar in the graph represent the number of paintings found to contain lead antimonate pigment divided by the total number of paintings of that period examined. Data from Laboratorium für Konservierung und Naturwissenschaftliche Untersuchung von Kunstwerken, Doerner-Institut, Munich, courtesy Dr. Frank Preusser.

2.32 ***Terminal Dates in the Period 1500 to the Present.*** We now know from the writings of Biringuccio and Piccolpasso (see 2.22) that lead antimonate yellow was synthesized for use in ceramic glazes, enamels, and glass at least as early as 1540 when the first edition of *Pirotechnia* was published. As mentioned, we do have one example of lead antimonate yellow in maiolica dating from c. 1510. Undoubtedly, the pigment was in use even earlier but we have no definitive literary source to link *giallolino* with antimony. For example, Leonardo da Vinci (1452–1519) refers to *giallo* in Part II, Folio 176 of the Codex Vaticanus Urbinas 1270, *Treatise on Painting*, and to *gialorino* in Part III, Folio 161; both of these words McMahon (Leonardo, 1956) translates simply as yellow. The latter was translated as Naples yellow by Rigaud in the late nineteenth century (Leonardo, 1877). We do, however, know that Leonardo's recipe for yellow in the Codex Atlanticus was realgar or orpiment dissolved in *aqua fortis* (Pedretri, 1964, p. 56, note 62).

We owe our knowledge of the terminus a quo of lead antimonate yellow in European painting to research done at the Doerner-Institut of the Bayerische Staatsgemäldesammlungen, Munich. In his analysis of samples from 262 paintings executed between c. 1818 and 1885, Kühn (1969) found that 155 (59%) contained the pigment although not always in the yellow passages. A subsequent report (Kühn & Zocher, 1970) indicated the presence of lead antimonate yellow in the yellow paint of 117 of 386 (30%) authenticated paintings from the period 1600–1900: 6 of 153 (4%) from the seventeenth century, 19 of 35 (54%) from the eighteenth century, and 92 of 198 (46%) from the nineteenth century. The most recent compiled figures for that laboratory (Dr. Frank Preusser, private communication) have been plotted as a histogram (frequency distribution) in fig. 2, from which it can be seen that the observed terminus a quo for lead antimonate yellow in European painting is c. 1630.

It is instructive to compare the histograms of lead-tin yellow and lead antimonate yellow (Kühn, 1973) as they occur in yellow paint (and in green paint as mixtures with azurite, Prussian blue, or cobalt blue). Lead-tin yellow was found in about 80% of yellow paints of the fifteenth and sixteenth centuries but apparently suffered a loss of popularity from c. 1626 onward until its observed terminus ad quem of 1750. An opposite trend is observed for lead antimonate yellow during the same period. On the basis of these studies, it appears that lead antimonate yellow enjoyed its greatest popularity between roughly 1750 and 1850 after which it was gradually replaced by lead chromate and cadmium sulfide yellows.

Since lead antimonate yellow is made with relative ease from readily available materials, a terminus ad quem cannot, strictly speaking, be assigned, even though its manufacture and use have been limited in this century. G. Siegle and Co. (now BASF Farben & Fasern AG, Bereich Siegle) is the only manufacturer of Naples yellow listed in the *Colour Index* (1971). The supplier states that manufacture has now ceased (private communication, 14 June 1975). The Belgian firm J. Blockx Fils, S.P.R.L., also used lead antimonate yellow in oil colors until 1974 when their suppliers stopped manufacturing it (private communication, 14 February 1975). In the United States of America, the Shepherd Chemical Co. and Hercules, Inc. (now Ciba-Geigy) have both supplied lead antimonate yellows at least until very recently (Preuss, 1974, p. 91). *Raw Materials Index* (1975) does not list a supplier for this pigment.

Lead antimonate yellow is still available from at least three artists' colormen. This has been confirmed by x-ray diffraction analysis of the following samples: *jaune de Naples* oil color in a tube labeled Fragonard from A. Drouant, Paris, which is a mixture of $Pb(SbO_3)_2$ and $Pb_2Sb_2O_7$ as well as barium sulfate; two samples of Naples yellow dry pigment in light and dark varieties from L. Cornelissen and Son, London; and *giallo di Napoli*, a dry pigment sample from Maimeri and Company, Milan.

The respective terminal dates of the lead antimonate and lead-tin yellows could be of considerable value in authentication studies. As Jedrzejewska (1959) has commented, "Even the most ingenious recent fakers were unlikely to have known about the existence of the stannate." In a somewhat different case, lead antimonate yellow was found in Etruscan-style wall paintings on terracotta which were proved to be modern forgeries on the basis of thermoluminescent and stylistic analyses (Fleming, et al., 1971).

3.0 GENERAL PIGMENT PROPERTIES

3.1 Color and Spectral Reflectance

3.11 ***Visible.*** The standard color designations (Kelly & Judd, 1976) for Naples yellow are brilliant yellow, light yellow, moderate yellow, pale yellow, and grayish yellow. The dry pigments in our possession were generally found to have, approximately, Munsell hue 5Y and Munsell value/chroma 9/8, 8.5/8, 8.5/10, and 8.5/12 (Munsell Color Company, 1973), all of which lie within the designation brilliant yellow. The oil paint tends to a Munsell value of 2.5Y, and one contemporary version of imitation lead antimonate yellow oil color was 10YR 8/6, an orange yellow.

The spectral reflectance of lead antimonate yellow, lead-tin yellows I and II, and cadmium yellow in linseed oil is shown in fig. 3. The lead antimonate yellow sample was supplied by H. C. von Imhoff, Musée d'art et d'histoire de Fribourg, Switzerland. The sample No. 551101 dates from c. 1920, was distributed by Standard Lack, Zürich, and contains zinc white. The lead-tin yellow samples were prepared by B. Mühlethaler and W. Pulver, Chemische-Physikalisches Laboratorium, Schweizerisches Landesmuseum, as part of a pigment research project (see von Imhoff, 1975; codes 550 101 and 550 102). Their synthesis was based on recipes 272 and 273 in the Bolognese manuscript (Merrifield, 1967, pp. 528–529). The cadmium yellow is contemporary, from Wilhelm Düll, Munich. A UV-visible spectrophotometer with integrating sphere was used with bandwidth 2 nm constant, specular reflectance component excluded, and reference barium sulfate.

As with that of lead-tin yellows (Kühn, 1968, p. 11), the color of lead antimonate yellow depends upon the temperature at which it is manufactured, a fact that was recognized by Brunner and Ehrmann (Brunner, 1837, pp. 381–382). They noted that three different temperatures produced three different shades, that lemon yellow or even sulfur yellow was produced with the strongest heat, that it was difficult

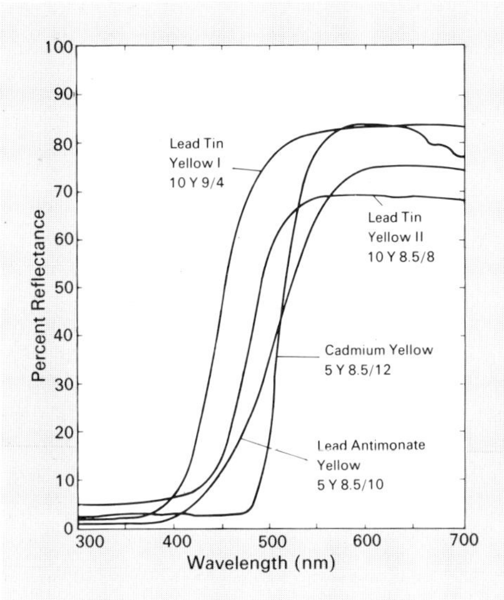

Fig. 3. Spectral reflectance curves of paints prepared in linseed oil with lead antimonate yellow, lead-tin yellow I and II, and cadmium yellow.

to reproduce the same shade, and that the color tended to orange if the temperature remained below the melting point of the salt which was one of its ingredients (*id est* 801°C). We now know that true lead antimonate yellow is not formed at moderate temperatures (see 4.1).

Doerner (1962, p. 61) and Mayer (1970, p. 94) note that the manufacturers produce a "light" and a "dark" Naples yellow, Mayer stating that "about six shades, from a greenish yellow to a comparatively pinkish orange yellow" are produced, while Wehlte (1975, p. 83) describes the pigment simply as a "slightly pale, medium yellow, sometimes with a reddish or flesh-coloured tinge." A very early reference to *giallolino* (Ferrante, 1599, p. 121) says it imitates the yellow color of broom (*Genista*). Lead antimonate yellow has excellent hiding power and was therefore less popular for watercolors (Wehlte, 1975, p. 84).

3.12 ***Infrared Reflectance.*** Heiber (1968, p. 148), using an infrared image converter, found that Naples yellow exhibited a higher reflectance of the near infrared (800–1200 nm) than did twenty-four other pigments examined and was only less reflective than lithopone. When applied as a hiding layer over a tempera underdrawing, lead antimonate yellow was found to be more transparent than lead white but less so than cadmium yellow, ranking tenth in transparency in the infrared among sixteen pigments (Heiber, 1968, p. 147).

3.2 Hiding Power and Tinting Strength

The refractive index of lead antimonate yellow varies from 2.01 to 2.28 (Gettens & Stout, 1966), a value slightly higher than that of lead white, but not as high as that of titanium white (2.5 to 2.9). Its hiding power, therefore, may be considered very good. Wehlte (1975, p. 84) considers its tinting strength also to be fairly good. Because of its opacity, the pigment has perhaps been less popular in watercolor than in oil. Wehlte further comments that "on brown underpainting Naples yellow, carefully graded to a thin veil, assumes a bluish-gray shade due to the 'turbid-medium effect.' The French painter Prud'hon knew how to use this effect ingeniously in the shadows of his flesh tones."

3.3 Permanence

Lead antimonate yellow is alkali-fast and is insoluble in organic solvents. Because of the presence of lead, it is darkened by atmospheric hydrogen sulfide and is therefore more useful in oil than in watercolor. The pigment is light fast (*Colour Index*, 1971, vol. 3, pp. 3,278–3,279; Doerner, 1962, p. 61; Wehlte, 1975, p. 84). Naples yellow was included in the 1886 list of durable pigments established by the German Society for the Advancement of Rational Painting (Levison, 1976, p. 17). The particular sample judged in the Russell and Abney report to be unstable after exposure to an equivalent of 480 years in a gallery illuminated by daylight was actually a mixture of zinc white and cadmium yellow (Brommelle, 1964).

3.4 Compatibility

According to Wehlte (1975, p. 84), lead antimonate yellow is compatible with all pigments except barium yellow in fresco and with all vehicles except those containing acid preservatives. The belief that lead antimonate yellow is affected by contact with iron was voiced by Dossie (1764) who wrote of the danger of discoloration if the pigment were used with a metal palette knife. Field (1835) also warned against mixture with any pigment containing iron such as ochre or Prussian blue. Remington and Francis (1954, p. 57) give this belief as the reason for the discontinuation of genuine lead antimonate yellow. Church (1901) remarked that mixtures of lead antimonate yellow and organic pigments deteriorated but neither he nor Doerner (1962, p. 61) could find any evidence that mixture of the compound with iron pigments caused discoloration. Stockler (1838, p. 7) also felt that it could be mixed with all earth and metallic pigments without fear of ill effects. Wehlte (1975, p. 81), however, notes that "experienced manufacturers know that Naples yellow should always be run on porphyry rollers and never on steel rollers." It is possible that residual chlorides or other impurities were responsible for the pigment changing to gray or grayish green through the formation of iron chloride or other compounds.

3.5 Chemical and Physical Properties

Lead antimonate yellow is insoluble in water, 5% Na_2CO_3 and organic solvents; partially soluble in 3N HCl (leaving a lead chloride precipitate) and 4N NaOH; and soluble in concentrated HNO_3 (leaving an antimony oxide residue) and H_2SO_4 (*Colour Index*, 1971, vol. 3, pp. 3,278–3,279; Doerner, 1962, p. 61; Palache, et al., 1951, vol. 2, p. 1,018; Plesters, 1956, p. 147).

The specific gravity of the pigment has been given as 6.6 (Wehlte, 1975, p. 84), a relatively heavy powder; with natural bindheimite values were reported at 5.16 and 5.28 (Mason & Vitaliano, 1953, p. 104). A synthetic analogue of bindheimite, described in detail by Friedrich and Marx (1962) and closely related to the pigment, was found to have a density of 7.32 g/cm^3, a melting point greater than 1,300°C, and a Vickers hardness (diamond-pyramid hardness) of 700 kg/mm^2 or 5.5–6 on the Mohs scale. De Wild (1929, p. 57) remarked upon the "great hardness" of the pigment.

3.6 Oil Absorption and Grinding Qualities

Doerner (1962, p. 61) states that lead antimonate yellow requires very little oil, about 15%, while Wehlte (1975, p. 84) regards the pigment's oil absorption as normal, requiring 25–35% poppy oil. Lead antimonate yellow has a strong drying effect on oil vehicles. Ruhrmann (1934) has listed Naples yellow among the pigments which do not break up an emulsion of water particles dispersed in oil and can therefore form a "van Eyck type" film which dries with a glossy finish. Because of its great hardness, the pigment is gritty and difficult to grind. Doerner (1962, p. 62) claims that it "needs very little grinding with the medium" and "that if too finely ground, it becomes heavier and more earthy in texture."

3.7 Toxicity

Sax (1975) states that lead antimonate has a high toxic hazard rating. This is particularly true in cases of major or prolonged ingestion or inhalation. It is only moderately toxic in contact with the skin.

4.0 COMPOSITION

4.1 Chemical Composition

Following Rose's investigations (1916, p. 308), the formula most often given for lead antimonate yellow in the pigment literature is $Pb_3(SbO_4)_2$ in which the mass ratio of antimony to lead is 0.39. Rose pointed out that Sb/Pb ratios found in the early recipes ranged from 0.18 to 1.33 but that some of the antimony could be lost through sublimation or dissolution during manufacture. He therefore selected a formula with a low Sb/Pb ratio.

The second most frequently cited formula is $Pb(SbO_3)_2$, having a high Sb/Pb ratio of 1.18, quite the opposite from that found in actual analyses (König, 1956, p. 69). Furthermore, the infrared spectrum of this compound given by Nyquist and Kagel (1971, pp. 200–201) bears no resemblance to lead antimonate yellow spectra. Other formulae which may be encountered are $Pb_8(SbO_4)_2$ (Rabaté, 1947, p. 262), $Pb(SbO_4)_2$ (Wehlte, 1975, p. 84; Plesters, 1956, p. 147) and $PbSb_2O_4$ (Remington and Francis, 1954, p. 57).

A pigment prepared according to a method of A. H. Church from 1 part potassium antimony tartrate, 2 parts lead nitrate, and 5 parts sodium chloride heated at 600°C for $2\frac{1}{2}$ hours has been analyzed by x-ray secondary-emission spectrometry, atomic absorption spectrophotometry, neutron activation analysis, and x-ray diffraction. It was found to contain 61.4% lead and 22.2% antimony, that is a Sb/Pb ratio of 0.36 (O'Shea, 1981).

It is now known from x-ray diffraction studies (see 5.31) that lead antimonate yellow is the isostructural anhydrous analogue of the mineral bindheimite with a crystal structure of the pyrochlore type. A correct theoretical formulation is therefore $Pb_2Sb_2O_7$, in which the Sb/Pb ratio is 0.59. In their analysis of a synthetic bindheimite produced during the cooling down of a lead refining furnace, Friedrich and Marx (1962, p. 73) found 62.8% PbO and 34.5% Sb_2O_5 by weight or a Sb/Pb ratio of 0.45. The deficiency of antimony and/or excess of lead is reflected in their proposed formula which, we suggest, is also appropriate for lead antimonate yellow:

$$Pb_y^{II} Sb_{2-x}^{V} O_7 \text{ where } 0 \leqslant x \leqslant 1 \text{ and } 2 \leqslant y \leqslant 3.$$

A wide range of composition is found in analyses of natural bindheimites, most of which contain significant amounts of Fe_2O_3 (limonite), As_2O_3, Bi_2O_3, Ag_2O_3, SiO_2 (quartz) or $PbCO_3$ (cerussite) (Shannon, 1920, p. 90; Montoya, 1969, p. 1727; Mason and Vitaliano, 1953, p. 104). The analyses of two bindheimite specimens from northern Italy have been reported by Natta and Baccaredda (1933, p. 278). The minerals from Oneta (Gormo), Valle Seriano, and Camerata Cornello, Val Brembana contained, respectively, 2.20% and 3.75% CuO as well as significant amounts of CaO, MnO, Fe_2O_3, and Al_2O_3. While there is no evidence that natural bindheimite was ever used as a pigment, a comparison between minor constituents in paint and in the mineral could be helpful in establishing this point.

Brunner (1837, p. 380; Dingler, 1828) found iron oxide, alum, silica, and calcium carbonate, usually in small amounts, in Naples yellow samples that he studied; he regarded these substances as accidental inclusions. We have found Al, Si, K, Cl, Ca and Fe by electron beam X-ray microanalysis in the lead antimonate yellow layer of a Matthias Stomer painting from 1630–1632 (see 5.33 and fig. 12). The presence of these elements is consistent with a recipe calling for alum and salt with the possible addition of a silicate filler.

Modern antimony yellow used in ceramics has tin oxide, aluminum oxide, and iron oxide as secondary constituents (Ahmed, 1962, p. 124). Detailed discussion of lead-antimony-oxygen and lead-antimony-tin-oxygen systems can be found in Hennig and Kohlmeyer (1957), Tagai and Shimohira (1963), Shimohira and Tagai (1964), Gerlach, et al., (1968), Urazov and Speranskaya (1956) and Kuznetsov and Koz'min (1958).

It would be expected from a survey of the literature that even if pure oxides of lead and antimony were used to synthesize lead antimonate yellow, more than one phase could be produced depending upon the ratios of the ingredients and the time and temperature of their calcination. In our x-ray diffraction studies of lead antimonate yellow samples, unidentifiable lines were frequently encountered that may result from such mixed oxides. In one sample, a mid-nineteenth-century Winsor & Newton pigment, elongated hexagonal prisms were observed by scanning electron microscopy (fig.10A). An oxide of lead and antimony in the hexagonal system is $PbSb_2O_6$ (Coffeen, 1956); this compound has been synthesized by heating equal amounts of $PbCO_3$ and Sb_2O_3 at 450°C for 50 hours, 750°C for 100 hours, and 1040°C for 72 hours (Magnéli, 1941).

4.2 Sources

4.21 ***Early Sources.*** In their isotope analyses of ancient glasses containing lead antimonate yellow, Brill, et al. (1973) discovered that the lead in XVIIIth dynasty glass-cored vessels from Tell el-Amarna was of local origin, probably from ore deposits in the Eastern Desert of Egypt, along the Red Sea. One such deposit, at Um Gheig, was found to have similar $^{208}Pb/^{206}Pb$ and $^{207}Pb/^{206}Pb$ ratios to those found in the artifacts. Furthermore, their results implied that the lead used for similar contemporaneous vessels from Nuzi and Tell al-Rimah as well as for later artifacts from Nimrud and Hasanlu was "of some recognizable type, probably coming from a highly localized source [and] was in use for many centuries for glass-making and related activities" (*op cit.*, p. 79). Lucas (1962, p. 195) states that neither antimony nor antimony ores occur in Egypt but that the material could have been obtained from Asia Minor, Persia or the Greek islands Mytilene and Chios. Partington (1975, p. 256) names sites in Kurdistan as sources of stibnite (Sb_2S_3), notably Mt. Takht-i-Suleiman in northwestern Iran. There are also important deposits of antimony oxides at Constantine, Algeria (Palache, et al., 1944, p. 273).

In the sixteenth century, there would undoubtedly have been many sources of lead, but on the subject of antimony, Biringuccio (1966, p. 92) is specific. He writes that "a smelted kind in cakes" was brought to Venice from Germany for use in the manufacture of bells. Piccolpasso (1934, p. 33) claims that this variety is the best for use in glazes. Antimony mines near the cities of Massa in Maremma, Sovana in the region of Siena, and an estate called Selvena in the region of Santa Fiore are mentioned by these two early writers. While there may have been other Italian sources, Biringuccio, who was from Siena, and Piccolpasso, who was from Castel Durante to the east (near Urbino), possibly wrote only of mines known locally. According to Palache, et al. (1944, p. 273), stibnite is found in limestone rocks at Pareta, Tuscany, and at Cetine di Cotormiano near Siena as well as in Westphalia, Saxony, Austria, Rumania, and Czechoslovakia. Smith and Gnudi (Biringuccio, 1966, p. 91) note that the word *antimony* in Biringuccio's time always meant the sulfide; the term *regulus of antimony* was later adopted for the metal.

The theories that Naples yellow was a volcanic product from the district of Naples or that its yellow color was derived from sulfur (Pomet, 1694) were frequently iterated in the seventeenth and eighteenth centuries. A summary of such references was supplied by de Bondaroy (1766), who states that he and Pott (1753) had separately formed the opinion, based on experiments they conducted, that Naples yellow was a manufactured pigment. Dossie (1764, pp. 98–99) likewise pointed out that "the different parcels of it vary too much from each [other] in specific qualities to allow the supposition of their being native earths of the same kind . . . found in the neighbourhood of Naples." Nevertheless, the opinion of d'Arclais de Montamy (de Massoul, 1797, p. 137) and others, that the pigment came from the slopes of Vesuvius, remained influential.

Laurie in his 1891 Cantor Lecture (Parry & Coste, 1902, p. 176) suggested that "possibly a search on Mount Vesuvius might result in the rediscovery of the original Naples yellow," and Blockx (1926) wrote that he intended to investigate the matter until he discovered that Naples yellow was by that time imported to Italy from France and Germany. Wehlte (1975, p. 84) states that "reports of former natural deposits in the Gulf of Naples have not been confirmed."

Merrifield (1967, vol. 1, pp. clx–clxii) based her theory that there were two varieties of Naples yellow, one artificial and one a native mineral, upon the following passage from Cennino Cennini written c. 1390:

> There is a yellow colour called giallolino, which is artificial and very compact. It is as heavy as a stone, and difficult to break. This colour is used in fresco, and lasts forever (that is on walls and on tempera pictures). It must be ground like the preceding with water. It is difficult to grind; and before grinding, as it is very difficult to pulverize, it should be broken in a bronze mortar, in the same way as the lapis amatite. When employed in painting, it is a very beautiful yellow; and with this colour and other mixtures which I will describe to you, you may paint beautiful foliage and herbage. And I have been informed that this colour is a real stone, produced in volcanoes; and it is for this reason that I said it is formed artificially, but not in the chemical laboratory.

A comprehensive survey of the minerals of Vesuvius is to be found in *Mineralogia Vesuviana* (Zambonini, 1910). Bindheimite, $Pb_2Sb_2O_6$ (O, OH), the natural mineral analogue of lead antimonate yellow, is not among the 276 minerals discussed there. Nor are any occurrences reported in southern Italy in *Dana's System of Mineralogy* (Palache, et al., 1951, vol. 2, p. 1,019), although ones in two northern locales are mentioned (see 4.1). Stibnite, Sb_2S_3, which would have been an ingredient of synthetic lead antimonate, also does not occur on Vesuvius. However, realgar, As_2S_2, and orpiment, As_2S_3, are found there, as is massicot, PbO, which "may easily be confused with bindheimite or other yellow ocherous lead or antimony minerals" (Palache, et al., 1951, vol. 1, p. 516). Realgar and orpiment were found as pigments on Pompeiian murals but lead antimonate yellow was not (Augusti, 1965, p. 69). Cennini (Thompson, 1933, pp. 28–29) described the use of realgar and orpiment as well as *giallorino* but it is still possible that the last was confused with one of the yellow mineral pigments which may have been mined on Vesuvius and thereby associated with Naples yellow.

A more compelling argument is that *giallolino* was linked by its color with a Neapolitan yellow tuff which was used in the construction during the twelfth century of the Castel dell'Ovo on the Bay of Naples (Rossi-Manaresi, 1976). This stone was available locally in quantity, it underlies the castle itself, and "because of its softness was very easy to quarry and fabricate with simple hand tools" (Lewin & Charola, 1978, p. 703). It consists of some analcime and sanidine imbedded in a matrix of the zeolite mineral chabazite. This material could have been Cennini's *giallorino* and the basis for all subsequent references to a volcanic source for Naples yellow.

A final possible explanation is that synthetic lead antimonate yellow was made primarily in Naples, and the pigment made there may have been of better quality than that found elsewhere (Mérimée, 1839). De Massoul (1797, p. 137) states that "Till the present time, the supply was drawn from Naples: formerly one person furnished the whole." Is it only coincidence that of the five earliest occurrences of lead antimonate yellow from the seventeenth century, four are in paintings by artists who studied or resided in Naples (see 6.0)? Matthias Stomer (1600–after 1650), for example, was in Naples in 1631 and apparently stayed on in Sicily and Southern Italy. Claude Lorrain (1600–1682) worked in Naples during the period c. 1618–1620, while Johann Heinrich Schönfeldt (1609–1682) also painted in the city as a young man. Luca Giordano (1634–1705) came from Naples.

4.22 ***Nineteenth- and Twentieth-Century Sources.*** It may prove difficult to reconstruct the history of lead antimonate yellow manufacturing in the eighteenth and early nineteenth centuries. We have not found any specific reference to the discovery of a particular process which made the pigment more appropriate for oil painting. It is quite probable that the development of improved milling technology was solely responsible for the manufacture of a finer pigment suitable for graphic artists.

Interesting information was furnished by ten European artists' colormen. Four have no records as to when they ceased supplying lead antimonate yellow, therefore making it likely that its replacement occurred in the mid-nineteenth century. It is of some significance that three of the four are British, since the colormaker George Field is reputed to have initiated in the second quarter of the nineteenth century the substitution of a mixture of cadmium yellow, a white pigment, and iron oxide (Hurst, 1922, p. 202). However, C. Roberson and Company supplied lead antimonate yellow until 1885 or possibly somewhat later. Through the courtesy of G. F. Roberson-Park, director of the company, the authors are pleased to have obtained a sample of pigment prepared in the mid-1800s. In addition to Roberson, Church (1901) names Edouard of Paris and Schönfeld of Düsseldorf as suppliers of the genuine pigment in the late nineteenth century. In Italy, Ferrario Belle Arte supplied both lead antimonate yellow and a substitute mixture, between 1932 and 1939. This history of continued sale suggests that there was still a demand for lead antimonate yellow, although it is clear that Jacques Blockx had a poor opinion of the color. Although his firm supplied the pigment until 1974, Blockx wrote that Naples yellow did not merit the importance many artists attached to it (Blockx, 1926).

4.3 Preparation

Lead antimonate yellow is synthesized from the calcination of a lead constituent and an antimony constituent often in the presence of other ingredients which, over the years, have included wine lees (tartar), alum, common salt, ammonium chloride, and potassium nitrate. The lead component has been introduced in the form of metallic lead (Pb), litharge (PbO), minium (Pb_3O_4), lead white [$2PbCO_3 \cdot Pb(OH)_2$], and lead nitrate [$Pb(NO_3)_2$], while metallic antimony (Sb), antimony oxide (Sb_2O_3), potassium antimonate ($KSbO_3$), antimony sulfide (Sb_2S_3) (the mineral stibnite), and tartar emetic have been used for the antimony. In one case (Brunner, 1837, p. 381), the lead-antimony alloy of nineteenth-century printers' type was used with salt and saltpetre to produce an inferior variety. Kühn (1968, p. 15, footnote) has observed that "admixtures of tin, apparently intentional, can sometimes be detected."

Many of the early lead antimonate recipes, some of which are listed in table 2, are vague with respect to the temperature at which the pigment is to be prepared and the length of time the ingredients are to be roasted. Indeed, some of them will not yield lead antimonate yellow at all. The earliest recipe — requiring lead white, potassium antimonate or antimony oxide, alum, and ammonium chloride (de Bondaroy, 1766) — and the similar recipes of Marcucci (1816), Bachhoffner (1837), and Stockler (1838) call for moderate heat, a dull red heat or a dark red heat usually in a covered crucible for from three to eight hours. A yellow will be obtained in such a reducing environment, but it is a heterogeneous mixture of lead and antimony oxides, chlorides and oxychlorides and, hence, not genuine lead antimonate yellow.

Mérimée (1839) specifically recommended that Passeri's (1758) ingredients (lead, antimony, *allume di feccia*, and common salt) be placed in the *least* heated part of a potter's oven in order to prevent the fusion and de-oxidation of the lead. He noted that four of Passeri's recipes do not call for salt, which Mérimée felt would give a portion of lead oxychloride, and he quotes Guimet as saying that only the oxides of lead and antimony are necessary.

The confusion surrounding these early recipes justifies the observation of Rabaté (1947, p. 262) that Naples yellow has been frequently confused with Mérimée's yellow, in which the lead antimonate is accompanied by lead oxychloride (Reclus, 1956, p. 377) or with Cassel yellow (Turner's yellow, Paris yellow, *jaune de Vérone*, *jaune Véronèse*, *jaune de Montpellier*, mineral

Table 2. **INGREDIENTS USED IN LEAD ANTIMONATE YELLOW RECIPES WITH THEIR PRESUMED MODERN NOMENCLATURE**

Date	*Source*	*Lead (Pb) Component*	*Antimony (Sb) Component*	*Other Ingredients*	
1559	Piccolpasso, 1934, pp. 33, 37–38.	$1\frac{1}{2}$–6 lbs. lead. Pb	1–4 lbs. antimony.[a] Sb? stibnite, Sb_2S_3?	1 oz.–1 lb. burnt wine lees.	$\frac{1}{2}$ oz.–$\frac{1}{2}$ lb. salt. NaCl
1758	Passeri, 1758, pp. 103–104.	$1\frac{1}{2}$–6 lbs. lead. Pb	1–4 lbs. antimony.[a] Sb? stibnite, Sb_2S_3?	1 oz.–1 lb. *allume di feccia*; *feccia*.[b]	NaCl
1766	de Bondaroy, 1766.	12 oz. ceruse. $2PbCO_3 \cdot Pb(OH)_2$	3 oz. *antimonium diaphoreticum*.[c]	1 oz. alum.[d]	1 oz. sal ammoniac. NH_4Cl
1786	Delalande; cited in Dingler's *Polytechnisches Journal*, 1828, p. 226.	3 parts lead, completely oxidized and sieved. PbO	1 part antimony, oxidized and sieved. Sb_2O_3.		
1797	de Massoul, 1797.	Spanish White.	diaphoretic antimony.[c]	calcined alum.	sal ammoniac. NH_4Cl
		Minium. Pb_3O_4	—	quick lime.	white pot ash. KOH
1813	Marcucci, 1816, p. 66.	12 parts *biacca di Plaiter*. $2PbCO_3 \cdot Pb(OH)_2$	12 parts *antimonio diaforetico*.[c, e] Sb_2O_3	2 parts *allume brugiato*.[f] Al_2SO_4	$\frac{1}{2}$ part sal ammoniac. NH_4Cl
1828	Dingler's *Polytechnisches Journal*, 1828, p. 224–5; after Passeri.	1 lb. 8 oz. calcined lead.[g] PbO	1 lb. calcined antimony.[c, g] Sb_2O_3	1 oz. of the alkali of cream of tartar. $K_2C_4H_4O_6 \cdot \frac{1}{2}H_2O$	1 oz. salt. NaCl
1828	Dingler's *Polytechnisches Journal*, 1828, p. 225; after Fougeroux de Bondaroy.	12 parts lead white. $2PbCO_3 \cdot Pb(OH)_2$	3 parts potassium antimonate. $KSbO_3$	1 part alum.[d]	1 part sal ammoniac. NH_4CL
1828	Dingler's *Polytechnisches Journal*, 1828, p. 227.	8 oz. minium. Pb_3O_4	12 oz. antimony. Sb	4 oz. zinc oxide. ZnO	—
		1 lb. pure lead oxide, minium or litharge. (PbO; Pb_3O_4)	1 *quent*[h] *antimon diaphoreticum*. $KSbO_3$	—	1 oz. sal ammoniac. NH_4Cl
1830	Mérimée, 1837; after Guimet; also cited in note by Ehrmann to Brunner, 1837.	2 parts pure minium. Pb_3O_4	1 part potassium antimoniate or diaphoretic antimony. $KSbO_3$ or Sb_2O_3	—	—

Date	Source	*Lead (Pb) Component*	*Antimony (Sb) Component*	*Other Ingredients*	
1837	Brunner, 1837, p. 380.	2 parts powdered lead nitrate. $Pb(NO_3)_2$	1 part finely powdered tartar emetic. $K(SbO)C_4H_4O_6\cdot \frac{1}{2}H_2O$	—	3 parts powdered cooking salt. NaCl
		equal parts of lead and antimony are made into an alloy and powdered. Pb	Sb	$1\frac{1}{2}$ parts saltpetre. KNO_3	3 parts cooking salt. NaCl
1837	Bachhoffner, 1837, p. 115.	12 oz. carbonate of lead, ceruse. $2PbCO_3\cdot Pb(OH)_2$	2 oz. common sulphuret of antimony. $Sb_2S_5 + Sb_4O_6 + S$?	$\frac{1}{2}$ oz. calcined alum.[d]	1 oz. muriate of ammonia, sal ammoniac. NH_4Cl.
1838	Stockler, 1838, p. 7.	12 parts lead white. $2PbCO_3\cdot Pb(OH)_2$	3 parts antimony oxide. Sb_2O_3	1 part alum.[d]	1 part sal ammoniac. NH_4Cl
1846	"Bailiff Thomas" and "Bailiff John"; cited in Miller, 1974, p. 120.	75 parts lead white. $2PbCO_3\cdot Pb(OH)_2$	14 parts oxide of antimony. Sb_2O_3	4 parts sulp. potass. of alumine alum.[d]	7 parts Mur. of ammonia. NH_4Cl
1861	J. Stenhouse and G. Hallet; patent; cited in Rose, 1916, p. 308.	2 parts minium. Pb_3O_4	8 parts antimony oxide. Sb_2O_3	1 part zinc oxide. ZnO	—
		3 parts lead oxide. PbO	3 parts antimony oxide. Sb_2O_3	1 part zinc oxide. ZnO	—
		2 parts lead oxide. PbO	1 to 4 parts antimony oxide. Sb_2O_3	1 part zinc oxide. ZnO	—
1871	Salvétat; cited in Rose, 1916, p. 309.	42 parts minium. Pb_3O_4	12 parts potassium antimonate. $KSbO_3$	9 parts crystallized boracic acid. H_3BO_3	12 parts sand. SiO_2
		9 parts minium. Pb_3O_4	6 parts potassium antimonate. $KSbO_3$	1 part calcined sodium carbonate. Na_2CO_3	3 parts zinc oxide. ZnO
1907	Cramer and Hecht; cited in Rose, 1916, p. 309.	73 parts lead nitrate. $Pb(NO_3)_2$	33 parts antimony oxide. Sb_2O_3	12 parts alumina. Al_2O_3	100 parts table salt. NaCl
1922	Hurst, 1922, p. 202	2 parts lead nitrate. $Pb(NO_3)_2$	1 part tartar emetic. $K(SbO)C_4H_4O_6\cdot \frac{1}{2}H_2O$		4 parts common salt. NaCl
1956	Reclus, 1956, p. 377.	2 parts lead nitrate. $Pb(NO_3)_2$	1 part emetic. $K(SbO)C_4H_4O_6\cdot \frac{1}{2}H_2O$		4 parts sodium chloride. NaCl

Date	Source	*Lead (Pb) Component*	*Antimony (Sb) Component*	*Other Ingredients*	
1971	Society of Dyers and Colourists, 1971, vol. 4, p. 4675.	lead nitrate. $Pb(NO_3)_2$	tartar emetic. $K(SbO)C_4H_4O_6 \cdot \frac{1}{2}H_2O$	alumina. Al_2O_3	sodium chloride. NaCl
		lead nitrate. $Pb(NO_3)_2$	antimony oxide. Sb_2O_3	potassium nitrate. KNO_3	zinc oxide. ZnO
		red lead. Pb_3O_4	antimony oxide. Sb_2O_3	potassium nitrate and lead nitrate. KNO_3 and $Pb(NO_3)_2$	zinc oxide. ZnO
		lead white and lead nitrate. $2PbCO_3 \cdot Pb(OH)_2$ and $Pb(NO_3)_2$	antimony oxide. Sb_2O_3		lead chloride. $PbCl_2$

a. In their translation of Biringuccio's *Pirotechnia* published in 1540, Smith and Gnudi point out that the term antimony at this time always referred to the sulfide although metallic antimony was probably being imported from Germany. (Biringuccio, 1966, p. 91, note). "The ore that I know is found in Italy in various places. From Germany they bring to Venice some of the smelted kind in cakes for the use of those masters who make bells .... It also serves to make yellow colours for painting earthenware vases and for tinting enamels, glasses and other similar works just as one would wish they were intrinsically yellow by virtue of the fire. There are several mines of this antimony ore in the region of Siena, among which is one near the city of Massa in the Maremma and another large quantity near another city called Sovana." (Biringuccio, 1966, p. 92.) See also Piccolpasso, 1934, p. 33 and footnotes 96 and 96b, p. 79.
b. "It should be noted with us the lees of wines are obtained in the month of November and of December more than at other times, because then wines undergo a change." (Piccolpasso, 1934, p. 29). This substance is crude potassium hydrogen tartrate (tartar, cream of tartar, lees, argol) which is deposited as a fine crystalline crust during fermentation of grape juice; it is called *faecula* in latin. Passeri uses *allume di feccia* in one recipe but *feccia* in the others. The term was apparently translated as alum by some and cream of tartar by others (Dingler's *Polytechnisches Journal*, 1828, p. 225), but only the latter was meant by Passeri.
c. The author of Dingler's *Polytechnisches Journal*, 1828, p. 224 reads Passeri's *Antimonio* as calcined antimony and de Bondaroy's *antimonium diaphoreticum* as potassium antimonate. However Marcucci (see note e) specifically defines *antimonio diaforetico* as the oxide of antimony.
d. Alum is presumably potassium aluminum sulphate, $K_2SO_4 \cdot Al_2(SO_4)_3 \cdot 24H_2O$.
e. "Antimonio diaforetico (o ossido di antimonio bianco per via di nitro)."
f. "Allume brugiato (ossio solfato di allumina, levatagli una porzione di acqua di cristallizzazione)."
g. Six further recipes for lead antimonate yellow are given after Passeri; the second incorrectly specifies 4 instead of 2 lbs. antimony.
h. A *quent* is equivalent to a dram (1.77 g), but this quantity seems to be too little in the context of this recipe.

yellow, Kassler yellow), which are exclusively lead oxychloride ($PbCl_2 \cdot 7PbO$).

De Massoul (1797, pp. 138–139) was perhaps the earliest to establish the importance of high temperature and long firing times; and this theme was taken up in the first rational discussion of lead antimonate yellow, that by Brunner and Ehrmann (1837). Brunner prepared his pigment from lead nitrate, tartar emetic, and common salt, the ingredients being purified by recrystallization and then finely powdered. The mixture was heated for two hours at a temperature such that the salt fused (*id est*, greater than 801°C). After cooling, the salt was rinsed out and the somewhat hard mass softened to a more or less fine powder. Overheating produced a very hard mass which would not soften in water and was difficult to grind. Ehrmann, repeating this process at three different temperatures, produced three different shades.

In the earliest known recipes, those of Cipriano Piccolpasso (see 2.22) for maiolica, the temperatures encountered in firing the glazes would be sufficient to produce lead antimonate yellow. We have obtained the genuine pigment by heating three parts of lead oxide and one part of antimony oxide at 950–1000°C for five hours. Natta and Baccaredda (1933) and Baccaredda (1936) have prepared a synthetic anhydrous analogue of bindheimite by precipitation from a neutral solution of potassium pyro-antimonate by lead acetate and by sintering lead acetate and antimonic acid at low red heat and purifying the product by washing with boiling water (Palache, et al., 1951, p. 1,019).

The procedures of five manufacturers have been described in some detail (British Intelligence Objectives Sub-Committee, n.d.). All of them require high temperature furnacing although there is considerable variation in the ingredients used. The companies mentioned (Deutsche Gold u. Silber Scheideanstalt, I. G. Farbenindustrie A. G., G. Siegle and Co., Riedel de Haen A. G., Kali Chemie A. G.) were producing a pigment primarily for use in ceramics or as iron enamel.

4.4 Adulteration and Sophistication (Imitation Lead Antimonate Yellow)

By 1850, Naples yellow was sold in a variety of shades; in England an orange-yellow variety known as French Naples yellow was on sale, reputedly capable of giving clear tints but not so useful in producing opaque green in mixtures as was the traditional variety which had a greenish tinge (Winsor & Newton, 1850). Possibly French Naples yellow was brightened with other pigments. A description of the late nineteenth-century production of Naples yellow by the French manufacturer Pinondel is instructive:

> Naples yellow is obtained by the calcination of oxychlorate of lead for the light tints and the addition of salt of Antimony for middle tints. As Antimony gives an earthy tone when it is used in too strong a quantity, we add Cadmium in the deeper shades and Jaune brilliant, but this addition takes place after the calcination of the oxychlorate of lead. As these two products do not act on each other the Naples Yellow as produced is little less permanent than that prepared from Antimony and as it is brighter, artists prefer it. As to its action on other colours, it is very slight, nevertheless we ought to avoid if possible the mixture of white lead, it is better to use the lighter qualities than reduce with white

(quotation from a letter from Pinondel of Paris to Roberson, 19 December 1885, supplied by G. F. Roberson-Park, C. Roberson and Company Limited, 13 March 1975).

Not only does this process involve the addition of cadmium yellow to produce the desired tint, but Pinondel is, in fact, describing a lead oxychloride pigment (Mérimée's yellow) and not genuine or lead antimonate Naples yellow.

The artists' pigment currently sold under the name Naples yellow is generally composed of cadmium sulfide or selenide and sometimes a yellow ochre or a red iron oxide with a white pigment; lead white, titanium white, and zinc white are all used by the firms who communicated with us. In one case lithopone is used in watercolor (see chapter on Cadmium Yellows, Oranges and Reds). Chromium-titanium yellow, nickel-titanium yellow, and yellows based on (parachloro-) orthonitraniline are also used as substitutes.

5.0 IDENTIFICATION — CHARACTERIZATION

5.1 Characterization by Light and Scanning Electron Microscopy

5.11 ***Light Microscopy.*** The microscopy of lead antimonate yellow has been noted by McCrone and Delly (1973, p. 496; *Particle Atlas* ref. no. 443). The compound is isotropic with refractive index 2.01–2.28 (Gettens & Stout, 1966, p. 148a), such that in Aroclor 5442 (refractive index 1.66), it exhibits moderately high relief comparable to that of lead-tin yellow and litharge. The values given for bindheimite are generally lower: 1.72–1.79 (Montoya, 1969, p. 1,727) for a bismuthian bindheimite; 1.84–1.87 (Palache, et al., 1951, p. 1,018); 1.850–1.899 (Winchell, 1965, p. 56); 1.905, 1.928 (Mason & Vitaliano, 1953, p. 104). An anisotropic variety, uniaxial negative, with $\omega = 2.08$ and $\epsilon = 1.82$ has also been reported (Winchell and Winchell, 1951, p. 235).

Lead antimonate yellow shows anomalous birefringence and remains visible between the crossed polars of a polarizing microscope. Undulose extinction has been observed at 1,000x magnification with a first-order red compensator. Fused agglomerates of smaller particles are typical, and small needles may also be observed (Hennig and Kohlmeyer, 1957, pp. 66, 68; Friedrich & Marx, 1962, p. 74). In our analyses, these needles have been shown to consist of lead and antimony, but they have also been linked to massicot, which may be easily confused with bindheimite (Palache, et al., 1951, p. 516). A mid-nineteenth-century Winsor & Newton sample included such needles (fig. 4); their characteristics under the scanning electron microscope are discussed below (see 5.12).

The finely divided particles of a milled lead antimonate yellow ($\leqslant 1$ μm) are a pale greenish yellow in plane polarized light (fig. 5A) and retain a pale yellow color in their boundaries between crossed polars (fig. 5B). Somewhat larger particles appear lemon yellow. The fused

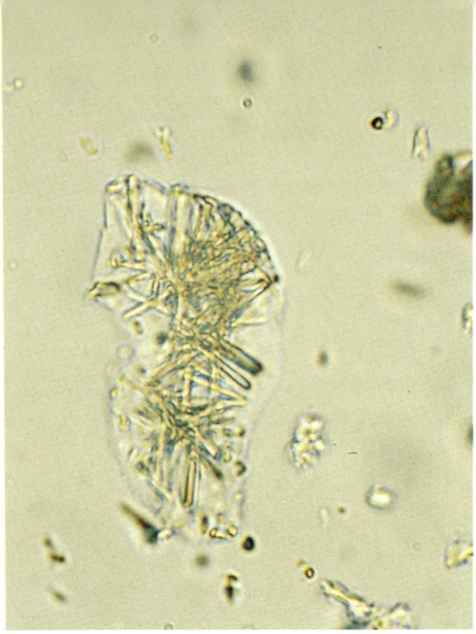

Fig. 4. Photomicrograph of mid-nineteenth-century sample of lead antimonate yellow provided by Winsor & Newton. Transmitted plane polarized light, Aroclor 5442 mount. Magnification 1000x.

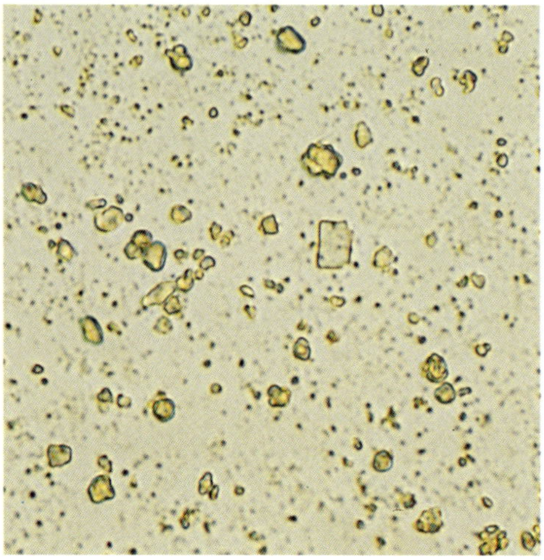

Fig. 5. Photomicrograph of lead antimonate yellow sample 551 101, from H. C. von Imhoff, Switzerland. Pigment produced in England c. 1920 and distributed by Standard Lack, Zurich. A. By transmitted plane polarized light in Aroclor 5442.

B. Same area between crossed polars. Magnification 1000x.

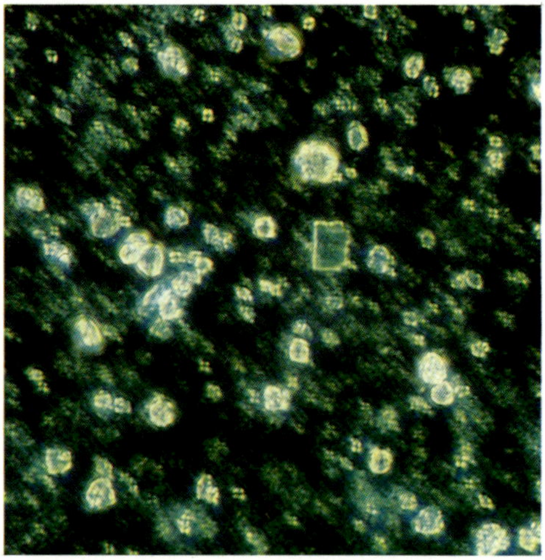

lumps tend to a deeper orange-yellow (fig. 6 and 7) which is also observed in bindheimite (fig. 8). Isotropic distorted octahedral crystals were observed in a sample synthesized from lead and antimony oxide (fig. 9) which was determined to be genuine lead antimonate yellow on the basis of x-ray diffraction.

Birefringent yellow particles are often observed in the manufacturer's samples of lead antimonate yellow; these possibly derive either from uncombined ingredients or from various lead and antimony oxides, chlorides, or oxychlorides which may have formed (see 4.3). The particles are easily distinguishable from lead antimonate yellow because they go to extinction upon rotation of the microscope stage. Massicot (PbO), for example, which at first glance may appear similar to lead antimonate yellow (de Wild, 1929, p. 57), is anisotropic. A compensator can be used to determine whether a weakly birefringent sample is isotropic lead antimonate yellow or anisotropic lead-tin yellow (Butler, 1977, p. 4), although in practice, these two pigments are not readily identified by light microscopy.

Lead antimonate yellow is weakly fluorescent, blue-white in ultraviolet (365 nm) and yellow in blue-violet light (McCrone & Delly, 1973, p. 868).

5.12 ***Scanning Electron Microscopy.*** The most consistent morphological characteristics observed in lead antimonate yellow samples by scanning electron microscopy are pseudo-hexagonal plates which appear more frequently as rounded, equant grains (McCrone & Delly, 1973, p. 735). Distorted octahedral plates were found in a sample we synthesized by calcining three parts of lead oxide (PbO) with one part of antimony oxide (Sb_2O_3) at 950–1,000°C for five hours (fig.9 and 10B). A mid-nineteenth-century

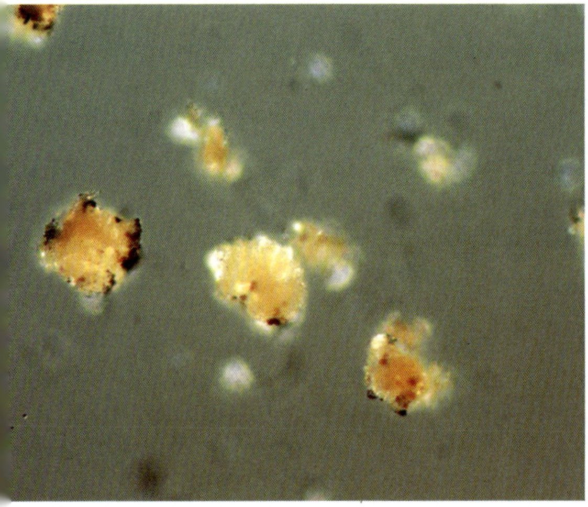

Fig. 6. Photomicrograph of particles of lead antimonate yellow in sample from *Portrait of Charles Churchill* by Sir Joshua Reynolds, 1755. Transmitted light, in Aroclor 5442 between partially crossed polars, analyzer at 12° to polarizer. Magnification 425x.

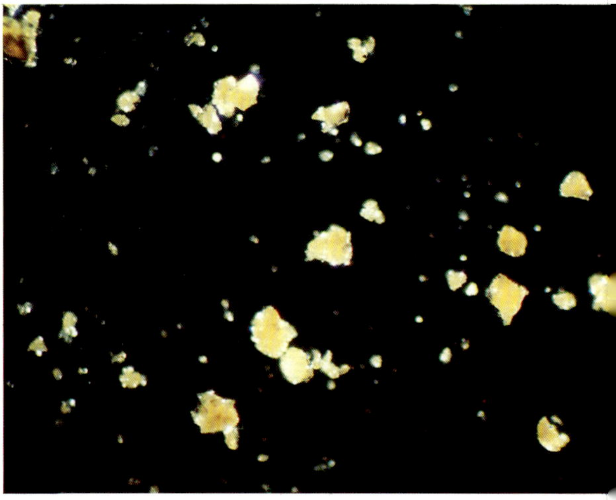

Fig. 7. Photomicrograph of lead antimonate yellow sample from *Portrait of Charles Churchill* by Sir Joshua Reynolds, 1755, courtesy National Gallery of Canada, Ottawa, accession no. 29. Between crossed polars; sample mounted in Aroclor 5442. Magnification 150x.

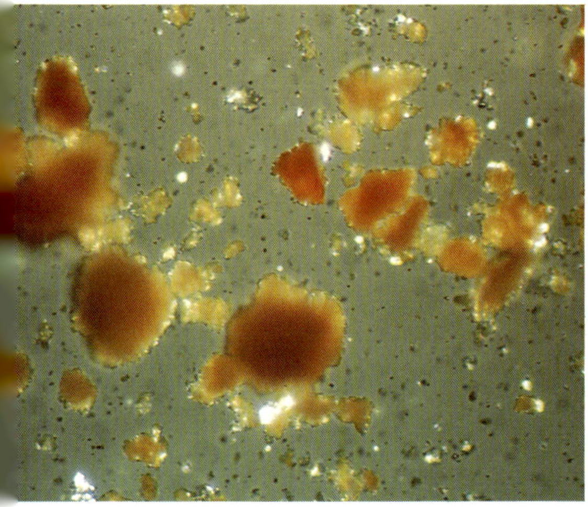

Fig. 8. Bindheimite sample from Pershing Co., Majuba Hill, Nevada. Courtesy National Mineral Collection, Geological Survey of Canada, catalogue no. 15131. Photomicrograph of sample in Aroclor 5442 between partially crossed polars by transmitted light, analyzer at 12° to polarizer. Magnification 150x.

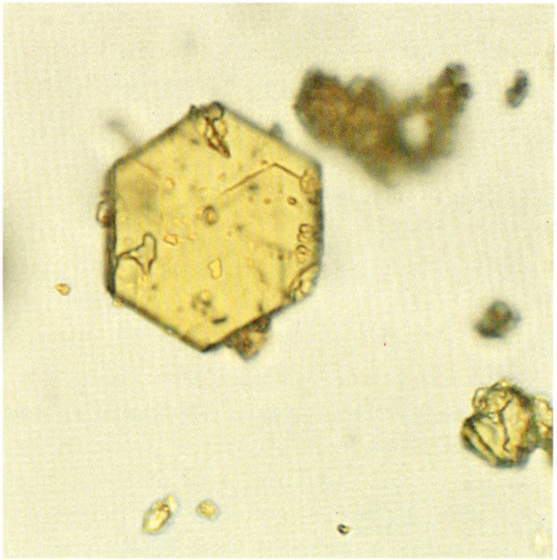

Fig. 9. Photomicrograph of lead antimonate yellow sample synthesized by calcining three parts lead oxide (PbO) with one part antimony oxide (Sb_2O_3) at 950–1,000°C for five hours. Different crystal from that in fig. 10B. Sample mounted in Aroclor 5442; viewed by transmitted plane polarized light. Magnification 1000x.

sample from Winsor & Newton exhibited lead-antimony rods of hexagonal cross section, growing from a central core. This may be an admixture of the phase $PbSb_2O_6$ which belongs to the hexagonal system (fig. 10A). More frequently, however, irregular, angular particles with a wide size range are observed.

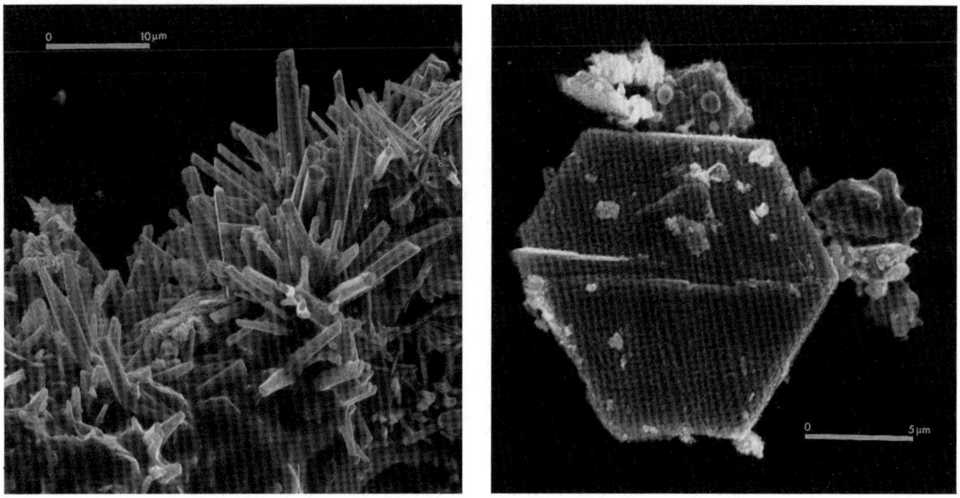

Fig. 10. Scanning electron micrographs of samples of lead antimonate yellow dispersed on a carbon planchet. Magnification 1400x. A. Winsor & Newton Ltd., mid-nineteenth century. B. Crystal of lead antimonate yellow from sample synthesized by calcining three parts lead oxide with one part antimony oxide at 950–1,000°C for five hours. Magnification 3000x. Etching parallel to crystal edges in B was produced after prolonged electron bombardment. Isometric x-ray maps of the crystal for lead and antimony indicated that it was a compound of lead and antimony.

5.2 Chemical Identification

Although lead antimonate yellow is only slightly soluble in strong acids, with care lead and antimony can be brought into solution and identified by microchemical tests.

5.21 ***Iodide and Chromate Tests for Lead.*** A few well-crushed pigment particles can be heated slowly with a drop of concentrated aqua regia and evaporated to dryness. The procedure may be repeated to ensure that more of the sample is brought into solution. Two drops of dilute acetic acid (5%) are then added to the yellow residue, preferably while the sample is still warm from the evaporation.

To test for lead, one drop of the supernatant solution is added to a depression slide containing a small drop of an aqueous solution (5% by weight) of potassium iodide. The formation of lead iodide as a yellow precipitate indicates the presence of lead (Plesters, 1956, p. 143; Chamot & Mason, 1931, p. 190).

A confirmatory test involves the addition of a second drop of the supernatant solution to a drop of saturated potassium chromate or potassium dichromate. If lead is present, a yellow lead chromate precipitate forms (Plesters, 1956, p. 143).

5.22 ***Rhodamine B Test for Antimony.*** Small amounts of antimony may be detected in the presence of a large quantity of tin by means of Rhodamine B (Feigl & Anger, 1972, p. 108). Antimony will slowly come into solution in concentrated hydrochloric acid. To accomplish this, a few well-crushed pigment particles are allowed to stand in one drop of hydrochloric acid (35%) for at least one hour and preferably much longer. A drop of this solution is then added to a larger drop of Rhodamine B reagent (tetraethylrhodamine; aqueous solution 0.01% by weight) on a white porcelain spot plate. The presence of antimony is indicated by a change in the color of the solution from pink to violet, with the subsequent formation of a purple precipitate if a sufficient quantity of antimony is present.

In the authors' experience, barium-strontium chromate and related chromate compounds were also found to give a positive reaction to this test. They can, however, be easily distinguished from lead antimonate yellow as they give a negative test for lead; lead dichromate does not produce the false positive result for antimony.

5.3 Instrumental Methods of Analysis

5.31 ***X-ray Diffraction.*** It has been demonstrated by x-ray diffraction studies that lead antimonate yellow is essentially identical with the natural mineral bindheimite and its synthetic analogue (König, 1956, p. 70; Riederer, 1972,

p. 11). The mineral was named after the German chemist Johann Jacob Bindheim, who first analyzed a Siberian variety in about 1792. X-ray diffraction must be regarded as the definitive technique for studying the pigment and the best means of differentiating it from lead-tin yellow II, with which it may be confused. Lead-tin yellow I, which was used more frequently than lead-tin yellow II, is easily distinguished from lead antimonate yellow by x-ray diffraction. In Kühn's x-ray diffraction analyses of lead-tin yellows from thirty-four works of art, only five samples (15%) were of type II (1968, p. 10). X-ray powder diffraction patterns of lead antimonate yellow, lead-tin yellow II, and bindheimite obtained by the authors are seen in fig. 11.

Bindheimite has an isometric crystal structure of the pyrochlore type [type $E8_1$; space group O_h^7-Fd3m; diffraction symmetry $4/m\ \bar{3}\ 2/m$; cell contents $Pb_{16}Sb_{16}O_{56}(H_2O)_8$] and has been reviewed by Palache, et al. (1951, pp. 1,017–1,020). The unit cell parameter of a synthetic analogue of bindheimite studied by Friedrich and Marx (1962, p. 74) was established at a = 10.47 Å, in close agreement with values obtained by Natta and Baccaredda (1933) for natural bindheimite, 10.40, 10.42, and 10.43 Å.

The value 10.47 Å is very close to Rooksby's (1964, p. 22) value of 10.694 ± 0.003 Å for cubic lead-tin oxide which, in practical terms, means that the difference in the *d*-spacings is of the order of only .06–.07 Å for the most intense lines of lead antimonate yellow and lead-tin yellow II; these lines are:

lead antimonate yellow:

3.016 2.617 1.849 1.578

lead-tin yellow II:

3.087 2.681 1.893 1.614.

In table 3, our x-ray diffraction analyses of lead antimonate yellow and a bindheimite specimen from Nevada, and lead-tin yellow II are compared with the Friedrich and Marx (1962) data for a synthetic analogue of bindheimite and the Mason and Vitaliano (1953) data for another natural bindheimite from Nevada. The x-ray diffraction patterns of the first three specimens are given in fig. 11: A, B, and C.

As demonstrated in fig. 11D, E, and F, one must be particularly careful in identifying lead antimonate yellow and lead-tin yellow II in mixtures with lead white. In these cases, the compounds can only readily be distinguished by the 3.016 Å line of lead antimonate yellow and the 3.087 Å line of lead-tin yellow II. Other lines are either partially or wholly coincident with lines of the lead white pattern.

Gerlach, et al. (1968) have published *d*-spacings of ten compounds of lead and antimony oxides, none of which match the bindheimite pattern. The two most intense lines of their $3PbO{\cdot}Sb_2O_5$ compound (the formula proposed by Rose, see 4.1), like those of the $3PbO{\cdot}Sb_2O_5$ porcelain enamel studied by Shimohira and Tagai (1964), were close to those of bindheimite, but the less intense lines cannot be matched. Ahmed (1962) produced diffraction patterns of $2PbO{\cdot}Sb_2O_5$ fired at 700, 800, 900, 1,000, and 1100°C, indicating a transition that starts at approximately 800°C and is completed at 900°C. This is consistent with his differential thermal analysis which shows exothermic activity that begins at 600°C and ends at 850°C with a peak at 650°C.

Friedrich and Marx have reported the preparation of a cubic form of $Pb_2Sb_2O_7$ (cited in Rooksby, 1964, p. 20), but Rooksby found that pure $Pb_2Sb_2O_7$, prepared by firing the appropriate mixture of lead and antimony oxides in air at 900°C, is slightly deformed from the cubic and exhibits rhombohedral symmetry. This structure is, however, "closely related to the cubic pyrochlore type, and is derived from it by a small relative extension of one of the three trigonal axes" (Rooksby, 1964, p. 20); such an extension can occur when other ingredients are incorporated into the compound. "The addition of potassium oxide, for example, to give a composition such as $KPbSb_2O_{6.5}$ effects such a change of structure. This may well have been the composition of the old artists' colour known as Naples yellow, which was made by igniting mixtures of lead nitrate and 'tartar emetic' (potassium antimony tartrate)" (Turner and Rooksby, 1959, p. VIII/23).

5.32 ***Infrared Spectrophotometry.*** Lead antimonate yellow has a distinct infrared spectrum in the region 1,400–200 cm^{-1} (fig. 12A) which permits the pigment to be distinguished from lead-tin yellow, chrome yellow, cadmium yellow, and chromium-titanium yellow. The characteristic absorptions in a lead antimonate yellow spectrum are listed in table 4, the general shape being a close match among historical samples (fig. 12D). A sample synthesized by calcining three parts of lead oxide (PbO) with one part of antimony oxide (Sb_2O_3) at

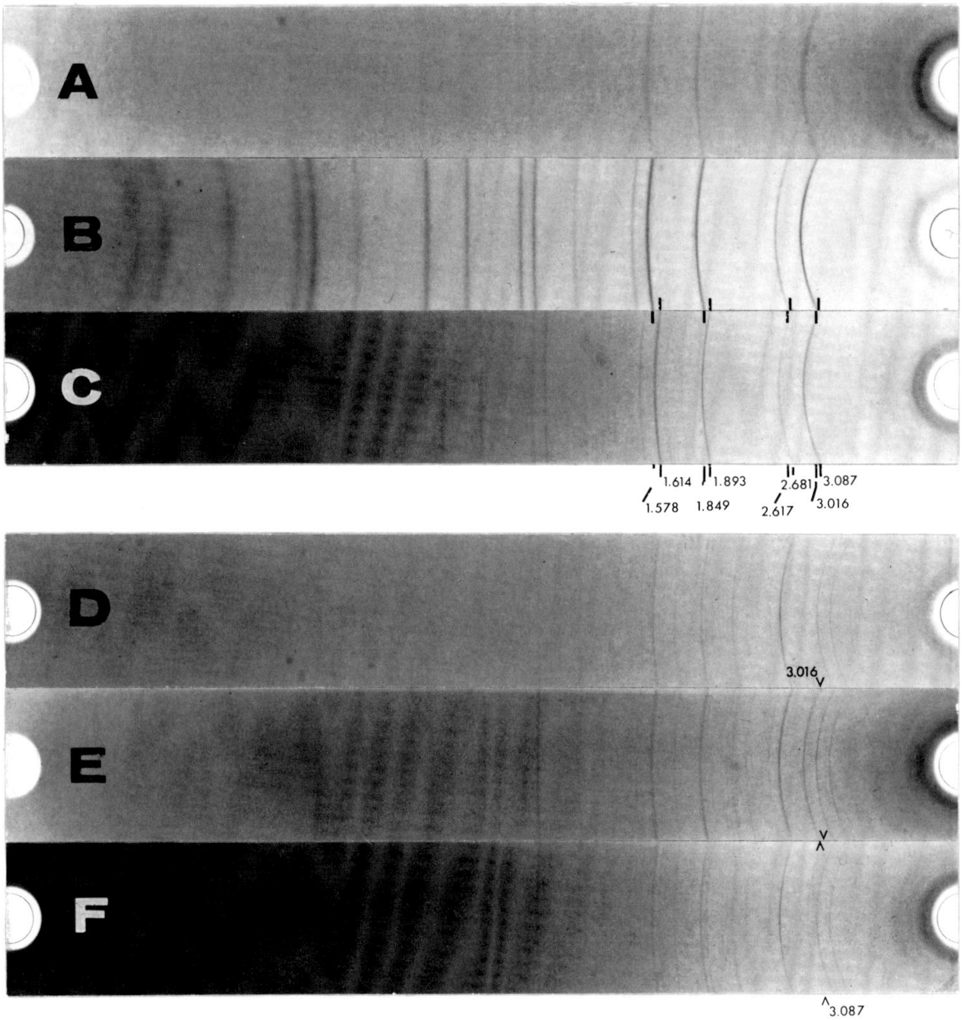

Fig. 11. X-ray powder diffraction patterns of lead antimonate yellow, lead-tin yellow II, and bindheimite. A. Bindheimite from Pershing Co., Majuba Hill, Nevada, courtesy National Mineral Collection, Geological Survey of Canada, catalog number 15131. B. Lead antimonate yellow, No. 551 101, containing zinc oxide, produced in England c. 1920 and distributed by Standard Lack, Zurich. C. Lead-tin yellow II, No. 550 102 synthesized by B. Mühlethaler and W. Pulver, Chemisch-Physikalisches Laboratorium, Schweizerisches Landesmuseum after Recipe 273 in the Bolognese manuscript (Merrifield, 1849, pp. 528–529) from 2 parts calcined lead and tin (Recipe 272), 2 ½ parts minium, and ½ part sand at 900–950°C. D. Lead white, basic carbonate, No. 182 101, Schminke. E. Lead white, basic carbonate mixed with lead antimonate yellow, No. 551 101. F. Lead white, basic carbonate mixed with lead-tin yellow II, No. 550 102. The pigment samples (B–F) were provided through the courtesy of H. C. von Imhoff, Musée d'art et d'histoire de Fribourg, Switzerland (von Imhoff, 1975). Debye-Scherrer 114.6 mm diameter camera, Kodak Industrex M film.

950–1000°C for five hours gave a spectrum (fig. 12B) with the same absorptions and additional weak peaks. These same peaks are found in the bindheimite spectrum (fig. 12C), but there they are broader, shoulders less distinct, and peaks due to impurities more pronounced.

Many lead antimonate yellows as formerly supplied contain silica or carbonates, basic lead sulfate, $PbSO_4 \cdot PbO$ (fig. 12D), titanium dioxide (particularly rutile, TiO_2), and barium sulfate ($BaSO_4$); sometimes these substances are present in small amounts but often they are admixed in considerable proportions. The pre-

Table 3. **X-RAY POWDER DIFFRACTION DATA FOR LEAD ANTIMONATE YELLOW, BINDHEIMITE, AND LEAD-TIN YELLOW II (CuK_{α} = 1.5418 Å)**

Lead Antimonate Yellow[a]		*Lead-tin Yellow II*[b]			*Bindheimite, synthetic*[c]			*Bindheimite,*[d] *Lovelock, Nevada*	
d(Å)	*I*	*d(Å)*	*I*	*hkl*	*d(Å)*	*I*	*hkl*	*d(Å)*	*I*
		9.571							
		6.172	7	111					
					5.97	5	112		
		3.345*							
		3.227	10	311					
					3.15	5	311		
		3.087	100	222					
3.016	100				3.01	100	222	3.03	100
					2.88	1	320		
2.812*		2.749*							
		2.681	44	400					
		2.639*							
2.617	39				2.61	18	400	2.62	60
2.473*		2.456	13	331					
2.394	9				2.40	5	331	2.41	10
		2.137*							
		2.062	3	511,333					
					2.04	1	510,431		
					2.01	2	511,333		
		1.893	35	440					
1.849	39				1.85	24	440	1.85	80
					1.77	1	530		
		1.618*							
		1.614	28	622					
1.578	38				1.58	14	622	1.58	80
		1.548	9						
1.512	9				1.51	5	444	1.51	30
1.308	5				1.31	3	800	1.31	20
		1.230	10						
1.201	13				1.20	3	662	1.20	40
1.17	11				1.17	3	840	1.17	40
1.07	8				1.07	1	844	1.07	40
1.01	9				1.008	2	666	1.00	40
					0.926	1	880	0.925	10
					0.885	2	10.62	0.884	50
					0.873	4	884	0.872	40
					0.828	4	12.40	0.827	30
					0.799	2	10.66	0.798	40
					0.789	1	12.44	0.788	40

*Lines attributed to zincite (ZnO).

a. Diffraction data obtained on Philips PW1050/76 Diffractometer, Ni filter, 40 kV, 30 mA. Relative intensities determined by peak areas. Intensities not given for lines due to impurities. Sample 551 101 obtained from H. C. von Imhoff, Musée d'art et d'histoire de Fribourg, Switzerland.

b. Diffraction data obtained as in a. Relative intensities determined by peak heights. Intensities not given for lines due to impurities. Sample 550 102 obtained from H. C. von Imhoff.

c. Products of a lead refining furnace of Stolberger Zinc AG. JCPDS Powder Diffraction File 18-687 (1982), Friedrich and Marx (1962).

d. Mason and Vitaliano (1953).

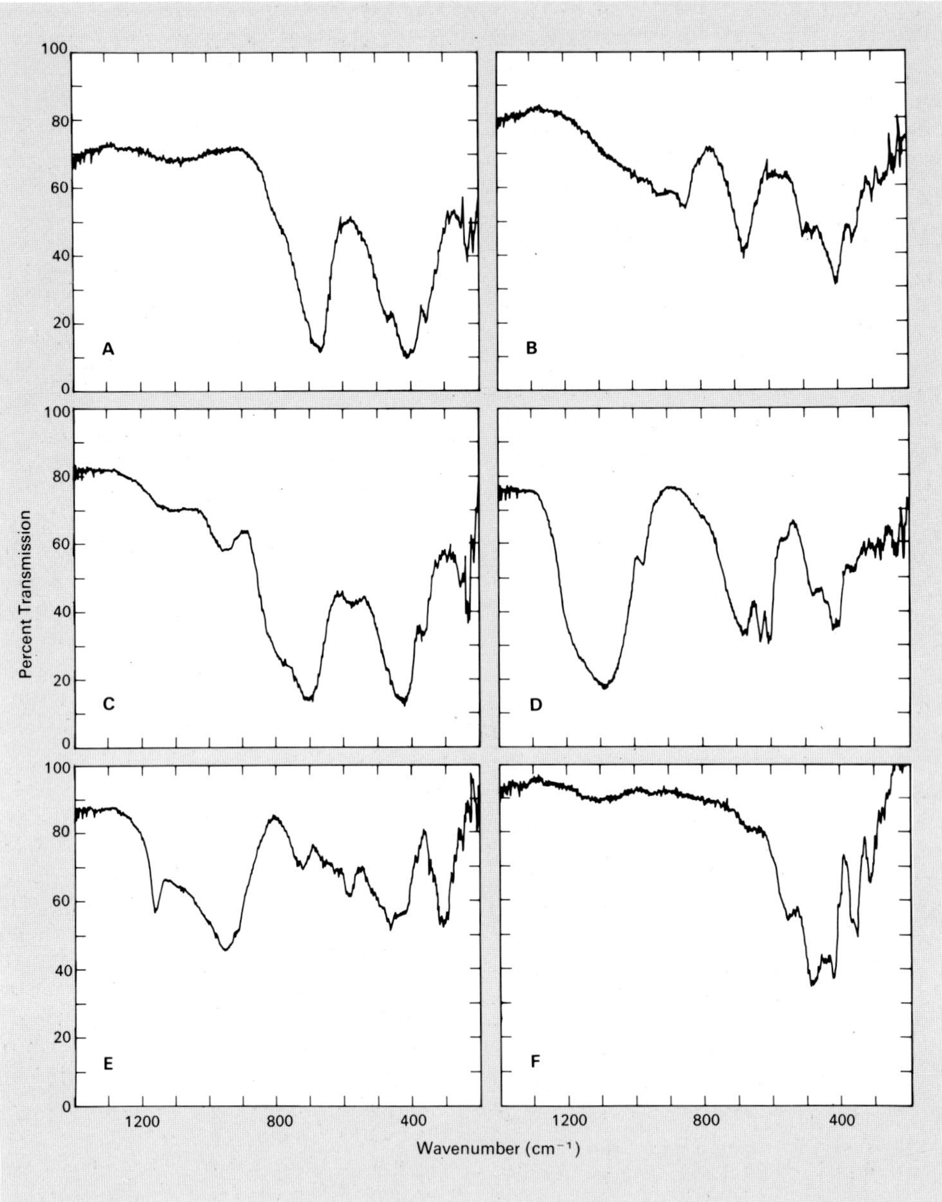

Fig. 12. Infrared spectra of lead antimonate yellow, bindheimite, and lead-tin yellow obtained using a diamond cell micro-sampling device. A. Lead antimonate yellow, sample No. 551 101, courtesy H. C. von Imhoff. B. Lead antimonate yellow synthesized by calcining three parts lead oxide (PbO) with one part antimony oxide (Sb_2O_3) at 950–1,000°C. C. Bindheimite from Pershing Co., Majuba Hill, Nevada, courtesy National Mineral Collection, Geological Survey of Canada; catalogue no. 15131. D. Lead antimonate yellow, Winsor & Newton Ltd., mid-nineteenth century. Contains lead white, basic sulfate ($2PbSO_4 \cdot PbO$). E. Lead-tin yellow II, no. 550 102, and F. Lead-tin yellow I, No. 550 101 courtesy H. C. von Imhoff. Both were synthesized by B. Mühlethaler and W. Pulver, Chemisch-Physikalisches Laboratorium, Schweizerisches Landesmuseum, after Recipes 272 and 273 in the Bolognese manuscript (Merrifield, 1849).

sence of titanium dioxide causes some difficulty because its spectrum has strong, broad peaks overlapping all but the 467 cm^{-1} peak of lead antimonate yellow. Fortunately, titanium dioxide is readily detectable by x-ray secondary-emission spectrometry and would not be expected in paintings executed before the 1920s. The spectrum of lead-antimony-titanium yellow is similar, but on careful examination, slight shifting and broadening of peaks may be observed to distinguish it from that of lead antimonate yellow.

For completeness, the spectra of lead tin yellow I and II are given (fig. 12E and F), since the spectral region below 900 cm^{-1} was not available to Kühn at the time his paper on these pigments was written (Kühn, 1968, p. 16).

5.33 ***Electron Beam Microanalysis.*** An electron probe microanalyzer or a scanning electron microscope with a lithium-drifted silicon x-ray detector and x-ray energy spectrometer may be used to great advantage to study pigments both as particles and in polished cross sections. The morphology and the elemental composition of lead antimonate yellow samples have each been studied in this way.

With an operating potential of 20 kV, an electron beam will excite the L lines of lead and, in addition, the M series, the main lines of which are $M\alpha_{2,1}$ = 2.339, 2.345 keV, $M\beta$ = 2.442 keV. For antimony, the principal lines excited are $L\alpha_{2,1}$ = 3.595, 3.604 keV, $L\beta_{1,4,3}$ = 3.843, 3.886, 3.932 keV, $L\beta_{2,15}$ = 4.100 keV, $L\gamma_1$ = 4.347 keV (Johnson & White, 1970).

In addition, elements characteristic of other ingredients used in manufacturing the pigment — such as aluminum, sulfur, potassium (alum), silicon (sand), sodium, chlorine (salt and ammonium chloride), and iron — as well as elements found in extenders — for example, calcium, barium, sulfur ($CaSO_4$, $BaSO_4$), zinc (ZnO), and titanium (TiO_2) — are frequently encountered. In the earliest example we have studied, a lead antimonate yellow layer from *Arrest of Christ* (1630–1632) by Matthias Stomer, considerable quantities of iron were found together with aluminum, silicon, sulfur, potassium, and calcium (fig. 13). The latter element is the principal source of interference with the antimony $L\alpha$ x-rays (Ca $K\alpha$ = 3.690 keV).

The scanning capability of an electron beam instrument allows one to map the distribution of lead and antimony both in cross sections and in individual particles (fig. 13C). Brill was able to show in this way (Brill, 1968, p. 56; Ibrahim, et al., 1976, p. 239) that the yellow pigment of the *opus sectile* panels from Kenchreai was $PbSnO_3$ and that the antimony also found in the glass was present as the white opacifier $Ca_2Sb_2O_7$. In Brill's investigation, x-ray diffraction did not produce conclusive evidence for this finding that has interesting consequences for research on the transition from antimony to tin-based colorants in the second to fourth centuries A.D.

Table 4. **INFRARED SPECTROPHOTOMETRY DATA ON LEAD ANTIMONATE YELLOW**

wavenumber (cm^{-1})	*approximate relative intensity*	*peak shape*
408 ± 7	100	—
666 ± 10	95 ± 10	—
467 ± 7	65	distinct shoulder
355 ± 5	65 ± 10	distinct shoulder
795 ± 10	10	broad, slight shoulder

Fig. 13 illustrates the kind of information that can be obtained by scanning across a given sector of a sample, in this case the polished cross section from *Arrest of Christ* by Matthias Stomer, 1630–1632. A secondary electron image of the sample is produced in 13A with scanning electron microscope (magnification 400x; accelerating voltage 20 kV). Fig. 13B is a photomicrograph of the same area (magnification 64x). Fig. 13C indicates the concentration of various elements across the line X–Y in fig. 13A. The profiles are intended to be qualitative only and are plotted at different expansions. A small antimony contribution is observed in the calcium scan. The particular elements that are present in each feature in the photomicrograph are clear from such traces. In fig. 13D, the antimony (Sb) peaks are observed without interference of calcium in an x-ray energy spectrum of the lead antimonate layer. The presence of iron (Fe), lead (Pb), aluminum (Al), and silicon (Si) is also revealed.

5.34 ***X-radiography.*** A radiograph was made of four paints prepared from dry lead antimonate yellow, lead-tin yellows I and II, and lead white; all were mixed to usable consistencies in linseed oil and spread as .06 mm thick

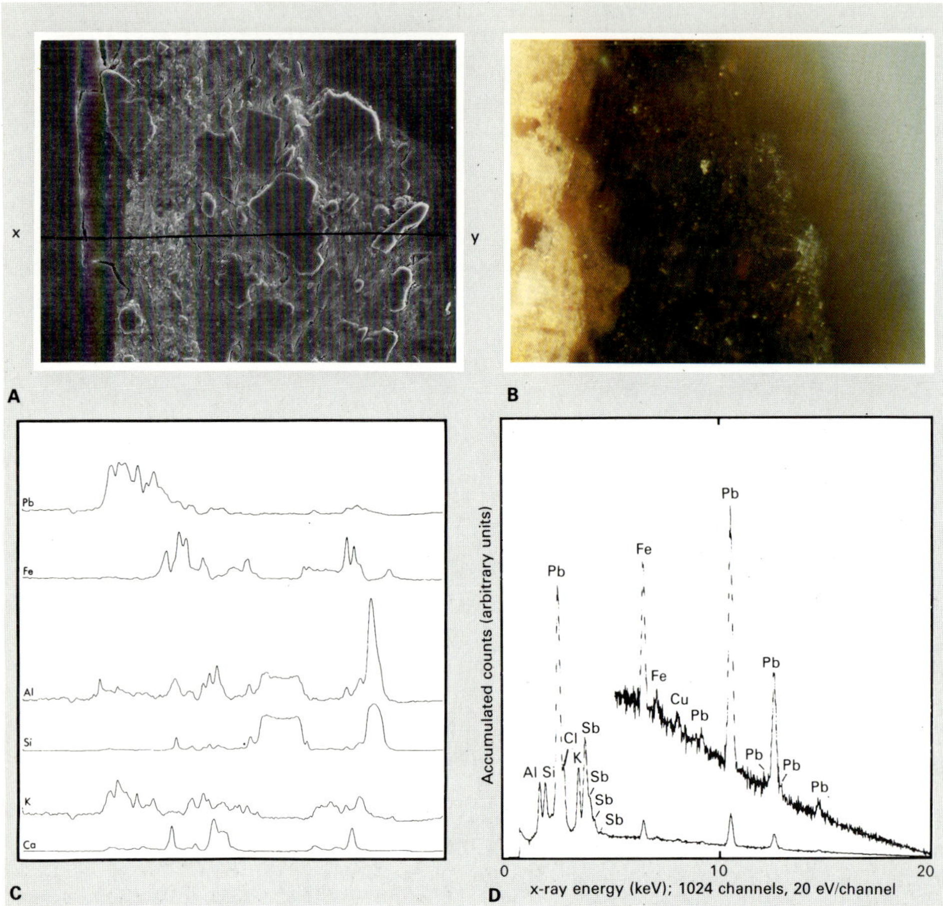

Fig. 13. Microscopy of a polished cross section from *Arrest of Christ* by Matthias Stomer, 1630–1632. Courtesy National Gallery of Canada, Ottawa, Accession no. 4094. A. Secondary electron image produced by the scanning electron microscope. Magnification 200x; accelerating voltage 20 kV. B. Photomicrograph of same area. Magnification 200x. C. Profiles of the concentration of various elements across the line X–Y in A. The profiles are intended to be qualitative only. D. Antimony is detected in this x-ray energy spectrum of the lead antimonate yellow layer.

films on a methacrylate sheet. This rather arbitrary experiment indicated that, for paint layers of equal thickness, lead antimonate yellow can be expected to be equivalent to lead white in its opacity to x-rays and somewhat less opaque than the lead-tin yellows.

5.35 ***Mass Spectroscopy: Lead Isotope Ratios.*** The technique of lead isotope analysis (Brill & Wampler, 1965; Stós-Fertner & Gale, 1978) has been used in an effort to locate the source of the lead component in ancient glass and ceramic glazes. Brill, et al. (1973) assembled samples of lead antimonate yellow and lead-tin yellow, in addition to lead white, for study by this method. As has been demonstrated for lead white, such studies could have a considerable bearing on provenance, dating, and authentication problems. The four stable isotopes of lead are ^{204}Pb, ^{206}Pb, ^{207}Pb, and ^{208}Pb; data on their relative abundance (ratios) in ancient artifacts and mining regions have been published in the three aforementioned references.

5.36 ***X-ray Fluorescence Spectrometry.*** Lead antimonate yellow may be detected in situ on paintings and ceramics by x-ray secondary-emission (fluorescence) spectrometry using radioisotope excitation and lithium-drifted silicon and germanium detectors (Kennedy, et al., 1970; Hanlan, 1971). Either $^{241}_{95}Am$ or $^{125}_{53}I$ sources may be used to excite the lead L lines and antimony K lines. The radiation emitted by the sources, Np L and Te K respectively, does not interfere with the analysis; however, artifactual x-rays arising in the collimator material, notably tin, must be considered when distinguishing lead-tin from lead antimonate yellow. In addition, the principal Compton scatter of the Te Kα line lies near the Sn Kα line.

The main lines for lead are $L\iota = 9.183$ keV, $L\alpha_{2,1} = 10.448, 10.550$ keV, $L\beta_{4,1,2,3} = 12.304, 12.612, 12.621, 12.791$ keV, $L\gamma_1 = 14.762$ keV; for antimony, $K\alpha_{2,1} = 26.106, 26.355$ keV, $K\beta_3 = 29.674$, $K\beta_1 = 29.721$ keV, $K\beta_2 = 30.388$ keV (Johnson & White, 1970).

While this technique is appealing because it is rapid and non-destructive, it registers predominantly surface contributions; the possibility that lead antimonate yellow may occur as an overpaint must be considered. Other pigments containing antimony are the little used antimony white (König, 1956, p. 49) and Mérimée's yellow (Reclus, 1956, p. 377).

5.4 Criteria for Positive Identification

Light microscopy, whether crystallographic or microchemical, is often the only available method of analysis. Lead antimonate yellow is the sole inorganic yellow pigment, of those compounds usually regarded as artists' pigments, that is isotropic and has a high refractive index (2.01–2.28). (Cobalt yellow, with a refractive index of only 1.72–1.76, exhibits normal isotropism and also frequently has a characteristic dendritic appearance.) However, because lead antimonate yellow exhibits anomalous birefringence, it is easily confused with lead-tin yellow.

Identification of lead and antimony in a yellow paint by x-ray secondary-emission spectrometry or microchemical tests suggests that lead antimonate yellow is present, providing the sample is free of chlorine (Mérimée's yellow) and titanium (modern lead-antimony-titanium yellow). The presence of tin certainly calls for corroboration by another method before classification. X-ray diffraction and infrared spectroscopy are definitive techniques for the positive identification of lead antimonate yellow. In conjunction with the study of cross sections by light microscopy and electron beam microanalysis, a procedure which is almost nondestructive to particulate samples involves the following: infrared spectroscopy with a diamond cell microsampling device; x-ray primary emission spectroscopy in a scanning electron microscope; x-ray diffraction with a Gandolfi-Debye-Scherrer camera; light microscopy in Aroclor 5442 or other suitable media.

The detection of lead in a yellow pigment, without confirmation of the presence of antimony, is not sufficient to identify lead antimonate yellow. Lemon lead oxide (massicot), lead-tin yellow, or a yellow organic lake, precipitated on or mixed with lead white, could be present (Jedrzejewska, 1959).

6.0 NOTABLE OCCURRENCES

During ancient times, lead antimonate was used in glass and ceramic glazes. These are described in chronological order under ancient occurrences. Identifications of the pigment in easel paintings are also listed. According to Birren (1965) the pigment has been reported in the palettes of Anton Raphael Mengs, Goya, Jacques-Louis-David, Jean-Auguste-Dominique Ingres, Eugène Delacroix and Paul Cézanne.

Lead Antimonate Yellow in Paintings — Notable Occurrences

Artist or School *Title, Date, Collection*	*Analyst and Method* *of Identification*
Matthias Stomer, *Arrest of Christ*, 1630–1632, National Gallery of Canada, Ottawa, 4094	Hanlan (1971), Authors, XRD, XRF
Johann-Heinrich Schönfeld, *Belshazzar's Feast*, 1633, *BSTG 10 786	BSTG, EMS, Micr.
Claude Lorrain, *Seaport at Sunrise*, 1674, BSTG 381	BSTG, EMS, Micr.
Eglon Hendrik van der Neer, *Genre Scene*, third quarter, seventeenth century, BSTG 6 449	BSTG, EMS, Micr.

Artist or School *Title, Date, Collection*	*Analyst and Method* *of Identification*
Luca Giordano, *Esther and Ahasuerus*, second half, seventeenth century, *BSTG, 6 250	BSTG, EMS, Micr.
Adriaen van der Werff, *Entombment of Christ*, 1703, BSTG 224	BSTG, EMS, Micr.
Herman van der Mijn, *Garden Flowers*, 1715, BSTG 904	BSTG, EMS, Micr.
Jan van Huysum, *Fruits, Flowers and Insects*, 1735, BSTG 2 077	BSTG, EMS, Micr.
Thomas Bardwell, *William Crowe*, 1746, Castle Museum, Norwich	Tally and Groen (1975), LMA, XRD, XRF
Thomas Bardwell, *Mary Phipps*, 1749, Pembroke College, Oxford	Tally and Groen (1975), LMA, XRD, XRF
August Querfurt, *Decampment*, 1749, BSTG 6 518	BSTG, EMS, Micr.
Christian Wilhelm Ernst Dietrich, *Ideal Landscape with Storm*, 1750, BSTG 6 283	BSTG, EMS, Micr.
Thomas Bardwell, *Earl of Pomfret*, 1753, Ashmolean Museum, Oxford	Tally and Groen (1975), LMA, XRD, XRF
Sir Joshua Reynolds, *Portrait of Charles Churchill*, 1755, National Gallery of Canada, Ottawa 29	Ruggles (1971), p. 21: XRF, XRD; Taylor (1972), pp. 11, 5
Canaletto (Giovanni Antonio Canal), *Interior of St. Mark's Church*, c. 1760, Montreal Museum of Fine Arts. 871	Authors, scanning electron microscopy: date from Rosenfeld, 1976, p. 388
Thomas Bardwell, *James Poole*, 1765, Castle Museum, Norwich	Talley and Groen (1975), LMA, XRD, XRF
Johann Philipp Hackert, *Forest Landscape with Animals*, 1766–1767, BSTG 12 907	BSTG, EMS, Micr.
Ferdinand von Kobell, *Mountain Scenes with Two Wanderers and Dog*, 1768, BSTG 6 189	BSTG, EMS, Micr.
Angelika Kaufmann, *Madonna and Child*, 1774, BSTG 6 365	BSTG, EMS, Micr.
Christian George Schüz, *River Landscape*, 1777, BSTG 6 533	BSTG, EMS, Micr.
Ferdinand von Kobell, *Alpine Landscape*, c. 1780, BSTG 6 190	BSTG, EMS, Micr.
Christian George Schüz The Elder, *The Pleasure Garden of the Favorites near Mainz*, 1784, BSTG 6 564	BSTG, EMS, Micr.
John Opie, *Portrait of an Elderly Woman*, c. 1790, BSTG 8 406	BSTG, EMS, Micr.
Josef Hauber, *The Painter Ferdinand Kobell*, c. 1795, BSTG 1 146 WAF	BSTG, EMS, Micr.
Johann Georg von Dillis, *The Triva Castle*, 1797, BSTG 9 392	BSTG, EMS, Micr.
Johann Georg von Dillis, *The Dillis Family before the Forest Ranger's House in Grungiebing*, c. 1800, BSTG 10 422	BSTG, EMS, Micr.
Friedrich Heinrich Füger, *Penitent Magdalene*, 1808, BSTG 288 WAF	BSTG, EMS, Micr.
Max Josef Wagenbauer, *Pasture*, 1810, BSTG 5 533	BSTG, EMS, Micr.
Josef Anton Koch, *Heroic Landscape with Rainbow*, 1812, BSTG 447 WAF	BSTG, EMS, Micr.
Domenicho Quaglio, *A Chapel in the Woods*, 1817, BSTG 1 615	BSTG, EMS, Micr.
Wilhelm von Kobell, *The Isar Valley with Castle Harlaching*, 1819, BSTG 9 212	BSTG, EMS, Micr.
Peter von Hess, *Italian Peasant Family at Tivoli*, 1820, BSTG 355 WAF	BSTG, EMS, Micr.

Artist or School Title, Date, Collection	*Analyst and Method of Identification*
Joseph Stieler, *Elector Maximilian Joseph as King Maximilian I of Bavaria*, 1822, *BSTG 1 021	BSTG, EMS, Micr.
Johann Georg von Dillis, *Landscape at Tegernsee*, 1825, BSTG 210 WAF	BSTG, EMS, Micr.
Lorenzo Quaglio II, *Peasant Couple before the Tavern*, 1826, BSTG 14 187	BSTG, EMS, Micr.
Ernst Fries, *Neckar Landscape*, 1828, BSTG 10 132	BSTG, EMS, Micr.
Ludwig Richter, *Landscape near Olevano*, 1829, BSTG 1 246 L	BSTG, EMS, Micr.
Theodore Leopold Weller, *Italian Peasants Returning Home*, 1831, BSTG 220 WAF	BSTG, EMS, Micr.
Carl Joseph Begas, *The Artist's Son*, 1835, BSTG 10 412	BSTG, EMS, Micr.
Heinrich Burkel, *In the Campagna, Rome*, 1837, BSTG 134 WAF	BSTG, EMS, Micr.
Peter von Hess, *King Otto's Entry into Athens*, 1839, BSTG 353 WAF	BSTG, EMS, Micr.
Franz von Rohden, *Holy Family*, prior to 1846, BSTG 821 WAF	BSTG, EMS, Micr.
Wilhelm von Kobell, *Isar Landscape*, 1849, BSTG 9 213	BSTG, EMS, Micr.
Moritz von Schwind, *Honeymoon Trip*, 1850/1852, BSTG 11 577	BSTG, EMS, Micr.
Arnold Böcklin, *Nymph at the Spring*, 1855, BSTG 11 538	BSTG, EMS, Micr.
Henri Coroënne, *The Dauphin Being Separated from His Mother*, 1859, BSTG 160 WAF	BSTG, EMS, Micr.
Albert Emil Kirchner, *Verona, Seen from Giusti Giardino*, 1862, BSTG 11 624	BSTG, EMS, Micr.
Anselm Feuerbach, *Garden of Ariosto*, 1863, BSTG 11 523	BSTG, EMS, Micr.
Arnold Böcklin, *Amaryllis*, 1866, BSTG 11 535	BSTG, EMS, Micr.
Franz von Lenbach, *The Vega of Granada*, 1868, BSTG 11 458	BSTG, EMS, Micr.
Arnold Böcklin, *Springtime*, c. 1870, BSTG 11 527	BSTG, EMS, Micr.
Wilhelm Xylander, *Dutch Landscape*, 1877, BSTG 11 650	BSTG, EMS, Micr.
Auguste Renoir, *Chysanthemums*, 1882, Art Institute of Chicago 1933.1173	Gettens and FitzHugh (1974), pp. 8, 11: XRD
Hans von Marées, *Self-Portrait*, 1883, BSTG 7 868	BSTG, EMS, Micr.
Gabriel von Max, *Monkeys as Art Judges*, 1889, BSTG 7 781	BSTG, EMS, Micr.
Theodor Heine, *Thuringian Landscape*, 1893, BSTG 12 945	BSTG, EMS, Micr.
Maurice Denis, *Italian Evening Landscape*, 1898, BSTG 8 656	BSTG, EMS, Micr.
Friedrich August Kaulbach, *Portrait of the Artist's Second Wife*, 1900, BSTG 8 149	BSTG, EMS, Micr.
Odilon Redon, *Rose Window in a Cathedral, c. 1912,* BSTG 13 080	BSTG, EMS, Micr.
Carl Max Schultheiss, *Flight into* Egypt, 1933, BSTG 9 934	BSTG, EMS, Micr.

*Bayerische Staatsgemäldesammlungen. Courtesy Dr. Frank Preusser

Micr. = Light Microscopy; EMS = Emission Spectroscopy; LMA = Laser Microprobe Analyzer; XRD = X-ray Diffraction; XRF = X-ray Fluorescence Spectrometry

Lead Antimonate Yellow in Antiquity — Notable Occurrences

Origin *Object or Sample, Date, Collection*	*Analyst and Method* *of Identification*
Egyptian, XVIIIth dynasty, Fragment of yellow glass from the site of Amenhotep II's (1450–1425 B.C.) palace at Thebes, 1580–1314 B.C. Victoria and Albert Museum, London. C.8-1946	Charleston, 1960, p. 3, note 1, analysis by H. P. Rooksby; Turner and Rooksby, 1959, pp. VIII/19–22, XRD. Pyrochlore-type substance akin to $Pb_2Sb_2O_7$
Egyptian, XVIIIth dynasty, Opaque yellow glass of a model coffin, 1580–1314 B.C. British Museum, London. No. 66654	Bimson and Shore, 1966, p. 106, XRD
Tell el-Amarna (Egypt: 27.44 N 30.55 E), Canes of yellow opaque glass, XVIIIth dynasty, 1580–1314 B.C. University College, London; Corning Museum of Glass. Pb-404 and Pb-405	Brill, et al., 1973
Tell el-Amarna (Egypt: 27.44 N 30.55 E), Yellow opaque glass threading from cored vessel, XVIIIth dynasty, 1580–1314 B.C. University College, London. Box no. UC 22938 and 22937; Corning Museum of Glass. Pb-453 and Pb-454	Brill, et al., 1973
Tell al Rimah (Iraq), Yellow opaque threading from cored vessel fragment, 1500 B.C. TR 3623; Institute of Archaeology, London; Corning Museum of Glass. Pb-409	Brill, 1968, p. 59, XRD: H. Holland, B. Gullo, J. Geiger, Corning Glass Works; also Brill, 1973
Tell al Rimah (Iraq), Cane from fused mosaic vessel, glass completely weathered, 1500 B.C. TR 65-24-44, 49; Corning Museum of Glass. Pb-410	Brill, 1968
Nuzi (Iraq: 35.22 N 44.18 E), Yellow opaque glass threading from cored vessel, 1450–1375 B.C. Semitic Museum, Harvard University, Cambridge. Nuzi no. M 100/1; Corning Museum of Glass 1206; Pb-408	Brill, et al., 1973; refers to Barag, D. In Oppenheim, et al., *Glass and Glassmaking in Ancient Mesopotamia*, Corning 1970; also Brill, 1968
Tell el-Amarna (Egypt: 27.44 N 30.55 E), Thin yellow rod (diameter 3 to 5 mm), 1370 B.C. Petrie Collection, University College, London	Turner and Rooksby, 1959, p. VIII/19–22, XRD. Pyrochlore-type substance akin to $Pb_2Sb_2O_7$
Nuzi (Iraq), Glaze from faience plaque, 1450–1350 B.C. G 52; 30-2-170 D; Corning Museum of Glass. 1208	Brill, 1968
Nippur (Iraq: 31.10 N 45.11 E), Badly weathered shard with a yellow pigmented glaze or paint, 1300–1000 B.C. S.26; location WA6, Loc. 4, Fl. 10–11 NE end	Hedges, 1976, p. 212, 213, XRF.
Egyptian, XXth dynasty, Yellow thread on a fragment of a small vase. Ramses III (1196–1166 B.C.), 1200–1085 B.C. W. E. S. Turner; provided by Z. Iskander, Antiquities Chemical Laboratory, Cairo	Turner and Rooksby, 1959, p. VIII/19–22, XRD. Pyrochlore-type substance akin to $Pb_2Sb_2O_7$
Hasanlu (Iran: 36.02 N 46.18 E), Segmented "frit" bead of quartz and yellow pigment, 1000–850 B.C. HAS S-170; Corning Museum of Glass. 713	Brill, 1968
Hasanlu, Small yellow glass bead, 1000–850 B.C. HAS S-103; Corning Museum of Glass. 731	Brill, 1968
Nineveh (Iraq: 36.24 N 43.08 E), Fragment of glazed bricks with portions of polychrome floral or geometric designs in green, yellow, and white; from the site of the Nabu temple. 800–600 B.C. 1930. 716a, b, d, e, f, h, j	Hedges and Moorey, 1975, p. 26, 31, XRF, XRD. (See also Partington, 1935, p. 283)

Origin *Object or Sample, Date, Collection*	*Analyst and Method of Identification*
Khorsabad (Iraq: 36.28 N 43.13 E), Cakes of pigment in Sargon II's palace, 722–705 B.C.	G. Contenau, *Manual d'archéologie orientale* (1931), cited in Partington, 1975, p. 292
Nimrud (Iraq: 36.06 N 43.19 E), Yellow glass inlay from ivory plaque, 700–600 B.C. ND 10, 240; Corning Museum of Glass. Pb-429	Brill 1968
Rhodes, Lumps of yellow cullet from bead factory debris, probably 300–100 B.C. Corning Museum of Glass. Pb-420 and 421	Brill, 1968
Alexandria, Mosaic plaque of glass of six different colors, 100 B.C. – 100 A.D.	Brill and Moll, 1963, pp. 146–147, 148, electron microprobe
Caesarea (Israel: 32.30 N 34.54 E), Opacifier in yellow glass of surface finds, late Hellenistic–early Roman. British Museum Research Laboratory, London	Bimson and Werner, 1967, p. 264: XRD.
Roman, Fragment of a *millefiori* bowl, 100 B.C.–100 A.D.	Brill and Moll, 1963
Brough, Westmorland, England, Dragonesque brooch, second century A.D. Ashmolean Museum, Oxford. Reg. No. 1927.872	Bateson and Hedges, 1975, XRD
Southern France, British cock brooch, second century A.D. Ashmolean Museum, Oxford. Reg. No. 1927.365	Bateson and Hedges, 1975, XRD
Early Celtic, Colorant and opacifier used in the yellow glass of a Castle Newe armlet, 50–150 A.D. The British Museum, London. Reg. No. 1946 4-2, 2	Brailsford, 1975, p. 81, XRD
Bégram, Afghanistan, Concentric sealing-wax of red and opaque yellow small rings of glass, each set constituting a circular unit of *millefiori* work; these have been let, in pairs alternating with units of green and blue rings, into the rim of a colorless glass plate, 100–300 A.D. Musée Guimet, Paris	Turner and Rooksby, 1959, p. VIII/19–22, XRD. Pyrochlore-type substance akin to $Pb_2Sb_2O_7$
Caerleon (Wales: 51.37 N 2.57 W), Threading from snake-thread vessel, 140–260 A.D. 54.389 A, 491; Corning Museum of Glass. 1119	Brill, 1968
Sardis (Turkey: 38.28 N 28.02 E), Mosaic tessera, 200–400 A.D.	Brill, 1968
Dod Law, near Wooler, Northumberland, England, Glass bangle fragment with an opaque surface layer of red and yellow in stripes, late first–second centuries. National Museum of Antiquities of Scotland	Newton, 1971, p. 15 and Stevenson, 1954, p. 209, 216, XRD. (E. T. Hall and F. Schweizer, Oxford)
Traprain Law, Scotland, Three pieces of broken glass bangles with color similar to that of Dod Law sample, late first–second centuries. National Museum of Antiquities of Scotland	Newton, 1971, p. 15, XRD. (M. Bimson, British Museum)
Forgery, "Etruscan" wall paintings on terracotta, c. 1960	Fleming, et al., 1971, p. 164 and Plate 28c (*hors texte*), XRD

XRD X-ray diffraction

XRF X-ray fluorescence

ACKNOWLEDGMENTS

The authors wish to thank Dr. Frank Preusser, formerly at the Doerner-Institut, Munich, now at the J. Paul Getty Conservation Institute, for supplying the list of notable occurrences of lead antimonate yellow in paintings of the Bayerische Staatsgemäldesammlungen. Joyce Plesters, National Gallery, London, contributed her valuable advice and helpful literature references. Useful information with respect to ancient occurrences was provided by Dr. Robert H. Brill, Corning Museum of Glass, and by Mavis Bimson, British Museum Research Laboratory, London. J. V. G. Mallet, Victoria and Albert Museum, provided us with a yellow *maiolica* fragment which established an early date for lead antimonate yellow. The mineralogical advice of Dr. Joel D. Grice, Mineral Sciences Division, National Museum of Natural Sciences, Ottawa, is also gratefully acknowledged. Pigment and mineral samples on which our research was based were obtained from H. Christoff von Imhoff, Musée d'art et d'histoire de Fribourg; Elisabeth West FitzHugh, Freer Gallery of Art, Washington; the late Richard D. Buck, then with the Intermuseum Laboratory, Oberlin; the Geological Survey of Canada, Energy, Mines and Resources, Ottawa; and the Restoration and Conservation Laboratory, National Gallery of Canada, Ottawa, where the Stomer cross section was prepared by J. MacGregor Grant. Translations from the German were kindly undertaken by our associates Ursus Dix and Barbara Klempan, and those from the Italian by the late Margaret McMunn, Multilingual Services Division, Secretary of State, Ottawa. Many of the references were located by Geraldine Wallis and Maureen Clark of our library and by Nancy Logan of the Canada Institute for Scientific and Technical Information. The draft manuscripts were prepared by Louise Coderre and Michèle Le Blanc. The following manufacturers were very helpful in supplying information and samples: BASF, Farben & Fasern (Bereich Siegle), J. Blockx Fils, L. Cornelissen and Son, Ferrario Belle Arti, La Pébéo, Reeves and Sons, C. Roberson and Company, George Rowney and Company, Koninklijke Talens, Günther Wagner Pelikan-Werke, and Winsor & Newton. The collaboration of our colleagues at the Canadian Conservation Institute is also gratefully acknowledged. The infrared spectra were prepared by R. Scott Williams. Marilyn E. Laver tested the reliability of the microchemical tests. Raymond H. Lafontaine prepared the reflectance spectra. Neil T. Adair, Wilfred Bokman, Elizabeth A. Moffatt and Gregory S. Young contributed to the x-ray diffraction, radiography, and scanning electron microscopy work. For his helpful advice and interest, we owe our thanks to Dr. Kenneth J. Macleod, Director, Conservation Research Services, Canadian Conservation Institute.

REFERENCES

N. Ahmed, "X-ray Study of the System Lead Monoxide — Antimony Pentoxide — Stannic Oxide," *Pakistan Journal of Scientific and Industrial Research* 5 (1962), 124–128.

G. B. Armenini, *De veri precetti della pittura* (Ravenna, 1586), 124–126.

S. Augusti, "Analysis of the Material and Technique of Ancient Mural Paintings," in *Application of Science in Examination of Works of Art*, Proceedings of the Seminar, 7–16 September 1965, Museum of Fine Arts (Boston, 1967), 67–70.

M. Baccaredda, "Piroantimoniati di metalli bivalenti: $Ca_2Sb_2O_7$, $Cd_2Sb_2O_7$, $Pb_2Sb_2O_7$," *Gazzetta chimica italiana* 66 (1936), 539–543.

H. Bachhoffner, *Chemistry as Applied to the Fine Arts* (London, 1837), 115.

J. D. Bateson and R. E. M. Hedges, "The Scientific Analysis of A Group of Roman-age Enamelled Brooches," *Archaeometry* 17, no. 2 (1975), 177–190.

M. Bimson and A. F. Shore, "An Egyptian Model Coffin in Glass," *The British Museum Quarterly* 30, no. 3–4 (1966), 105–109.

M. Bimson and A. E. Werner, "Two problems in Ancient Glass: Opacifiers and Egyptian Core Material," *Annales du 4e Congrès international d'etude historique du verre*, Ravenna and Venice, 13–20 May 1967 (Liège, 1969), 262–266.

V. Biringuccio, *The Pirotechnia*, trans. C. S. Smith and M. T. Gnudi (Cambridge, 1966).

F. Birren, *History of Color in Painting* (New York, 1965), 31–76.

F. D. Bisagno, *Trattato della pittura* (Venice, 1642), 106–107, 117–119.

J. Blockx, *A Compendium of Painting* (1881; Antwerp, 1926), 74–75.

A. D. F. de Bondaroy, "Mémoire sur le giallolino ou jaune de Naples," in *Histoire de l'académie royale des sciences* (Paris, 1766), 303–314.

A. D. F. de Bondaroy, "Sur le giallolino ou jaune de Naples," in *Histoire de l'académie royale des sciences* (Paris, 1769), 60–64.

C. B. [Claude Boutet?], *Traité de mignature, pour apprendre aisément à peindre sans maitre*, 3rd ed. (Paris, 1684), 36.

J. Brailsford, *Early Celtic Masterpieces from Britain in the British Museum* (London, 1975), 81.

R. H. Brill, "The Scientific Investigation of Ancient Glasses," in *Proceedings, Eighth International Congress on Glass 1968* (1968), 47–68.

R. H. Brill and S. Moll, "The Electron-beam Probe Microanalysis of Ancient Glass," in *Recent Advances in Conservation*, contributions to the IIC Rome Conference, 1961, ed. G. Thomson (London, 1963), 145–151.

R. H. Brill, W. R. Shields, and J. M. Wampler, "New Directions in Lead Isotope Research," in *Application of Science in the Examination of Works of Art* (Boston, 1973), 73–83.

R. H. Brill and J. M. Wampler, "Isotope Ratios in Archaeological Objects of Lead," in *Application of Science in the Examination of Works of Art*, Proceedings of the Seminar, 7–16 September, 1965, Museum of Fine Arts (Boston, 1967), 155–166.

British Intelligence Objectives Sub-Committee, *The German Manufacture of Certain Inorganic Pigments* (London, n. d.), 165–173, 221–222.

N. S. Brommelle, "The Russell and Abney Report on the Action of Light on Water Colours," *Studies in Conservation* 9, no. 4 (1964), 140–152.

K. Brunner, "Ueber das Neapelgelb." (Dingler's) *Polytechnisches Journal*, vol. 77, pt. 63 (Stuttgart, 1837), 379–383 (including a commentary by Ehrmann).

M. H. Butler, "Applications of the Polarizing Microscope in the Conservation of Works of Art," *ZEISS Information* 22, no. 84 (1977), 4–5.

E. M. Chamot and C. W. Mason, *Handbook of Chemical Microscopy, vol. 2: Chemical Methods and Inorganic Qualitative Analysis* (New York, 1931).

R. J. Charleston, "Lead in Glass," *Archaeometry* 3 (1960), 1–4 (esp. note 1).

A. H. Church, *The Chemistry of Paints and Paintings*, 3rd ed. (London, 1901), 158–159.

W. W. Coffeen, "Antimonates as Additives to Barium Titanate Dielectric Bodies," *Journal of the American Ceramic Society* 39, no. 4 (1956), 154–158.

Colour Index, 3rd ed. (Bradford, Yorkshire, England, Research Triangle Park, North Carolina, 1971), 3,278–3,279; vol. 4, 4,675; vol. 5, ix, 5,630.

Dingler, "Ueber die Bereitung des Neapelgelbes (Mineralgelbes)," *Polytechnisches Journal*, vol. 58, pt. 28 (1828), 224–228.

M. Doerner, *The Materials of the Artist and Their Use in Painting with Notes on the Techniques of the Old Masters*, trans. by E. Neuhaus (New York, 1962), 61–62.

R. Dossie, *The Handmaid to the Arts*, vol. 1, 2nd ed. (London, 1764), 98–99.

C. L. Eastlake, *Methods and Materials of Painting of the Great Schools and Masters*, vol 1. (New York, 1960), 420, 439.

J. Elsum, *The Art of Painting after the Italian Manner* (London, 1704), 114.

Encyclopaedia Britannica, Articles on Pottery by G. Savage, Maiolica, Urbino maiolica, and Castel Durante ware (Chicago, 1974).

M. Farnsworth and P. Ritchie, "Spectrographic Studies on Ancient Glass: Egyptian Glass, Mainly of the Eighteenth Dynasty, with Special Reference to Its Cobalt Content," *Technical Studies in the Field of the Fine Arts* 6, no. 3 (1938), 155–173.

F. Feigl and V. Anger, *Spot Tests in Inorganic Analysis*, trans. R. E. Oesper, 6th ed., rev. (Amsterdam, 1972).

I. Ferrante, *Dell'historia naturale*, vol. 4 (Naples, 1599), 121.

G. Field, *Chromatography: or a Treatise on Colours and Pigments, and of Their Powers in Painting, & c.* (London, 1835), 78–79.

S. J. Fleming, *Authenticity in art: The Scientific Detection of Forgery* (New York, 1976).

S. J. Fleming, H. Jucker, and J. Riederer, "Etruscan Wall-paintings on Terracotta: A Study in Authenticity," *Archaeometry* 13, no. 2 (1971), 143–167.

G. Friedrich and R. Marx, "Untersuchungen an synthetischem Bindheimit—$Pb_2Sb_2O_7$," *Erzmetall* 15 (1962), 72–75.

J. Gerlach, U. Hennig, R. Kurz, and F. Pawlek, "Beitrag zur Kenntnis des Systems Blei-Antimon-Sauerstoff," *Metallwissenschaft und Technik* 22, no. 1 (1968), 15–24.

R. J. Gettens and E. W. FitzHugh, "Malachite and Green Verditer," *Studies in Conservation* 19, no. 1 (1974), 2–23.

R. J. Gettens and G. L. Stout, *Painting Materials, A Short Encyclopedia* (1942; New York, 1966), see also R. J. Gettens and G. L. Stout, "Trial Data on Painting Materials — Pigments and Inert Materials," *Technical Studies in the Field of the Fine Arts* 8, no. 1 (1939), 12–60.

J. F. Hanlan, "The Use of EDX for Fragile and Refractory Objects," *Developments in Applied Spectroscopy* 10 (1971), 15–32.

N. W. Hanson, "Some Painting Materials of J. M. W. Turner," *Studies in Conservation* 1 (1954), 162–173.

R. D. Harley, *Artists' Pigments: c. 1600–1835, a Study in English Documentary Sources*, 2nd ed (London, 1982), 90–91.

R. E. M. Hedges, "Pre-Islamic Glazes in Mesopotamia-Nippur," *Archaeometry* 18, no. 2 (1976), 209–213.

R. E. M. Hedges and P. R. S. Moorey, "Pre-Islamic Glazes at Kish and Nineveh in Iraq, *Archaeometry* 17, no. 1 (1975), 25–43.

M. Heiber, "The Use of an Infra-red Image Converter for the Examination of Panel Paintings," *Studies in Conservation* 13, no. 3 (1968), 145–149.

H. Hennig and E. J. Kohlmeyer, "Untersuchungen über das gegenseitige Verhalten von Blei-Antimon-Oxyden im schmelzflussigen Zustande," *Erzmetall* 10, no. 2 (1957), 64–71.

G. H. Hurst, *Painters' Colours, Oils and Varnishes*, 6th ed. (London, 1922), 202.

L. Ibrahim, R. Scranton, and R. Brill, *Kenchreai, Eastern Port of Corinth, Part II, The Panels of Opus Sectile in Glass* (Leiden, 1976), 227–255.

H. C. von Imhoff, *Research Project on Pigment Identification and Controlled Natural Alteration through Age*. ICOM paper 75/4/4. Committee for Conservation, The International Council of Museums, 4th Triennial Meeting, Venice, 1975, 3 pp.

R. Jacobi, "Über den in der Malerei verwendenten gelben Farbstoff der alten Meister," *Angewandte Chemie* 54, no. 1/2 (1941), 27–30.

JCPDS Powder Diffraction File. JCPDS International Center for Diffraction Data, Swarthmore, Pennsylvania, 1982.

H. Jedrzejewska, "Yellow Lead Pigments and Their Possible Use in Detecting Fakes," *Studies in Conservation* 4, no. 2 (1959), 78.

G. G. Johnson, Jr. and E. W. White, *X-ray Emmission Wavelengths and keV Tables for Nondiffractive Analysis* (Philadelphia, 1970).

K. L. Kelly and D. B. Judd, *Color Universal Language and Dictionary of Names* (Washington, 1976).

P. L. Kennedy, R. W. Tolmie, J. Hanlan, and J. M. Grant, *X-ray Fluorescence Spectra of Artists' Pigments Used for Paintings* (Ottawa, 1970), 9, 17, samples I-76 and I-77.

R. König, *Anorganische Pigmente und Röntgenstrahlen* (Stuttgart, 1956), 49, 69–70, 120.

H. Kühn, "Lead-tin yellow," *Studies in Conservation* 13, no. 1 (1968), 7–33.

H. Kühn, *Die Pigmente in den Gemälden der Schack-Galerie* (Munich, 1969).

H. Kühn, "Terminal Dates for Paintings Derived from Pigment Analysis," in *Application of Science in the Examination of Works of Art*, ed. W. J. Young (Boston, 1973), 199–205.

H. Kühn and C. Zocher, "Feature Cards for the Storing of Technical Data which Result from the Scientific Examination of Works of Art," *Studies in Conservation* 15, no. 2 (1970), 102–121.

V. G. Kuznetsov and P. A. Koz'min, "Structure of the Phase with the Composition $Pb_3Sb_2O_{8.47}$," *Zhurnal Neogranicheskoi Khimii* 3 (1958), 2,361–2,365 (cited in *Chemical Abstracts* 55 [1961], 1,134).

A. P. Laurie, *The Materials of the Painters Craft* (London, 1910), 127, 208.

A. P. Laurie, *The Pigments and Mediums of the Old Masters* (London, 1914), 12, 50.

E. Lemaire, "Le jaune de Naples," *La nature*, no. 3073 (15 September 1941), 313–314.

H. W. Levison, *Artists' Pigments, Lightfastness Tests and Ratings* (Hallandale, Florida, 1976), 17–18.

S. Z. Lewin and A. E. Charola, "Scanning Electron Microscopy in the Diagnosis of 'Diseased' Stone," *Scanning Electron Microscopy/1978* 1 (1978), 695–704.

G. P. Lomazzo, *A Tracte Containing the Artes of Curious Paintinge Caruinge and Buildinge*, trans. R. Haydocke (Milan, 1584; Oxford, 1598; Farnborough, England, 1970).

A. Lucas, *Ancient Egyptian Materials and Industries*, 4th ed. rev. (London, 1962).

W. C. McCrone and J. G. Delly, *The Particle Atlas*, 4 vols., 2nd ed. (Ann Arbor, 1973), 496, 735, 868.

A. Magnéli, "Crystal Structure of $PbSb_2O_6$ and Isomorphous Compounds," *Arkiv för kemi mineralogi och geologi* 15B, no. 3 (1941).

L. Marcucci, "Del giallo di Napoli (o ossido gialli di piombo mescolato con ossido di antimonio)," *Saggio analitico-chimico sopra i colori minerali* (Rome, 1813, 1816), 66–67.

B. Mason and C. J. Vitaliano, "The Mineralogy of the Antimony Oxides and Antimonates," *Mineralogical Magazine* 30 (1953), 100–112.

C. de Massoul, *A Treatise on the Art of Painting and the Composition of Colours* (London, 1797), 137–139.

R. Mayer, *The Artist's Handbook of Materials and Techniques*, 3rd ed. rev. (New York, 1970).

M. J. F. L. Mérimée, *The Art of Painting in Oil, and in Fresco: Being a History of the Various Processes and Materials Employed, from Its Discovery by Hubert and John van Eyck, to the Present Time*, trans. W. B. Sarsfield Taylor (London, 1839), 99–103.

M. J. F. L. Mérimée, *De la peinture à l'huile, ou des procédés matériels employés dans ce genre de peinture, depuis Hubert et Jan van Eyck jusqu'à nos jours* (Paris, 1830), 110–114.

M. P. Merrifield, *Ancient Practice of Painting*, vol. 1 (London, 1849), clx–clxii.

M. P. Merrifield, *Original Treatises, Dating from the XIIth to XVIIIth Centuries on the Arts of Painting, in Oil, Miniature, Mosaic and on Glass; of Gilding, Dyeing, and the Preparation of Colours and Artificial Gems*, 2 vols., intro S. M. Alexander (London, 1849; New York, 1967), 1: clvi–clxiii, 2: 528–529.

J. J. Miller, II, *English Yellow-Glazed Earthenware* (Washington, 1974), 5–6, 119–120.

J. Montoya, "Mineralogical Notes: Bismuthian Bindheimite, Mammoth Mine, Eureka, Utah," *The American Mineralogist* 54, no. 11–12 (1969), 1,726–1,728.

Munsell Color Company, *Munsell Book of Color, Neighbouring Hues Edition, Matte Finish Collection* (New York, 1973).

G. Natta and M. Baccaredda, "Tetrossido di antimonio e antimoniati," *Zeitschrift für Kristallographie, Mineralogie und Petrographie* 85 (1933), 271–295.

R. G. Newton, "A Preliminary Examination of a Suggestion That Pieces of Strongly Coloured Glass Were Articles of Trade in the Iron Age in Britain," *Archaeometry* 13, no. 1 (1971), 11–16.

R. A. Nyquist and R. O. Kagel, *Infrared Spectra of Inorganic Compounds 3800-45 cm⁻¹* (New York, 1971), spectrum no. 304, 200–201.

A. L. Oppenheim, R. H. Brill, D. Barag, and A. von Saldern, *Glass and Glassmaking in Ancient Mesopotamia* (Corning, New York, 1970).

M. L. O'Shea, "A Historical and Chemical Analysis of the Artists' Pigment Naples Yellow," (Ph. D. Thesis, Reed College, 1981).

C. Palache, H. Berman, and C. Frondel, *The System of Mineralogy*, 7th ed. (New York, vol. 1, 1944; vol. 2, 1951).

E. J. Parry and J. H. Coste, *The Chemistry of Pigments* (London, 1902), 176–179.

J. R. Partington, *A History of Chemistry*, vol. 2 (London, 1961), 77–78.

J. R. Partington, *Origins and Development of Applied Chemistry* (London, 1935; New York, 1975), 139, 256, 282–284, 292.

G. Passeri, "Istoria delle pitture in majolica fatte in Pesaro, e ne' luoghi," in *Nuova raccolta de'opuscoli scientifici, e filologici*, vol. 4 (Venice, 1758), 103–104. See also: M. Couret, "Method of Preparing Naples Yellow," in *The Repertory of Arts and Manufacturers* 3 (1795), 349–350.

C. Pedretti, *Leonardo da Vinci on Painting; A Lost Book (Libro A)* (Berkeley, 1964), 56–58.

C. Piccolpasso, *Li tre libri dell'arte del vasaio*, trans. B. Rackham and A. Van de Put (London, 1934).

J. Plesters, "Cross-sections and Chemical Analysis of Paint Samples," *Studies in Conservaton* 2, no. 3 (1956), 110–157.

P. Pomet, *Histoire générale des drogues* (Paris, 1694), 93.

G. della Porta, *Natural Magick*, ed. and preface D. J. Price (New York, 1957), 185.

M. J. Pott, *Lithéognosie ou examen chymique des pierres et des terres en général* (Paris, 1753).

H. P. Preuss, *Pigments in Paint* (Park Ridge, New Jersey, 1974), 91.

[H. Rabaté] "A propos du jaune de Naples," *Peintures, pigments, vernis* 23, no. 9 (1947), 261–262.

B. Rackham, *Italian Maiolica* (London, 1952), 15.

B. Rackham, "Italian Maiolica: Notes on Its Materials from Piccolpasso," *Studies in Conservation* 2, no. 1 (1955), 41–44.

E. Raehlmann, *Uber die Farbstoffe der Malerei* (Leipzig, 1914), 24.

Raw Materials Index, Pigment section, National Paint and Coatings Association (Washington, 1975).

M. Reclus, "Les pigments minéraux colorés," in *Chimie des peintures, vernis et pigments*, vol. 2, G. Champetier and H. Rabaté, eds. (Paris, 1956), 377.

J. S. Remington and W. Francis, *Pigments, Their Manufacture, Properties and Uses*, 3rd ed. (London, 1954), 57.

E. L. Richter and H. Härlin, "A Nineteenth-century Collection of Pigments and Painting Materials," *Studies in Conservation* 19, no. 2 (1974a), 76–82.

E. L. Richter and H. Härlin, "The Pigments of the Swiss Nineteenth Century Painter Arnold Böcklin," *Studies in Conservation* 19, no. 2 (1974b), 83–87.

J. Riederer, *Pigment Analysis with Mineralogical Methods*. ICOM paper 15/72/2. Committee for Conservation, The International Council of Museums, Madrid, 2–8 October 1972, 20 pp., esp. 11.

H. P. Rooksby, "Opacifiers in Opal Glasses through the Ages," *G.E.C. Journal of Science & Technology* 29, no. 1 (1962), 20–26.

H. P. Rooksby, "A Yellow Cubic Lead Tin Oxide Opacifier in Ancient Glasses," *Physics and Chemistry of Glasses* 5, no. 1 (1964), 20–25.

F. Rose, *Die Mineralfarben und die Durch Mineralstoffe Erzeugten Farbungen* (Leipzig, 1916), 306–310.

M. N. Rosenfeld, "Problems of Iconography in Italian Paintings," *Apollo* 103, no. 171 (May 1976), 384–391.

R. Rossi-Manaresi, "Causes of Decay and Conservation Treatments of the Tuff of Castel dell'Ovo," in *2nd International Symposium on the Deterioration of Building Stone* (Athens, 1976), 233–248.

M. Ruggles, "A Reynolds Revived," *The National Gallery of Canada, Bulletin* 17 (1971), 16–23.

W. Ruhrmann, "Das Bindemittel der alten Meister," *Technische Mitteilungen für Malerei* 50 (1934), 43–47, 52–56, 60–67, 74–76, 81–84. Abstract by G. L. Stout, *Technical Studies in the Field of the Fine Arts* 3, no. 3 (1935), 180–183.

N. I. Sax, *Dangerous Properties of Industrial Materials*, 4th ed. (New York, 1975), 859–860.

E. V. Sayre, "Summary of the Brookhaven Program of Analysis of Ancient Glass," in *Application of Science in Examination of Works of Art*, Proceedings of the Seminar, 7–16 September 1965, Museum of Fine Arts (Boston, 1967), 145–154.

E. V. Shannon, "The Occurrence of Bindheimite as an Ore Mineral," *Economic Geology* 15 (1920).88–93.

T. Shimohira and H. Tagai, "Orientation of Crystals of Lead Antimonate and Titanate on the Surface of Aluminum Porcelain Enamel," *Yogyo kyokai shi* 72, no. 8 (1964), 121–125.

R. B. K. Stevenson, "Native Bangles and Roman Glass," Society of Antiquarians of Scotland, *Proceedings* 88 (1954), 208–221.

J. K. Stockler, *Praktisches Hülfsbuch des Kunstfreundes* (Pesth and Leipzig, 1838), 7.

Z. Stös-Fertner and N. H. Gale, "Chemical and Lead Isotope Analysis of Ancient Egyptian Gold, Silver and Lead," in *Proc. 18th International Symposium, Archaeometry and Archaeological Prospection* (Oxford, 1978).

H. Tagai and T. Shimohira, "Studies on the Low-melting Enamel of the System PbO-Sb_2O_3," *Yogyo kyokai shi* 71, no. 4 (1963), 71–79.

M. K. Talley, Jr. and K. Groen, "Thomas Bardwell and His Practice of Painting: A Comparative Investigation between Described and Actual Painting Technique," *Studies in Conservation* 20, no. 2 (1975), 44–108.

J. M. Taylor, "An Improved X-ray Macroprobe Technique for Cross-section Analysis of Museum Objects," *International Institute for Conservation of Artistic and Historic Works*, 10th Congress (Lisbon, 1972).

D. V. Thompson, trans., *Il libro dell'arte, the Craftsman's Handbook of Cennino d'Andrea Cennini* (New Haven, 1933; New York, 1960).

D. V. Thompson, *The Materials and Techniques of Medieval Painting* (New York, 1956), 179–180.

D. V. Thompson and G. H. Hamilton, eds., *An Anonymous Fourteenth-century Treatise: De arte illuminandi, The Technique of Manuscript Illumination* (New Haven, 1933), 1, 28.

W. E. S. Turner and H. P. Rooksby, "A Study of the Opalizing Agents in Ancient Glasses throughout Three Thousand Four Hundred Years," *Glastechnische Berichte* 32K, no. 8 (1959), 17–28.

W. E. S. Turner and H. P. Rooksby, "A Study of the Opalizing Agents in Ancient Glasses throughout Three Thousand Four Hundred Years, Part II," *Advances in Glass Technology. Proceedings.* Part II (1963), 306–307.

G. G. Urazov and E. I. Sperenskaya, "Reaction of Lead Oxide with the Oxides of Antimony," *Zhurnal Neorganicheskoi Khimii* 1 (1956), 1,418–1,429 (cited in *Chemical Abstracts* 51 [1957], 927).

A. Ure, *A Dictionary of Arts, Manufacturers, and Mines*, vol. 2, 3rd ed. (London, 1843), 878.

Leonardo da Vinci, *A Treatise on Painting*, trans. J. F. Rigaud (London, 1877), 90.

Leonardo da Vinci, *Treatise on Painting* (Codex Urbinas Latinus 1270), trans. A. P. McMahon, vol. 1 (Princeton, 1956), 81, 200; vol. 2, pt. III, fol. 161, pt. II, fol. 76.

K. Wehlte, *The Materials and Techniques of Painting; with a Supplement on Colour Theory*, trans. U. Dix (New York, 1975).

A. M. de Wild, *The Scientific Examination of Pictures; An Investigation of the Pigments Used by the Dutch and Flemish Masters from the Brothers Van Eyck to the Middle of the 19th Century*, trans. L. C. Jackson (London, 1929), 56–58, pl. 21, charts *hors texte*.

A. N. Winchell and H. Winchell, *Elements of Optical Mineralogy, An Introduction to Microscopic Petrography*. Part II. *Descriptions of Minerals*, 4th ed. (New York, 1951), 235.

H. Winchell, *Optical Properties of Minerals, A Determinative Table* (New York, 1965), 56.

Winsor & Newton, *The Art of Landscape Painting in Oil Colours*, 3rd ed. (London, 1850), 10–11.

F. Zambonini, "Mineralogia Vesuviana," in *Atti della reale accademia delle scienze fisiche e matematiche* 14, series 2, no. 6 (Naples, 1910).

Carmine—Cochineal Carmine and Kermes Carmine

HELMUT SCHWEPPE and HEINZ ROOSEN-RUNGE

NOMENCLATURE IN VARIOUS LANGUAGES

English: cochineal
German: Cochenille, Nopalschildlaus, Koschenille
Russian: кошениль
French: cochenille
Italian: cocciniglia
Spanish: cochinilla, grana cochinilla, zacatillo

English: kermes
German: Kermes
Russian: кермес
French: kermès, kermes vermilio
Italian: chermes
Spanish: quermes

1.0 INTRODUCTION

Carmine, a red colorant that has been employed as a dye and lake pigment since antiquity, is prepared principally from two species of scale insects, cochineal or kermes. Owing to the introduction of synthetic organic colorants of greater lightfastness, the commercial use of carmine has diminished although it may still be obtained.

Kermes is one of the oldest organic colorants known; it is mentioned in both the Old and New Testaments of the *Bible* (1931). This early use of kermes is discussed in Parnell (1846, pp. 10–11) and Laurie (1911, pp. 269–272). The colorant was used in ancient Egypt and the near East and is mentioned in the writings of Dioscorides and Pliny. It was also employed in miniature painting during late antiquity and the Middle Ages. Today kermes is rarely encountered except perhaps where native craft practices persist. In the *Colour Index* (1971), it is listed as Natural Red 3, no. 75460.

Cochineal, native to the New World, was used by the Aztecs for dyeing and painting and was brought to Europe in the sixteenth century following the Spanish conquest. When hot aqueous extract of cochineal is precipitated with iron-free alum, carmine lake is obtained. The *Colour Index* (1971) lists cochineal as Natural Red 4, no. 75470.

Owing to its fugitive nature, carmine is now used principally in cosmetics and in student-grade artists' paints. It is the only organic pigment presently permitted in eye makeup preparations in the United States (Dunn, 1973).

The chemical sections in this monograph are by H. Schweppe. Discussions of the pictorial techniques of the medieval illuminators, the reconstruction of terminology used during the Middle Ages, as well as the sections on the history of art are contributed by the late H. Roosen-Runge. The section on the history of carmine preparation was written by both authors. Throughout, the characteristics of cochineal carmine, the better known and more commonly encountered pigment in recent history, will be considered first, followed by discussion of the much older kermes carmine.

HELMUT SCHWEPPE, head, central laboratory for analytical services, Applications Departments, BASF Aktiengesellschaft, Ludwigshafen am Rhine.

HEINZ E. ROOSEN-RUNGE (deceased), professor of the history of art, Kunsthistorisches Institute of the University of Würzburg.

1.1 Brief Definition of Pigment

Carmine is the currently accepted generic name for two closely related colored substances that have been employed under various designations from ancient times to the present. There are two primary species of wingless female scale insects that provide the pigments cochineal and kermes. When chemically treated, these slightly different substances yield respectively the distinctive compounds, carminic acid and kermesic acid (see fig. 1).

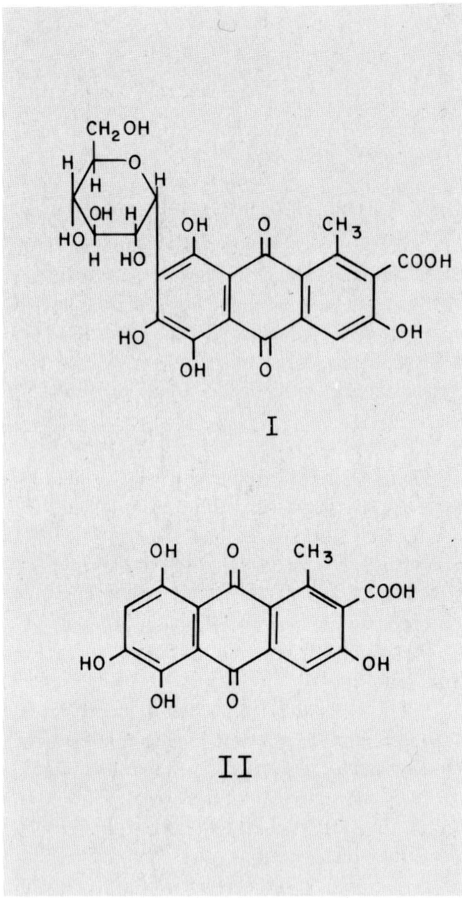

Fig. 1. Formulas of carminic acid (I) and kermesic acid (II).

The definitive structural formula of carminic acid (I) is that determined by Overeem and van der Kerk (1964), later confirmed by Bhatia and Venkataraman (1965). Formula II is the definitive structure of kermesic acid according to Gadgil, et al. (1968).

The source of cochineal carmine, the cochineal or nopal female scale insect (*Dactylopius coccus* L. Costa), is indigenous to the New World and is found in Mexico, Central America, and regions of South America. The host plants are cacti from two principal genera, *Opuntia* and *Nopalea*, represented by prickly pear and torch thistle, respectively. Cochineal carmine pigment, a protein-containing aluminum-based lake of carminic acid, was introduced into Europe about 1523 after the conquest of Mexico by the Spanish, who recognized its similarity to their native kermes. The most common carmine pigment is prepared by precipitating a hot aqueous extract of cochineal with alum that is free from iron. Cochineal scarlet is obtained by precipitating with tin salts.

Kermes female scale insect (*Kermes vermilio* Planch.) has been categorized into several species found in different areas of Europe and the Orient. The host plants are scarlet oak (*Quercus coccifera* L.), possibly the holm oak (*Quercus ilex* L.), roots of the perennial knawel (*Scleranthus perennis* L.), and strawberries as well as broad bean. A crimson-red lake is obtained by the precipitation of the extract of the kermes scale insect with iron-free alum. Besides kermesic acid, according to Dimroth and Scheurer (1913), kermes also contains a small amount of a second dye of unknown constitution, flavokermesic acid (about 3 g in 5 kg of kermes).

1.2 Current Terminology

1.21 ***Pigment.*** The modern name for the colorant is carmine. When the origin is known, the material can be referred to either as cochineal carmine or kermes carmine.

In practice, the colorants are usually encountered in the form of a lake. To achieve additional precision in characterization of the pigment, one may identify the inorganic constituents used to prepare the lake, principally citing calcium, tin, perhaps magnesium, in addition to aluminum.

1.22 ***Raw Materials.***

Cochineal: A distinction is made between the cultivated Mesteck cochineal (*grana fina, mestica, mestèque* cochineal) and the wild or wood cochineal (*grana silvestra*). The commercially available Mesteck cochineal (from Mestec, a province in Honduras) is coated with a whitish dust; the wood cochineal is covered with a type of down (Born, 1936). *Dactylopius coccus* Costa

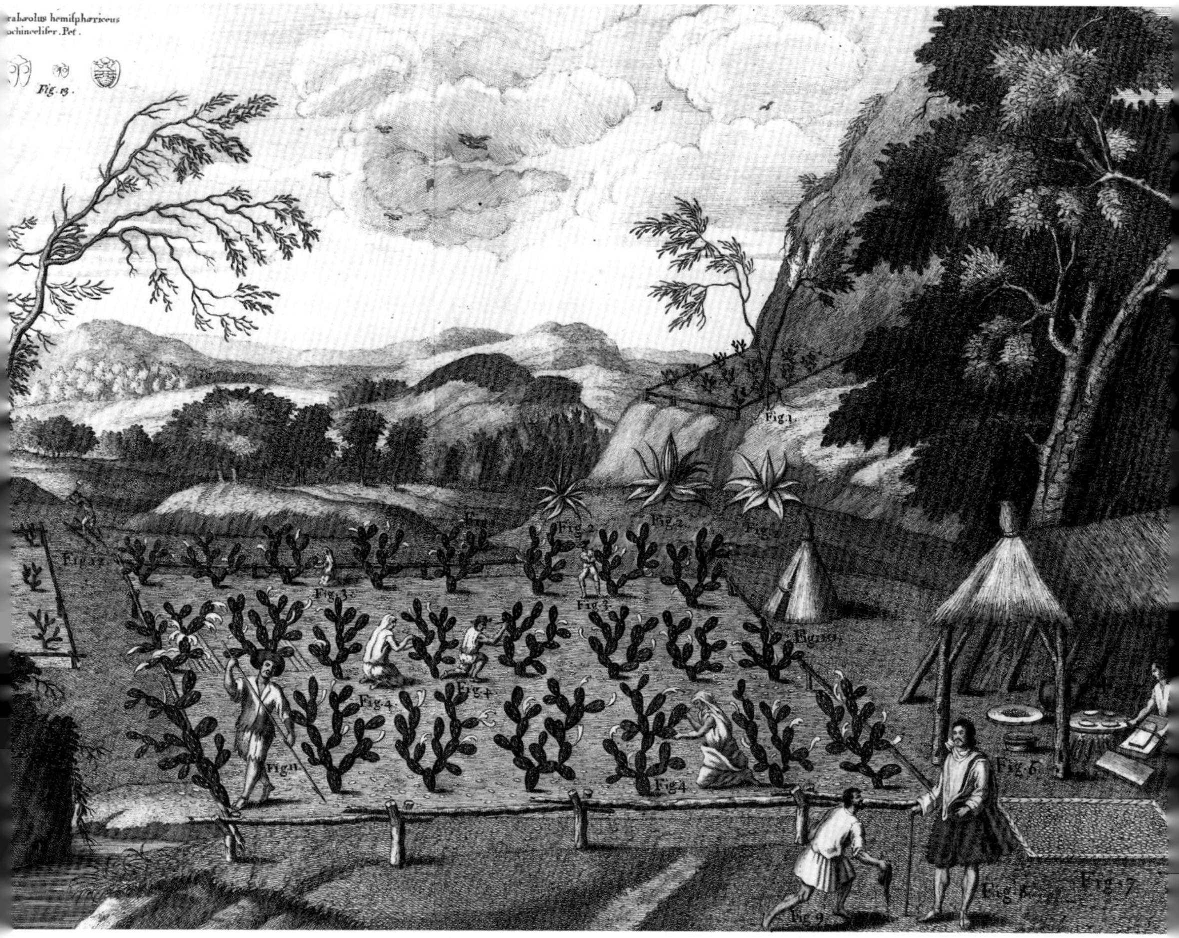

Fig. 2. Engraving of a Mexican *nopalery* as seen by a traveler in the eighteenth century. The work is entitled *The Manner of Propagating, Gathering and Curing ye Grana or Cochineal, Done by an Indian in the Bishoprick of Guaxaca in the Kingdom of Mexico in America.* From Sloane (1725), courtesy Hunt Institute for Botanical Documentation, Carnegie-Mellon University, Pittsburgh.

(earlier known as *Coccus cacti* L. or *Coccus cacti coccinelliferi*) was the most widespread of the cultivated cochineal species. Another Mesteck cochineal, *Dactylopius confusus* Cockerell, which occurs to this day in the feral form on *Opuntia exaltata* in the area of Cuzco, Peru (Fester, 1953), was cultivated during the Inca period. Fig. 2 illustrates a typical *nopalery* (plantation) as seen by a traveler in early eighteenth-century Mexico (Sloane, 1725). The wild cochineal is only half the size of the cultivated variety and has a lower content of dye. However, it propagates more rapidly than Mesteck cochineal, providing six harvests per year. Two other wild cochineal species found in north-western Argentina are *Dactylopius ceylonicus* Green and *D. tomentosus* Lam.

Kermes: The kermes scale insect (*Kermes vermilio* Planch.) lives on the host tree, the kermes or scarlet oak (*Quercus coccifera* L.) growing in the Near East, Spain, southern France, southern Italy, and on several Greek islands, especially Crete. This insect must have been the principal source for the production of a pigment in the ancient world. A second type of kermes, *Kermococcus illicis* L. (previously *Coccus illicis* L.) is to be found on the holm oak (*Quercus ilex* L.) in the Mediterranean region (Born, 1936). Kermes pigment cannot be made from this species, although many authors erroneously state otherwise.

Polish kermes (*Porphyrophorus polonicus* L.) lives on the roots of the perennial knawel (*Scleranthus perennis* L.), which is a common plant on the sandy soils of Eastern Europe. The insect was collected in Poland, Lithuania, the Ukraine, Prussia and Saxony. According to Hofenk-de Graaf and Roelofs (1972), Polish kermes contains both kermesic and carminic acid.

The scale insects, *Coccus uvae-ursi*, living on the bearberry (*Arctostaphylos uvae-ursi*), and *Coccus fragariae*, living on strawberries, have long been collected in Russia and used for the preparation of coloring matter. A different species, the scale insect *Porphyrophorus hamelii*, is found in Armenia. These three insects are considerably larger in size than the Polish kermes (Born, 1936). Pubetz (1872) also mentions the scale insect, *Coccus arabae*, living on the fava (broad) bean and cultivated in southern France since 1851. It is possible that it is identical to *Coccus fabae*.

1.3 Lac, a Related Colorant

Before closing these introductory remarks, we should remind the reader of a third important species of scale insect, *Laccifer lacca* Kerr (previously *Coccus laccae*), which lives on the twigs of certain host trees that grow in India and Southeast Asia, e.g. *Butea frondosa* Roxb., *Ficus religiosa* (L.), and *Zizyphus jujuba* Lam (Thomson, 1957). This insect produces the lac resin, better known by its refined product, shellac. It is found as "stick lac" on twigs of the above-mentioned trees which become encrusted with a reddish mammilate resin having a glossy fracture. From the resin, a red coloring matter can be extracted which is related chemically to cochineal and kermes.

In India this red dyestuff has been used since ancient times to dye silk (Ure, 1853, vol. 2, p. 10). Indian lac resin was imported for dyeing into Spain and Provence as early as 1220, the resin being macerated in water to extract the coloring matter and the aqueous solution later evaporated to form cakes known as lac dye (Laurie, 1911, p. 267). The dyestuff, CI Natural Red 25 (*Colour Index*, 1971) contains laccaic acid A as the main colored component, a compound that has a hydroxyanthraquinone chemical structure related to that of carminic and kermesic acid (see 1.1).

An adequate discussion of the utilization of lac would require a separate monograph. However, because an investigator attempting to distinguish the various reds derived from scale insects needs to keep in mind the historically important colorant from lac, some of its characteristics will be described in the analytical section.

2.0 **HISTORY**

2.1 Archaic, Obsolete and Misrepresentative Names

2.11 ***Cochineal.*** The Aztecs used a red substance for dyeing and painting that they called *nochezli* (Born, 1936). Following the conquest of Mexico, the Spaniards recognized the similarity of *nochezli* to their kermes and called it *grana cochinilla* (the Latin term *grana* and a diminutive form of *coccinus* for scarlet color). The French form of the word, *cochenille*, was soon incorporated into the English and German languages.

Extraction with alum solution and precipitation of the extract with soda produced a product that was called Florentine, Viennese, Venetian, Parisian, and Kugel lake — all largely fanciful names. In the case of Florentine lake, for example, the reference is to the early production of pigment in Florence, but much confusion can arise concerning the precise character of the colorant involved because kermes was undoubtedly employed before the arrival of cochineal.

Listed in the 1956 edition of the *Colour Index* are other names with which CI Natural Red 4 may also be associated: *cochenille*, *kermenz*, *kirmanjie*, *koscherilje*, globe lake, karmesin lake, *krappcarmin*, Munich lake, new red lake, and *Purpurcarmine*.

2.12 ***Kermes.***

2.121 ***Antiquity:*** In Greek and Roman antiquity, various terms were used for kermes, a colorant also known in the Orient (Born, 1947). The kermes pellets were considered by some to be fruits of the host plant; by others, small animals. Theophrastus (370–287 B.C.) used the term κοκκοσ (berry), and Dioscorides in the first century A.D. used this same word, describing how the κοκκοσ was harvested by scraping it with the fingernails from the host plant, an obvious reference to the kermes scale. The term *coccus* for kermes is similarly found in Pliny's writings in the first century A.D., although he also knew it as *granum* (grain)

(*Naturalis historia* IX, 141; XII, 8; XXIII, 3; XXXVII, 204). However, Pliny seems to describe an animal with the word *scolecium* (small worm) (*Naturalis historia* XXIV, 4).

Saint Jerome (Eusebius Hieronymus) (c. 348–420 A.D.) used the terms *baca* (berry), *granum*, and *vermiculus* (small worm). He also knew the term *coccus*, which he identifies as the Biblical *tholath* and equates with *vermiculus*, thus again suggesting an animal origin: "pro cocco iuxta Latinum eloquium apud Hebraeos tholath, id est vermiculus scribitur." Isidore of Seville (c. 570–636 A.D.) in his encyclopedia *Originum sive Etymologiarum* noted that the κοκκοσ of the Greeks was also called *vermiculum* or *rubrum* and that it was obtained from tree branches in the woods (Inst. XIX, 28, 1) (Blümner, 1912, p. 248; Roosen-Runge, 1967, vol. 2, p. 40).

As the use of the colorant spread afield, the synonyms were frequently exchanged, misunderstood and changed in meaning. Thus, the basic roots, from the Greek κοκκοσ as well as the Latin words *granum* and *vermiculum*, were all early terms for the description of crude kermes. From these root words, the numerous variations were derived.

2.122 ***The Middle Ages:*** The term *kermes*, which is said to have its origin in the Sanskrit *kirmidja* (engendered by a worm; from *krim*, Persian for worm), is found widely incorporated in many European languages (Kurdian, 1941). The French *cramoisi*, the German *Karmin*, *Karmoisin*, and *Karmesin*, as well as the English *crimson* are derived from it. The French *vermillion* and *vermeil*, the English *vermilion* and *vermeil*, synonyms frequently used during the Middle Ages for *kermes*, can be traced to the ancient *vermiculum* (little worm). The last apparently referred to a color that was similar to cinnabar in hue. The ways in which these and other terms used in various early manuscripts are reviewed below.

Vermiculum-Vermilion: Kermes appeared as *vermiculo* in chapter 43 of the *Capitulare de villis* (the instruction for the management of the royal Carolingean estates); the term indicates a raw material used for dyeing purposes as noted by Pertz (1835); Brühl (1971); Ploss (1952); Roosen-Runge (1967, vol. 2, p. 41). This substance, whatever it might have been, was ordered with other materials such as flax, wool, wood, madder wool cards, etc., for delivery to the women's workhouses. Notably the material appears under an antique name, obviously reflecting an old tradition.

At approximately the same date, the collection of manuscripts *Compositiones ad tingenda musiva* of the capitular library in Lucca, Codex Lucensis 490 (Hedfors, 1932, p. 40), also mentions a *vermiculum* (C 25–D 11). It is apparent from the text of this recipe, which describes the dyeing of pelts, that the term must refer to crude kermes, since digested urine is added to the *vermiculum* to dye the pelts that were pretreated with galls and alum. In this passage, we find the recommended use of reagents that were discussed repeatedly during the later Middle Ages in descriptions of the preparation of crude kermes, although the results of such treatments were variously described. Also mentioned are several other names for raw materials that must have been handed down through various traditions no longer known. A Byzantine tradition may be involved, for a number of the instructions for pigment preparation in the manuscript from Lucca must have been written originally in Greek; unfortunately the original text has been Latinized in a rather barbaric manner. The substance used for dyeing is called *vermiculum* in the passage mentioned and *coccus* in another. Again we can assume that crude kermes is meant, for the *coccus* is ground and leached in urine (Hedfors, 1932, p. 40, 165; Roosen-Runge, 1967, vol. 2, pp. 41, 94).

The terms found in the Lucca manuscript were carried into the high Middle Ages. They are used repeatedly in the copies and variations of these recipes that were combined in the texts of the *Mappae Clavicula*, compiled beginning in Carolingean times. This compendium is mentioned as early as 821 in the book catalogue from Reichenau, which lists one volume, "Mappae Clavicula de efficiendo auro, vol. I." A collection of recipes and directions, which combined art technology with general technical and chemical instructions, the *Mappae Clavicula* was copied and amended repeatedly during the Middle Ages. The most complete text known, a specimen from the twelfth century, is preserved in the Corning Glass Museum, New York. This version probably originated in northern France c. 1170–1180 and must have been a direct copy of an almost contemporaneous text from southeastern England. English words inserted in the Latin text were copied so faithfully that the southeastern dialect remains recognizable

(Phillips, 1847; Smith & Hawthorne, 1974; Roosen-Runge, 1967, vol. 1, pp. 19, 160). Both the *vermiculum* and *coccus* terms occur in this twelfth-century manuscript. However, the text is often linguistically superior to that of the Lucca manuscript, an indication that it contained the original draft of the recipes from which the Lucca version was copied in an inferior form. Thus, we conclude that crude kermes must have been known under these names far into the Middle Ages.

The term *vermiculum* for crude kermes recurs into the fifteenth century. Jean Le Begue, a licentiate in the law and Greffarius Generalium Magistrorum of the Paris mint, prepared a compendium on the painting techniques of book illumination, drawing from contemporary and older texts. He referred to tracts by Theophilus and Heraclius, whom he called "admirabilis et doctissimus magister de omni scientia picturae artis" and "sapientissimus vir." The text of Le Begue's compendium was reprinted by Merrifield (1849, vol. 1, pp. 16–321), and Roosen-Runge (1973, columns 1465, 1468) has reviewed its contents and the probable dates of the tracts by Theophilus and Heraclius that it includes. Le Begue used *vermiculum* in the same sense in which it was used by Isidore, the Lucca manuscript, and the Klosterneuburg fragment (Ganzenmüller, 1956).

Coccarin-Coccus: As mentioned, in addition to the Latin recipe texts, the Lucca manuscript contains a series of instructions which are Greco-Byzantine in origin. The expression *coccarin*, for example, may have been derived from the Greek term κοκκαγψορ for kermes (κοκκοσ) (Hedfors, 1932, p. 165; Roosen-Runge, 1967, vol. 2, p. 24). As he earlier said of κοκκοσ (*vermiculum*), Isidore of Seville in his seventh-century *Originum sive Etymologiarum* describes *coccarin* as "Nascitur sicut superius dictum est, in foliis cedri" (libri XX.) According to Hedfors (1932, p. 165) *querci* has degenerated into *cedri*, and the text refers to the kermes living on the kermes oak (Roosen-Runge, 1967, vol. 2, p. 24, 93). The fact that *coccarin* in this recipe is to be combined with a *vermiculum* to obtain a *vermiculum* is highly characteristic of the confusing manner in which the Lucca manuscript juggles the names for both raw materials and the mixtures prepared from them; this inconsistency of usage perhaps indicates various origins for the recipes combined in this collection. If *coccarin* describes crude kermes, since its origin from a deciduous tree is clearly specified, then the *vermiculum* mentioned as another basic component cannot have the same meaning. It must refer to a prepared kermes. The combination must have yielded still another substance, the preparation of which was therefore titled "Alia compositio vermiculi," to distinguish it from the "normal *vermiculum*" (Roosen-Runge, 1967, vol. 2, p. 43).

Characteristic of the traditionalism of the Middle Ages is the reappearance of the *coccarin* term used in the Lucca manuscript in the twelfth-century *Mappae Clavicula* text. The *coccarin* of the Lucca recipe R 31–S 4 is also repeated by the comment "nascitur in foliis cedri" in a parallel recipe of another Carolingean Mappae Clavicula fragment from Klosterneuburg, discovered by W. Ganzenmüller (1956). The term appears again in the Corning manuscript in the instructions CLXX with the same meaning (Roosen-Runge, 1967, vol. 2, pp. 24, 42, 94). In the Klosterneuburg fragment, a "vermiculum terrenum," a soil-produced *vermiculum*, is mentioned and described as "qui in folia cernina nascitur." The mention of growth on oak leaves excludes the possibility that the insect referred to may be the Polish kermes, which grows on the roots of a shrub, an assumption that could be suggested by the term *terrenum*. However, this writer must be referring to crude kermes; the word *terrenum* therefore might only imply that this *vermiculum* grew on a plant.

Granum-Lacca: *Granum*, another antique name for kermes, is found in writings by Pliny and Hieronymus, and also was used in the Middle Ages. Both *crana* and *grana* are used in the recipes R 31–S 4 of the Lucca manuscript (Roosen-Runge, 1967, vol. 2, p. 93) and CLXXV of the *Mappae Clavicula* of the twelfth century.

While the names mentioned in the earlier paragraphs can definitely be shown to refer to kermes, the same can only be surmised for *lacca*, which is found in the Lucca manuscript and the Corning manuscript of the *Mappae Clavicula*. However, there is good reason to believe that the term indicates kermes in both the Lucca manuscript instructions for the preparation of a *russeum* R 28–31 and the *Mappae Clavicula* recipe for *alia lazurin* (CLXXII) (Roosen-Runge, 1967, vol. 2, p. 57, color photomicrographs M 101–102). When kermes is prepared according to these directions (Roosen-Runge, 1967, vol. 2, pp. 42, 43, 93, 94), the result of the experiment is a pigment rust-red in color, matching the term *russeum* used in the recipe of the Lucca manuscript.

The name *lacca* was retained into the late Middle Ages, but not for kermes itself, rather for a carmine-red pigment obtainable from kermes or raw wool dyed with kermes. Such recipes are given by Le Begue in his *Experimenta de coloribus* (Merrifield, 1849, vol. 1, pp. 51, 53), in the fourteenth century artists' book by Cennino Cennini (Chapter 44) (Ilg, 1888; Thompson, 1933), as well as in the mid-fifteenth-century ms. 165 (recipe 137–139) of the library of San Salvatore, Bologna (Merrifield, 1849, vol. 1, pp. 455, 457; Roosen-Runge, 1973, columns 1466, 1467, 1471).

The Italian name for the dye was *grana*, from which comes the English expression "to dye in grain." Shakespeare, in *Twelfth Night*, has given testimony to the high regard held for the colorant, when he has Olivia say of her complexion, "'Tis ingrain sir! 'twill endure wind and weather."

Kermes was thus referred to by widely differing names in the early, high, and late Middle Ages. During this period the term *vermiculum* applied to a number of more or less red dyes for which kermes was the raw material. Some of these were called *lacca* and *lacha* as well. In these cases, however, the substances were described as colorants for dyeing, with the exception of the *lacca* recipe of the *Mappae Clavicula* CCLIII which, according to its title, was to be used as a pigment for the painting of wood or walls. Recipes R 1 31–S 4 (*Mappae Clavicula* CLXXV) and Lucca manuscript S 4–10 (*Mappae Clavicula* CLXXVI, section 1) describe a dry pigment (Roosen-Runge, 1967, vol. 2, p. 44).

In descriptions of pelt dyeing, the use of vermiculum as a dye is stated expressly; hence, the parallel recipes should also be ones for dyeing. It will be the goal of further research to demonstrate these recipes' use in the contemporary textile arts or the dye trade. The colorants can be prepared as pigments as well, and used for painting in combination with the binders of the Middle Ages — *clarea* (white of egg), *ichthyocollon* (isinglass), and gum arabic.

The color of kermes carmine has long been known as carmoisine red or crimson (Roosen-Runge, 1973). Carmoisine red is darker than scarlet and ordinary carmine and is less vivid. *Vermiculum* is a warm orange-red pigment prepared with vinegar or lemon juice (Roosen-Runge, 1967, vol. 2, p. 46).

2.123 *Polish Kermes:* The Polish scale insect *Porphyrophorus polonicus* (L.) (previously *Margarodes polonicus* L.) was important for the preparation of dye in Central and Eastern Europe. It was known under the names Polish Kermes, German Kermes, Root Kermes, Root Cochineal, *cremesino xavo*, *cremese*, and *chremesino tedesco* (Rosetti, 1548).

2.124 *Other names:* In Germany, during the twelfth to fourteenth centuries, kermes was collected with religious ceremony on the Feast of John the Baptist and was therefore called *Johannisblut* (Saint John's blood) (Ure, 1853, vol. 2, pp. 2–3). Additional obsolete names for CI Natural Red 3 are *kerms*, *alkermes*, scarlet corns, and kermes berries or grains. These names commonly used in the past appear in the second, 1956, edition of the *Colour Index*. A pigment known in the nineteenth century as kermes mineral was finely ground antimony sulfide (Ure, 1853, vol. 2, p. 2).

2.2 History of Use

2.21 ***Cochineal.*** The wingless scale insect involved in the production of carmine is a native of North, Central, and South America. We know that the colorant derived from it was used to dye textiles at least at the time of the old Peruvian Paracas culture (700 B.C.) (Fester, 1953). The cultivation of cochineal apparently spread from Mexico (Oaxaca) to Brazil, Peru, and Haiti (1777); and later to the East Indies (1789), Malta and Java (1828), Algiers (1834) and the Canary Islands (1835) (Pubetz, 1872). Edelstein (1958) describes the fascinating story of Nicholas Joseph Thiery de Menonville (Avocat de Parliament and Botanist to the King of France) who traveled to Guaxaca, Mexico in 1777 to purchase cacti with live cochineal and smuggle them out of the country to Haiti. In Ure's *Dictionary of Arts* (1853, vol. 1, pp. 449–452) a description is given of the raising and harvesting of the insect. Fig. 3 shows some of the primitive utensils used in the *nopalaries* of Mexico during the late nineteenth century.

When the Spaniards landed in Mexico in 1512, they found the Aztec Indians dyeing textiles and painting with this coloring matter (Born, 1936). Following the conquest, cochineal was brought to Europe and began to replace the domestic kermes both for dyeing and as an artists' pigment. The earliest date for the dye's introduction into Europe is given as 1518. Mathioli is cited for the statement that cochineal

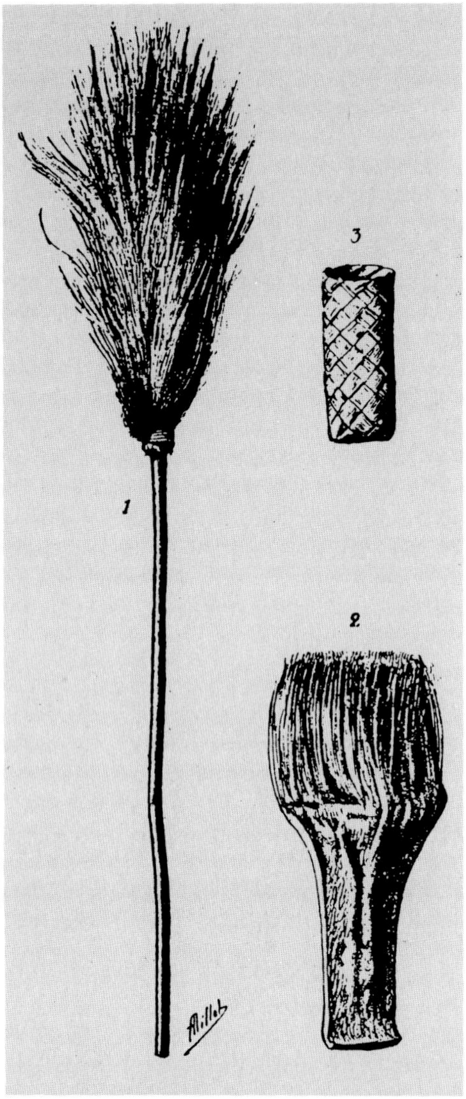

Fig. 3. Native utensils used in working with cochineal (from Diquet, 1909).
1. An *escobeta*, a soft brush used to clean away contaminating debris, such as the first moult, without harming the insects.
2. *Zilhuastle* or *chilihuostl*, a stiff brush made from an Agave leaf, used to remove the insects from the plants at the time of harvest.
3. Small container, fabricated from palm leaves, used to transfer live insects to the individual plants.

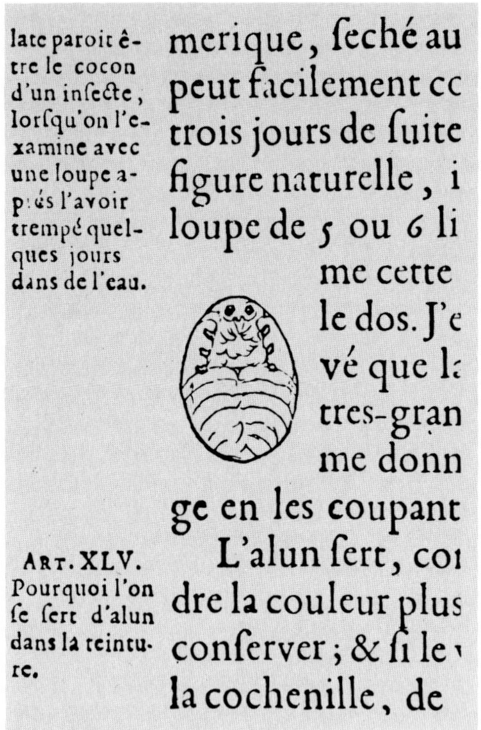

late paroit ê-
tre le cocon
d'un inſecte,
lorſqu'on l'e-
xamine avec
une loupe a-
prés l'avoir
trempé quel-
ques jours
dans de l'eau.

merique, ſeché au
peut facilement cc
trois jours de ſuite
figure naturelle, i
loupe de 5 ou 6 li
me cette
le dos. J'e
vé que l:
tres-gran
me donn
ge en les coupant

ART. XLV.
Pourquoi l'on
ſe ſert d'alun
dans la teintu-
re.

L'alun ſert, coı
dre la couleur plus
conſerver; & ſi le
la cochenille, de

Fig. 4. Perhaps the earliest illustration of the cochineal insect, from Nicolaas Hartsoeker, *Essay de dioptrique* (Paris, 1694). Hartsoeker was a manufacturer of lenses, microscopes, and telescopes who demonstrated that the red dye was derived from the dried bodies of female insects full of red eggs, not from red berries as had been previously thought (Partington, 1961).

was quite common in 1549 (Laurie, 1911, p. 271).

Fig. 4 shows what is considered to be the earliest illustration of the insect, from a 1694 book on optics by Nicolaas Hartsoeker, a manufacturer of lenses, microscopes, and telescopes who lived near Paris. The author demonstrated that the dye source was the dried bodies of female cochineal insects which were full of red eggs; Leeuwenhoek is said to have confirmed Hartsoeker's finding in 1704 but without acknowledging his source (Partington, 1961).

2.22 ***Kermes.*** The history of this ancient dye is related by Ure (1853, vol. 2, pp. 2–3) who remarks that "good kermes is plump, of a deep red colour, of an agreeable smell and a rough and pungent taste." It has been known in the East since the days of Moses and, as noted above, was employed by dyers in ancient Greece and Rome. The Spaniards paid half of their tribute to the Romans with kermes grains. That produced from the neighborhood of Emerita in

Lusitania (Portugal), was considered the best quality; that from Sicily, the worst. From time immemorial kermes has been used to dye silk in India.

Mention of the dye occurs in Exodus, 26:1 in the directions for making the curtains of the Temple (Laurie, 1911, p. 270). That the color produced by the kermes insect was the scarlet of Moses seems reasonably clear; the word *scarlet* in the books of Moses referred to a worm, and by analogy kermes would literally be translated as "wormdye." The word has been variously interpreted to mean either "double dyed" or "the best scarlet" and seems to have indicated a bright deep red dye. Altogether the terms seem to point to a species of *Kermococcus*, doubtless *K. vermilio*. It is so understood by the Septuagint and Vulgate (Bodenheimer, 1960). Professor Tychsen tells us that *tola*, the ancient Phoenician name for the dye, was used by the Hebrews and is found in the Syrian translation of Isaiah 18:1. After the captivity, the Jews more commonly used the Armenian word *zeheri*. These remarks are only a fragment of the highly illuminating discussion by Parnell (1846, pp. 10–14).

Levy (1959) has described the use of kermes for the dyeing of wool in old Mesopotamia. Even in Homer's time, kermes was an important commercial article and the only source of coloring matter for scarlet dye. The origination of such usage is attributed to the Phoenicians (*Bayer Farben Revue*, 1972).

Discoveries in southern France prove that kermes served as a dye for textile fibers as far back as the Neolithic period (Forbes, 1956). A detailed description of the use of these colorants for dyeing from antiquity to the early Middle Ages has been provided by Brown (1943).

Pfister (1935) concluded that Polish kermes was of far more than regional importance in antiquity. For example, the presence of Polish kermes on Sassanid textiles indicates that Persia had to import dyes from the regions now belonging to southern Russia, where the Scythians once lived.

2.3 Terminal Date

Natural cochineal carmine is still produced today, but owing to its fugitive nature it is used mostly for cosmetics. Its price during 1970 ranged from $42.25/lb. to $52.50/lb. (Dunn, 1973). Synthetic carminic or kermesic acid is much too difficult to prepare to be of commercial interest.

3.0 GENERAL PIGMENT PROPERTIES

3.1 Color and Spectral Reflectance

Fig. 5 shows a sample of "carmine nacarat," a very pure carmine, in the form of a palette-knife coat. Deep red carmine lakes, obtained by dissolving the best grades of carmine in a small amount of ammonia and precipitating with alum, tend to have a bluish shade. When inert white additives such as starch, blanc fixe, or Georgia kaolin are added before the precipitation, lighter colored lakes are obtained (Gentele, 1909).

Scarlet-colored carmine lakes are obtained with tin (II) salts (cochineal scarlet); the addition of magnesium sulfate to a cochineal extract produces a rose red to crimson-colored lake upon addition of sodium carbonate solution. Carmines can also acquire a weakly violet nuance when they are exposed to ammonia vapor for a short time (Gentele, 1909).

Fig. 6 shows the variations in hue of several types of carmine paints coated on watercolor paper at incomplete hiding. (In the paint industry, when paint is applied to a half white and half black background and the division of the two areas is invisible, the condition is known as complete hiding; if the division is seen, it is incomplete hiding.) The spectral reflectance curves of several of these representative hues are

Fig. 5. Paintout of very pure carmine, *carmin nacarat*. The vehicle is Medium III from Lucas (Dr. Fr. Schoenfeld and Co., Düsseldorf), based on a mixture of a cyclohexanone and an acrylic resin.

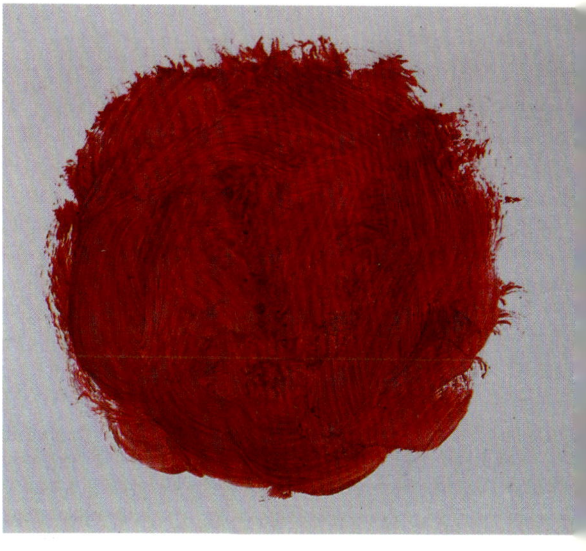

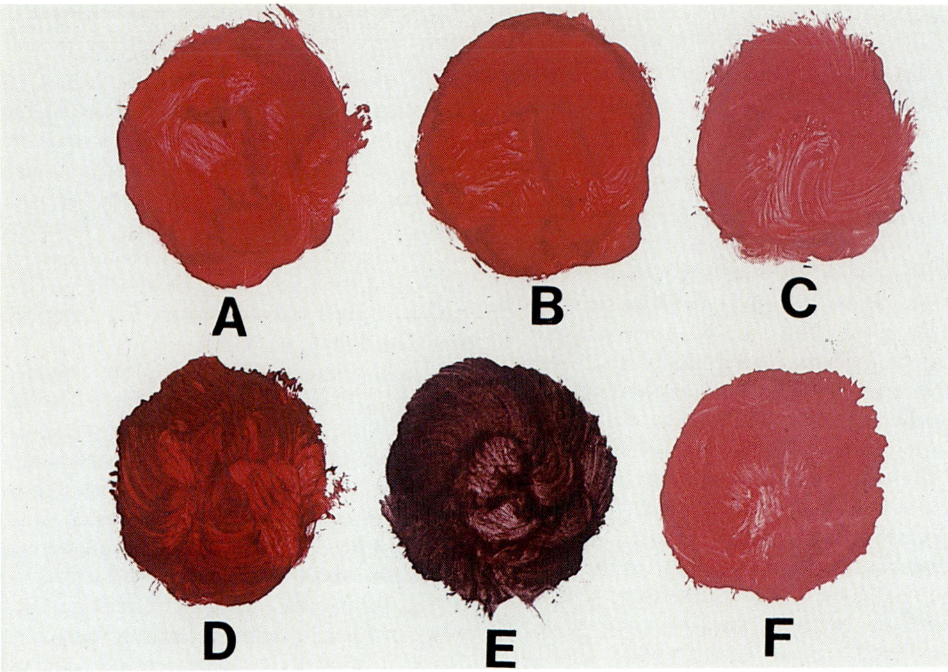

Fig. 6. Paintouts of various carmine-based pigments. A, B, C. Different preparations of cochineal carmine. D. Carmine lake light (aluminum base). E. Carmine lake dark (aluminum base). F. Carmine lake (tin base). Vehicle same as that in fig. 5.

Fig. 7. Spectral reflectance curves of representative carmine lake pigments painted at incomplete hiding over white watercolor paper. Colors specified in terms of Munsell notation Hue/Value/Chroma. (Compare with figs. 6 and 8.)

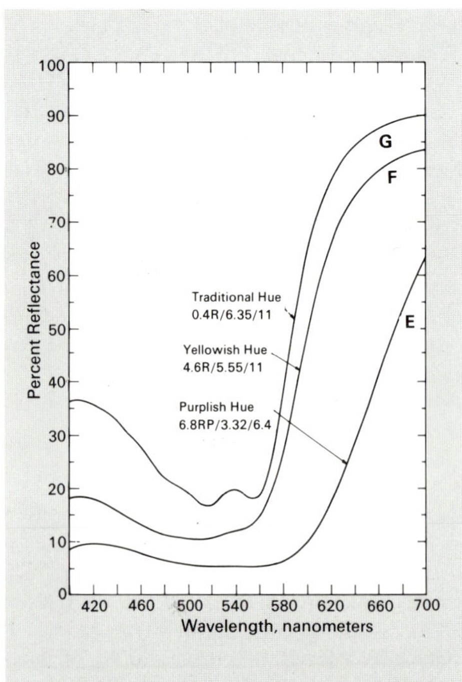

shown in fig. 7 and the location of the hues on the CIE chromaticity diagram is shown in fig. 8. In the latter figure one can see that the hue can vary from a dark purplish red (Sample E, dominant wavelength about 497c — a purple hue that is the complement to 497 nm green) to a strong red or deep yellowish pink (Sample F, dominant wavelength about 612 nm). These designations are the ISCC-NBS standard color names (Kelly & Judd, 1976). The influence of pH in changing the absorption curves of carmine in solution may be seen in fig. 9 (see section 3.5).

3.2 Hiding Power and Tinting Strength

Carmine lakes tend to be translucent when used in varnishes or oil colors; they are thus employed principally as glazes in these mediums. However, the pigments are reasonably opaque when used in tempera and glue-size vehicles.

3.3 Permanence

Carmine lakes have little resistance to light and weather; they fade readily even under incandescent illumination. Oil paints and watercolors based on modern carmine lakes exhibit a light-

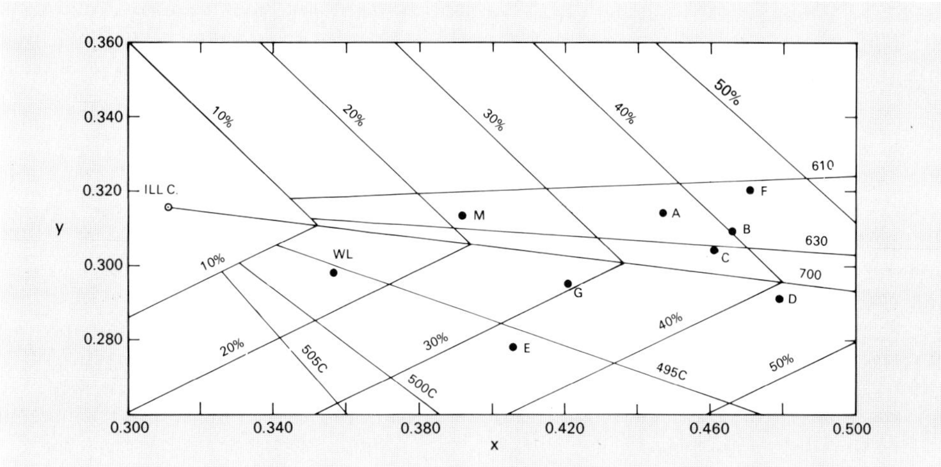

Fig. 8. Location of various cochineal carmine hues on CIE chromaticity diagram. Samples A–F same as in fig. 6. G. Masstone of carmine lake in animal glue. M. Masstone of carmine lake in oil. WL. Chromaticity of lead white in oil.

Fig. 9. Spectral absorption curves for aqueous solutions of carminic acid at various pH values.

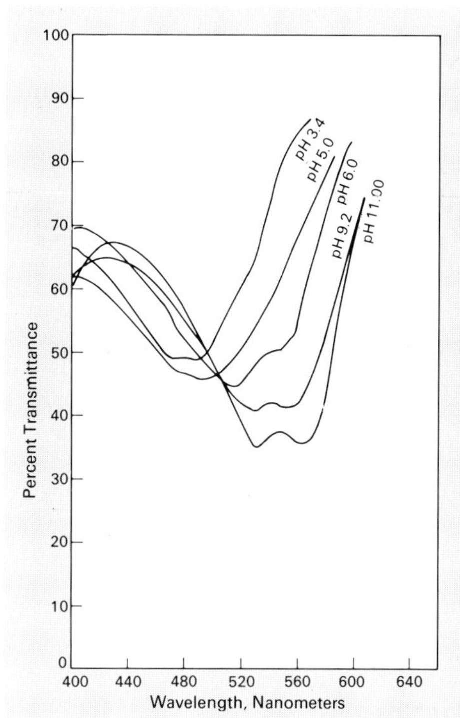

fastness equivalent to between 1 and 2 on the British Standard 1006:1971 blue-wool scale (Feller, 1963). Padfield and Landi (1966) found that cochineal and kermes, when dyed on wool, tended to exhibit higher lightfastness: equivalent to BS1006 Class 3 with an aluminum mordant; between Class 5 and 7 with a tin mordant. Duff, et al. (1977) reported that a textile dyed with the tin mordant fell in BS1006 Class 4, more fugitive than when dyed with alizarin mordanted with aluminum or tin (Class 5-6). Dyeings with lac are generally considered to have better fastness than those with cochineal and kermes.

A comment regarding the long-term stability of the raw material is made by Ure (1853, vol. 1, p. 450): "Cochineal keeps for a long time in a dry place. Hellot says that he has tried some 130 years old, which produced the same effect as new cochineal."

3.4 Compatibility

Carmine is not affected by hydrogen sulfide (Schultz, 1931; Seufert, 1955), but would be affected by basic or acidic conditions. Thus, although carmine lakes are suitable for watercolor, size, oil and egg tempera, their resistance to the alkalinity of lime and waterglass leaves much to be desired.

3.5 Chemical Properties

The chemical difference between cochineal carmine and kermes carmine is slight (see 1.1 and 5.2), and the same fundamental hues are formed with the various metallic ions. Carmine is a dark purplish amorphous-appearing substance which forms lakes of varied hues with different metals — with mercury or tin, a scarlet; with zinc or aluminum, a crimson; with iron, a gray or purplish color. It also forms a dull violet color with barium, green with uranium, and purple with chromium (Rawson, et al., 1915, p. 111; Dunn, 1973).

Carmine is insoluble in alcohol and practically insoluble in water (Gentele, 1909). It dissolves in mineral acids producing an orange-red color; in sodium carbonate solution it turns a deep red and in strongly alkaline solutions, a deep violet. Fig. 9 shows the changes in the spectrophotometric absorption curves as the pH of the solution is varied over the range 3.4 to 11.0. Carminic acid, obtained in a pure form via its lead salt, forms red needles which darken at 130°C and carbonize at 250°C.

Kermesic acid undergoes similar reactions. It is most easily distinguished from carminic acid, however, by its solubility in ether. It is also much more sparingly soluble in cold water. Kermesic acid is soluble without decomposition in concentrated sulfuric acid giving a violet-red solution, which on addition of boric acid becomes clear blue. If flavokermesic acid is present as an impurity, a dull bluish-violet results. Remarks of further interest concerning the chemical behavior of this material can be found under 5.2. Rawson, et al. (1915) cite a peculiarity of kermes: it possesses a pleasant aromatic smell which is frequently imparted to cloth dyed with kermes.

3.6 Oil Absorption and Grinding Qualities

Carmine lakes require from 46 to 100% oil for grinding (Seufert, 1955; *Raw Materials Index*, 1975).

3.7 Toxicity

Carmine is used in cosmetics and medical studies. When injected into the blood stream, the dye has been known to cause illness owing to the lack of sanitary conditions during its preparation (see reference to Roueché, 1971 in 4.2).

4.0 COMPOSITION

4.1 Chemical Composition

4.11 ***Carminic Acid.*** Carminic acid, 7-α-D-glucopyranosyl-9,10-dihydro-3,5,6,8-tetrahydroxy-1-methyl-9,10-dioxo-2-anthracenecarboxylic acid ($C_{22}H_{20}O_{13}$), is the main colored component in cochineal (see fig. 1). In Polish kermes it is probably present together with kermesic acid. Carminic acid crystallizes from methanol/water as red prisms that decompose at 130°C with no melting point. The specific optical rotation is $[\alpha]_{645}^{15} = 51.6°$ (water). The crystals, difficult to dissolve in cold water (0.13 g/100 ml) and in alcohol, are also difficult to dissolve in ether. Insoluble in benzene and chloroform, they are soluble in pyridine (3.34 g/100 ml) and in methanol. Oxidation in the air produces deoxycarminic acid.

Dry cochineal generally contains between 10 and 14% carminic acid. Frey (1931) concluded from this low yield that a second dye, neocarminic acid, must be present in the mother liquor, the investigation of which he described. We know from more recent experience that secondary components frequently appear in the thin layer chromatographic analysis of commercial carmine samples, as noted in table 3.

Following publication of the initial study of cochineal dye by Pelletier and Caventou (1818), carminic acid itself was isolated for the first time by de la Rue (1847). Dimroth and Kämmerer (1920) postulated a preliminary structural formula; Ali and Haynes (1959) determined the constitution of the side chain at C_7. Overeem and van der Kerk (1964) finally determined the precise constitution (formula I, fig. 1), subsequently confirmed by Bhatia and Venkataraman (1965).

A marked color change for carminic acid in solution takes place in a pH range of 5–6, orange on the acid side becoming purplish and violet on the alkaline side (see fig. 9). As mentioned, carminic acid forms various colored lakes with inorganic cations (see 3.5). The addition of hydrogen potassium oxalate and tartaric acid also produces colored precipitates.

4.12 ***Kermesic Acid.*** Kermesic acid, 9,10-dihydro-3,5,6,8-tetrahydroxy-1-methyl-9,10-dioxo-2-anthracenecarboxylic acid ($C_{16}H_{10}O_8$) (fig. 1), in the form of brick-red needles decomposing at 250°C, is the main colored component in *Kermes vermilio* Planch. It is considered to be the main component in Polish kermes together with a lesser amount of carminic acid. Kermesic acid is readily soluble in methanol, ethanol, and glacial acetic acid. It is less soluble in ether (but more soluble in ether than is carminic acid), sparingly soluble in cold water, and insoluble in benzene and chloroform. It dissolves in concentrated sulfuric acid to yield a violet-red or blue color in a 10% solution of boric acid in concentrated sulfuric acid. Violet is the solution color in alkali (Dimroth, 1910). In the two types of kermes mentioned, kermesic acid occurs at a concentration of approximately 1%, accompanied by about 0.06% flavokermesic acid ($C_{13}H_8O_6$). The latter forms yellow needles which are soluble in concentrated sulfuric acid, giving a red brown color; when used with tin mordant it will dye wool dull orange (Dimroth & Scheurer, 1913).

The constitution of kermesic acid was determined by Dimroth (1910), and Dimroth and Fick (1915). Later the structure was revised by Overeem and van der Kerk (1964) and once more by Gadgil, et al. (1968) (formula II, fig. 1).

4.2 Sources

This section will primarily discuss the types and sources of the dried cochineal and kermes insects, the raw materials for the colorants. A list of current suppliers of prepared cochineal, carmine, and carminic acid is given in table I.

Table 1. **LIST OF CURRENT SUPPLIERS OF COCHINEAL, CARMINE AND CARMINIC ACID**

Carl Roth GmbH & Co. KG Schoemperlenstr. 1–5 D-7500 Karlsruhe 21 Germany	carmine and cochineal
E. Merck Postfach 4119 D-6100 Darmstadt Germany	carmine and carminic acid
Fluka AG Buchs Switzerland	carmine and carminic acid
G. Siegle & Co. GmbH D-7000 Stuttgart 30 Germany	carmine
Horace Cory & Co. Ltd. London England	carmine
J. T. Baker Chemical Co. 222 Red School Lane Phillipsburg, N.J. 08865 U.S.A.	carminic acid
H. Kohnstamm & Co., Inc. 161 Avenue of the Americas New York, N.Y. 10013 U.S.A.	carmine and carmine lake
General Color Co. Newark, N.J. U.S.A.	carmine
Eastman Kodak Co. Eastman Organic Chemicals Rochester, N.Y. 14650 U.S.A.	carminic acid and carmine alum lake

The cochineal insect is found in Mexico, Central America, and regions of South America and may be considered as native to the New World because it was not known before the discovery of the Western Hemisphere. Since then, both the host plant (cacti) and the insect have been successfully introduced into other countries (fig. 10). Some dates for the spread of cochineal culture to the East Indies, Malta and Java, Algiers, and the Canary Islands are given in 2.21.

There are two principal forms of dried cochineal. One is the silver variety, which displays a grayish red color because the furrows of its body are covered with a fine down or white bloom. The so-called black variety is dark reddish brown and devoid of any down (Leggett, 1944). The granules measure about 2.5–3 mm in diameter (fig. 11), generally having a wrinkled, frequently striated surface. The male insect, which does not yield any coloring matter, is much smaller and, being furnished with wings, does not remain attached to the plants (de Puyster, 1919).

The females are collected two to four times a year shortly before they lay eggs and are killed with hot water, steam, or dry heat. It is said that a plantation will yield up to 400 kilograms of insects in a good year. Dunn (1973) states that each insect contains about 10% carminic acid, and Gardner (1901) cites the figure of 70,000 dried insects per pound of cochineal. As mentioned earlier, the yield of the colorant from the wild varieties was considerably lower (see 1.22).

Cultivated cochineal is considered to be relatively rich in coloring matter, containing from 10 to 14% colorant by weight. The colorant exists in the dried insect — principally in the eggs — as a glycoside of carminic acid, from which the true coloring constituent, carminic acid, is readily produced (Rawson, et al., 1915, p. 110). A saponifiable wax, coccerin, also exists in cochineal, the amount varying from 0.5 to 4%. Carmine pigment is prepared by extracting cochineal with water and precipitating an aluminum salt of carminic acid with alum, giving a product which still contains nitrogenous material (Mayer & Cook, 1943; Joshi, 1962).

Cochineal takes on different appearances depending upon the way in which it is killed

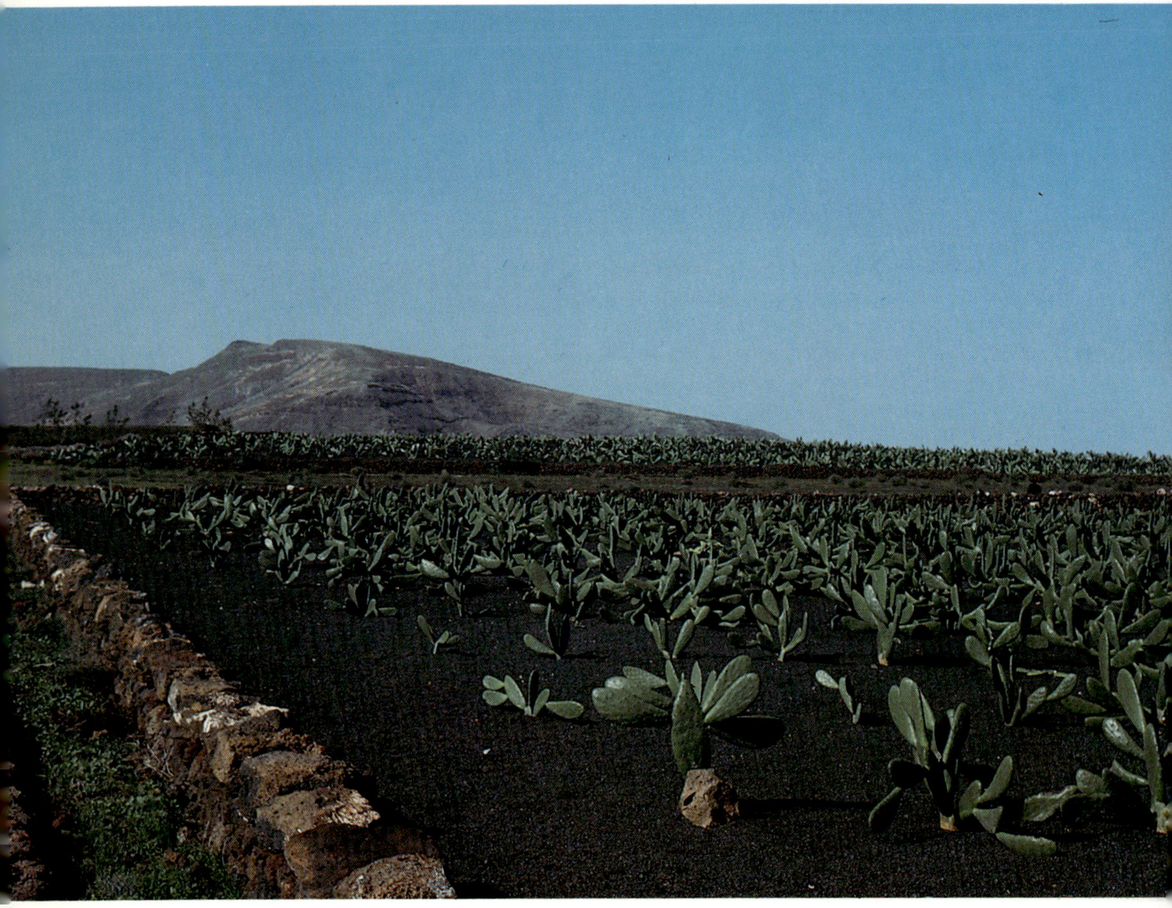

Fig. 10. Photo of *nopalery* on Lanzarote (Canary Islands) taken in 1979, courtesy of F. B. Mortimer-Ford, Horace Cory & Co., Ltd., London.

(Pubetz, 1872): *regenerida* is a shiny silver-white cochineal killed by the heat of the sun; *jaspeada*, a spotted ash-gray cochineal, is obtained by killing with steam. Some dye is leached out while the insect is in contact with hot water and steam; this settles on its surface on drying, imparting a foxy color. Insects placed in hot pans or on hot stones become dark to black in color and therefore were called *cochineal nigra* or *zaccatille*. The dead cochineal were sifted into categories of larger and smaller whole insects or shells. Fragments were sold as cochineal dust or *granilla*.

The amounts of cochineal produced in the past were considerable. According to available documents, the state of Oaxaca, Mexico, produced 27,000 tons of cochineal during the 100 years from 1758 to 1858 (Born, 1936). Diquet (1909) provides a description of cochineal culture in Mexico together with pictures of *nopalaries* and a bibliography of writings extending from 1666 to 1882. Guatemala reached the peak of its export production in 1858 with 922 tons, while the Canary Islands were leading in cochineal export in 1869 with 2,717 tons. Approximately 250 tons were harvested in the Canary Islands, and more than 100 tons in northern Argentina (in Marmol and the surrounding country) in 1927. China exported approximately 640 tons in 1924 (Ullmann, 1930).

The following types of cochineal were commercially available during the nineteenth century when the commodity was at the height of its use (Girardin, 1868–1869):

(1) Honduras cochineal, which was shipped direct to England, included three varieties: (a) black or *Zaccatille* cochineal, which was blackish or pink with traces of a white coating; (b) gray, spotted, or silver-colored cochineal, which

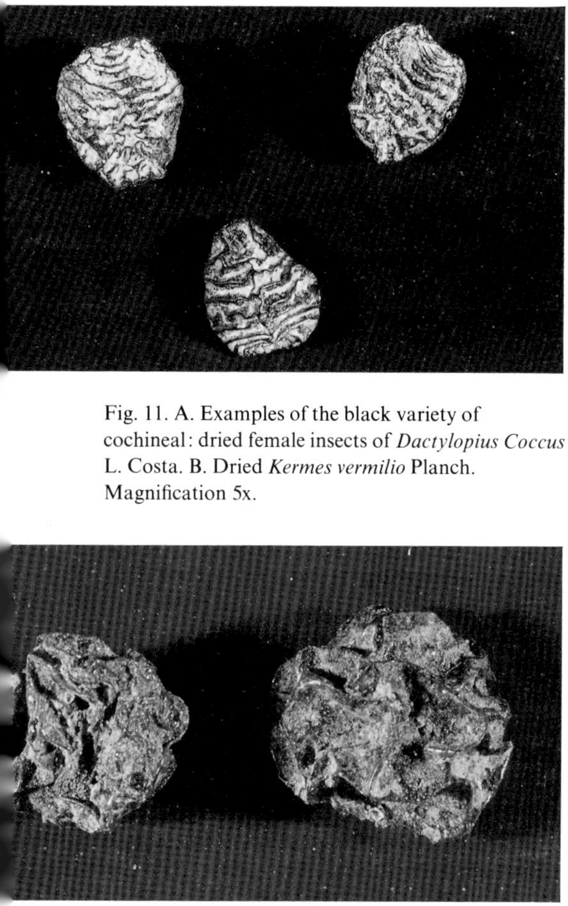

Fig. 11. A. Examples of the black variety of cochineal: dried female insects of *Dactylopius Coccus* L. Costa. B. Dried *Kermes vermilio* Planch. Magnification 5x.

had a whitish, lustrous powdery coating, (c) red cochineal which had a whitish dust deep in the folds.

(2) Vera Cruz cochineal, imported via Bordeaux and Le Havre, was available in the same three types as was the Honduras. The grains were light and shriveled, with a lustrous silvery gray color.

(3) Canary cochineal, also called Teneriffe cochineal, was shipped from the Canary Islands via Cadiz and Marseille.

(4) Java cochineal, shipped direct to Amsterdam and Rotterdam, was small in size and of inferior quality.

Wright (1963) reviews the history of cochineal production and describes contemporary attempts to revive *nopaleries* in Mexico. He says that a plantation often contained 50,000 host plants. The town of Nochisttan on the main highway from the capital to Oaxaca City derives its name from the word *Nochezttan* meaning: "the place where there is much cochineal." Several other authors have also given detailed descriptions of cochineal production: Pubetz (1872, pp. 400–409), Rupe (1900, pp. 170–172), Born (1936, pp. 228–240), Karr (1936) and Wright (1963).

Reference to the present practice of drying cochineal in the streets and dooryards of Peru and the consequent contamination of the dyestuff with salmonellae bacteria appeared in the *New Yorker* (Roueché, 1971). In one instance the natural dyestuff had been used for a dye-transit-time test on a number of patients in the city of New York, infecting them with salmonellae. The story of how the cause of the disease was eventually traced to the contaminated cochineal is a fascinating one.

Cochineal ammoniacale (also known as prepared or ammoniacal cochineal) was the ammonium salt of carminic acid, obtained by digesting powdered cochineal with ammonia and evaporating the excess. It was sold in dark brown slabs.

Kermes, which contains only about one-tenth as much coloring matter as cochineal, was commercially available in the form of round, brownish red, lustrous but slightly dusty, pea-sized kernels (figs. 11 and 12). Each kernel had a tiny aperture filled with dark red, crumb-like particles. When the kernels were ground, a red powder soluble in water and alcohol was obtained.

Kermes was harvested mainly in the eastern Mediterranean; Venice was the port of entry for the product from the East and the Orient. It is also known that the Arabs had brought the colorant to Spain from Persia. During the Middle Ages, kermes was produced in southern France, Portugal, Morocco, Algeria and Crete. Little is known, however, about the amounts harvested. Marseille became the most important port in southwestern Europe for the export of kermes.

Polish or root kermes had recognized drawbacks; it contained less coloring matter, and its collection was laborious as well as tedious, for the "kermes acorns" were only located by lifting out each plant by its roots. The insect was then removed, the roots cleaned, and the shrub replanted. Digging up the plants often killed them and thus replacements were required (Leggett, 1944). The culture of the insects in nineteenth-century Poland has been extensively discussed by Kowecka (1957). The procurement

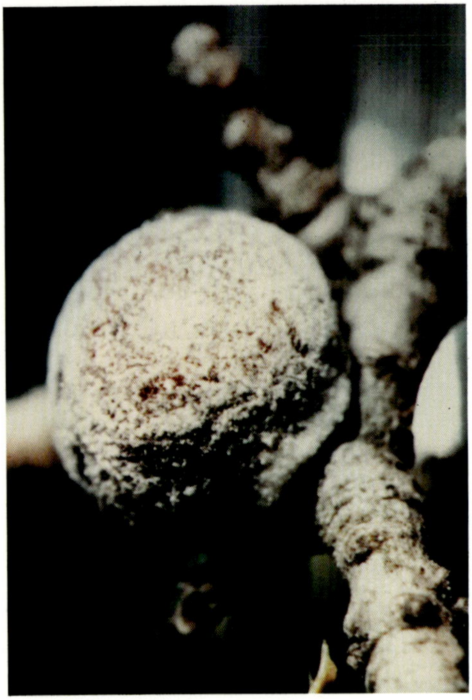

Fig. 12. Species of Kermes scale insect: A. *Kermes vermilio* Planch. which yields dye. B. *Kermes ilicis* L., black, berry-like.
Samples collected by D. Matile-Ferrero, Museum National D'Histoire Naturelle, Paris, from Quercus coccifera at Sérignon, Vaucluse, France, June 18, 1978. Magnification 6x.

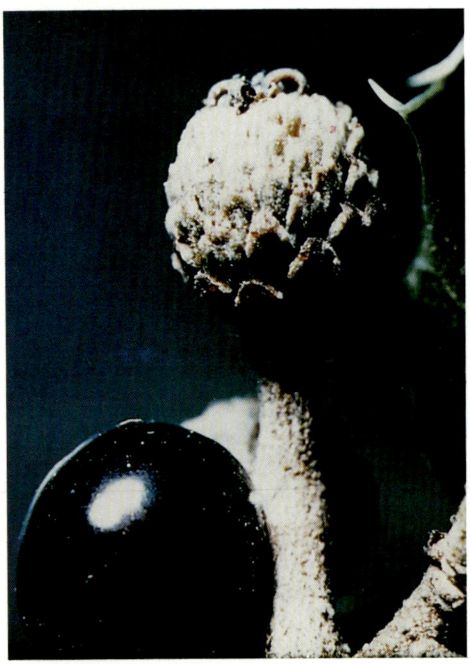

of an authentic sample of kermes today is a difficult assignment unless a site could be found where the local natural dyes are still prepared in the old tradition.

4.3 Preparation

4.31 ***Preparation of Carminic Acid from Cochineal.*** Dimroth (1913) reported that the extracts of commercial cochineal with hot water (30 liters/kg) were generally evaporated to one liter, reacted with 400 ml concentrated sulfuric acid with constant cooling, and allowed to crystallize from alcohol (yield 50 g/kg). If the crystallization is poor, the violet-colored lead lake can be made and then reacted with a calculated amount of sulfuric acid in methanol (approximately 7 liters/kg lead lake), resulting in the formation of lead sulfate and carminic acid in the methanol. After the lead sulfate is filtered off, the filtrate can be evaporated under vacuum until crystals are formed. Recrystallization from water/methanol (1:6.5) yields about 100 g carminic acid per kg of lead salt.

4.32 ***Preparation of Kermesic Acid from Kermes.*** To prepare kermesic acid, powdered kermes (1 kg) can be thoroughly extracted with ether (Dimroth, 1910). Approximately 20 g wax and 120 g fat are thereby removed while the dyestuff remains mostly undissolved. The ether is driven off, and the powder remaining is soaked with 700 ml of approximately 14% ethereal hydrochloric acid and allowed to stand overnight. The colorant present in the kermes as the salt is converted to the acid form by this procedure. Next, the acidified material is extracted with ether in a Soxhlet extractor until the solvent run-off is almost colorless, which requires approximately 100 hours. The ether extract, which still contains extracted fats and waxes in addition to the coloring matter, is evaporated to dryness and the residue is treated with methanol. The dye dissolves, whereas most of the fats and waxes remain behind when the solution cools. The preparation is filtered and 10–12 ml 2N aqueous sodium acetate solution is added to the filtrate, which precipitates the kermesic acid in the form of a brown-red sodium salt. This is separated by means of a Büchner funnel, the flavokermesic acid remaining in the filtrate. The precipitate is then transferred to a glass beaker, hot water added, and the mixture dissolved by the addition of sodium hydroxide, then treated with an excess of boiling hydrochloric acid. The kermesic acid crystallizes as small, brick-red needles (Perkin & Everest, 1918). The following day, these are separated by filtration and washed with dilute hydrochloric acid, then with water. They decompose at about 250°C without melting.

4.33 *Preparation of Cochineal Carmine and Carmine Lakes.* The "carmine of Madame Cenette," famous for its beauty, is said to have been produced in the following manner (Rupe, 1900, p. 173; de Puyster, 1919): 1 kg powdered cochineal is boiled in a copper kettle for two hours with at least 25 liters of soft river water. Then 90 g sodium nitrate is added, and after three minutes of additional boiling, 120 g oxalic acid are added and the mixture boiled for another ten minutes. The kettle is taken from the fire and allowed to stand for four hours to permit its contents to clarify. The clear liquid is then decanted into a glass vessel, and this is allowed to stand for three weeks. The separated carmine is filtered off and dried in the shade.

M. Dechan (cited by de Puyster, 1919) collected a number of standard recipes for the preparation of cochineal carmine. Other recipes for the preparation of carmine lakes may be found in Roosen-Runge's treatise (1967, vol. 2, pp. 40–46).

Carmine (cochineal carmine) or crimson pigment is currently the aluminum or aluminum-calcium lake of carminic acid which contains protein in varying amounts. It is obtained by the precipitation of the hot, aqueous extract with iron-free alum and is an intensely red fine powder. The best and purest type is called *carmine nacarat*. Wagner (1939) assumes that, during the preparation of carmine, the alkali salt of carminic acid is formed instead of a lake and that the aluminum hydroxide that precipitates during the slightly alkaline treatment is active only as an adsorbent that makes the colorant less soluble.

Burnt Carmine or roasted carmine (French: *carmine brulé*) is a brownish, dark purple-to-violet or lilac-brown color obtained by heating the carmine. Carmine violet is obtained by precipitating aqueous cochineal extract with lead acetate in acetic acid solution. Madder carmine is a madder lake whose hue has been altered with carmine.

Scarlet lake is usually a mixture of carmine (crimson lake) and vermilion, but alizarin lake may replace the carmine (Church, 1915, p. 208). Mixed lakes are prepared by the precipitation of carmine on freshly prepared alizarin lakes (cochineal-madder lake, scarlet red), on cinnabar (Munich lake), or on red-wood lakes treated with cochineal decoctions (cochineal red-wood lake, lively red).

4.34 *Preparation of Kermes-Carmine during the Middle Ages.* Processes for the preparation of carmine from kermes during this period are described by Roosen-Runge (1967, vol. 2, pp. 43–44), based on instructions in fifteenth-century treatises. One recipe, from Le Bégues' "Experimenta de coloribus," no. 11 (Merrifield, 1849, vol. 1, pp. 50–51), has been translated as follows:

> To make fine lake. — Take the ashes of oak,[1] and make a ley, and boil in it clippings of fine scarlet of rubea de grana[2] until the colour is extracted from the clippings, and then strain the ley with the colour through a linen cloth. Afterwards take some more ley, similar to what you first took, and heat it; and put into it some finely powdered roche alum, and let it stand until the alum is dissolved. Then strain it through the strainer with the other liquor or ley in which the clippings were put, and immediately the ley will be coagulated, and make a lump or mass, which you must stir well. Remove it afterwards from the vase, and lay it on a new hollow brick, which will absorb the ley, and the lake will be left dry. You must afterwards take it off the brick and keep it for use.

As a reconstruction of this recipe, Roosen-Runge carried out the preparation in the following manner:

> 300 mg kermes insects are first pulverized as finely as possible in a mortar. Then 5 ml of a saturated lye from oak ashes is added, and both again triturated finely to a smooth liquid. The liquid is strained through a fine linen cloth so that the shell particles are removed. Then an additional 5 ml saturated oak ash lye is heated, and 150 mg alum powder added to this liquid. After the alum has dissolved in the lye, the latter is again strained through a fine linen cloth into the kermes liquid obtained in the first operation. The precipitate separates immediately (the recipe says "subito dictum lessivium stringetur"), producing a colored mass from the colorless clear liquid. The pulpy colored mass is set on a dry shard of a fired clay pot, which absorbs the liquid, leaving a dark red substance that can easily be scraped off.

---

1. The Turkey oak, the *cerro* of the Italians.

2. Strictly speaking, *rubea* means madder, and *grana*, kermes; but as it appears that at this period kermes was widely used in dyeing scarlet, and as the recipes for making this "Lacca di Cimatura" generally employ the clippings of cloth dyed with kermes, it is probable that the kermes, and not madder, was meant in the present case.

A similar method, in the course of which a lye from hardwood ashes (*weid eschen*, see Ploss, 1952, p. 218 under "Weitasche"), and then alum, were added to the crushed kermes, is described in a workshop book formerly at the Municipal Library of Strasbourg (Cod. A VI, 19). The manuscript, which is only available in a copy, probably dates from the fifteenth century, but its contents are possibly older ("the late Middle Ages" was the date postulated recently by E. Ploss, 1972, p. 695). The text of this manuscript is found in Berger (1912, p. 174) and Borradaile (1966, p. 34), recipe no. 29 "Wellent ir machen schoen fin paris rot" (If you want to make a beautiful fine Paris red).

4.4 Adulteration, Sophistication and Substitution

The most common fraud described in the early twentieth century involved partially extracting the cochineal with water for a short time; the insects were then dried and sold as black cochineal. The cochineal could also be placed in a sack and shaken with talc or lead sulfate and sold as white or silver cochineal. The latter deception was easily detected by grinding the cochineal and mixing them with water, whereupon the talc or lead sulfate fell to the bottom. Good cochineal should not give over 5 to 6% of ash on ignition (de Puyster, 1919).

Napier (1869) describes nineteenth-century adulteration practices as follows:

> Another adulteration of cochineal consists in taking part of the coloring substance by a rapid ebullition in water, then steeping the insects into a concentrated logwood or peachwood liquor, drying, and covering them with ground chalk or talc.
>
> This adulteration is detected by means of lime-water, which completely precipitates the coloring matter of cochineal, and leaves the solution clear. If logwood or peachwood be present, the solution remains purplish red after the addition of lime-water.

With the introduction of synthetic pigments, attempts have been made to replace carmine with organic pigments similar in hue. Among these are Eosin (lead lake) (C.I. 45380), Lake Red C® (C.I. 15585), Lithol Red RS® (C.I. 15630), and Lithol Rubine® pigments (C.I. 15825, 15850, 15865) in the form of their calcium, barium, or manganese lakes. Even Helio Fast Carmine B and G® (C.I. pigment red 96 and 95) are suitable for this purpose (*Colour Index*, 1971). More or less applicable substitutes for carmine are the inorganic pigments cinnabar, red lead, chrome red, and molybdate red.

Kaolin, blanc fixe, and starch are used as substrates for carmine lakes in addition to aluminum hydroxide, calcium carbonate, and zinc oxide. Technical grade carmines frequently contain up to 60% lactose as diluent.

Sophistication of carmines with starch can be detected by boiling a sample in water, filtering it, and adding several drops of iodine solution to the cold filtrate. If starch is present, the solution turns blue to violet.

As mentioned, madder carmine is madder lake toned with carmine. Scarlet lake is usually a mixture of carmine lake and vermilion, but alizarin lake may replace the former (Church, 1915, p. 208). Ure (1853, vol. 1, p. 451) described the detection of vermilion thus: "Quantities of carmine were treated with a solution of caustic potash which completely dissolved them, with the exception of a beautiful red powder not acted on by potash and concentrated acid, and which was recognized to be red sulphuret of mercury or vermilion."

5.0 IDENTIFICATION — CHARACTERIZATION

5.1 Characterization by Optical and Chemical Microscopy

Lakes of carmine, alizarin, and lac tend to be isotropic and of low refractive index. Usually the reddish mass or individual particles appear translucent and "sandy" or granular within the grains rather than transparent as do, for example, natural malachite or azurite particles. Often the collection of reddish particles is not uniform, some being distinctly translucent, others more like particles of reddish-stained clay (see fig. 13).

Microscopic examination usually can only reveal the presence of an organic red (purplish red or scarlet) which is probably a lake. In such cases an essentially isotropic reddish "stain" is present that may not be dispersed into particles of well-defined morphology. Other samples may show rounded particles of low profile in Canada balsam or negative profile in mounting media of refractive index about 1.6. Occasionally substances such as chalk, barytes, or starch grains may accompany the reddish material, but their presence only provides indirect evidence concerning the character of the original colorant.

Occasionally one may extract the colorant from even microscopic-sized particles and obtain a spectrophotometric absorption curve of

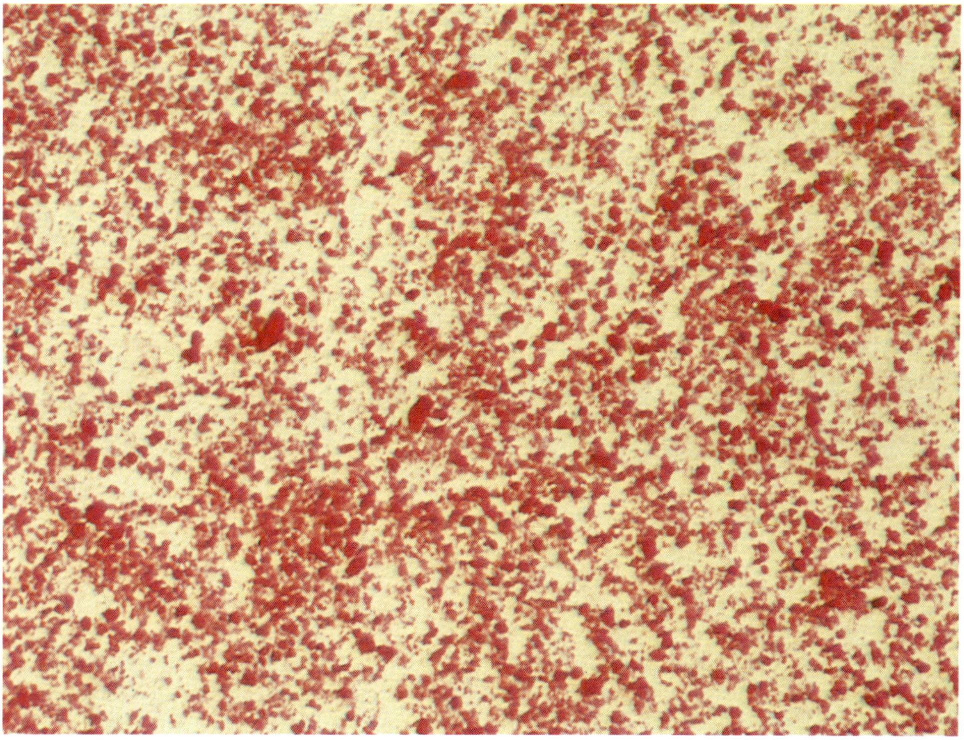

Fig. 13. Photomicrograph of modern cochineal-carmine lake artists' pigment. Magnification 650x.

the colored solution such as described in 5.32. Sometimes the colorant can be extracted with an organic solvent, and isolated by evaporating the solvent to dryness. An infrared spectrogram can then be obtained as Kühn (1968) has described in the case of alizarin. In the latter procedure, it is advisable to gain skill and experience on small specimens of known materials before attempting to perform the test on a precious sample from the work in question.

Besides the two methods just described which, if successful, can provide conclusive evidence for the specific colorant involved, there is a modern technique of microscopy that holds the promise of providing an objective method for characterizing organic colorants with reasonable confidence: microspectrophotometry. The procedure has been described by Kirby (1977). However, her identification of kermes in five Italian paintings of the fifteenth to sixteenth centuries and of cochineal in two Dutch paintings of the seventeenth century remains tentative, owing to the similarities in the curves.

5.2 Chemical Identification

5.21 ***Spot Tests.*** One of the simplest initial tests that can be carried out is observation of the fluorescence of the pigment under ultraviolet radiation. Cochineal and kermes carmine fluoresce a vivid pink; madder lake (purpurin) a dull orange, and alizarin and lac lake do not fluoresce at all (Pfister, 1935). The fluorescence under near ultraviolet is observable in the solutions of the colorants in concentrated sulfuric acid (98%) (Pfister, 1935; Schweppe, 1977).

Two to ten drops of concentrated sulfuric acid are added to 0.1–1.0 mg pigment in a small procelain dish; after approximately one minute the color of the solution and the fluorescence are observed. Coarse particles and samples containing binders can be ground and dispersed in the sulfuric acid with a glass rod. The mixture can subsequently be diluted with water, and the color of the solution is observed as well as any flocculation.

A 10% solution of boric acid in concentrated sulfuric acid is added to a second sample of pigment, and the color of the solution as well as the fluorescence is observed (Schweppe, 1977). Small samples of pigment can also be boiled with dilute hydrochloric acid (1:10), dilute ammonia

Table 2. **COLOR REACTIONS OF COCHINEAL-CARMINE, KERMES-CARMINE, AND LAC LAKE**

	(CC) Cochineal-carmine	*(KC) Kermes-carmine*	*(LL) Lac Lake*
sulfuric acid	fuchsin	dull red-violet	fuchsin
uv-fluorescence of sulfuric acid solution*	orange-red	orange-red	orange-red
10% boric acid dissolved in concentrated sulfuric acid	dark blue	brown-violet	dark blue
uv-fluorescence of boric acid–sulfuric acid solution	—	—	—
dilute ammonia (1:10)	intense fuchsin	fuchsin	—
dilute HCl (1:10)	intense orange-red	intense orange-red	intense orange-red
2% sodium hydroxide solution	intense fuchsin	intense fuchsin	intense fuchsin
2% alum solution	fuchsin	fuchsin	fuchsin

*Orange ultraviolet fluorescence indicates the presence of madder lakes, rose madder lakes, madder carmine, or purpurine lakes.

(1:10), or 2% sodium hydroxide solution, and the colors of the solutions observed (Pfister, 1935). The results of the various color reactions are compiled in table 2.

If one wished to distinguish between the natural madder lakes and synthetic alizarin lakes, a small sample can be boiled with a 2% alum solution in water. In madder lakes the purpurin content dissolves yellowish red and displays an orange ultraviolet-induced fluorescence. Alizarin lakes do not dissolve in alum solution.

5.22 *Microanalytical Distinction of Cochineal Carmine and Kermes Carmine*

The pigment to be tested, approximately 10–15 ng, is transferred to the depression of a porcelain spot plate and triturated in a drop of concentrated sulfuric acid with a glass rod. Cochineal carmine (CC), kermes carmine (KC), and lac lake (LL) may yield magenta (CC, LL) or red-violet (KC). When 1–3 mg boric acid are then added and ground with a glass rod in the drop of sulfuric acid, kermes (KC) yields a brown-violet; the other two, (CC) and (LL), a dark blue or a gray-blue. Then, 2 ml water is added to dilute the sulfuric acid, the solution is added to 1 ml ether and shaken vigorously in a microseparatory funnel.

In the ether phase, kermes (KC) turns orange; the other two, (CC) and (LL), remain in the aqueous phase. In the presence of madder lakes, however, after phase separation, the upper (ether) layer will be yellow (alizarin) to orange (purpurin, pseudopurpurin). Because kermesic acid colors the ether layer orange-red, carmine has to be detected in the presence of madder lakes by thin-layer chromatography (TLC) or by infrared spectroscopy.

The orange-colored ether layer is separated and transferred into a small test tube. When 1 ml water and 1–3 drops of a 1% uranyl acetate solution are added, the water layer turns olive after shaking and the ether layer is colorless (Schweppe, 1977).

If the aqueous phase remains orange-red after shaking with ether, it is extracted by shaking with 1 ml pentyl alcohol; (CC) and (LL) are orange colored in the pentyl alcohol phase, which can be separated in a microseparatory funnel. The alcohol extract is shaken with 1 ml water, and 1–3 drops of a 1% uranyl acetate solution are added. If only a very small amount of material is available, a 0.5% solution of uranyl acetate should be used and a blank prepared for comparison. The aqueous layer will turn green in the presence of (CC) and (LL). Approximately 0.5 ng (CC) or (LL) can be detected with this reaction, but a distinction between (CC) and (LL) is not possible.

Identification of madder will be described in detail in a subsequent monograph. When no other means seem available, some of the extracted material can always be isolated by evaporating the extraction solvent to dryness. Then an infrared spectrum can be taken of the isolated material and compared to that of samples of known materials prepared in the same manner (see 5.32).

5.23 *Determination of the Inorganic Components of Carmine Lakes.* The principal metallic ions in the inorganic component of a lake can be indicated by a preliminary examination by emission spectrographic analysis or x-ray fluorescence. The paragraphs that follow describe a traditional wet-chemical method of analysis, a procedure that does not involve the use of elaborate or advanced equipment.

Carmine is completely soluble in ammonia solution; hence, one method of isolating the inorganic components of lakes is to treat the sample with ammonium hydroxide, filter off the residue, which consists of the hydrated oxides of aluminum (white), iron (brown), and chromium (gray-green), and wash the residue with a little of the ammonia solution. Calcium ions remain in solution.

An alternative procedure for isolating hydrated oxides entirely free from organic matter is to dissolve the sample in concentrated sulfuric acid, dilute the solution with water, and extract successively with ether and isopentyl alcohol. The acid solution is evaporated nearly to dryness, any residual organic matter is oxidized by repeatedly adding a drop of perchloric acid and reheating, and the solution is again evaporated to dryness. The residue is dissolved in the minimum volume of 20% hydrochloric acid, diluted with an equal volume of water, and filtered. Excess ammonia solution is added to the filtrate, the hydrated oxides of aluminum, iron, and chromium are isolated by filtration, and the filtrate is reserved for determination of calcium.

To separate aluminum, boiling 1N sodium hydroxide is poured over the residue on the filter. Acetic acid is added to the filtrate until the solution has a pH value of about 5; a few drops of a saturated solution of morin (3, 5, 7, 2′, 4′-pentahydroxyflavone) in methanol are added, and the solution diluted with the same volume of alcohol. Aluminum displays an intensive green fluorescence.

To the remaining hydroxides after the removal of the aluminum hydroxide, a drop of a 5% solution of potassium ferrocyanide and a drop of 10% sulfuric acid is added. Blue coloration due to Prussian blue indicates the presence of iron.

To test for chromium, a part of the hydroxide residue should be set aside for fusion with Na_2CO_3 and $NaNO_3$ (1:1). The cooled fused mass can be taken up with a few drops of 10% sulfuric acid, and a few drops of a saturated solution of diphenylcarbazide in ethanol are added. Red-violet coloring indicates chromium.

For the identification of calcium, a few drops of a 5% solution of ammonium oxalate are added to the filtrate of the hydroxide residues. A white sediment indicates calcium. The test with chlorindazon C (1-6′-chloroindazol-3′-ylazo-2-hydroxynaphthalene-3-carboxylic acid) is especially suitable for the detection of calcium in very small samples of lakes (Schweppe, 1969). In this case a few drops of a 0.1% solution of chlorindazon C in dimethylformamide: ammonia:ethanol (5:45:50) are added to the filtrate of the hydroxide precipitation with ammonia. In the presence of calcium the solution is colored red-violet, and after a short time flakes appear (detection limit: 10 ng Ca; limit of concentration: $1{:}10^6$). By addition of ether, the sensitivity of the calcium test can be increased (detection limit: 5 ng Ca; limit of concentration: $1{:}10^8$). In this case the red-violet flakes will form in the boundary layer between ammonia and ether.

Before the hydroxide precipitation with ammonia, a part of the hydrochloric acid solution of the inorganic components should be removed if identification of tin and phosphate is desired. The recognition of small amounts of tin is done with the cacotheline test for stannous ions (Gutzeit, 1929). To the hydrochloric acid solution a small amount of zinc dust is added in order to transform the tin into the bivalent form. A few drops from the reduced solution are placed on filter paper which has been previously impregnated with a saturated solution of cacotheline in water. A violet stain indicates stannous compounds. Trivalent vanadium, titanium, and uranium as well as molybdenum and tungsten compounds of lower valency interfere with the test by the formation of colored precipitates.

For the detection of phosphate, a part of the hydrochloric acid solution is evaporated to dryness, and the residue dissolved in a little nitric acid. After dilution of the residue with water, an excess of ammonium molybdate solution is added and boiled off. A yellow precipitation of ammonium molybdophosphate indicates the presence of phosphate (detection limit: 50 ng phosphate).

5.24 ***Chromatography — Determination of the Organic Components of Carmine Lakes.*** Hydroxyanthraquinones such as carminic acid and kermesic acid can be readily differentiated by paper chromatography (PC) (Paris & Rousselet, 1958; Shibata, et al., 1950), by thin-layer chromatography (TLC) (Boutroup, 1970; Hofenk-de Graaf & Roelofs, 1972; Masschelein-Kleiner, et al., 1967, 1968, 1969, 1970; Schweppe, 1975) and by gas chromatography (GC) (Furuya, et al., 1966). Thin-layer chromatography (TLC) is frequently preferred over other methods for the identification of organic artists' colorants because it is simple to perform and offers the possibility of analyzing samples in the microgram range.

Carmine lakes must be hydrolyzed and the free hydroxyanthraquinones must be separated from the inorganic cations prior to the TLC comparison. Hydrolysis is carried out with mineral acids, and the separation of the hydroxyanthraquinones from the inorganic cations is accomplished by extraction with organic solvents that are not miscible with water. Although a number of possibilities are mentioned in the literature (Hofenk-de Graaf & Roelofs, 1972; Masschelein-Kleiner, 1967; Masschelein-Kleiner & Heylen, 1968; Roelofs, 1972), only one simple method of hydrolysis will be described which, from our own experience, also provides useful additional information in the form of color reactions.

A 0.1–1.0 mg sample of pigment is placed in a small porcelain dish, 2–10 drops concentrated (98%) sulfuric acid are added, and the coarse pigment or paint particles are crushed with a glass rod. The color of the solution and the uv-fluorescence are compared with the corresponding behavior of carminic acid and kermesic acid (see table 2). Then, several milligrams of boric acid are added to the sulfuric acid, the mass ground with a glass rod, and the change in color observed and compared with the known sample. Subsequently, 5 ml water is added, and the acidified aqueous solution is extracted with 5 ml n-pentyl or isopentyl alcohol in a micro-separatory funnel. The pentyl alcohol phase is shaken several times with 1-ml portions of water until the aqueous phase is only weakly acid. Then, the pentyl alcohol phase is evaporated to dryness, and the residue taken up in several drops of acetone and used for a TLC comparison with known samples of carminic acid, kermesic acid, and the laccaic acids. Acetylcellulose (with 10% acetyl content) without binder (e.g., MN-300-Ac from Macherey & Nagel) may be used as the adsorbent. The solvent system is ethyl acetate:tetrahydrofuran:water (6:35:47) (Masschelein-Kleiner, 1967; Masschelein-Kleiner & Heylen, 1968).

A good separation of substituted hydroxyanthraquinones such as those in carminic acid, kermesic acid, and the laccaic acids, as well as of their secondary components, can also be achieved on micropolyamide F-1700 ready-to-use plates (Schleicher & Schüll), with solvents (A) methanol-formic acid (95:5) (Schweppe, 1975, 1977) and (B) methyl ethyl ketone-formic acid (7:3) (private communication, Dr. K. Bredereck). These plates are available as polyester support foils covered with polyamide sintered on both sides. Separation requires only short periods of time because of the high development rates. Approximately 20–30 minutes

Table 3. **hR_f VALUES* IN THE THIN-LAYER CHROMATOGRAPHY OF HYDROXYANTHRAQUINONES OCCURRING IN MADDER AND SOME TRADITIONAL RED PIGMENTS FROM SCALE INSECTS**

Layer Material: micropolyamide F-1700 (Schleicher & Schüll)
Solvent Systems: (A) methanol-formic acid (95:5)
(B) methyl ethyl ketone-formic acid (7:3)†

The chromatography was performed under standard conditions (saturated chamber).

	hR_f Values	
	in Solvent A	*in Solvent B*
alizarin	50	66
purpurine	39	62
carminic acid	32	45
kermesic acid	14	33
flavokermesic acid	17	37
laccaic acids, A, B, C, D, E, and other hydroxyanthraquinones occurring in lac lake**	—	27, 34, 45 51, 55, 60, 70

* $hR_f = R_f \times 100$.

** Samples of pure laccaic acids A, B, C, D, E were not on hand. Therefore a connection with the hR_f values is impossible. Solvent A is unfit for separating laccaic acids.

† This solvent mixture suggested by Professor Dr. K. Bredereck, formerly of the University of Stuttgart, Federal Republic of Germany.

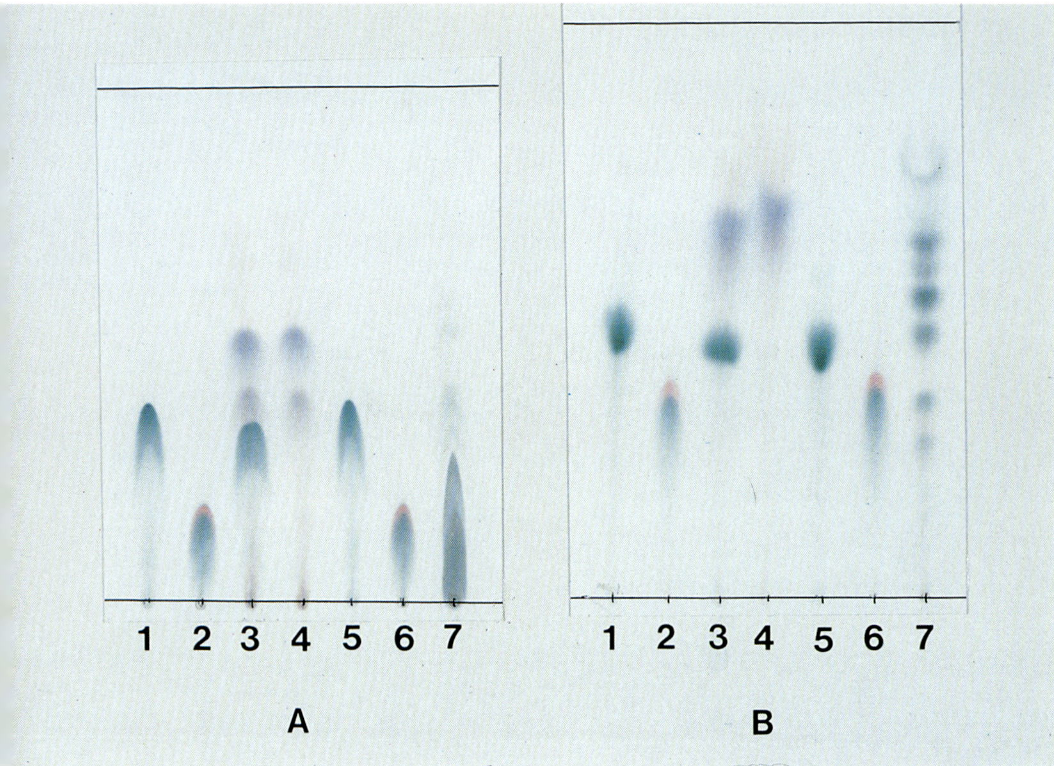

Fig. 14. Distinguishing between red artists' pigments (hydroxy-anthraquinones) by thin-layer chromatography. Layer material: Micropolyamid F-1700 (Schleicher & Schüll). Solvent systems: A. Methanol-formic acid (95:5). B. Methyl ethyl ketone-formic acid (7:3). Color reaction: Uranyl acetate (1% in water) + methanol (1:1). Pigments: 1. cochineal carmine (hydrolyzed); 2. kermes carmine (hydrolyzed); 3. madder carmine (= cochineal carmine + madder lake; hydrolyzed); 4. madder lake (hydrolyzed); 5–7 control substances: 5. carminic acid; 6. kermesic acid with flavokermesic acid (red spot); 7. laccaic acids (constituents in lac lake).

are all that is necessary with the solvent systems described.

For optimum differentiation, the following comparisons between unknown and known materials are performed with the developed chromatogram in sequence: 1) inherent color, 2) fluorescence color under uv-light, 3) change in color with ammoniacal vapor, 4) color reactions with calcium, aluminum or uranium salts.

The R_f values of carminic acid, kermesic acid, and their secondary components which are of unknown composition, as well as those of colorants comprising of laccaic acids and several other hydroxyanthraquinones are compiled in table 3.

Fig. 14 shows two thin-layer chromatograms of carminic acid, kermesic acid, and several other hydroxyanthraquinones on micropolyamide F-1700 with the solvent systems described above.

5.3 Instrumental Methods of Analysis

5.32 *Infrared Spectroscopy.* Flieder (1968), Masschelein-Kleiner and Heylen (1968), and van't Hul-Ehrnreich (1970) have described the determination of natural organic pigments by the comparison of infrared spectra. Abrahams and Edelstein (1964, 1967) have described the identification of natural colorants on old textiles. Correlations between the location of absorption maxima in the infrared range and the constitutions of hydroxyanthraquinones can be found in the work of Bloom, et al., (1959).

An infrared spectrum may be prepared without hydrolysis if a pure cochineal carmine is available. Fig. 15A shows a spectrogram of carmine-nacarat lake. Sharper absorption bands result, however, when the test sample of a

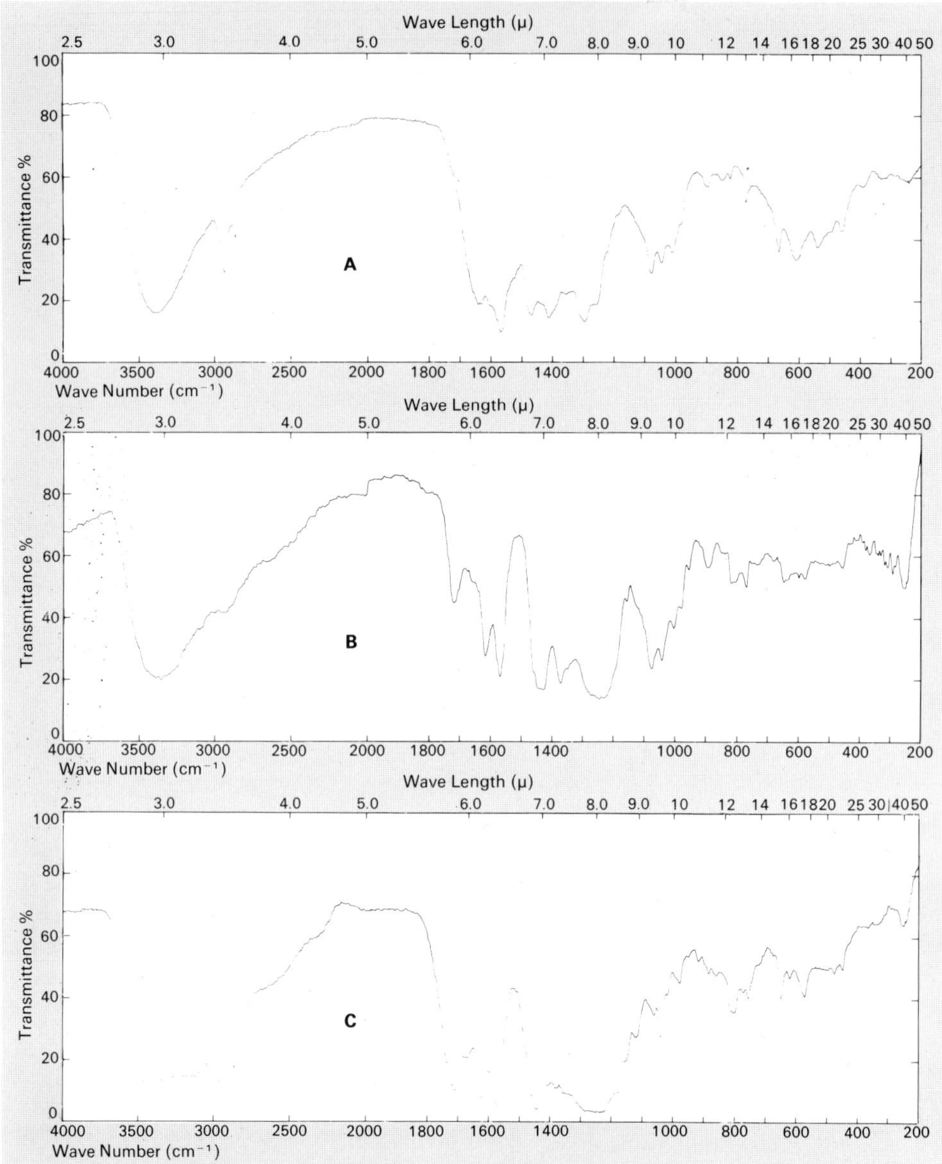

Fig. 15. Infrared curves of A. Carmine Nacarat. B. Carminic acid. C. Kermesic acid.

carmine lake is hydrolyzed, and the spectrum of the free carminic or kermesic acid is then obtained. The hydrolysis can be performed as described for the preparation of the free acids prior to thin-layer chromatography.

An infrared spectrum of carminic acid obtained from cochineal carmine by hydrolysis is shown in fig.15B. Fig.15C shows a spectrum of kermesic acid obtained from kermes carmine.

Recording the infrared spectrum with a 12 mm-diameter potassium bromide tablet requires a larger amount of material than thin-layer chromatography. Because the available amount of sample material will usually be small, a microtechnique for recording the spectrum on 2-mm tablets is appropriate. This method requires only approximately 10 ng pure substance.

5.33 ***Spectroscopy in the Visible Range.*** The measurement of the absorption spectrum of a pigment in the visible range is not disturbed by the presence of colorless secondary components. Saltzman, et al., (1936) were the first to use this method for the analysis of organic colorants on old textiles. They have detected Tyrian purple and cochineal on old Peruvian fabrics. Using the Saltzman method, Kashiwagi (1976) published the absorption spectra of cochineal and of red dyes found at Chancay and Tiahuanaco, claiming good correspondence with samples of known modern materials.

The absorption spectra of carminic and kermesic acid in the range 400–700 nm can be obtained with solutions in concentrated sulfuric acid — for example, a 10% solution of boric acid in concentrated sulfuric acid or in dilute sodium hydroxide solution. The various hydroxyanthraquinones frequently develop different colors in solution under these conditions and therefore have different absorption spectra.

Table 4 gives the wavelengths corresponding to the absorption maxima of carminic and kermesic acid in the visible range in various solvents (Dimroth & Fick, 1915). The weaker bands are indicated by the figures in parentheses. The spectral absorption curves of cochineal solutions at different pH values are shown in fig. 9.

Table 4. **PRINCIPAL ABSORPTION PEAKS OF COLORANTS IN SOLUTION (Dimroth and Flick, 1915)**

	Wavelengths in nm	
Solvent	*Carminic Acid*	*Kermesic Acid*
in concentrated sulfuric acid:	544 nm	(633) nm
	504	(543)
	(474)	501
		(470)
in concentrated sulfuric acid + 10% boric acid:	(624)	623
	(581)	(576)
	498	(529)
	(466)	493
		(463)
in potassium hydroxide:	571	567
	530	527
	(492)	(493)

An analysis can sometimes be made without taking a sample by recording the spectral reflectance curve in the visible range, a method described by Johnston and Feller (1963) and demonstrated for the distinguishing of alizarin and carmine. Reflectance curves of well-characterized known samples must of course be available.

5.34 ***Comparison of Analytical Methods.*** The thin-layer chromatographic comparison of the hydrolysis products is superior to the other described methods with respect to sensitivity in the identification of cochineal and kermes carmine. Perhaps the second most effective technique is measurement of the absorption spectrum in solution; if the spectrum is recorded in the visible range, colorless components will not interfere. If the quantity of material permits separation of interfering components (colorless as well), the resulting infrared spectrum of the individual colored constituents will provide the greatest amount of information. When only very small samples are available for analysis, however, success is sometimes possible only with a sensitive color reaction such as one of those described in table 2.

6.0 NOTABLE OCCURRENCES

Owing to the fugitive nature of common lakes prepared from cochineal and kermes, many instances of this colorant's loss in paintings and textiles may be postulated. A faded example would be decidedly notable, but the world must wait for the time when techniques of analysis can be refined sufficiently to establish for certain the identity of the organic colorant that has all but disappeared. Methods for positively identifying lakes have only lately been developed that possess suitable sensitivity for the small samples usually available from artistic and historic works; these are thin-layer chromatography, absorption spectroscopy in solution, and the recently described techniques of microspectrophotometry (Kirby, 1977). The results thus far, however, have yielded a scarcity of confident identifications of samples taken from paintings.

At the moment, it is the textile chemist who can provide the most extensive list of notable occurrences. For example, cochineal has been identified on textiles from Peru through the use of spectrophotometric curves. The analyses gave characteristic absorption peaks between 550–570 nm. Thirteen fibre specimens from late Maxca burials were analyzed, and five were identified as cochineal. In two samples the mordant was alum, in the other three, iron (Young, 1957a). Of a lot of eight samples submitted for analysis from a Chancay-style grave at Zapallan Peru, three were identified as cochineal, two with alum mordant and the third with iron (Young, 1957b).

The number of instances in which carmine has been unequivocally identified in paintings and painting materials has been limited, primarily because of the need to employ analytical methods other than the more routinely used chemical microscopy or x-ray diffraction. We can, however, point to the positive identification by Talley and Groen (1975) of carmine in two paintings by Bardwell: the portrait of Mary Phipps, c. 1749, at Pembroke College, and that of James Poole, c. 1765, at the Castle Museum. Thin-layer chromatography was used in these cases. In an earlier instance, we find Hanson (1954) identifying carmine, by chromatography and solubility, in a work box that belonged to Joseph Mallord William Turner, 1775–1851.

Birren (1965) lists the occurrences of carmine (C) and crimson (Cr) lakes in the palettes of the following artists: François Boucher, 1703–1770 (C); Anthony Raphael Mengs, 1728–1779 (Cr); Francisco de Goya, 1746–1828 (Cr); Gilbert Stuart 1755–1828 (Cr); Jean Louis David, 1748–1825 (Cr); Paul Cézanne, 1839–1906 (Cr); Pierre Bonnard, 1867–1947 (C); André Derain, 1880–1954 (C); Georges Braque, 1882–1963 (C); Raoul Dufy, 1877–1953 (C); Maurice Utrillo, 1883–1955 (Cr); Giorgio de Chirico, 1888–1978 (C). He also cites a treatise on painting techniques published by J. B. Corneille in 1648 where carmine is mentioned.

Of the forty-six watercolor artists who reported their pigments to Russell and Abney (1888), fourteen claimed the use of carmine and crimson lake.

The red coats of the British Army were dyed with this colorant until it was supplanted by the azo scarlets in 1878 (Dunn, 1973).

REFERENCES

D. H. Abrahams and S. M. Edelstein, "A New Method for the Analysis of Ancient Dyed Textiles," *American Dyestuff Reporter* 55 (1964), 19–25.

D. H. Abrahams and S. M. Edelstein, "A New Method for the Analysis of Ancient Dyed Textiles," *Archaeological Chemistry, A symposium, University of Pennsylvania* (Philadelphia, 1967), 15–24.

M. A. Ali and L. J. Haynes, "C-glycosol Compounds. Part III. Carminic Acid," *Journal of the Chemical Society* (1959), 1,033–1,035.

"Aus der Geschichte der Färberei XIII. Kostbare Insekten," *Bayer Farben Revue*, no. 21 (1972), 26–30.

E. Berger, *Quellen und Technik der Fresko-,Öl - und Temperamalerei des Mittelalters* (Munich, 1912; Walluf, 1973).

S. B. Bhatia and K. Venkataraman, "The Positions of the Carboxylic Group in Carminic Acid," *Indian Journal of Chemistry* 3 (1965), 92–93.

Bible, King James version (New York, 1931), Isaiah 1:18, Exodus 26:1, Hebrews 9:19.

F. Birren, *History of Color in Painting* (New York, 1965).

B. Bischoff, "Die Überlieferung des Theophilus-Rugerus nach den ältesten Handschriften (Quellengeschichtliche Untersuchungen zur *Schedula diversarum artium* des Theophilus)," *Münchner Jahrbuch der bildenden Kunst*, Folge 3, vol. 3/4 (1952/1953), 145–149.

H. Bloom, L. H. Briggs, and B. Cleverley, "Physical Properties of Anthraquinone and Its Derivatives. Part I. Infrared Spectra," *Journal of the Chemical Society* (1959), 178–185.

H. Blümner, *Technologie und Terminologie der Gewerbe und Kunst bei Griechen und Römern*, vol. 1 (Leipzig and Berlin, 1912).

F. S. Bodenheimer, "Animals in the Life of the Ancient Middle East," in *Animals and Man in Bible Lands* (Paris, 1960), 76, 80, 138, 157, 200.

W. Born, "Der Scharlach," *Ciba-Rundschau* 7 (November 1936), 218–228; "Die Cochenille," 228–240.

W. Born, "Le kermes et la cochenille de Pologne," *Les cahiers Ciba* 1 (1947), 330–338.

V. Borradaile and R. Borradaile, *The Strasburg Manuscript* (London, 1966), 34.

J. Boutrop, *Möglichkeiten zur Analyse von Färbungen auf Wolle mit Cochenillerot unter besonderer Berücksichtigung chromatographischer Verfahren* (Krefeld, 1970).

B. Brown, "Kermes — An Ancient Dyestuff," *Textile Colorist* 65 (1943), 187–189.

C. Brühl, *Capitulare de villis Cod. Guelf. 254 Helmst.* (Stuttgart, 1971).

A. H. Church, *The Chemistry of Paints and Painting*, 4th ed. (London, 1915).

Colour Index, vol. 2, 2nd ed. (Bradford, Yorkshire, England, and Lowell, Massachusetts, 1956).

Colour Index, vols. 1–5, 3rd ed. (Bradford, Yorkshire, England, and Research Triangle Park, North Carolina, 1971).

O. Dimroth, "Uber den Farbstoff des Kermes," *Berichte der Deutschen Chemischen Gesellschaft* 43 (1910), 1,387–1,401.

O. Dimroth, "Uber die Carminsäure," *Liebig's Annalen der Chemie* 399 (1913), 1-35.

O. Dimroth and R. Fick, "Uber den Farbstoff des Kermes," *Liebig's Annalen der Chemie* 411 (1915), 315–338.

O. Dimroth and H. Kämmerer, "Uber die Carminsäure," *Berichte der Deutschen Chemischen Gesellschaft* 53B (1920), 471–480.

O. Dimroth and W. Scheurer, "Uber den Farbstoff des Kermes," *Liebig's Annalen der Chemie* 399 (1913), 43–61.

M. L. Diquet, "Histoire de la Cochenille an Mexique," *Journal, Société des americanistes de Paris* 6 (1909), 75–99.

C. R. Dodwell, trans., *Theophilus, De diversis artibus* (London, 1961).

D. G. Duff, R. S. Sinclair, and D. S. Stirling, "Light-induced Colour Changes of Natural Dyes," *Studies in Conservation* 22 (1977), 161–169.

M. J. Dunn, "Carmine Lake," *Pigment Handbook*, vol. 1, ed. T. Patton (New York, 1973), 697–698.

S. M. Edelstein, "Spanish Red — Thiery de Menonville's Voyage à Guaxaca," *American Dyestuff Reporter* 47, no. 1 (1958), 1–8.

R. L. Feller, "Standards of Exposure to Light," *Bulletin of American Group. The International Institute for Conservation* 4, no. 1 (1963), 10–12.

G. A. Fester, "Einige Farbstoffe südamerikanischer Kulturvölker, *Isis* 44 (1953), 13–16.

F. Flieder, "Mise au points des techniques d'identification des pigments et des liants inclus dans la couche picturale des enluminures de manuscrits," *Studies in Conservation* 13 (1968), 49–86.

R. J. Forbes, *Studies in Ancient Technology*, vol. 4 (Leiden, 1956), 102.

Frey, "Zur Kenntnis des Carmins und der Neocarminsäure" (Diss., Techn. Hochschule, Zurich, 1931).

T. Furuya, S. Shibata, and H. Iizuka, "Gas-liquid Chromatography of Anthraquinones," *Journal of Chromatography* 21 (1966), 116–118.

D. D. Gadgil, A. V. Rama Rao, and K. Venkataraman, "Structure of Kermesic Acid," *Tetrahedron Letters*, No. 18 (Pergamon Press, 1968), 2,223–2,227.

W. Ganzenmüller, "Ein unbekanntes Bruchstück der Mappae Clavicula aus dem Anfang des Jahrhunderts," in *Beiträge zur Geschichte der Technik und der Alchemie*, rpt. (Weinheim, 1956), 336–349.

F. B. Gardner, *The Painters' Encyclopaedia* (New York, 1901).

J. G. Gentele, *Gentele's Lehrbuch der Farbenfabrikation. 3. Band: Die Lackfarben*, ed. A. Buntrock (Braunschweig, 1909), 27–37.

J. P. L. Girardin, *Chimie générale et appliquée* (Paris, 1868–1869).

G. Gutzeit, "Sur une méthode d'analyse qualitative rapide. I. De quelques reactions spécifiques et spéciales des cations et anions les plus usuels," *Helvetica Chimica Acta* 12 (1929), 713–740.

N. W. Hanson, "Some Painting Materials of J. M. W. Turner," *Studies in Conservation* 1 (1954), 162–173.

H. Hedfors, *Compositiones ad tingenda musiva* (Uppsala, 1932).

J. H. Hofenk-De Graaf and W. G. T. Roelofs, "On the Occurrence of Red Dyestuffs in Textile Materials from the Period 1450–1600; Origin, Chemical Constitution, Identification," *Plenary Meeting ICOM Committee for Conservation*, October 2–5, 1972, Madrid (ICOM Report 5/72/3).

E. H. van't Hul Ehrnreich, "Infrared Microspectroscopy for the Analysis of Old Painting Materials," *Studies in Conservation* 15 (1970), 175–182.

A. Ilg, *Das Buch von der Kunst oder Traktat von der Malerei des Cennino Cennini da Colle di Valdelsa* (Vienna, 1888).

R. M. Johnston and R. L. Feller, "The Use of Differential Spectral Curve Analysis in the Study of Museum Objects," *Dyestuffs* 44 (1963), 277–286.

B. S. Joshi, "Some Natural Anthroquinone Colouring Matters" in: *Recent Progress in the Chemistry of Natural and Synthetic Colouring Matters and Related Fields*. Eds. T. S. Gore, B. S. Joshi, S. V. Sunthankar, and B. D. Tilak. (New York, 1962), 127–136.

A. E. Karr, "Cochineal and Its Colors," *Textile Colorist* 58 (1936), 672–674.

K. M. Kashiwagi, "An Analytical Study of Pre-Inca Pigments, Dyes, and Fibers," *Bulletin of the Chemical Society of Japan* 49, 5 (1976), 1,236–1,239.

K. L. Kelly and D. B. Judd, *Color, Universal Language and Dictionary of Names* (Washington, D.C., 1976).

J. Kirby, "A Spectrophotometric Method for the Identification of Lake Pigment Dyestuff," *National Gallery Technical Bulletin* 1 (1977), 35–48.

E. Kowecka, "The Problem of the Production of Plant Colorants in Poland in the First Half of the Nineteenth Century," *Kwartalnik historii kultury materialnej, Warszawa* 5 (1957), 50–66.

H. Kühn, "A Study of the Pigments and the Grounds Used by Jan Vermeer," *Report and Studies in the*

History of Art, National Gallery of Art, Washington (1968), 155–175.

H. Kurdian, "History of Kirmiz," *Journal of the American Oriental Society* 61 (1941), 105–107.

A. P. Laurie, *The Materials of the Painter's Craft* (Philadelphia, 1911), 267, 269–272.

W. F. Leggett, *Ancient and Medieval Dyes* (Brooklyn, 1944), 69–89.

M. Levey, *Chemistry and Chemical Technology in Ancient Mesopotamia* (London, 1959).

L. Masschelein-Kleiner, "Microanalysis of Hydroquinones in Red Lakes," *Mikrochima Acta* 6 (1967), 1080–1085.

L. Masschelein-Kleiner and J. B. Heylen, "Analyse des Laques rouges anciennes," *Studies in Conservation* 13 (1968), 87–97.

L. Masschelein-Kleiner and L. Maes, "Etude technique de la Tapisserie Tournaisienne au XVe et XVIe siècles, Les colorants," *Bulletin de l'Institut royal du patrimoine artistique* 12 (1970), 269–272.

L. Masschelein-Kleiner, N. Znamensky-Festraets, and L. Maes, "Etude technique de la Tapisserie Tournaisienne a XVe siècle, Les colorants," *Bulletin de l'Institut royal du patrimoine artistique* 11 (1969), 34–41.

F. Mayer and A. H. Cook, *The Chemistry of Natural Coloring Matters* (New York, 1943).

M. Merrifield, *Original Treatises on the Arts of Painting*, vol. 1 (London, 1849).

J. Napier, *Chemistry Applied to Dyeing* (Philadelphia, 1869), 369.

J. C. Overeem and G. J. M. van der Kerk, "Revised Structures for Cochenillic Acid and for the Insect Pigments Carminic and Kermesic Acids," *Recueil des travaux chimiques des pays-bas* 83 (1964), 1,023–1,035.

T. Padfield and S. Landi, "The Light-fastness of the Natural Dyes," *Studies in Conservation* 11 (1966), 181–196.

R. Paris and R. Rousselet, "Caractèrisation des colorants d'origine végétale à l'aide de la chromatographie sur papier," *Annales pharmaceutiques françaises* 16 (1958), 747–756.

E. A. Parnell, *A Practical Treatise on Dyeing and Calico-Printing* (New York, 1846), 10–14.

J. R. Partington, *A History of Chemistry*, vol. 2 (London, 1961), 451–454.

Pelletier and Caventou, *Annales de Chemie et de Physique* (2), 8 (1818), 250–286.

A. G. Perkin and A. E. Everest, *The Natural Organic Colouring Matters* (London, 1918).

G. H. Pertz, *Monumenta Germaniae historica*, 1 (Hannover, 1835).

R. Pfister, "Teinture et alchimie dans l'Orient héllénistique," *Seminarium Kondakovianum VII, Institut Kondakov* (Prague, 1935), 1–50.

T. Phillipps, "*Mappae Clavicula*, A Treatise on the Preparation of Pigments during the Middle Ages," *Archaeologia* 32 (1847), 183–244.

E. Ploss, *Das Bamberger Malerbüchlein, zur Verwandtschaft der spätmittelhochdeutschen Farberezept. Festchrift für Herman Heimpel zum 70 Geburtstag*, vol. 3 (Göttingen, 1972), 693–703.

E. Ploss, "Ein Malerbüchlein aus dem Bamberger Karmeliten Kloster," *100. Bericht des historischen Vereins für die Pflege der Geschichte des ehemaligen Fürstbistums* (Bamberg, 1964), 331–346.

E. Ploss, "Studien zu den deutschen Maler- und Färberbüchern des Mittelalters" (Ph.D. diss., Munich, 1952).

A. Pubetz, "Die Farbstoffe, ihre Anwendung und Erkennung: Vegetabilische und animalische Farbstoffe," in *Praktisches Handbuch der gesammten Färberei und Druckerei* (Berlin, 1872), 400–409.

B. de Puyster, "Use of Organic Dyestuffs for Lake Manufacture, III Cochineal," *Color Trade Journal* 5 (1919), 39–43; "IV Kermes, Indian Yellow, Sepia," *Color Trade Journal* 4 (1919), 98–100.

Raw Materials Index, Pigment section (Washington D. C., 1975).

C. Rawson, W. M. Gardner, and W. F. Laycock, *A Dictionary of Dyes, Mordants and Other Compounds* (London, 1915).

W. G. T. Roelofs, "Thin Layer Chromatography; An Aid for the Analysis of Binding Materials and Natural Dyestuffs from Works of Art," *Plenary Meeting ICOM Committee for Conservation*, October 2–5, 1972, Madrid (ICOM Report 9/72/3).

H. Roosen-Runge, "Artikel Farbe, Farbmittel der abendländischen ma. Buchmalerei," *Reallexikon zur deutschen Kunst* 6 (Stuttgart, 1973), 1,463–1,492.

H. Roosen-Runge, *Farbgebung und Technik frühmittelalterlicher Buchmalerei*, vols. 1 and 2 (Munich, 1967).

G. Rosetti, *The Plictho of Gioanventura Rosetti*, trans. S. M. Edelstein and H. C. Borghetty (1548; Cambridge, Massachusetts, 1969).

B. Roueché, "Annals of Medicine," *New Yorker* (September 4, 1971), 66–81.

W. de la Rue, *Annalen der Chemie und Pharmacie* 64 (1847), 1.

H. Rupe, *Die Chemie der natürlichen Farbstoffe* (Braunschweig, 1900), 170–195.

W. J. Russell and W. De W. Abney, *Report on the Action of Light on Water Colours* (London, 1888), 73–77.

M. Saltzman, A. M. Keay, and J. Christensen, "The Identification of Colorants in Ancient Textiles," *Dyestuffs* 44 (1963), 241–251.

G. Schultz, *Farbstofftabellen*, vol. 1 (Leipzig, 1931), 498–500, 510–512; also vol. 2 (Leipzig, 1931), 7.

H. Schweppe, "Nachweis natürlicher organischer Künstlerpigmente in Mikromengen," *International Symposium on Microchemical Techniques*, May 22–27, 1977, Davos. *Mikrochimica Acta*, 2 (Vienna, 1977), 583–596.

H. Schweppe, "Nachweis von Farbstoffen auf alten Textilien," *Zeitschrift für analytische Chemie* 276 (1975), 291–296.

H. Schweppe, "Neues Spurenreagenz zum Nachweis von Calcium," *Zeitschrift für Analytische Chemie* 244 (1969), 310–312.

G. Seufert, *Farbnamenlexikon von A bis Z* (Berlin, 1955).

S. Shibata, M. Takito, and O. Tanaka, "Paper Chromatography of Anthraquinone Pigments," *Journal of the American Chemical Society* 72 (1950), 2,789–2,790.

H. Sloane, *A Voyage to the Islands Madera Barbadoes, Nieves, S. Christophers and Jamacia*, vol. 2 (London, 1725).

C. S. Smith and J. G. Hawthorne, "*Mappae Clavicula* — A Little Key to the World of Medieval Techniques, An Annotated Translation based on a Collation of the Sélestat and Phillips-Corning Manuscripts," *Transactions of the American Philosophical Society*, n.s. vol. 64, pt. 4 (1974), 1–128.

M. K. Talley, Jr., and K. Groen, "Thomas Bardwell and His Practice of Painting: A Comparative Investigation between Described and Actual Painting Technique," *Studies in Conservation* 20 (1975), 44–108.

D. V. Thompson, trans., *Il libro dell'arte, the Craftman's Handbook of Cennino d'Andrea Cennini* (New Haven, 1933).

R. H. Thomson, *Naturally Occurring Quinones* (London, 1957, 2nd ed. 1971).

F. Ullmann, *Enzyklopädie der technischen Chemie*, vol. 5 (Berlin, 1930), 148.

A. Ure, *A Dictionary of Arts, Manufacturers, and Mines*, 4th ed., 2 vols. (London, 1853).

H. Wagner, *Die Körperfarben*, 2nd ed. (Stuttgart, 1939), 333–335.

N. P. Wright, "A Thousand Years of Cochineal: A Lost but Traditional Mexican Industry on Its Way Back," *American Dyestuff Reporter* 52 (1963), 635–639.

W. J. Young, "Analyses of Textile Dyes," appendix III, pp. 53–54 in S. K. Lothrop and J. Mahler, *Papers of the Peabody Museum of Archaeology and Ethnology*, vol. 50, no. 2 (Cambridge, Massachusetts, 1957a).

W. J. Young, "Analyses of Textile Dyes," appendix IV, pp. 33–34 in S. K. Lothrop and J. Mahler, *Papers of the Peabody Museum of Archaeology and Ethnology*, vol. 50, no. 2 (Cambridge, Massachusetts, 1957b).

Appendices

Terminology and Procedures Used in the Systematic Examination of Pigment Particles with the Polarizing Microscope

ROBERT L. FELLER and MICHAEL BAYARD

In spite of the many sensitive and highly sophisticated analytical techniques available today, it is imperative that powders and pigments be examined under the microscope. This is the simplest and most reliable way to evaluate the relative homogeneity or heterogeneity of specimens and to detect the presence of traces of materials that might otherwise be overlooked. Microscopic examination narrows down the possibilities quickly and inexpensively. With heterogeneous specimens, its use is indispensable. At the very least, examination under a polarizing microscope can frequently establish unequivocally what a sample *does not* contain.

The use of standard terminology when reporting analytical results will do much to improve the effectiveness of record keeping and communication. However, polarized-light microscopy has been employed in a number of disciplines for more than one hundred years and, as a consequence, the practice has acquired a variety of terminologies derived from the different areas of specialization. It is appropriate, therefore, to review and give definition to the terms used throughout this volume to describe the characteristics of pigment particles seen under the mineralogical microscope — birefringence, refractive indices, and particle size — and to discuss as well certain other morphological and crystallographic characteristics of pigments.

It is beyond the scope of this brief appendix to present a comprehensive treatise on the use of the mineralogical microscope to characterize crystalline materials at magnifications of 60 to 1,000x. The reader must rely on standard texts for extensive theoretical and practical explanations of such subjects as the characteristics of the six crystallographic classes, the reasons for birefringence, and the determination of optic sign (Stoiber & Morse, 1972; Bloss, 1961; Winchell & Winchell, 1964; and McCrone & Delly, 1978). Nonetheless, many of these topics have been cited in the text because such information is considered pertinent to the characterization of crystalline materials. Reference to these properties will prove increasingly useful as the analyst's sophistication, skill, and experience develop.

We will take up the terms in the order of a systematic examination, beginning with the simplest and most obvious morphological properties — such as size, shape, and surface character — and proceeding to the more advanced aspects of optical character such as refractive index and birefringence. It is the authors' hope that, for those who do not have special training in the basic microscopists' techniques and procedures, the addition of this discussion of terminology will enhance the usefulness of the sections of each chapter devoted to microscopic examination.

CHECKLIST OF PROPERTIES

The analyst's examination of a dispersed sample of paint with the aid of a polarizing microscope should proceed by consideration of an orderly sequence of properties for each individual pigmentary component. These attributes should be defined or described as objectively as possible. The following constitutes a basic checklist:

1. Morphological Properties
A. Homogeneity
B. Size — coarse, medium, fine, very fine

C. Shape — equant, tabular, lamillar, acicular; rounded, angular
D. Surface character or structure — smooth or rough
E. State of aggregation — dendritic, spherulitic, stellate, globular, fibrous
F. Crystal form and system

2. **Optical Properties**

A. Color by transmitted and by reflected light
B. Pleochroism
C. Refractive index — low, medium, or high; relief
D. Birefringence — weak, moderate, strong, extreme
E. Extinction — parallel, symmetrical, oblique
F. Optic sign or sign of elongation — positive or negative
G. Interference figure

3. **Properties Determined by Ancillary Methods**

A. By microchemical tests (see McCrone 1952)
B. By dispersion staining
C. By use of a hot/cold stage

In the paragraphs that follow, a brief discussion will be given of terms and concepts relating to the morphological and optical properties listed above.

Morphological Properties

Homogeneity

One of the most useful initial observations to make is an evaluation of the homogeneity of a dispersed sample of the pigments in a particular paint. One would wish to know whether a green, for example, is based primarily on a single green-hued pigment, perhaps mixed with a white, or if it is based on a mixture of blue and yellow. Does a white pigment appear to be a mixture of titanium white, zinc white and barium sulfate or of lead white and chalk? Does a relatively pure sample direct from the manufacturer contain fillers or inerts? Such features markedly affect the analytical results.

Particle Size

There should be general agreement concerning the terminology used in describing the size of individual particles viewed under the micro-scope at 60 to 1000x. Size designations are generally based on the diameter of a particle or an aggregate, although, in the case of needle-shaped particles, both length and width may be cited. Size is reported in terms of the CGS system, generally micrometers (10^{-6} meters), which is the same as the older term, micron.

The classification scheme in table 1 is similar to that suggested in 1956 by the Paint Research Station in England (*Pigment Particles*, 1956). The proposed three inclusively defined divisions for the sizes of individual pigment particles fall in approximately equal steps on a logarithmic scale. (Setting the division at 3.162 micrometers instead of 3 would make the divisions exactly equal on a scale of logarithms to the base 10; thus, $\log 1 - \log 3.162 = 0.5$; $\log 3.162 - \log 10 = 0.5$, etc.) The employment of a logarithmic scale to separate the classes is appropriate if particle sizes tend to follow the log-normal distribution, as suggested by Irani and Callis (1963).

The resolution achievable with the optical microscope is 0.3 micrometer; this sets the lower limit for particles that can be detected by optical methods. The "very fine" classification in table 1 thus will rarely be employed by the analyst using the optical microscope, but should prove useful for pigments examined with an electron microscope.

According to the American Society for Testing Materials Standard D185-45, and related paint standards, special note is taken of the percentage of particles above 44 micrometers, usually determined in paint laboratories by sieving through a 325-mesh screen. Many specifications state that no more than 0.5 to 3% of the pigment sample shall be above this size (Patton, 1964). The upper limit in many European standards of paint formulation, on the other hand, is about 60 micrometers (Nylen, 1965). Thus, there seems little need to specify an upper limit in table 1 for "coarse"; most modern pigment particles in a well-prepared paint are likely to be smaller than 44 to 60 micrometers. On the other hand, particles of pigment made of coarsely ground

Table 1. **PROPOSED CLASSIFICATION OF INDIVIDUAL PARTICLES BY SIZE**

Paint Research Station Classes (1956)		*Proposed Classes*	
		above 10 μm*	coarse
2 to 30 μm*	coarse	10 to 3	large
5 to 1	medium	3 to 1	medium
1 to 0.1	fine	1.0 to 0.3	fine
0.2 to 0.01	very fine	below 0.3	very fine

*1μm = 1 micrometer = 10^{-6} meters or 10^{-4} centimeters (obsolete: 1 micron). 1 cm = 10,000 μm. 1 inch = 25,400 μm.

mineral may be as large as 60 micrometers or greater.

The very wide range in size that one can expect to find in samples of commercially prepared pigments can be seen in various graphs presented by Patton to describe the distribution of particle sizes (Patton, 1973). If the data seen in fig. 6 in the chapter of Barium Sulfate are indeed representative, then the range in size that would encompass 90% of a given sample of pigment may have at least a tenfold spread in dimension. In many instances, the standard deviation about a mean value of particle size (which theoretically would encompass 68% of the sample) could very well span a tenfold range in dimension. This being the case, the distribution of particle sizes in a modern pigment may frequently extend across two of the classes proposed. It may therefore be impractical to attempt to characterize an entire sample of pigment as belonging solely in one of the classes given in table 1. Table 1 will serve primarily to standardize the microscopist's terms used to describe the size of individual particles or isolated agglomerates of particles, lending more precise meaning to statements such as: "Clinging to the large particles of X were many fine particles of Y."

If the ocular of the microscope is fitted with a measuring reticule, it will be convenient and less ambiguous for the microscopist simply to report the approximate dimensions of particles in terms of micrometers. When a particle is needle- or bladelike, of course, both the long and the short dimensions can be cited.

Shape

Equant particles are those that are roughly symmetrical, having a length-to-width ratio of no more than five to one, and a thickness commensurate to width. Quartz and calcite are commonly encountered as equant particles. The ideal or extreme example of such a particle is, of course, a sphere. If the thickness of an otherwise equant particle is considerably reduced, the term *tabular* can be applied or the particle can be called a platelet; a reduction in one dimension of the particle to about $\frac{1}{10}$ the other would be necessary for assignment to this class. If there is extreme reduction in thickness, the term *lamillar* (Winchell & Winchell, 1964) can be applied, or the particle may be called simply a flake. It is said to be acicular (needle-shaped) if the length-to-width ratio is greater than five to one and its thickness is similar to its width. Strontium chromate usually provides a familiar example of *acicular* crystals; certain varieties of zinc white are also so classified. A variation on acicular when width and thickness are not closely alike, is a blade; a chunkier particle can be said to be *columnar* (Winchell & Winchell, 1964). Many terms used in mineralogy can be adopted to describe the shape of pigment particles. For example, standard systems for describing sphericity-roundness have been proposed by petrologists (high or low sphericity, angular, subangular, rounded) (Brewer, 1964; El-Hinnawi, 1966).

Definitions of shape can seldom be rigorous; individual grains in a sample of strontium chromate will certainly be found with length-to-width ratios less than five to one. Nonetheless, it may frequently be useful to assign an assemblage of particles to one of the above classifications to which the majority conform. Equant, tabular, lamillar, acicular, and columnar are adjectival terms that all analysts should recognize, particularly in the description of individual particles. The nouns platelet, needle, and blade are less formal terms.

Surface

There are two main descriptions of surface texture : *smooth* and *rough*. A rough surface may be irregular or may have specific patterns such as parallel striations, conchoidal fracture, crystalline cleavage planes, or etch pits associated with crystallinity. The various morphological characteristics are not independent; thus, classification of a particle with respect to a particular crystal system usually is dependent upon its possessing definite shape and surface characteristics.

State of Aggregation

The state of aggregation or agglomeration of particles should be noted. The term *dendritic* refers to a branching structure of crystal growth, frequently composed of blade-shaped crystals. Certain commercial preparations of aureolin have been found in a dendritic form (see fig. 5, chapter on Cobalt Yellow). *Spherulites* are associations in which many acicular units radiate in three dimensions from a central point; the structure frequently becomes apparent under crossed polars. *Stellate* agglomerates are two-dimensional versions of the spherulitic form. Spherical agglomerations without detectable radiate order can be referred to simply as *globular*. In the case of *fibrous* associations, such as asbestos, the individual particles are bound together in a parallel array.

Crystal Form and System

The crystallographic properties of a pigment are among the most useful for identification. Unfortunately, it is rare to find text-book-perfect crystal forms in actual samples, although occasionally rather good indications of crystal form can be observed in coarsely ground minerals. Ordinarily, however, the particles are either too small to permit adequate observation of their characteristics, or have been subjected to grinding or other operations that have destroyed the crystals and obscured their crystallographic form. Furthermore, well-defined crystals may never have been formed at all during manufacture, or they may have been altered through processing. Even when crystal faces are well defined there are often distortions in appearance owing to the optical microscope's ability to present only a projection of a three-dimensional object onto a two-dimensional plane. There are also distortions and variations in *crystal habit*, that is, in the way the crystal formed. It is not uncommon for a crystal to grow preferentially in one direction at the expense of others. Within any such distortions, however, the interfacial angles tend to have constant values that can be used in characterization.

When crystals are well-formed, it should be possible to identify them as belonging to one of the six crystal systems to which the chemical substance usually has already been unequivocally assigned on the basis of x-ray diffraction analysis. These six systems, diagrammed in fig. 1, are defined as follows:

Cubic: Three mutually perpendicular axes with equal atomic spacing.

Tetragonal: Two mutually perpendicular axes having equal spacing, with different spacing along the axis perpendicular to the other two.

Hexagonal: Three axes 120° apart with equal atomic spacing and different spacing along the axis perpendicular to the plane of the other three.

Orthorhombic: Three mutually perpendicular axes with unequal atomic spacing.

Monoclinic: Two perpendicular axes with unequal spacing, the third nonperpendicular axis with different spacing.

Triclinic: Three oblique axes with unequal spacing.

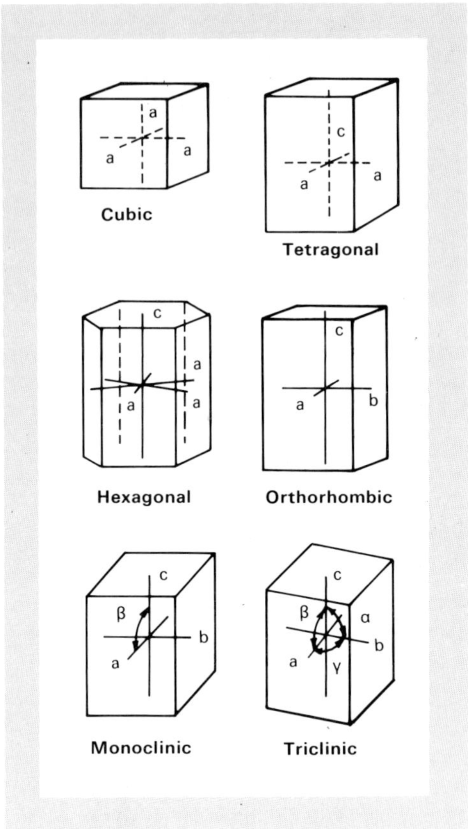

Fig. 1. The six crystal systems.

The cleavage of crystals is a useful property to note; some substances, such as calcite, cleave as near-perfect crystals, while others, such as quartz, show only conchoidal fracture. The degree of perfection of a given crystal may be described as either *euhedral*, completely bounded by characteristic faces; *subhedral*, bounded only partially by characteristic faces; or *anhedral*, exhibiting no crystallographic faces. Most minerals tend to cleave more perfectly in one or two of their crystal planes than in others and are, therefore, characterized as subhedral.

Referring to details of crystal habit or form, one may define a particle with two parallel and opposite faces as *pinacoid* (Hurlbut, 1952). A form with 3, 4, 6, 8, or 12 faces whose axes are usually aligned along one of the principal crystallographic axes is considered a prism, or is said to be *prismatic*. A form composed of 3, 4, 6, 8, or 12 nonparallel faces that meet in a point is *pyramidal*. For example, one finds the mineral handbooks describing malachite [$CuCO_3\cdot$

$Cu(OH)_2$] and crocoite ($PbCrO_4$) as prismatic, zincite (ZnO) pyramidal, and rutile (TiO_2) "prismatic . . . with dipyramid terminations common" (Hurlbut, 1952).

Optical Properties

Color

The color of particles observed with unpolarized light should always be reported in terms of its hue, value, and possibly chroma as observed by transmitted or reflected light. The hue of the transmitted light depends on the absorption characteristics of the particle: if yellow light is absorbed, the transmitted color is complementary blue.

The intensity of the color observed by transmitted light is dependent not only on the absorption characteristics of the chemical substance but also on the extent of the optical path length or on the number of particles in an agglomeration. As a consequence, some materials that are highly colored in massive samples may appear very pale, if not almost colorless, when viewed as small grains under the microscope. The analyst must always be alert to the possibility of not perceiving or fully appreciating the color of minerals when they are observed as thin flakes in transmitted light.

The color observed by reflected light depends on the refractive index of the particle relative to that of the surrounding medium, as well as on the particle's size and spectral absorption characteristics. If the refractive index difference is small there will be little light reflected, and the analyst will be well advised to refer to the color of the light transmitted. If the particular sample viewed consists of very large particles or is densely populated with absorbing particles—or if they are much higher in refractive index than that of the surrounding medium, so that little light is transmitted—then the color of the reflected light will be most informative.

There is a standard color nomenclature available, defined in the dictionary by Kelly and Judd; there are also standard abbreviations for describing colors, for example, rO for reddish orange and yG for yellowish green (Kelly & Judd, 1976). These authors describe a now rarely encountered comparison ocular, or Abbé-type camera lucida, by which microscopic samples may be compared with the color of Munsell chips.

The apparent distribution of color across a particle may be affected by varying thickness; moreover, with particles 3 to 0.3 micrometers in diameter, the refractive index difference between the particle and the mounting medium may result in varying color perceptions. As an example of the latter, medium-sized particles of cadmium yellow can appear light yellow in the center and orange at the periphery. There also may be some alteration of color of very small particles owing to secondary chromatic aberration in the optical system of the microscope. Finally, if reference is to be made to published color data, or if colors are to be matched using a comparison microscope, it is necessary to use illumination of similar color temperature in making visual evaluations.

Today, spectrophotometric curves can be determined for individual particles, as has been described by Kirby (1977). This technique will allow colors seen under the microscope to be characterized far more objectively than was possible in the past.

Pleochroism

The color of particles by transmitted light may sometimes vary with the crystallographic direction in which light passes through the crystal. As a result, when a crystal is viewed in plane-polarized light (that is, when the polarizer, but not the analyzer, is in place), the color may change as the stage of the microscope is rotated. When a fragment shows a two-color change, it is technically said to be *dichroic*; when it can exhibit three or more changes in color it is referred to by the more general term *pleochroic*. The first behavior would be expected in the case of uniaxial crystals; the second, in biaxial (see Birefringence). The polarizers used in most modern microscopes provide an extreme example of pleochroism; in one direction the analyzer almost completely absorbs the plane-polarized light transmitted by the polarizer (the field is dark), and at 90° to this the analyzer becomes more or less transparent to all wavelengths (the field is bright). As with all polarization phenomena, the optical path (thickness of the crystal particle) will influence the pleochroic colors observed.

Refractive Index

When light passes from a less dense to a more dense medium, its velocity is slowed down; a beam of light will be bent or refracted. The ratio of the velocity of light in air (ideally, the velocity in a vacuum) to the velocity in various sub-

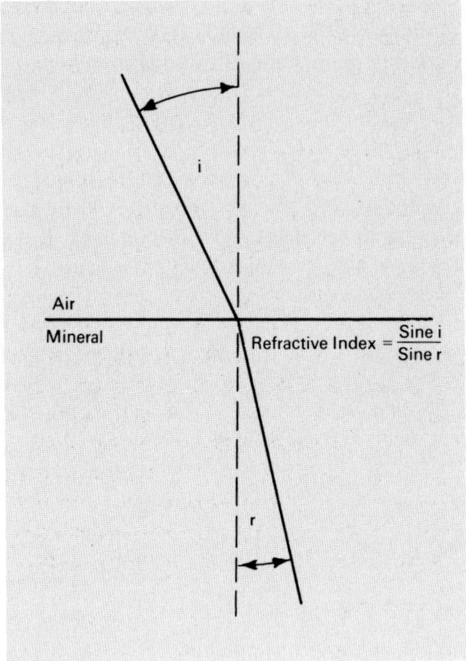

Fig. 2. Refraction of light ray entering medium of higher refractive index.

stances is expressed as the index of refraction or *refractive index*, N, of the substance. It is measured by the angle at which the ray is bent, specifically; it is the ratio of the sine of the angle of incidence of the light ray to the angle of refraction, N = sine i/sine r (fig. 2). The *refringence* of a substance in a given situation refers to its power to produce refraction and is designated by its index of refraction relative to that of the surrounding medium.

In the cubic crystal system, the light travels at the same speed in all directions within the crystal, which is said to be *isotropic*. Light does not travel at the same speed in all directions within substances that belong to the other five crystal classes; these are said to be *anisotropic*. When such crystals at random orientation are observed under crossed polars, they are seen to be *birefringent* (see below).

In passing through crystals of the tetragonal and hexagonal systems, a beam of light can be divided into two rays which are plane polarized at 90° to one another and which travel through the crystal at different speeds; as a consequence, such crystals have two refractive indices. Crystals of the orthorhombic, monoclinic, and triclinic classes possess three refractive indices.

When the refractive index of a colorless crystal particle is very close to that of the mounting medium, a beam of light will pass through the particle with little deviation in path. There will be little refringence, and one will scarcely see the particle in the microscopic field. On the other hand, if there is a considerable difference between the refractive indices of pigment and medium, the light will be bent considerably at the borders of the particle. The borders will appear dark; colorless, nonabsorbing particles will be seen in *relief* (fig. 3). A difference in refractive index between the particle and mounting medium of 0.04 or less is considered by Stoiber and Morse to result in low relief, a difference of 0.04 to 0.12 in moderate, and above 0.12 in high (Stoiber & Morse, 1972). Winchell (1928) has suggested five categories of relief, but three may be sufficient for the less-experienced analyst. If the refractive index of the mounting medium is higher than that of the mineral grain, the relief is said to be negative; if

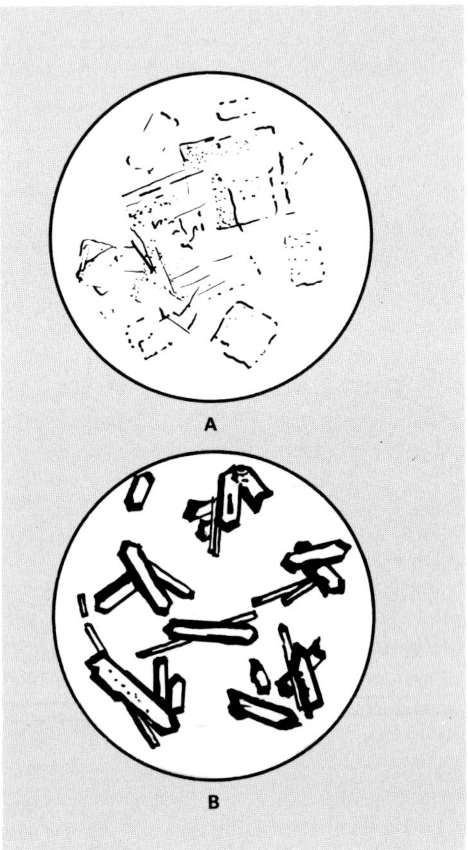

Fig. 3. Examples of low (A) and high (B) relief.

the grain is higher in refractive index than the medium, relief is said to be positive. The notations of relief can be conventionally designated as (+) or (−) and L, M, or H.

The refractive index or indices of a particle are seldom measured precisely during routine examinations. Instead, the values are simply compared relative to that of the mounting medium. Canada balsam has a refractive index of about 1.53, for example; Fisher Scientific Company's "Permount," about 1.52. Because the appearance of particles is so strongly influenced by the refractive index of the surrounding medium, the mounting medium or its refractive index must always be specified when photomicrographs are taken for publication or for the record.

Fig. 4. Explanation of formation of the bright Becke line (B) when microscope is focused above crystals mounted in media of refractive index lower or higher than that of the crystal. Arrows indicate direction of light near edges of the crystal. As focus of the microscope is raised from the crystal plane to x, the bright line moves toward the substance of higher refractive index.

More than thirty years ago, Aroclor 5442 was introduced as a mounting medium; its index, 1.66, divides the known number of artificial inorganic minerals approximately in half. The Aroclors, based on chlorinated aromatic compounds, have since been found to be carcinogenic and must be handled with considerable care if not abandoned altogether. With this medium, or a replacement of equally high index, it is possible to visually recognize four refractive index categories: low, below 1.55; medium, 1.55–1.66; high, 1.66–2.0; and very high, above 2.0 (McCrone & Delly, 1978). The first three can be determined rather easily based on observation of the Becke line, to indicate whether the index of the particle or the mounting medium is higher, and observation of the relief, to note the relative difference between indices of particle and medium. Estimations of refractive index above 2.0 will prove difficult in normal mounting media.

A simple diagram (fig. 4) will explain the well-known Becke-line technique for estimating whether the refractive index of a particle is

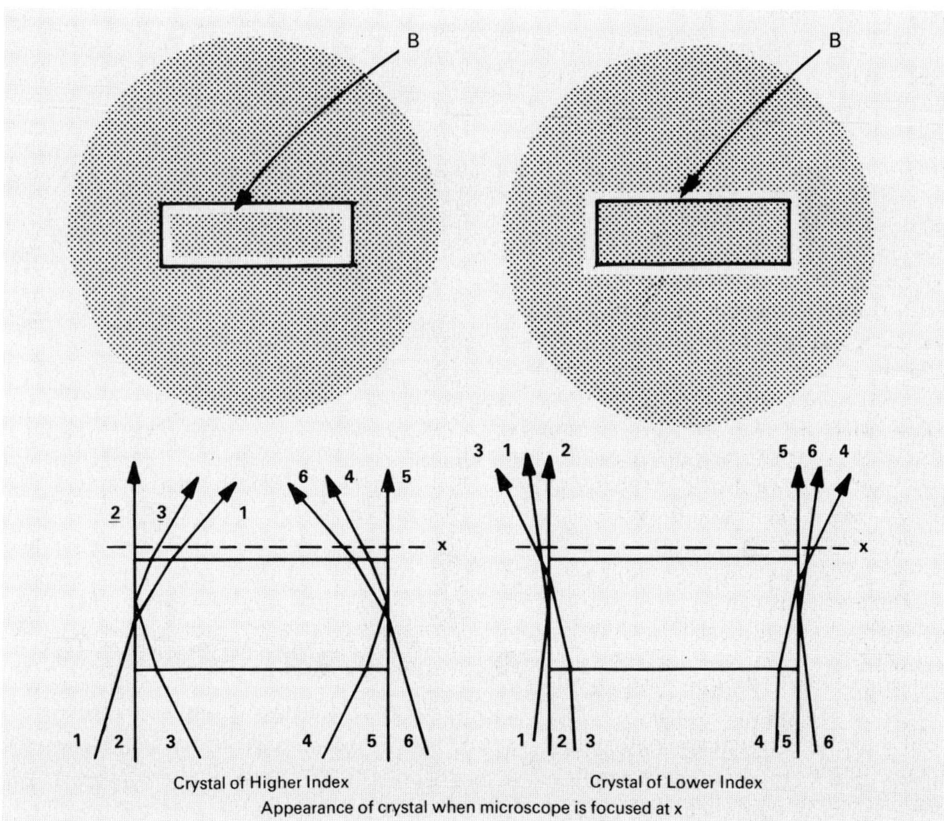

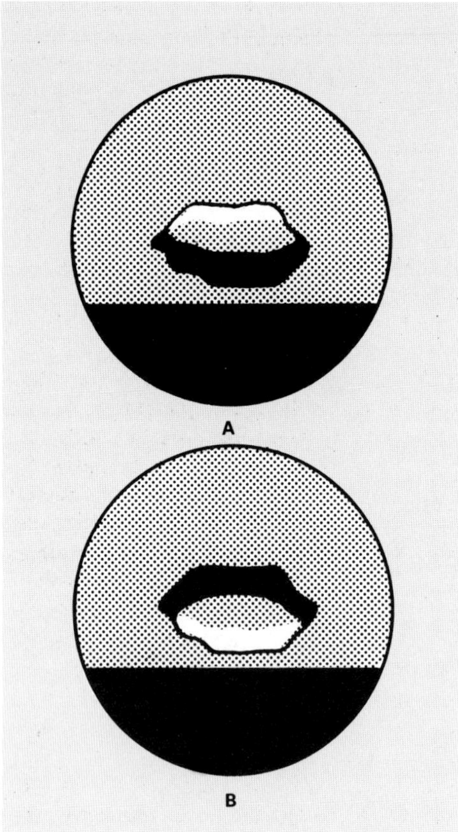

Fig. 5. Determination of relative refractive index by the method of oblique illumination or the Schroeder van der Kolk method. A. Refractive index of particle is higher than that of surrounding medium. B. Refractive index of particle is lower than that of surrounding medium. The stop has been inserted part way up from the bottom.

above or below that of the mounting medium. There is, however, an additional, highly reliable, technique which accomplishes the same purpose: the oblique illumination or Schroeder van der Kolk method. This technique involves the gradual or partial insertion of an opaque stop into the light path below the stage (usually below the polarizer). If the shadows on the particles appear on the same side as the stop, the refractive index of the particle is usually greater than that of the surrounding medium; if the shadow is opposite, the refractive index of the particle is usually lower than that of the medium, as illustrated in fig. 5. We are forced to say "usually" because the opposite effect may be observed in certain optical systems. Nonetheless, the proper rule for any given microscope can be quickly determined simply by carrying out a trial procedure with a substance known definitely to have a refractive index higher or lower than that of the mounting medium.

The use of oblique substage illumination can often enhance appearance and reveal much concerning the characteristic form and surface of particles; therefore some of the photomicrographs in the text have been taken with the light that passes through the specimens obliquely shielded to various degrees. Oblique illumination tends, however, to make particle relief appear higher than the microscopist might ordinarily expect. Hence whenever oblique illumination has been used in the photomicrographs accompanying this text, the fact is always stated.

As an aid in estimating the relative indices of particles based on their relief, a set of substances of known refractive index, mounted in the normally utilized mounting medium, can be prepared for ready reference. A possible set of isotropic reference materials could include potassium alum, refractive index 1.456; sodium chloride, 1.544; ammonium chloride, 1.640; potassium iodide, 1.667; sodium iodide, 1.775; lithium iodide, 1.955; and diamond, 2.42.

When there is a need to determine an index precisely, the immersion method employing liquids of known refractive index is used. If a hot stage is available, the range of indices determined in any one liquid may be extended because its refraction can be varied by raising or lowering the temperature. Thus, perhaps only four or five liquids will be all that are needed to span the normal range of indices.

Birefringence

Birefringence refers to the difference in refractive index dependent upon the path in which plane-polarized light passes through a crystal. The term isotropic describes a particle exhibiting no birefringence. Noncrystalline, amorphous, materials are isotropic, but the only crystals that are isotropic are those of the cubic system; here the perpendicular axes are the same length, and the crystal is characterized by a single refractive index. All other crystal systems are anisotropic, exhibiting two or more different refractive indices. Those exhibiting only two indices are the tetragonal and hexagonal systems; they have one axis that will produce no double refraction or polarization and are hence known as uniaxial. The other anisotropic crystal systems — orthorhombic, monoclinic, and triclinic — exhibit three different refractive indices and have two axes showing no double refraction or polarization: they are designated as biaxial.

One usually detects or observes birefringence with the polars fully crossed, the field in the microscope is then dark (black). If anisotropic crystals are present they are ordinarily seen in various degrees of brightness; many particles will blink or twinkle as the stage of the mineralogical microscope is rotated. Individual particles will usually go to extinction at various orientations as the stage is rotated. Their behavior should be observed closely. (See section below describing Extinction.)

Under crossed polars, the brighter the gray-to-white appearance of a dispersion of crystals as the stage is rotated, the higher their birefringence. If, instead of a grayish-to-white appearance, the particles exhibit vivid interference colors (described below), the birefringence is likely to be strong to extreme. Uniaxial crystals can falsely appear to be isotropic under crossed polars if they happen to be lying in such a way that polarized light enters the particle in the direction of the optic axis.

Within crystals of the tetragonal and hexagonal system (those characterized as uniaxial), the light transmitted through the crystal is split into two beams that are polarized in planes perpendicular to one another. One beam, called the ordinary (O) ray, is not refracted; it travels with the same velocity in all directions and its refractive index is designated as ω. Light traveling through the crystal along this path is named the *optic axis* of the crystal and is not polarized. The other ray is refracted and travels with a velocity that depends on its path through the crystal. This is called the extraordinary (E) ray; its refractive index designated as ϵ. The difference between the two refractive indices is called the birefringence, B. Thus in uniaxial crystals: $B = \epsilon - \omega$.

In order to describe anisotropic crystals objectively, it is useful to refer to a standard *scale of birefringence*. Table 2, based on the standard work of Winchell (1928), presents the terms used in optical mineralogy, together with a list of representative minerals; an alternative list of minerals familiar to workers in the field of artists' pigments has been added.

It must be remembered that the observed degrees of birefringence will be dependent in part on the thickness of the particles, that is, the length of the path through which the light has passed. Hence, comparison of the birefringence of an unknown with a standard, or comparisons between two minerals, must always be made using particles of similar thickness. When thin, pigment-sized particles are considered, the "very weak" category in table 2 would not be observable. It is proposed that the term weak apply to birefringences below .01; moderate, .01–.03; strong, .03–.06; and extreme, greater than .06.

Whichever refractive index, ω or ϵ, is greater

Table 2 **SCALE OF BIREFRINGENCE (After Winchell, 1928)**

Rank	*Familiar Minerals*	*Representative Minerals (Winchell, 1928)*	*Birefringence*			*Interference Color*	
			Familiar Minerals	*Representative Minerals*	*Descriptive Terminology**	*5 μm*	*30 μm*
1	apatite	leucite	ca. .003	.001	(very weak)	black	dark gray
2	quartz	orthoclase	.009	.004–.009	weak	dark gray	light gray-white
3	barite	hypersthene	ca. .012	.010–.019	moderate	light gray	yellow-red 1°
4	glauconite	augite	ca. .019	.019–.028	(rather strong)	light gray-white	blue-green 2°
5		tourmaline		.028–.037	strong	white	yellow-red 2°
6	anhydrite or talc	muscovite	.035–.050	.037–.055	(very strong)	white-light yellow	blue-red 3°
7	calcite	titanite	.17	>.055	extreme	light red	gray-white

*A simpler scale of birefringences is recommended by the authors: weak <0.01, moderate 0.01–0.03, strong 0.03–0.06, and extreme >0.06. The terms in parentheses are those suggested by Winchell in ranking seven degrees of birefringence.

determines what is known as the optic sign of the mineral (described below). Winchell considers it convenient to have symbols for the greater and smaller indices; the higher index is designated as N_g and the lower as N_p, the subscripts referring to "great" and "petty", or small (Winchell & Winchell, 1964). When the light travels through the crystal at right angles to the optic axis, the index for the slower ray (higher refractive index) is labeled N_g and that of the faster, N_p.

The direction in which there is no double refraction is called the optic axis. Three crystal systems, the orthorhombic, monoclinic, and triclinic, exhibit two such axes and are called biaxial. One of the two optic axes, designated as α, lies in the plane of the minimum index of refraction; the other, situated in the plane of the maximum index, is designated as γ. The intermediate index of refraction, β, lies in a plane which is perpendicular to that in which the two optic axes lie and is known as the optic normal. The birefringence of biaxial systems is defined as $B = \gamma - \alpha$.

The angle formed by the two optic axes is called the *optic angle* and is designated as 2V. When observed with the microscope, the apparent optic angle, designated as 2E, is larger than 2V because the rays are bent away from the perpendicular when they emerge from the crystal into air. The true angle can be calculated from the apparent angle by the formula:

$$\text{Sin V} = \frac{\text{Sin E}}{\beta}$$

Biaxial crystals, with two optic axes, have two optic angles, which add up to 180°. Unless specified otherwise, 2V refers to the acute angle.

Considerable training and experience is needed to determine the three refractive indices of biaxial crystals and to report the birefringent effects precisely. The subject is beyond the scope of the present discussion.

Extinction

The property of *extinction* referred to above can be used to describe certain characteristics of crystals objectively. As mentioned, at certain positions with respect to crossed polars, a uniaxial substance will go to extinction (become dark). When the stage is rotated through 360°, the crystals ordinarily will go to extinction at four positions at right angles to each other. If this occurs when an edge of the crystal is oriented parallel to the polarization directions of the microscope (usually these are aligned in the vertical and horizontal directions of the field), the extinction is termed *parallel.* It is *symmetrical* if an extinction direction bisects a prominent interfacial angle of the crystal and is *oblique* if the extinction axis intersects the long direction of a crystal at an angle (fig. 6). If one is certain of the crystallographic orientation of the specimen, the extinction angle can be used as an identifying property.

There are two other types of extinction that are simple to observe as the stage is rotated (polars crossed): *undulose*, which produces a rippling or band pattern of extinction within the same grain, and *spherulitic*, which results in the appearance of a stationary cross on the particle.

Interference Colors

When anisotropic materials are viewed between crossed polars using a "white" light source, colors known as interference colors are frequently seen. Simply expressed, the reason for this phenomenon is as follows. When the plane-polarized white light from the polarizer (below stage) is passed through a biaxial material, the waves are split into two plane-polarized components that are at right angles to one another. Because one of the components travels more slowly than the other, the waves in that component will become retarded, out of phase with the other waves. Recalling the laws of wave motion, the reader will realize that, at some particular wavelength, the two sets of waves emerging from the crystal may be completely out of phase and therefore effectively cancel one another; light of another wavelength, after passing through the crystal, may be exactly in phase with the waves emerging from the component at 90° to the first train of waves. The net result is that certain wavelengths of light, upon emerging from the crystal, may cancel one another and not be seen, while other wavelengths may reinforce one another. The transmitted light is thus often colored, the consequence of having lost some of the wavelengths and having others reinforced. Under crossed polars, cancelling occurs when the waves are retarded by multiples of the wavelength, $n\lambda$; reinforcement takes place when the differences in wavelength are of the order of $n\lambda + \lambda/2$. This statement may not be as one first expects, but the reader can find an excellent explanation in Stoiber and Morse (1972) as well as in other standard texts.

The resultant colors are known as *interference*

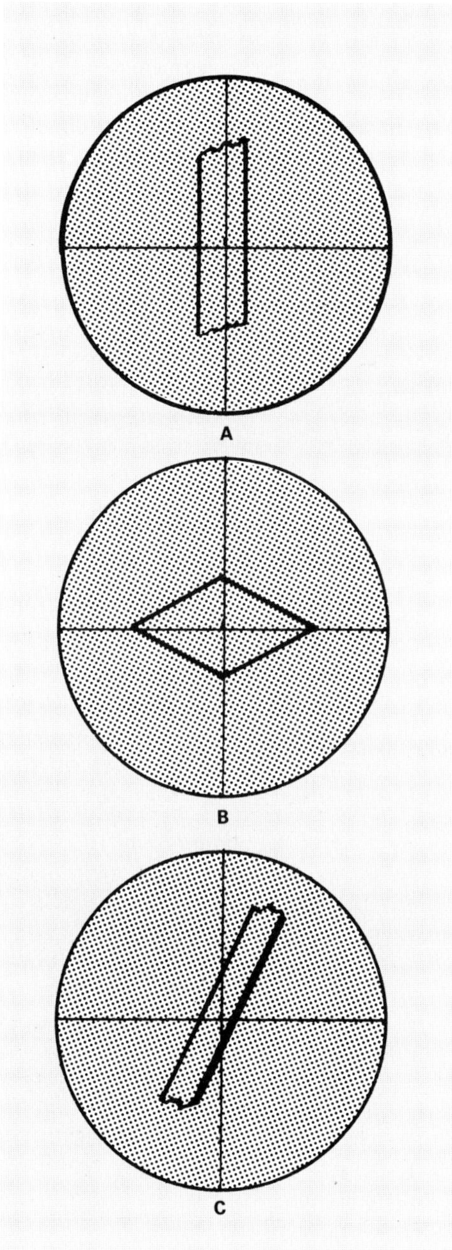

Fig. 6. Examples of parallel (A), symmetrical (B), and oblique (C) extinction.

or *retardation* colors and are those illustrated in the familiar Michel-Levy charts that accompany most texts on optical mineralogy. The retardation, Δ, is often given by the formula Δ = thickness x $(N_2 - N_1)$.

In other words, the retardation is equal to the thickness of the particle times the birefringence. The particular colors of retardation serve as a convenient indicator of the difference in refractive indices between the mineral's two observed crystallographic directions, providing thickness is known and is sufficiently great. Thus, retardation colors are of considerable use in typical thin-sections of minerals, which are customarily polished to a thickness of 30 micrometers. Retardation color is much more rarely of assistance in the case of particles that are only 5 micrometers thick, as is noted in table 2.

As steadily increasing retardations are achieved, passing through λ, 2λ, 3λ, the gamut of hues essentially repeat themselves, while at the same time decreasing in chroma. The first sequence, "first-order" colors (1°), terminates in a violet hue called "first-order red" (Red I). The subsequent series of colors are not as saturated and are described as second order, third order, etc. With experience, the microscopist can recognize the order of the color seen under crossed polars, at least up to about the fourth order. Retardation is often expressed in terms of wavelengths, the out-of-phase shift that is necessary to produce the color seen. For example, the retardation of first-order red is said to be about 550 nm.

Measurement of retardation, as mentioned, is often useful in characterizing optical properties of anisotropic crystals. It takes special training, however, to employ such observations with precision because the interference color is influenced not only by the birefringence (the difference in refractive indices) but also by the thickness of the particle and the relation between the direction of the light passing through the crystal and the direction of its optic axis or axes.

Optic Sign

In a uniaxial crystal, the velocity of light in the direction of the extraordinary or E ray may be greater or lesser than that of the ordinary or O ray. If the refractive index ϵ is greater than ω (i.e. if the velocity of the O ray is greater than that of the E ray), the optic sign of the uniaxial crystal is said to be positive; if the refractive index ϵ is less than ω, the optic sign is said to be negative. In the latter case, the ordinary (O) ray is slower, or retarded more (having a higher refractive index), and the crystal is said to be "uniaxial negative."

If the particle is a well-formed crystal mounted in a medium of known refractive index, then observation of the Becke line and comparison of the relative relief of the crystal in different orientations will disclose whether ϵ or ω is the higher index. Most texts illustrate this mode of determination with calcite crystals; these are seen in high relief in Canada balsam under plane-polarized light (analyzer removed) when the axis in the ω direction is vertical and are scarcely to be seen when oriented so that ϵ is vertical. Calcite is uniaxial negative; its particles often exhibit symmetrical extinction.

Alternatively, a compensator or accessory plate can be used to determine the direction that has the highest index in a well-characterized crystal. If the shape of the particle is not indicative of one or more of the crystallographic axes, then the compensator is used to observe an *interference figure* (usually difficult to achieve with small grains). Even without such a device, the procedure of mounting the particles in media of successively higher or lower refractive index will eventually show, for a crystal of the tetragonal or hexagonal system, a match for one index in all orientations. This refractive index will be ω.

As mentioned, the direction of the fast or slow ray in passing through a crystal is frequently determined by use of an accessory plate. The gypsum (selenite) or quartz (designated as sensitive-tint, Red I or λ), mica ($\lambda/4$), or quartz-wedge (I–III orders or 1° to 3°) plates are those most commonly employed. The sensitive-tint plate is regularly referred to in this text and the appearance it imparts is illustrated in many of the colored photomicrographs. When inserted in the accessory slot in the microscope, the mounted sliver of gypsum or quartz retards the plane-polarized light one full wavelength as it passes through the microscope.The field takes on a violet hue, the so-called first-order red in a Michel-Levy chart. Such an accessory is called, in general, a retardation plate; more specifically, this particular type is designated as the sensitive-tint plate, for the eye is especially sensitive to slight changes in hue in this range of interference colors. The direction of the slow ray in the mineral wafer mounted in the accessory plate is usually marked on the device as an arrow (see fig. 7).

If a crystal under examination gives an interference color under crossed polars which ranks lower than first-order white (a color of low retardation), a sensitive-tint plate is used. (When higher interference colors are observed, a quarter-wave plate is more appropriate.) The crystal under observation is turned 45° from extinction. When the appropriate accessory plate is inserted, the arrow indicating the slow direction in the accessory will, in most microscopes, be at an angle of 45° with respect to the optical planes of the crossed polars. Fig. 7 illustrates a typical change in color that will

Fig. 7. Example of crystals with positive (A) and negative (B) elongation when observed with the sensitive-tint accessory plate.

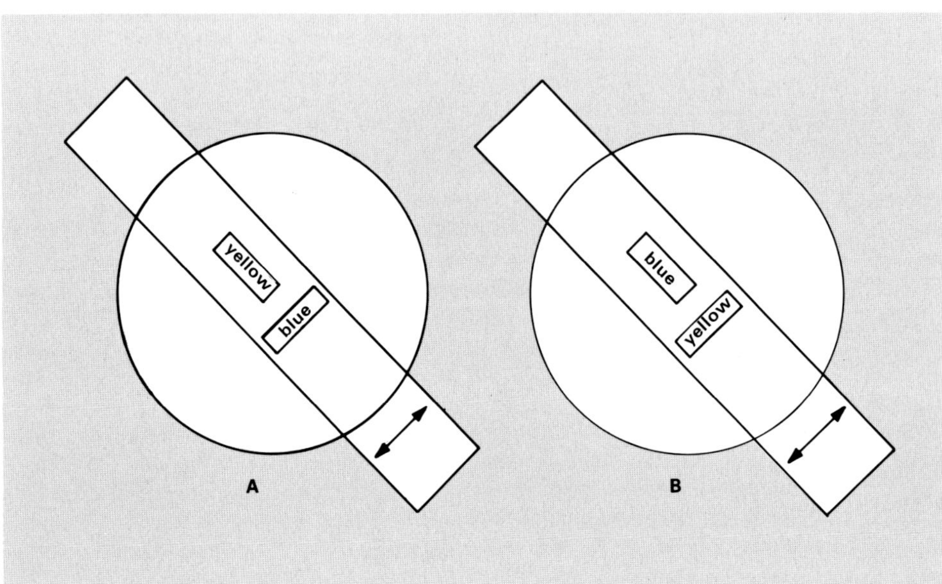

normally be seen when the sensitive-tint plate is in position with the polars crossed. If the particles are seen as blue in the NE/SW orientation, the retardation is said to be positive; if yellow, the retardation is negative.

Some uniaxial negative minerals commonly encountered in the examination of pigments are talc, glauconite, calcite, geothite, and hematite. Fewer uniaxial positive minerals are known: quartz and chalcedony are perhaps the examples most frequently encountered.

Determination of the optic sign of biaxial crystals (orthorhombic, monoclinic, and triclinic systems) requires considerable knowledge and experience; it is a job for the expert. There are three indices of refraction; the optic sign is considered positive if the difference in indices $\alpha-\beta$ is greater than that in $\beta-\gamma$; it is negative if $\alpha-\beta$ is less than $\beta-\gamma$. Either the observation of interference figures or a complete run-down of all three indices is needed to characterize biaxial substances thoroughly. Primarily, interference figures are employed, but the reader must be referred to standard texts on optical mineralogy for details of this subject.

Sign of Elongation

Unless well-trained and experienced, the analyst is not often sure of the crystallographic orientation of individual uniaxial crystals and therefore is not confident enough to determine whether the E or O ray is slowest. However, if a particle is elongated or needle-shaped, microscopists frequently find it useful to report the sign of elongation. If the slow ray (high refractive index) is parallel to the crystal's long direction, the particle is said to exhibit positive elongation. The elongation is negative if the long direction has the lower index. The sign of elongation does not have to correspond to the optic sign, although it frequently does.

Throughout this text we have referred to the sign of elongation for many of the pigments, for it is a property readily observed simply by employing a sensitive-tint plate to determine whether the long direction has a high or low index. When the particle is oriented with the longest axis in the same direction as the arrow on the plate and the refractive index is highest in this direction (the light travels at a slower rate), the particle will be seen in a blue or blue-green retardation color and is said to exhibit positive elongation; when the color is yellow or orange, the elongation is negative. Fig. 7 provides a ready reference to these rules. Strontium chromate needles exhibit positive elongation.

Interference Figures

The observation of so-called interference figures is usually necessary in order to characterize biaxial crystals. The study of uniaxial and biaxial interference figures is particularly applicable in mineralogical thin sections where thickness is great (usually 30 micrometers), grain size large, and a variety of crystal orientations are possible. It is rare for an interference figure to be of value in the characterization of artists' pigments, however, because of the small size of available particles and the difficulty of orienting them.

Ancillary Methods

The three ancillary microscopic techniques noted in the outline — microchemical tests, dispersion staining, and use of a hot/cold stage — are included for the sake of completeness. These procedures and other supplementary analytical techniques are usually employed to confirm the results of routine examination with the polarizing microscope. The tradition of using relatively simple and inexpensive microchemical tests in the examination of pigments is a long one, and such tests have been regularly described in the present text.

CONCLUSION

The individual particles of pigment in an artist's paint are rarely discernible with the naked eye; hence the extent and character of mixtures of finely divided materials can scarcely be understood without the aid of magnification. This appendix has been added to help explain the terms customarily used to describe the properties of pigment that can be observed under the microscope at magnifications of 60 to 1,000 diameters.

In the introduction to this volume, it was pointed out that one may wish to identify pigments at three broad levels of precision, by establishing the fundamental chemical type (lead white, iron oxide, chromate yellow), the variety (lead, strontium or barium chromate, kermes or cochineal carmine), and, in special situations, the specific correspondence or identity. One should also realize that there are various levels of certainty that may be achieved in the identification: *likely, most probable*, and *positive* (unequivocal). (Kelly and Judd [1976], for example, suggest that there are six levels of precision at which a color can be specified.)

The authors consider that, except when used by an experienced and well-trained microscopist, microscopic examination alone will rarely yield an unequivocal identification of a pigment. (The observation of microfossils in natural chalk is a familiar and notable exception.) This fact is reflected in the final section of each chapter in which it is emphasized that the basic criterion for positive identification of any crystalline compound is the x-ray diffraction pattern. All other factual evidence that can be accumulated regarding a particular pigment under investigation represents characterization essentially in terms of achieving a moderate or high probability of correctness. If one would consider "one chance of error in a million" to constitute unequivocal, positive identification, then perhaps it is fair to consider that one chance of error in ten to one hundred would represent a *likely* — one chance in one hundred to one thousand a *most probable* — identification.

By systematically observing the properties presented in the outline at the beginning of this chapter, the analyst can increase the relative certainty of identification. As a rough rule of thumb, one may suggest that a half-dozen qualitatively characterized properties in the checklist must agree with the properties of a known material before a likely identification can be assumed. Infrared spectra, microchemical tests, and spectrographic measurements are valuable supplementary techniques which will do much to increase the precision of the characterizations and raise the level of certainty to the order of one chance of error in a hundred or better. As mentioned, characterization by x-ray diffraction pattern constitutes most nearly an unequivocal, or positive identification of common crystalline materials.

These concluding remarks are intended to remind the reader that there are a number of levels of certainty at which one may wish to characterize a pigment. There are also various levels of specificity at which a pigment may be defined: chemical type, variety, or correspondence/identity. We trust that this brief review of the properties of crystalline materials observable with a polarizing microscope will help to improve the effectiveness and precision of microscopic examination as applied to the study of artists' pigments.

REFERENCES

F. D. Bloss, *An Introduction to the Methods of Optical Crystallography* (New York, 1961).

R. Brewer, *Fabric and Mineral Analysis of Soils* (New York, 1964), 21–24.

E. E. El-Hinnawi, *Methods in Chemical and Mineral Microscopy* (New York, 1966).

C. S. Hurlbut, Jr., *Dana's Manual of Mineralogy* (New York, 1952), 36–39.

R. R. Irani and C. F. Callis, *Particle Size: Measurement, Interpretation and Application* (New York, 1963).

K. L. Kelly and D. B. Judd, *Color, Universal Language and Dictionary of Names* (Washington, 1976).

J. Kirby, "A Spectrophotometric Method for the Identification of Lake Pigment Dyestuffs," *National Gallery Technical Bulletin* [1] (1977), 35–48.

W. C. McCrone "The Microscopical Identification of Artists' Pigments," J. 11C-Covodeder Group 7, Nos. 1 & 2 (1982), 11-34.

W. C. McCrone and J. G. Delly, *The Particle Atlas*, vol. 5, 2nd ed. (Ann Arbor, 1973) or W. C. McCrone, L. B. McCrone and J. G. Delly, *Polarized Light Microscopy* (Ann Arbor, 1978).

P. Nylen, *Modern Surface Coatings* (New York, 1965), 356.

T. C. Patton, *The Pigment Handbook* (New York, 1973), vol. 1.

T. P. Patton, *Paint Flow and Pigment Dispersion* (New York, 1964), 208–209.

Pigment Particles (Teddington, 1956), 6.

R. E. Stoiber and S. A. Morse, *Microscopic Identification of Crystals* (New York, 1972).

A. N. Winchell, *Elements of Optical Mineralogy; Part 1. Principles and Methods*, 3rd ed. (New York, 1928), 118.

A. N. Winchell and H. Winchell, *The Microscopic Characters of Artificial Inorganic Solid Substances: Optical Properties of Artificial Minerals* (New York, 1964).

Standard Specifications of Pigment Composition and Color

RUTH JOHNSTON-FELLER

Throughout these monographs, the Colour Index (CI) generic name and number and the five-digit CI constitution number have been used as a standard identification of each pigment. This system for characterizing pigments and dyes is the result of a continuing joint effort of the Society of Dyers and Colourists in Great Britain and the American Association of Textile Chemists and Colorists in the United States. The CI generic name is based on the most common method by which the colorant is applied, such as pigment (P), vat dye (Vat), acid dye (Acid), combined with a hue description such as red (R) or purple (P). The generic number following the name refers to the particular type of pigment. Thus, the generic name and number of natural barium sulfate, barytes, is CI Pigment White 22 (CI PW-22), whereas the synthetic barium sulfate, blanc fixe, has the generic name and number CI Pigment White 21 (CI PW-21).

The five-digit CI constitution number following the generic name and number designates the chemical constitution of each colorant when it is known. (Some modern pigments have no constitution number because their exact chemical structures are proprietary.) Blocks of numbers have been assigned according to the general chemical character of the colorant. Thus, all natural colorants are given CI numbers from 75000 to 75999; inorganic pigments are assigned numbers from 77000 to 77999, and phthalocyanine colorants are numbered 74000 to 74999. In the case of the barium sulfates, for example, both the barytes and blanc fixe have a CI number of 77120 because both consist of the compound barium sulfate, $BaSO_4$. The customary format for the CI designation used in this text is exemplified by that for barytes: CI Pigment White 22, No. 77120.

A standard nomenclature has also been employed to describe the visual color of pigments in masstone and in mixtures with white (tints). The standard color names are based on a division of the three-dimensional Munsell color space into 267 blocks which have been denominated according to a standard schedule of hue names with lightness (Munsell value) and saturation (Munsell chroma) word modifiers. The system, called the "Universal Color Language," has been devised jointly by a committee of the Inter-Society Color Council and the National Bureau of Standards. The latest publication, available from the Superintendent of Documents (United States Government Printing Office, cat. no. C13.10:440, $3.25), is the 1976 edition, *Color: Universal Language and Dictionary of Names* by K. L. Kelly and D. B. Judd, National Bureau of Standards Special Publication 440.

To determine the Universal Color Name, the Munsell notation of the particular color is first required. This may be obtained either by direct visual comparison with specimens in the *Munsell Book of Color* or by spectrophotometric measurement and subsequent appropriate calculations. The diagrams describing the system are then consulted in *Color: Universal Language and Dictionary of Names* to locate the name corresponding to the Munsell notation. Although not used in the monographs, a number which has been assigned to each block may also be cited. Use of the number facilitates easy reference to the dictionary of color names that have been garnered from many of the other color systems in existence. Thirty-one hue-range diagrams for the Munsell space are provided. In determining the universal name one must first find the proper Munsell hue diagram. One then locates the Munsell value on the vertical axis and the Munsell chroma on the horizontal axis. The block in which straight lines drawn through these factors intersect describes the universal color name and block number for the specimen under consideration. Thus, a typical brilliant red with Munsell notation of 3R 4/13 is called "strong red" and numbered 12. If the color is lightened (increased in value by addition of white), it becomes "deep pink," numbered 3. If the color is decreased in saturation (by addition of gray) it becomes "moderate red," numbered 15; if it is further decreased in saturation (by more addition of gray), it is named "grayish red," numbered 19.

Throughout the monographs, we have used this system to express the colors of the pigments described, both in their masstone and in mixtures with white. Thus, green earth in masstone has a color name of "dark grayish green", carmine lakes in transparent watercolors vary from "dark purplish red" to "deep yellowish pink", and aureolin in watercolor is "brilliant yellow".